Lecture Notes in Mathematics

Edited by A. Dold and B. Eckmann

854

Algebraic K-Theory Evanston 1980

Proceedings of the Conference
Held at Northwestern University
Evanston, March 24–27, 1980

Edited by E. M. Friedlander and M. R. Stein

Springer-Verlag
Berlin Heidelberg New York 1981

Editors

Eric M. Friedlander
Michael R. Stein
Department of Mathematics, Northwestern University
Evanston, Illinois 60201/USA

AMS Subject Classifications (1980): 12A35, 12A62, 13B20, 13D15, 14C35, 14C40, 14F15, 16A54, 18F25, 20G10, 20G35

ISBN 3-540-10698-7 Springer-Verlag Berlin Heidelberg New York
ISBN 0-387-10698-7 Springer-Verlag New York Heidelberg Berlin

Library of Congress Cataloging in Publication Data Main entry under title: Algebraic K-theory. (Lecture notes in mathematics; 854) Bibliography: p. Includes index. 1. K-theory--Congresses. 2. Algebraic number theory--Congresses. 3. Geometry, Algebraic--Congresses. I. Friedlander, E. M. (Eric M.), 1944-. II. Stein, Michael R., 1943- III. Northwestern University (Evanston, Ill.) IV. Series: Lecture notes in mathematics (Springer-Verlag); 854. QA3.L28 vol. 854 [QA169] 510s [512'.55] 81-5333AACR2 ISBN 0-387-10698-7

Printing and binding: Beltz Offsetdruck, Hemsbach/Bergstr.
2141/3140-543210

Introduction

A conference on algebraic K-theory, supported by the National Science Foundation, was held at Northwestern University during the period March 24-27,1980. These proceedings contain some of the papers presented at that conference as well as some related papers. On behalf of the participants, we thank the National Science Foundation for its financial support, Northwestern University for its hospitality, and the Northwestern mathematics department staff for its friendly assistance.

Eric M. Friedlander

Michael R. Stein

December 10, 1980

Table of Contents

THE DILOGARITHM AND EXTENSIONS OF LIE ALGEBRAS

Spencer Bloch (*)
Department of Mathematics
University of Chicago

One of the nice things about algebraic K-theory is that it is forever leading the researcher in new and unexpected directions. Several years ago I encountered in this connection the dilogarithm function

$$\int \log(1-t)\,\frac{dt}{t},$$

which was the key ingredient in a regulator map

$$K_2(X) \to H^1(X,\mathbf{C}^*)$$

for a Riemann surface X. I had various constructions for this map; a direct function theoretic approach when X was an elliptic curve in [1], a sheaf-theoretic approach in [3], and an interpretation via periods of integrals and generalized intermediate jacobians in [4]. All were complicated and in various ways unsatisfactory.

More recently Deligne found a simpler and more powerful construction based on interpreting $H^1(X,\mathbf{C}^*)$ as the group of line bundles on X with connections. This fitted nicely with an idea of Ramakrishnan that the dilogarithm could be interpreted as a single valued map

$$\mathbf{P}^1_{\mathbf{C}} - \{0,1,\infty\} \to H(\mathbf{Z})\backslash H(\mathbf{C}) \tag{0.1}$$

where H(R) for any ring R is the Heisenberg group of unipotent 3 × 3 upper triangular matrices. $H(\mathbf{Z})\backslash H(\mathbf{C})$ has a natural structure of principal bundle over $\mathbf{C}^* \times \mathbf{C}^*$ with fiber $\mathbf{C}^*$. Moreover this bundle has a standard (non-integrable) holomorphic connection ∇. A symbol $\{f,g\} \in K_2(X)$ corresponds to a map

$$(f,g): X \to \mathbf{C}^* \times \mathbf{C}^*$$

(ignore for the moment the problem of zeroes and poles of f and g as well as the question of whether f and g are well-defined) so we may associate to $\{f,g\}$ the bundle

$$(f,g)^*(H(\mathbf{Z})\backslash H(\mathbf{C}),\nabla) \in H^1(X,\mathbf{C}^*).$$

That this works and leads to a regulator map is the content of §1. Ramakrishnan's map (0.1) above gives the Steinberg relation

(*) Partially supported by the NSF

$$(f,1-f)^*(H(\mathbb{Z})\backslash H(\mathbb{C}),\nabla) = (e).$$

As an application we answer in (1.24) a question of Tate ([16], p.250] concerning torsion in K_2.

In communicating to me his construction, Deligne remarked (cryptically, as is his wont) that he had found it while thinking about Kac-Moody Lie algebras. Aspects of that relationship (Another new and unexpected direction!) are discussed in §2. Taking X = Spec(R) to be an affine Riemann surface, the regulator map leads to an extension

(0.2) $$0 \to H^1(X,\mathbb{C}^*) \to W \to SL(R) \to 1$$

It then turns out that one can associate to (0.2) a central extension of lie algebras

(0.3) $$0 \to H^1(X,\mathbb{C}) \to \mathfrak{g} \to sl(R) \to 0$$

given by the lie algebra cocycle

(0.4) $$A,B \longmapsto \mathrm{Trace}(A\cdot dB) \in H^1(X,\mathbb{C}).$$

In fact (0.3) is the universal central extension of $sl(R)$ in the category of lie algebras over $\mathbb{C}$. More generally, if k is a commutative ring with $\frac{1}{2} \in k$ and R is a k-algebra (commutative with 1) then

$$H_2(sl(R),k) \cong \Omega^1_{R/k}/dR = \text{Kähler 1-forms mod exact 1-forms},$$

and the cocycle (0.4) leads to the universal central extension (over k)

$$0 \to \Omega^1/dR \to \mathfrak{g} \to sl(R) \to 0.$$

The proof of this is a straightforward application of ideas of Steinberg and is given in §3. It would be exciting to know the groups $H_*(sl(R),k)$ for $* > 2$ and to understand their relation to higher regulators. (A technical point: it might be better to work with the relative groups $H_*(sl(R),sl(k);k)$. This is closer in spirit to the calculations which have been done for Kac-Moody algebras [9],[10],[12], and avoids a good deal of garbage which would otherwise appear in H_3.) In fact, based on the ideas in [9] and [12], it seems natural to conjecture when $k = \mathbb{C}$

$$\lim_{\substack{\to\\ n}} H_*(sl_n(R),sl_n(\mathbb{C});\mathbb{C}) \cong \lim_{\substack{\to\\ n}} H_*((U_n(\mathbb{C}),e)^{(X,x_0)},\mathbb{C})_{top.sp.}$$

$$\cong \bigotimes_g \lim_{\substack{\to\\ n}} H_*(\Omega U_n(\mathbb{C})/U_n(\mathbb{C}),\mathbb{C})_{top.sp.}$$

where $g = \dim H^1(X,\mathbf{C})$. Indeed, the second isomorphism is clear, because X is homotopic to a wedge of g circles so base point preserving maps from X to $U_n(\mathbf{C})$ look like g-tuples of based maps $S^1 \to U_n(\mathbf{C})$. Our calculation of $H_2(sl(R),sl(\mathbf{C});\mathbf{C})$ agrees with the conjecture.

On an algebraic level, the lie algebra cocycles one can write down suggest maps of complexes for each $r > 1$

$$\begin{array}{ccccccccc}
\bigwedge^{2r+2} sl_n(R) & \longrightarrow & \bigwedge_k^{2r+1} sl_n(R) & \xrightarrow{\partial} & \bigwedge^{2r} Sl_n(R) & \longrightarrow \cdots \longrightarrow & \bigwedge^{r+1} sl_n(R) & & \\
\downarrow & & \downarrow & & \downarrow & & \downarrow & & \\
0 & \longrightarrow & R & \xrightarrow{d} & \Omega^1_{R/k} & \longrightarrow \cdots \longrightarrow & \Omega^r_{R/k} & \longrightarrow & 0 ,
\end{array}$$

although at the moment I can only get the signs to work when $\Omega^2_{R/k} = (0)$. Such maps would give cohomology classes in

$$H^p(sl_n(R),H^q_{DR}(R/k)) \quad \text{for} \quad p+q = 2r+1,\ r > 1,\ p > q+3$$

as well as classes in $H^{p+1}(sl_n(R),\Omega^p_R/d\Omega^{p-1}_R)$. Thinking of R as the ring of functions on some space X, and $H^*(sl_n(R))$ as a rational approximation to the singular cohomology of the mapping space $SL_n(\mathbf{C})^X$, some of these classes might correspond to pull backs of indecomposable classes under the evaluation map $X \times SL_n(\mathbf{C})^X \to SL_n(\mathbf{C})$.

As must be clear by now I do not claim any tremendous originality for this work. In addition to the debt to Deligne, Ramakrishnan, and Steinberg mentioned above, I would like to acknowledge considerable inspiration from a Colloquium talk given by Kac at Chicago this year as well as from several conversations with D. Kazdan and C. Soulé.

§1. The Regulator Map

Define for any ring R, the Heisenberg group

(1.1) $$H(R) = \left\{ \begin{pmatrix} 1 & r_1 & r_2 \\ 0 & 1 & r_3 \\ 0 & 0 & 1 \end{pmatrix} \in M_3(R) \right\} .$$

Let $N = H(\mathbf{Z}) \backslash H(\mathbf{C})$

(1.2) $$\begin{pmatrix} 1 & a & b \\ 0 & 1 & c \\ 0 & 0 & 1 \end{pmatrix} \equiv \begin{pmatrix} 1 & a' & b' \\ 0 & 1 & c' \\ 0 & 0 & 1 \end{pmatrix} \iff \exists\, m,n,p \in \mathbf{Z} \text{ such that } a+m = a' \quad b+mc+n = b' \quad c+p = c' .$$

Clearly then we have a map

(1.3) $$\pi : N \to \mathbf{C}^* \times \mathbf{C}^*, \quad \pi \begin{pmatrix} 1 & a & b \\ 0 & 1 & c \\ 0 & 0 & 1 \end{pmatrix} = (e^{2\pi i a}, e^{2\pi i c})$$

identifying N as a principal bundle over $\mathbf{C}^* \times \mathbf{C}^*$ with fiber $\mathbf{C}^*$.

To write down a cocycle for N, let U_+ (resp U_-) $\subset \mathbf{C}^*$ be the complement of negative (resp. positive) real axis. Define branches $\log_+$ and $\log_-$ on U_+ and U_- by

(1.4) $$-\pi < \operatorname{Im} \log_+ < \pi \ , \ 0 < \operatorname{Im} \log_- < 2\pi .$$

There are sections, e.g. $s_{++}: U_+ \times U_+ \to \pi^{-1}(U_+ \times U_+)$

(1.5) $$s_{++}(\alpha,\beta) = \begin{pmatrix} 1 & \frac{1}{2\pi i}\log_+(\alpha) & 0 \\ 0 & 1 & \frac{1}{2\pi i}\log_+(\beta) \\ 0 & 0 & 1 \end{pmatrix}$$

and similarly for s_{+-}, s_{-+}, and s_{--}. On the other hand

$$\begin{pmatrix} 1 & \frac{1}{2\pi i}\log_+(\alpha) & 0 \\ 0 & 1 & \frac{1}{2\pi i}\log_+(\beta) \\ 0 & 0 & 1 \end{pmatrix} = A \begin{pmatrix} 0 & \frac{1}{2\pi i}\log_-(\alpha) & 0 \\ 0 & 1 & \frac{1}{2\pi i}\log_+(\beta) \\ 0 & 0 & 1 \end{pmatrix}$$

where

$$A = \begin{pmatrix} 1 & \frac{1}{2\pi i}(\log_+(\alpha) - \log_-(\alpha)) & \frac{1}{4\pi^2}(\log_+(\alpha) - \log_-(\alpha))\log_+(\beta) \\ 0 & 1 & 0 \\ 0 & 0 & 1 \end{pmatrix}$$

This gives for a transition cocycle

$$(1.6) \qquad c_{++,-+} = \beta^{\frac{1}{2\pi i}(\log_-(\alpha) - \log_+(\alpha))} = c_{+-,--} .$$

Note

$$(1.7) \qquad \frac{1}{2\pi i}(\log_-(\alpha) - \log_+(\alpha)) = \begin{cases} 0 & \text{on upper } \frac{1}{2} \text{ plane} \\ 1 & \text{on lower } \frac{1}{2} \text{ plane} \end{cases} ,$$

and hence represents a generating cocycle in $H^1(C^*,Z)$. Writing $\mathcal{O}^*_{C^*}$ for the sheaf of invertible analytic functions on C^*, the standard funtion on C^* is a section in $\Gamma(C^*, \mathcal{O}^*_{C^*})$. Formula (1.7) can be interpreted as saying that the class in $H^1(C^* \times C^*, \mathcal{O}^*_{C^*})$ of N is the (exterior) cup product of a generator for $H^1(C^*,Z) \cong Z$ and the tautological global section of $\mathcal{O}^*_{C^*}$.

The next step is to put a connection on this bundle. In down to earth terms this amounts to writing the logarithmic derivative of the cocycle (1.6) as a difference of 1-forms, one on each of the two intersecting open sets, e.g.,

$$(1.8) \qquad \frac{1}{2\pi i}(\log_-(\alpha) - \log_+(\alpha))\frac{d\beta}{\beta} = \omega_{++} - \omega_{-+} .$$

It is natural to take

$$(1.9) \qquad \omega_{++} = \omega_{+-} = -\frac{1}{2\pi i}\log_+(\alpha)\frac{d\beta}{\beta} \ ; \ \omega_{-+} = \omega_{--} = -\frac{1}{2\pi i}\log_-(\alpha)\frac{d\beta}{\beta} .$$

Here is another, rather more canonical construction of the connection. (Again the idea arose in conversation with Rama krishnan.) Consider the group

$$H^* = H_C \Big/ \left\{ \begin{pmatrix} 1 & 0 & n \\ 0 & 1 & 0 \\ 0 & 0 & 1 \end{pmatrix} \middle| \ n \in Z \right\}$$

and the exact sequence

$$0 \longrightarrow \mathbf{C}^* \longrightarrow H^* \xrightarrow{\varphi} \mathbf{C} \times \mathbf{C} \longrightarrow 0 .$$

Let $A = \varphi^{-1}(\mathbf{Z} \times \mathbf{Z}) \cong \mathbf{C}^* \cdot (\mathbf{Z} \times \mathbf{Z})$. It is not hard to show A is abelian and there is a natural character

$$\psi : A \to \mathbf{C}^*$$

which is the identity on $\mathbf{C}^*$ and trivial on $\mathbf{Z} \times \mathbf{Z}$. The space of sections of the line bundle over $\mathbf{C}^* \times \mathbf{C}^*$ discussed above can be identified with the space of holomorphic functions $f : H^* \to \mathbf{C}$ such that

$$f(ha) = \psi(a) f(h) , \quad h \in H^*, \ a \in A.$$

H^* acts on this space of functions by left translation, so the lie algebra $\mathrm{lie}(H^*)$ acts by derivations. The map

$$d\varphi : \mathrm{lie}(H^*) \to \mathbf{C} \times \mathbf{C}$$

is canonically split as a surjection of vector spaces, via

$$\frac{\partial}{\partial u_1} \longmapsto \begin{pmatrix} 0 & 0 & 0 \\ 0 & 0 & 1 \\ 0 & 0 & 0 \end{pmatrix}$$

$$\frac{\partial}{\partial u_2} \longmapsto \begin{pmatrix} 0 & 1 & 0 \\ 0 & 0 & 0 \\ 0 & 0 & 0 \end{pmatrix}$$

so vector fields on $\mathbf{C}^* \times \mathbf{C}^*$ act as derivations on sections. the reader can check (it is a good exercise!) that the connection thus obtained is the one given explicitly above.

Now given X an open Riemann surface, and f,g non-vanishing holomorphic functions on X, we consider $(f,g) : X \to \mathbf{C}^* \times \mathbf{C}^*$ and pull back the bundle with connection constructed above to X. The resulting bundle, denoted $r(f,g)$, will have an integrable connection (because dim X = 1), so we may view $r(f,g)$ as an element of the group $H^1(X, \mathbf{C}^*)$.

Proposition (1.10). (i) $r(f, g_1 g_2) = r(f, g_1) r(f, g_2)$

(ii) $r(f,g) = r(g,f)^{-1}$

(iii) $r(f_1 f_2, g) = r(f_1, g) r(f_2, g)$.

Proof. (i) The exact sequence

$$0 \longrightarrow \mathbf{C}_X^* \longrightarrow \mathcal{O}_X^* \xrightarrow{\text{dlog}} \Omega_X^1 \longrightarrow 0$$

shows that a 1-cocycle on X with values in $\mathbf{C}^*$ can be represented by an $\mathcal{O}_X^*$ cocycle u_{ij} together with a trivialization

$$du_{ij} \cdot u_{ij}^{-1} = \omega_i - \omega_j \ .$$

If we take an open cover $X = X_+ \cup X_-$ where

$$X_{+,f} = \{x \in X \mid f(x) \not\leq 0\} \ , \ X_{-,f} = \{x \mid f(x) \not\geq 0\} ,$$

we see from (1.6), (1.8), (1.9) that $r(f,g)$ is represented by

$$(1.11) \qquad g^{\frac{1}{2\pi i}(\log_-(f) - \log_+(f))} \ ; \ \omega_+ = -\frac{1}{2\pi i}\log_+(f)\frac{dg}{g} \ , \ \omega_- = \frac{-1}{2\pi i}\log_-(f)\frac{dg}{g} \ .$$

These formulae are multiplicatively linear in g, proving (i).

The $\mathbf{C}^*$ 1-cocycle represented by a pair $\{u_{ij}, \omega_i\}$ as above will be a coboundary if there exists an $\mathcal{O}^*$ 1-cochain v_i such that $u_{ij} = v_i v_j^{-1}$ and $dv_i \cdot v_i^{-1} = \omega_i$. Cover X by open sets

$$X_A = X_{a_1,f} \cap X_{a_2,g} \qquad a_1, a_2 = +,-$$

and take

$$v_A = e^{\frac{-1}{2\pi i}\log_{a_1}(f)\log_{a_2}(g)}$$

The pair $\{v_A v_B^{-1}, dv_A \cdot v_A^{-1}\}$ represents $r(f,g)f(g,f)$, proving (ii).

Finally, (iii) follows from (i) and (ii). Q.E.D.

Define functions $\ln_r(x)$ by

$$(1.12) \qquad \ln_1(x) = \frac{1}{2\pi i}\log(1-x) \ , \ \ln_r(x) = \frac{1}{2\pi i}\int \ln_{r-1}(x)\frac{dx}{x} \ .$$

Proposition (1.13) (Ramakrishnan): The assignment

$$x \longmapsto \begin{pmatrix} 1 & \ln_1(x) & \ln_2(x) \\ 0 & 1 & \ln_1(1-x) \\ 0 & 0 & 1 \end{pmatrix}$$

gives a well-defined map $\psi: \mathbf{P}^1 - \{0,1,\infty\} \to N$. Moreover, ψ is horizontal with respect to the connection (1.9).

Proof. The functions

$$\frac{1}{2\pi i}\log(1-x)\ ,\ \frac{-1}{4\pi^2}\int \log(1-x)\frac{dx}{x}$$

depend on the choice of a path from 0 to x. The map ψ being well-defined means that two different paths lead to transformations

$$\begin{aligned} \ln_1(x) &\longmapsto \ln_1(x) + m \\ \ln_2(x) &\longmapsto \ln_2(x) + \frac{m}{2\pi i}\log(x) + n \end{aligned} \qquad m,n \in \mathbb{Z}$$

with (and this is the point) the same integer m. The reader can check m = winding number of the difference of the two paths about 1.

To show ψ is horizontal we can work locally on $\mathbb{C}^* \times \mathbb{C}^*$. Fix A = (a,b); a,b = +,-, $U_A = U_a \times U_b \subset \mathbb{C}^* \times \mathbb{C}^*$, and consider the section σ_A

$$\sigma_A(\alpha,\beta) \equiv \begin{pmatrix} 1 & \frac{1}{2\pi i}\log_a(\alpha) & 0 \\ 0 & 1 & \frac{1}{2\pi i}\log_b(\beta) \\ 0 & 0 & 1 \end{pmatrix} \quad \text{(modulo left action of } H(\mathbb{Z}).)$$

Identifying the fibre with $\mathbb{C}^*$, one gets

(1.14 $\quad \psi(x) = \exp(2\pi i\cdot \ln_{2,a}(x))\cdot\sigma_A(1-x,x);\ \ln_{2,a}(x) = \frac{-1}{4\pi^2}\int \log_a(1-x)\frac{dx}{x}$.

The horizontality condition is most simply understood if one thinks in terms of the corresponding line bundle, trivialized over U_A by σ_A. The connection becomes

$$\nabla(1) = \omega_A = -\frac{1}{2\pi i}\log_a(\alpha)\frac{d\beta}{\beta}\ ;\ \nabla(f) = f\nabla(1) + df$$

so $f \circ \sigma_A$ is horizontal if $df\cdot f^{-1} = \frac{1}{2\pi i}\log_a(\alpha)\frac{d\beta}{\beta}$. Substituting

$$\alpha = 1 - x,\ \beta = x\ ,\ f = \exp(2\pi i\cdot\ln_{2,a}(x))$$

yields a solution to this differential equation. Q.E.D.

Corollary (1.15). If f and 1-f are non-vanishing holomorphic functions on the Riemann surface X, then r(f,1-f) = (e).

Proof. The bundle r(f,1-f) will have a horizontal section by (1.13). Q.E.D.

Let $k(X)$ be the field of meromorphic functions on the Riemann surface X. Using the Matsumoto description of $K_2(k(X))$, we deduce from (1.10) and (1.15)

<u>Corollary (1.16)</u>. r induces a map

$$r: K_2(k(X)) \to \varinjlim_{S \subset X \text{ finite}} H^1(X-S, \mathbb{C}^*).$$

Suppose now $x \in S$, and f and g are non-vanishing holomorphic functions on $X-S$ which are meromorphic at x. Define the <u>tame symbol</u>

$$(1.17) \qquad T_x(f,g) = (-1)^{\mathrm{ord}_x f \cdot \mathrm{ord}_x g} \left(\frac{g^{\mathrm{ord}_x f}}{f^{\mathrm{ord}_x g}}\right)(x) \in \mathbb{C}^*$$

Finally, let $\partial_x : H^1(X-S, \mathbb{C}^*) \to \mathbb{C}^*$ be the residue map from the Gysin sequence, so we have

$$(1.18) \qquad 0 \to H^1((X-S) \cup \{x\}, \mathbb{C}^*) \to H^1(X-S, \mathbb{C}^*) \xrightarrow{\partial_x} \mathbb{C}^* \to H^2((X-S) \cup \{x\}, \mathbb{C}^*) \to \cdots$$

<u>Proposition (1.19)</u>. Notation being as above, we have

$$\partial_x r(f,g) = T_x(f,g).$$

<u>Proof</u>. Both sides are multiplicative in f and g, skew-symmetric, and trivial when $g = 1 - f$ as well as when $f(x)$ and $g(x) \neq 0, \infty$. A simple manipulation reduces us to the case $f(x) \neq 0, \infty$, $\mathrm{ord}_x g = 1$, so $T(f,g) = f(x)^{-1}$. To calculate $\partial_x r(f,g)$ consider the diagram of complexes of sheaves (written vertically)

$$(1.20) \qquad \begin{array}{ccccccccc} 0 & \longrightarrow & \mathcal{O}_X^* & \longrightarrow & \mathcal{O}_X^*(x) & \longrightarrow & \mathbb{Z}_x & \longrightarrow & 0 \\ & & \downarrow \mathrm{dlog} & & \downarrow \mathrm{dlog} & & \downarrow & & \\ 0 & \longrightarrow & \Omega_X^1 & \longrightarrow & \Omega_X^1(x) & \xrightarrow{\text{residue}} & \mathbb{C}_x & \longrightarrow & 0 \end{array}$$

where $\mathcal{O}_X^*(x)$ denotes the sheaf of functions meromorphic at x and $\neq 0, \infty$ off x. The exact sequence of first (hyper) cohomology groups of these complexes is canonically identified with

$$H^1(X, \mathbb{C}^*) \to H^1(X-x, \mathbb{C}^*) \xrightarrow{\partial_x} \mathbb{C}^* .$$

Recall the data respresenting $r(f,g) \in H^1(X-x,\mathbf{C}^*)$ included differential forms (1.11),

$$\omega = -\frac{1}{2\pi i}\log(f)\,\frac{dg}{g}\,.$$

It follows from (1.20) that the presumption for calculating $\partial_x r(f,g)$ is

$$\exp(2\pi i \cdot \text{residue}_x(\omega)) = f(x)^{-1}. \qquad \text{Q.E.D.}$$

We can now construct the global regulator map from the diagram

(1.21)

$$\begin{array}{ccccccc}
 & & K_2(X) & \longrightarrow & K_2(k(X)) & \xrightarrow{\ \amalg T_x\ } & \coprod_{x\in X} \mathbf{C}^* \\
 & & \Big\downarrow r & & \Big\downarrow r\ (1.16) & & \Big\| \\
0 & \longrightarrow & H^1(X,\mathbf{C}^*) & \longrightarrow & \varinjlim\limits_{\substack{S\subset X\\ \text{finite}}} H^1(X-S,\mathbf{C}^*) & \xrightarrow{\ \amalg\partial_x\ } & \coprod_{x\in X} \mathbf{C}^*
\end{array}$$

The top row is the localization sequence in algebraic K-theory, and the bottom is the Gysin sequence in topology. We get finally

(1.22) $$r: K_2(X) \to H^1(X,\mathbf{C}^*).$$

As an application, let $\mu \subset \mathbf{C}^*$ be the group of roots of 1, and consider the diagram (cf. [1])

$$\begin{array}{ccccccc}
0 & \longrightarrow & \mathrm{Tor}_1(J(X),\mu) & \longrightarrow & \mathbf{C}(X)^*\otimes\mu & \longrightarrow & \coprod_{x\in X}\mu \\
 & & \Big\downarrow \alpha & & \Big\downarrow \text{symbol} & & \Big\downarrow \\
0 & \longrightarrow & \Gamma(X,\underline{K}_2) & \longrightarrow & K_2(\mathbf{C}(X)) & \xrightarrow{\ \text{tame}\ } & \coprod_{x\in X}\mathbf{C}^*
\end{array}$$

(Here X is a complete smooth curve over $\mathbf{C}$, and $J(X)$ = jacobian(X).)

<u>Corollary</u> (1.23). $\alpha: \mathrm{Tor}_1(J(X),\mu) \to \Gamma(X,\underline{K}_2)$ is injective.

<u>Proof</u>. The image of Tor_1 in $K_2(\mathbf{C}(X))$ consists of symbols

$$\{f, e^{\frac{2\pi i}{n}}\}$$

where all zeros and poles of f have multiplicity divisible by n. On the complement

X_f of the singular set of f, this class maps under the regulator map to the trivial line bundle with connection given by $\frac{1}{n}\frac{df}{f}$. The corresponding class in $H^1(X_f, \mathbf{C}^*)$ is trivial if and only if $f = g^n$ for some g. Q.E.D.

Corollary (1.24). Let F be the function field of a curve defined over an algebraically closed ground field k of characteristic 0. Let $\zeta \in k$ be a primitive n'th root of 1. Then the map

$$F^*/F^{*n} \longrightarrow {}_nK_2(F) \qquad f \longmapsto \{f,\zeta\}$$

is an isomorphism.

Proof. Tate shows in [16] that the map is surjective. But

$$\{f,\zeta\} = 1 \Rightarrow \text{tame}\{f,\zeta\} = 1 \Rightarrow (f) = nD$$

for some divisor D on the curve. We may assume $k = \mathbf{C}$ and apply (1.23) to get $f \in F^{*n}$. Q.E.D.

§2. Lie Algebras

We suppose now that X is smooth affine algebraic curve defined over $\mathbf{C}$, and $R = \mathbf{C}[X]$ is the ring of algebraic funtions on X. We write $E(R)$ and $St(R)$ for the elementary matrices and Steinberg group of R, respectively. The universal central extension is

(2.1) $$0 \to K_2(R) \to St(R) \to E(R) \to 1.$$

We "push out" this extension via the regulator map from §1

$$r: K_2(R) \to H^1(X, \mathbf{C}^*)$$

to get

(2.2) $$0 \to H^1(X, \mathbf{C}^*) \to W(R) \to E(R) \to 1.$$

Finally, in order to replace E(R) by something more palatable we can sheafify (2.2) for the Zariski topology on X

(2.3) $$0 \to \mathscr{H}^1(\mathbf{C}^*) \to \underline{\underline{W}} \to \underline{\underline{SL}} \to 0.$$

It is not hard to show [6], $\Gamma(X, \mathscr{H}^1) \cong H^1(X, \mathbf{C}^*)$ while $H^1(X, \mathscr{H}^1(\mathbf{C}^*)) \cong H^2(X, \mathbf{C}^*) = (0)$ (because X is affine). Taking sections of (2.3) over X therefore yields

(2.4) $$0 \to H^1(X, \mathbf{C}^*) \to \underline{\underline{W}}(R) \to SL(R) \to 1.$$

It is interesting to note that if we took sections of (2.3) over a <u>complete</u> curve X, we would get an exact sequence of H^1 terms

(2.5) $$0 \to \mathbf{C}^* \to H^1(X, \underline{\underline{W}}) \to H^1(X, \underline{\underline{SL}}) \to e .$$

The set $H^1(X, \underline{\underline{W}})$ cries out for a geometric interpretation. Speaking vaguely, it represents a $\mathbf{G}_m$-bundle M over the "space" $H^1(X, SL)$. The S-valued points of M are SL-bundles B on $X \times S$ together with trivializations of the Cartier divisor on S:

$$\pi_{2*}(c_2(B)) , \quad c_2(B) = \text{second chern class of } B \text{ on } X \times S.$$

In other words, the pull-back of M over S is the principal $\mathbf{G}_m$-bundle associated to $\pi_{2*}(c_2(B))$.

I want to consider the infinitesimal structure of (2.4). Recall that the connection ∇ on the bundle N in §1 was not integrable. Its curvature form K is computed by differentiating the forms ω in (1.9),

(2.6) $$K = \frac{1}{2\pi i}\frac{d\alpha}{\alpha} \wedge \frac{d\beta}{\beta}.$$

Suppose we conside the "thickened" Riemann surface $X[\varepsilon]$ with ring of functions $R[\varepsilon] = R + R\cdot\varepsilon$, $\varepsilon^2 = 0$. Given f,g $R[\varepsilon]^*$, we can pull back (N,∇) as in §1, only now since $\frac{\partial}{\partial\varepsilon}$ provides an independent tangent directon, the curvature $(f,g)^*K \neq 0$ in general. In fact

(2.7) $$(f,g)^*K = \frac{df}{f} \wedge \frac{dg}{g} \in \Omega^2_{R[\varepsilon]} = \Omega^1_R \wedge d\varepsilon \cong \Omega^1_R .$$

For any functor F:(rings) $\to$ (ab. groups), let t_F be the functor $t_F(R) = \mathrm{Ker}(F(R[\varepsilon]) \xrightarrow[\varepsilon \to 0]{} F(R))$. It is a result of Van der Kallen that the above procedure defines an isomorphism

(2.8) $$t_{K_2}(R) \xrightarrow{\cong} \Omega^1_R$$

for any ring R with $1/2 \in R$ [17]. We obtain

(2.9) $$0 \to \Omega^1_R \to t_{St}(R) \to sl(R) \to 0.$$

The problem that presents itself, however, is that (2.9) is not an exact sequence of lie algebras. In fact $t_{St}(R)$ doesn't have a natural lie algebra structure. (This is liked to the failure of excision for K_2. [5])

To see this, replace R in (2.9) by $R[\delta]/(\delta^2)$ and consider the kernel of the specialization $\delta \mapsto 0$

(2.10) $$0 \to \Omega^1_R\cdot\delta \oplus R\cdot d\delta \to \mathrm{Ker}(t_{St}(R[\delta]) \to t_{St}(R)) \to sl(R)\cdot\delta \to 0.$$

Thus the usual lie algebra construction

$$(1 + A\varepsilon)(1 + B\delta)(1 + A\varepsilon)^{-1}(1 + B\delta)^{-1} = 1 + [A,B]\varepsilon\delta$$

fails in this case because the middle term in (2.10) is not isomorphic to $t_{St}(R)$ (the left-hand term is too big).

Suppose however we consider the defintion of the pullback connection on $X[\varepsilon]$ via cocycle data, as in §1, for the complex

$$\begin{array}{ccc} \mathcal{O}^*_{X[\varepsilon]} & \longrightarrow & \Omega^1_{X[\varepsilon]} \\ \wr\Vert & & \wr\Vert \\ \mathcal{O}^*_X \cdot (1+\varepsilon\,\mathcal{O}_X) & \longrightarrow & \Omega^2_X \cdot \varepsilon \oplus \mathcal{O}_X \cdot d\varepsilon \oplus \Omega^2_X \end{array} .$$

From this point of view, it is natural to work rather with relative differentials, i.e., to set $d\varepsilon = 0$. The tangential data obtained by pulling back N over $X[\varepsilon]$ becomes a class

(2.11) $$s(f,g) \in \mathbf{H}^1(X, \mathcal{O}_X \to \Omega^1_X) \cong \Omega^1_R/dR.$$

If $f = 1 + f_1\cdot\varepsilon$, $g = g_0 + g_1\cdot\varepsilon$, an easy calculation from (1.9) yields

(2.12) $$s(f,g) = f_1\frac{dg_0}{g_0} .$$

It is important to realize (*) that the same construction can be carried out with $R[\varepsilon]$ replaced by $R \otimes_C A$ for any augmented artinian ring A, and yields an invariant in $(\Omega^1_R/dR) \otimes_C \mathrm{Ker}(A \to C)$. If we take, for example, $A = R[\varepsilon,\delta]/(\varepsilon^2,\delta^2)$, $f = 1 + a\varepsilon$, $g = 1 + b\delta$; $a,b \in R$, we get

(2.13) $$(adb) \otimes \delta\varepsilon \in (\Omega^1_R/dR) \otimes_C (\varepsilon,\delta,\delta\varepsilon).$$

It follows easily from (2.7) and (2.12) that the exact sequence of tangent funtors one gets by applying s (2.11) to $K_2(R[\varepsilon])$ is exactly the sequence obtained by reducing (2.9) modulo exact differentials

(2.14) $$0 \to \Omega^1_R/dR \to \mathfrak{g} \to sl(R) \to 0.$$

The key point is that (2.14) <u>is</u> an exact sequence of lie algebras. To see this, note that $\mathfrak{g}$ is a quotient of $t_{St}(R)$ and hence is commutative by [13]. We write $x^{r\varepsilon}_{ij}$, $h_{ij}(r\varepsilon)$ for the images in $\mathfrak{g}$ of the corresponding elements in $t_{St}(R) \subset St(R[\varepsilon])$ (notation as in [14]). $e_{ij}(r) \in sl(R)$, $i \neq j$, has entries r in the $(i,j)^{th}$ place and zeroes elsewhere, and $d_{ij}(r)$ is diagonal with r in the $(i,i)^{th}$ place and -r in the $(j,j)^{th}$. We define a group-theoretic section

(2.15) $$\sigma: sl(R) \to \mathfrak{g}, \quad \sigma(e_{ij}(r)) = x^{r\varepsilon}_{ij}, \quad \sigma(d_{ij}(r)) = h_{ij}(1 + r\varepsilon).$$

To define a lie algebra structure on $\mathfrak{g}$, we need to describe (2.14) in terms of a lie algebra cocycle. To do this in a natural way, we work with the ring $C[\varepsilon,\delta]/(\varepsilon^2,\delta^2)$ and consider the extension (whose existence was remarked above)

$$0 \to (\Omega^1_R/dR) \underset{C}{\otimes} (\varepsilon,\delta,\varepsilon\delta) \to \Gamma \to SL((\varepsilon,\delta,\varepsilon\delta)R) \to 1 .$$

Lemma (2.16). In Γ we have the identity for $r,s \in R$

$$[x_{ij}^{r\varepsilon}, x_{ji}^{s\delta}] h_{ij}(1 - rs\varepsilon\delta) = -rds \otimes \varepsilon\delta .$$

Proof. We use the computations in [13]. In particular (cf. op.cit. 1.2)

$$\langle s\delta, -r\varepsilon\rangle = x_{ij}^{r\varepsilon} \cdot x_{ji}^{s\delta} \cdot x_{ij}^{-r\varepsilon} \cdot x_{ji}^{-s\delta} h_{ij}(1 - rs\varepsilon\delta).$$

(The general definition is

$$\langle a,b\rangle = x_{ji}^{(-b(1+ab)^{-1})} x_{ij}^{a} x_{ji}^{b} x_{ij}^{(-a(1+ab)^{-1})} h_{ij}^{-1}(1+ab).$$

We have simplified, using $\varepsilon^2 = \delta^2 = 0$ and $h_{ij}(x) = h_{ji}(x)^{-1}$.) Moreover by (op.cit. 3.12),

$$\langle s\delta, -r\varepsilon\rangle = -rds \otimes \delta\varepsilon \in (\Omega^1_R/dR) \underset{C}{\otimes} (\varepsilon,\delta,\delta\varepsilon). \quad \text{Q.E.D.}$$

We now want to describe the lie algebra structure on $\mathfrak{g}$, using the section σ. Lemma (2.16) suggests we should have

(2.17)
$$[\sigma(e_{ij}(r)),\sigma(e_{ji}(s))] - \sigma([e_{ij}(r),e_{ji}(s)]) = -rds$$
$$[\sigma(e_{ij}(r),\sigma(e_{k\ell}(s))] = \sigma([e_{ij}(r),e_{k\ell}(s)]) , \quad k \neq j \text{ or } \ell \neq i.$$

Lemma (2.18). For $n > 2$, $B = (b_{ij}) \in sl_n(R)$, write $dB = (db_{ij})$ = the corresponding matrix of differentials. Then if k is a commutative ring and R is a k-algebra (commutative with 1) the map

$$\psi: \Lambda_k^2 sl_n(R) \to \Omega^1_{R/k}/dR \qquad \psi(A \wedge B) = -Tr(A \cdot dB)$$

is a two-cocycle for the k-lie algebra $sl(R)$, and hence defines a central extension of k-lie algebras

$$0 \to \Omega^1_{R/k}/dR \to L \to sl_n(R) \to 0 .$$

*For details, see [5].

Proof. In any case, ψ defines a 2-cochain, and we have

$$\begin{aligned}\partial\psi(A_0,A_1,A_2) &= \sum_{i<j} (-1)^{i+j+1}\psi([A_i,A_j],A_0,\ldots,\hat{A}_i,\ldots,\hat{A}_j,\ldots,A_2)\\ &= \psi([A_0,A_1],A_2) - \psi([A_0,A_2],A_1) + \psi([A_1,A_2],A_0)\\ &= -\mathrm{Tr}([A_0,A_1]dA_2) + \mathrm{Tr}([A_0,A_2]dA_1) - \mathrm{Tr}([A_1,A_2]dA_0) .\end{aligned}$$

Since we are computing modulo exact differentials, and $\mathrm{Tr}(AB) = \mathrm{Tr}(BA)$, we find $\mathrm{Tr}(A\cdot dB) = -\mathrm{Tr}(dA \cdot B) = -\mathrm{Tr}(BdA)$. Hence

$$\begin{aligned}\mathrm{Tr}([A_0,A_1]dA_2) &= -\mathrm{Tr}(A_2 d[A_0,A_1]) = -\mathrm{Tr}(A_2 dA_0 A_1 + A_2 A_0 dA_1 - A_2 dA_1 A_0 - A_2 A_1 dA_0)\\ &= \mathrm{Tr}(A_0 A_2 dA_1 + A_2 A_1 dA_0 - A_1 A_2 dA_0 - A_2 A_0 dA_1) .\end{aligned}$$

Substituting in the above yields $\partial\psi(A_0,A_1,A_2) = 0$. Q.E.D.

Remark (2.19). ψ is in fact a relative cocycle for $sl_n(R)$ rel. $sl_n(k)$.

Note when R is the ring of functions on an affine Riemann surface X, we have $\Omega^1_{R/\mathbf{C}}/dR \cong H^1(X,\mathbf{C})$. The discussion in this section can be summarized as follows: infinitesimal deformation (in the sense of (2.11)) of the extension

(2.4) $$0 \to H^1(X,\mathbf{C})^* \to \underline{\underline{W}}(R) \to SL(R) \to 1$$

leads to a central extension of $\mathbf{C}$-lie algebras

(2.20) $$0 \to H^1(X,\mathbf{C}) \to \mathfrak{g} \to sl(R) \to 0.$$

The cocycle for this extension is

(2.21) $$A \wedge B \longrightarrow -\mathrm{Tr}(A\cdot dB).$$

§3. Universality

The purpose of this section is to prove:

Theorem (3.1). Let k be a commutative ring with $1/2 \in k$, R a k-algebra (commutative with 1). Let $n \geq 5$ be an integer. Then the extension (2.18)

$$0 \to \Omega^1_{R/k}/dR \to L \to sl_n(R) \to 0$$

defined by the 2-cocycle $Tr(A \cdot dB)$ is the universal central extension of $sl_n(R)$ in the category of k-lie algebras. In particular,

$$H_{2,k\text{-lie alg.}}(sl_n(R),k) \cong H_{2,k\text{-lie alg.}}(sl_n(R),sl_n(k);k)$$
$$\cong \Omega^1_{R/k}/dR \ .$$

Proof (compare [14]). Define a k-vector space $\mathcal{M}$ with generators $\varepsilon_{ij}(r)$, $1 \leq i \neq j \leq n$, $r \in R$ and relations

$$\varepsilon_{ij}(ar + bs) = a\varepsilon_{ij}(r) + b\varepsilon_{ij}(s); \ r,s \in R; \ a,b \in k.$$

Let M be the free k-lie algebra on $\mathcal{M}$ subject to relations

$$(3.3) \qquad [\varepsilon_{ij}(r),\varepsilon_{k\ell}(s)] = \begin{cases} 0 & k \neq j \text{ and } \ell \neq i \\ \varepsilon_{i\ell}(rs) & j = k, \ \ell \neq i \ . \end{cases}$$

Clearly the association $\varepsilon_{ij} \mapsto e_{ij}$ defines a surjection

$$(3.4) \qquad \pi\colon M \to sl_n(R) \to 0.$$

Define elements $(i \neq j)$

$$(3.5) \qquad H_{ij}(r,s) = [\varepsilon_{ij}(r),\varepsilon_{ji}(s)], \ I_{ij}(r,s) = H_{ij}(r,s) - H_{ij}(s,r)$$
$$= H_{ij}(r,s) + H_{ji}(r,s) \ .$$

Clearly $I_{ij}(r,s)$ Ker π.

Lemma (3.6). $I_{ij}(r,s) \in$ Center(M).

Proof. By (3.3), it suffices to prove

$$[\varepsilon_{\ell m}(q), I_{ij}(r,s)] = 0$$

when one of the indices ℓ, m is distinct from i and j. Since $I_{ij}(\alpha,s) = I_{ji}(r,s)$, it suffices to consider the cases $\ell = i$, $m \neq i,j$, and $m = i$, $\ell \neq i,j$.

$$[\varepsilon_{im}(q),[\varepsilon_{ij}(r),\varepsilon_{ji}(s)]] = [[\varepsilon_{im}(q),\varepsilon_{ij}(r)],\varepsilon_{ji}(s)] + [\varepsilon_{ij}(r),[\varepsilon_{im}(q),\varepsilon_{ji}(s)]]$$

$$= -[\varepsilon_{ij}(r),\varepsilon_{jm}(qs)] = -\varepsilon_{im}(qrs)$$

$$[\varepsilon_{\ell i}(q),[\varepsilon_{ij}(r),\varepsilon_{ji}(s)]] = [[\varepsilon_{\ell i}(q),\varepsilon_{ij}(r)],\varepsilon_{ji}(s)] + [\varepsilon_{ij}(r),[\varepsilon_{\ell i}(q),\varepsilon_{ji}(s)]]$$

$$= \varepsilon_{\ell i}(qrs) .$$

The right-hand side is symmetric in r and s in both cases, so the lemma follows from (3.5). Q.E.D.

Lemma (3.7). $H_{ij}(q,rs) = H_{i\ell}(qr,s) + H_{\ell j}(qs,r)$.

Proof.

$$H_{ij}(q,rs) = [e_{ij}(q),[e_{j\ell}(r),e_{\ell i}(s)]]$$

$$= [[e_{ij}(q),e_{j\ell}(r)],e_{\ell i}(s)] + [e_{j\ell}(r),[e_{ij}(q),e_{\ell i}(s)]]$$

$$= [e_{i\ell}(qr),e_{\ell i}(s)] + [e_{j\ell}(r) - e_{\ell j}(qs)]$$

$$= H_{i\ell}(qr,s) + H_{\ell j}(qs,r). \qquad \text{Q.E.D.}$$

Lemma (3.8). $I_{ij}(r,s)$ is independent of i and j.

Proof. Setting $r = 1$ in (3.7) and then interchanging q and s gives

$$H_{ij}(q,s) = H_{i\ell}(q,s) + H_{\ell j}(qs,1)$$
$$H_{ij}(s,q) = H_{i\ell}(s,q)+H_{\ell j}(qs,1) .$$

Now subtract to get $I_{ij}(q,s) = I_{i\ell}(q,s)$. Independence of i results from $I_{ij}(r,s) = I_{ji}(r,s)$ (3.5). Q.E.D.

We now write

(3.9) $I(r,s) = I_{ij}(r,s).$

Lemma (3.10). The assignment

$$r\,ds \longrightarrow I(r,s)$$

defines a map

(3.11) $$\rho:\Omega^1_{R/k}/dR \to \text{Center } M \subset \text{Ker}(\pi:M \to sl_n(R)).$$

Proof. We must show

(i) $I(r,s)$ is k-bilinear and alternating.

(ii) $I(q,rs) = I(qr,s) + I(qs,r)$.

(iii) $I(1,r) = 0$.

proof of (i): k-bilinearity is clear, and I is alternating by (3.5).

proof of (ii): rewrite (3.7) and then switch i and j, r and s, getting

$$H_{ij}(q,rs) = H_{i\ell}(qr,s) + H_{\ell j}(qs,r)$$
$$H_{ji}(q,rs) = H_{\ell i}(qr,s) + H_{j\ell}(qs,r)\ .$$

Assertion (ii) follows by addition.

proof of (iii): $I(1,r) \underset{(i)}{=} -I(r,1) \underset{(ii)}{=} -2I(r,1) \Longrightarrow I(1,r) = 0.$ Q.E.D.

Write $\overline{M} = \text{Coker}(\rho)$, and let $\overline{H}_{ij}(r,s)$, $\overline{\varepsilon}_{ij}(r)$ denote the obvious elements in $\overline{M}$.

Lemma (3.12) Let $\mathcal{K} \subset \overline{M}$ be the subgroup generated by all $\overline{H}_{ij}(1,s)$. Then $\mathcal{K}$ is generated by all $\overline{H}_{ij}(1,s)$. As a k-vector space, $\overline{M}$ is generated by the $\overline{\varepsilon}_{ij}(r)$ and the $\overline{H}_{ij}(1,s)$.

Proof. Let $\mathcal{K}' \subset \mathcal{K}$ be the subgroup generated by all $\overline{H}_{ij}(1,s)$. Taking $q = 1$ in (3.7) and using $\overline{H}_{ij}(r,s) = -\overline{H}_{ji}(r,s) = -\overline{H}_{ji}(s,r)$, we get

$$\overline{H}_{i\ell}(r,s) \equiv \overline{H}_{j\ell}(r,s) \mod \mathcal{K}'\ .$$

Since $n > 4$, it follows that $\overline{H}_{ij}(r,s)$ is independent (mod $\mathcal{K}'$) of i and j. Because $\overline{H}_{ij}(r,s) = \overline{H}_{ji}(r,s)$, we get $\overline{H}_{ij}(r,s) \in \mathcal{K}'$.

Let $M' \subset \overline{M}$ be the linear span of the $\bar{\varepsilon}_{ij}(r)$ and the $\overline{H}_{ij}(1,s)$. We must show M' is closed under bracket. Using (3.3), the Jacobi identity, and $\mathcal{N}' = \mathcal{N}$, the problem reduces to computing

$$[\bar{\varepsilon}_{ij}(r),\bar{H}_{ij}(1,s)] \underset{(3.7)}{=} [\bar{\varepsilon}_{ij}(r),\bar{H}_{i\ell}(1,s)] + [\bar{\varepsilon}_{ij}(r),\bar{H}_{\ell j}(1,s)]$$

$$\underset{(\text{Jacobi})}{=} [\bar{\varepsilon}_{i\ell}(1),[\bar{\varepsilon}_{ij}(r),\bar{\varepsilon}_{\ell i}(s)]] + [\bar{\varepsilon}_{\ell j}(1),[\bar{\varepsilon}_{ij}(r),\bar{\varepsilon}_{j\ell}(s)]]$$

$$= -2\bar{\varepsilon}_{ij}(rs) \in M'. \qquad \text{Q.E.D.}$$

<u>Corollary (3.14)</u>. The map $\bar{\pi}:\overline{M} \to sl_n(R)$ induced from (3.4) is an isomorphism.

<u>Proof.</u> $\bar{\pi}$ is clearly surjective. To prove injectivity, it suffices to show that any vector space relation between the elements

$$\bar{\pi}(\bar{\varepsilon}_{ij}(r)) = e_{ij}(r), \quad \bar{\pi}(\overline{H}_{ij}(1,s)) = d_{ij}(s)$$

hold in $\overline{M}$. Such relations are generated by

$$d_{ij}(s) + d_{j\ell}(s) = d_{i\ell}(s)$$

so it suffices to show

$$\overline{H}_{ij}(1,s) + \overline{H}_{j\ell}(1,s) = \overline{H}_{i\ell}(1,s).$$

This follows from (3.7), using $\overline{H}_{i\ell}(1,s) = \overline{H}_{i\ell}(s,1)$. Q.E.D.

From (3.11) and (3.14) we find an exact sequence

(3.15) $$\Omega^1_{R/k}/dR \xrightarrow{\rho} M \xrightarrow{\pi} sl_n(R) \longrightarrow 0.$$

<u>Lemma (3.16)</u>. M is the universal central extension of $sl_n(R)$ in the category of k-lie algebras.

<u>Proof.</u> Let

$$0 \to A \to N \to sl_n(R) \to 0$$

be another central extension. Choose elements $E'_{ij}(r) \in N$ lifting the $e_{ij}(r)$. I claim first

(3.17) $$[E'_{ij}(r),E'_{\ell m}(s)] = 0 \quad , \quad i \neq m \text{ and } j \neq \ell.$$

In fact, $E'_{\ell m}(s) = [E'_{\ell p}(s),E'_{pm}(1)] + a$, $a \in A$, $p \neq i,j,\ell,m$. (Here is where we use $n \geq 5$.) Thus

$$[E'_{ij}(r),E'_{\ell m}(s)] = [[E'_{ij}(r),E'_{\ell p}(s)],E'_{pm}(1)] + [E'_{\ell p}(s),[E'_{ij}(r),E'_{\ell p}(s)]]$$

$$= 0$$

because both inner brackets lie in A. Now define

$$E_{ij}(r) = [E'_{i\ell}(r),\ E'_{\ell j}(1)] \equiv E'_{ij}(r) \mod A.$$

Note this is independent of ℓ because

$$0 = [E'_{\ell m}(1),[E'_{i\ell}(r),E'_{mj}(1)]] = -[[E'_{i\ell}(r),E'_{\ell m}(1)],E'_{mj}(1)] + [E'_{i\ell}(r),E'_{\ell j}(1)].$$

Define $\eta:M \to N$, $\eta(\varepsilon_{ij}(r)) = E_{ij}(r)$. To check this works, we must show

$$[E_{ij}(r),E_{\ell m}(s)] = \begin{cases} 0 & \ell \neq j \quad i \neq m \\ E_{im}(rs) & \ell = j \quad i \neq m \\ -E_{\ell j}(rs) & m = i \quad \ell \neq j \end{cases} .$$

The first of these follows from (3.17). The second is

$$[E_{ij}(r),E_{jm}(s)] = [E_{ij}(r),[E_{j\ell}(s),E_{\ell m}(1)]]$$

$$= [E_{i\ell}(rs) + a,\ E_{\ell m}(1)] = [E_{i\ell}(rs),E_{\ell m}(1)] = E_{im}(rs)$$

(Here $a \in A$.) The third identity is similar.

Since $M = [M,M]$, we deduce from the above that M is the universal central extension of $sl_n(R)$. Q.E.D.

We turn now to the proof of (3.1). We have

$$\begin{array}{ccccccc} & & \Omega^1_{R/k}/dR & \xrightarrow{\rho} & N & \longrightarrow & sl_n(R) & \longrightarrow & 0 \\ & & \vdots\downarrow & & \downarrow \eta & & \| & & \\ 0 & \longrightarrow & \Omega^1_{R/k}/dR & \longrightarrow & L & \longrightarrow & sl_n(R) & \longrightarrow & 0 \end{array}$$

Note $L = (\Omega^1_{R/k}/dR) \oplus sl_n(R)$, and the bracket is

$$[(0,A),(0,B)] = (Tr(A\cdot dB),[A,B]).$$

Thus

$$n\rho(rds) = n(I(r,s)) = n[\varepsilon_{ij}(r),\varepsilon_{ji}(s)] - n[\varepsilon_{ij}(s),\varepsilon_{ji}(r)].$$

Moreover

$$n(\varepsilon_{ij}(r)) = [(0,e_{i\ell}(r)),(0,e_{\ell j}(1))] = (0,e_{ij}(r))$$

whence

$$n\rho(rds) = (rds - sdr,0) = (2rds,0).$$

This establishes an isomorphism of extensions over $sl_n(R)$ between M and L and completes the proof of (3.1).

Bibliography

1. Bloch, S., Higher regulators, Algebraic K-theory, and zeta functions of elliptic curves, lecture notes, U.C. Irvine.

2. ________, Algebraic K-theory and zeta functions of elliptic curves, ICM, Helsinki, 1978.

3. ________, Appliations of the dilogarithm function in algebraic K-theory and algebraic geometry, Intl. Symp. on Algebraic Geometry, Kyoto (1977) 103-114.

4. ________, Lecture Notes on Algebraic Cycles, Duke University Lecture notes in mathematics IV, (1979).

5. ________, K_2 of artinian $\mathbb{Q}$-algebras, with aplications to algebraic cycles, Comm. in Algebra 3 (1975) 405-428.

6. ________, and Ogus, A., Gersten's conjecture and the homology of schemes, Ann. Ec. Norm. Sup. t. 7, fasc. 2 (1974).

7. Cartier, P., Théorie des groupes, fonctions théta, et modules des variétés abeliennes, Sem. Bourbaki, no. 338 (1968).

8. Deligne, P., Le Symbole modéré handwritten notes, dated July 25, 1979.

9. Garland, H., Dedekind's η-function and the cohomology of infinite-dimentional lie algebras, Proc. Nat. Acad. Sc. (U.S.A.) Vol. 72 (1975) 2493-2495.

10. ________, and Lepowsky, J., Lie algebra homology and the MacDonald-Kac formula, Inventiones Math. vol. 34 (1976) 37-76.

11. Katz, N., Nilpotent connections and the monodromy theorem: appliations of a result of Turrittin, Publ. Math. I.H.E.S. 39(1970).

12. Lepowsky, J., Generalized verma modules, loop space cohomology, and MacDonald-type identities, Ann. Ec. Norm. Sup. 4e serie, t. 12, (1979) 169-234.

13. Maazen, H., and Stienstra, J., A presentation for K_2 of split radical pairs preprint no. 31, Utrecht University (1976).

14. Milnor, J., Introduction to Algebraic K-theory, Ann. Math. studies no. 72, Princeton University Press, Princeton (1971).

15. Ramakrishnan, handwritten notes on the dilogarithm.

16. Tate, J., On the torsion in K_2 of fields, Algebraic Number Theory Symposium, Kyoto (1976).

17. van der Kallen, W.L.J., Sur le K_2 des nombres duanx, C.R. Acad. Sci. Paris 273 (1974) 1204-1207.

A SPECTRAL SEQUENCE FOR THE K-THEORY OF AFFINE GLUED SCHEMES

BARRY H. DAYTON

Northeastern Illinois University
Chicago, IL. 60625 USA

and

Queen's University
Kingston, Ontario
Canada

and

CHARLES A. WEIBEL*

University of Pennsylvania
Philadelphia, PA 19104 USA

Abstract An affine scheme is 'glued' if it is the colimit of a finite diagram of affine schemes. We first develop several recognition criteria for determining when an affine scheme is glued. Under mild hypotheses, for example, glued schemes are seminormal. We then investigate the K-theory of glued schemes and develop an Atiyah-Hirzebruch type spectral sequence which converges to the Karoubi-Villamayor K-theory of the glued scheme. This allows us to compute K_0 of some interesting rings and generalize a number of previous results in the literature.

AMS (MOS) Subject classification (1970) - Primary 18F25
Secondary 13D15, 14F15, 14M99, 18A30, 18G40

Key Words and Phrases: Algebraic K-Theory, Spectral Sequence, K_1-regular, Seminormal, Karoubi-Villamayor K-theory

---

*Second author supported by NSF Grant

This paper is motivated by the paper [20] of L. Roberts where he calculates the Karoubi Villamayor K-theory of some reducible affine curves and surfaces. The present authors have been of the opinion for some time that Roberts' method could be extended to cover a larger class of surfaces and some varieties of higher dimensions. The purpose of this paper is to present such a generalization.

We recall the method used by Roberts. The varieties in question are constructed by a glueing operation which results in a conductor square

$$(*)\qquad \begin{array}{ccc} A & \longrightarrow & \overline{A} \\ \downarrow & & \downarrow \\ A/I & \longrightarrow & \overline{A}/I \end{array}$$

where A is the coordinate ring of the variety, $\overline{A}$ its integral closure, and I the conductor of A in $\overline{A}$. We now restrict ourselves to curves of "type II" as defined in [20]. The components of $\operatorname{Spec} A$ are $\operatorname{Spec} A_i$, $\overline{A}_i = \underset{\sim}{k}[t]$ for some field $\underset{\sim}{k}$, so $\overline{A} = \prod_{i=1}^{n} \underset{\sim}{k}[t]$ with one factor for each component, $A/I = \prod_{i=1}^{m} \underset{\sim}{k}$ with one factor for each non-regular point of $\operatorname{Spec} A$, and $\overline{A}/I = \prod_{i=1}^{M} \underset{\sim}{k}$ with one factor for each point of $\operatorname{Spec} \overline{A}$ over a non-regular point of $\operatorname{Spec} A$. In this situation $\overline{A} \to \overline{A}/I$ will be a GL-fibration so there is an exact Mayer-Vietoris sequence

$$KV_{q+1}(A/I) \oplus KV_{q+1}(\bar{A}) \xrightarrow{\alpha_{q+1}} KV_{q+1}(\bar{A}/I) \to KV_q(A) \to KV_q(A/I) \oplus KV_q(\bar{A}) \xrightarrow{\alpha_q}$$

It thus suffices to determine the maps α_q, which are maps from a product $\prod KV_q(\underset{\sim}{k})$ $(= KV_q(A/I) \oplus KV_q(\bar{A}))$ to another such product. Roberts shows these maps can be described by a matrix α which is the incidence matrix of a bipartite graph G. The vertices of G are the non-regular points and the components of Spec A and the edges of G are inclusions of non-regular points in components, i.e. in 1-1 correspondence with points of Spec $\bar{A}$ over non-regular points.

At this point we part with Roberts' calculation and note that $\ker \alpha = H^0(G)$, $\operatorname{coker} \alpha = H^1(G)$ where $H^*(G)$ is the cohomology of the graph G [4, §4]. Thus we conclude that there is a short exact sequence (which in this case splits)

$$(**) \quad 0 \to H^1(G;KV_{q+1}(\underset{\sim}{k})) \to KV_q(A) \to H^0(G;KV_q(\underset{\sim}{k})) \to 0$$

Hence the number $M - m - n + 1$ (when G is connected) that Roberts obtains should be interpreted as the rank of $H^1(G)$. We will see (Example 4.7 below) that, even for surfaces, numerical invariants are not sufficient to describe $KV_*(A)$: the cohomology groups may not be direct sums of $KV_*(\underset{\sim}{k})$'s and the sequence (**), with dimensions raised by one, may not split.

We now interpret the graph G as the geometric realization of a category $\mathcal{G}$ whose objects are the non-regular points and the components of Spec A and whose morphisms are the inclusions. If we let F be the contravariant functor $\mathcal{G} \to$ ($\underset{\sim}{k}$-algebras) taking the value $\underset{\sim}{k}$ on the non-regular points, $\underset{\sim}{k}[t]$ on the components with morphisms $\underset{\sim}{k}[t] \to \underset{\sim}{k}$ representing the inclusions, then the inverse limit of F on $\mathcal{G}$, $\lim F(\mathcal{G})$, is seen to be the pullback in (*), i.e. $\lim F(\mathcal{G}) = A$.

The above suggests that the proper generalization of Roberts' calculation is to give the KV-theory of the limit $\lim A(\mathcal{G})$ of a $\underset{\sim}{k}$-algebra valued functor on a finite category $\mathcal{G}$ in terms of the cohomology of the realization $B\mathcal{G}$ of $\mathcal{G}$. In fact with appropriate hypotheses on $\mathcal{G}$ and A we are able to obtain a spectral sequence

$$E_2^{p,-q} = H^p(\mathcal{G};KV_q(\underset{\sim}{k})) \Rightarrow KV_{q-p}(\lim A(\mathcal{G}))$$

In §0 we summarize the finite category theory we will need and introduce the notion of a cellular category. We will need the category $\mathcal{G}$ above to be cellular in order to obtain the desired E_2-term for our spectral sequence.

In §1 we discuss limits of ring-valued functors on finite categories and give recognition criteria for such limits. For instance, under mild conditions these limits are seminormal.

Our method uses, as does Roberts', the Karoubi-Villamayor Mayer-Vietoris sequences of squares similar to (*). Certain maps must be GL-fibrations for these sequences to exist. The simplest condition that a map be a GL-fibration is that it be surjective with K_1-regular codomain. In §2 we discuss the K_1-regularity of rings $\lim A(\mathfrak{a})$. This is a non-trivial problem as, even though $\lim A(\mathfrak{a})$ may be seminormal, it may not even be K_0-regular. §3 is devoted to the proof of our main theorem. In §4 we give a number of applications including the KV-theory of the reducible varieties studied by Roberts in [20], the KV-theory of hyperplanes in [6], and the KV-theory of certain unions of linear subvarieties in $\mathbb{A}^n_k$.

In §5 we give a general spectral sequence which includes our spectral sequence of §3 and the classical Atiyah-Hirzebruch spectral sequence for topological K-theory. We are able to sharpen some calculations by comparing these two spectral sequences. We are also able to generalize statement (4) of [20, p. 51] to show that all the rings $\lim A(\mathfrak{a})$ to which our spectral sequence of §3 applies are K_0-regular.

By KV_* we mean the groups introduced in [12]. These are denoted K^h_i $i \geq 0$ in [9] and coincide with the classical K_0 and the negative groups of Bass [3] when $* \leq 0$.

§0. Bookkeeping Categories

If α,β are objects of a small category $\mathcal{C}$ we write $\alpha \leq \beta$ if there is a morphism $\alpha \to \beta$. We will call $\mathcal{C}$ essentially a poset if the objects of $\mathcal{C}$ form a partially ordered set (poset) and the only endomorphisms are the identities. If, in addition, there is at most one morphism between any two objects we call $\mathcal{C}$ (by abuse) a poset.

Suppose $\mathcal{C}$ is essentially a poset. An object α of $\mathcal{C}$ is called maximal if it is maximal under the ordering $\leq$. An object γ of $\mathcal{C}$ will be called mediate if $\gamma < \alpha$, $\gamma < \beta$ for distinct maximal elements α,β of $\mathcal{C}$. We say that a full subcategory $\mathcal{B}$ of $\mathcal{C}$ is generated by a set S of objects when $\beta \in \mathcal{B}$ iff $\beta \leq \sigma$ for some $\sigma \in S$.

By the cohomology $H^*(\mathcal{C};G)$ of the category $\mathcal{C}$ with coefficients in the abelian group G we mean $H^*(B\mathcal{C};G)$, where $B\mathcal{C}$ is the geometric realization of the nerve $N\mathcal{C}$ of $\mathcal{C}$ (see [18]). The topological space $B\mathcal{C}$ is particularly nice when $\mathcal{C}$ is essentially a poset, since $B\mathcal{C}$ is essentially a simplicial complex with one p-cell for every chain $\sigma_0 \to \cdots \to \sigma_p$ of arrows, none of which is an identity. (Essentially a complex means among other things that there may be more than one cell with a given set of vertices : consider the example of the category $\cdot \rightrightarrows \cdot$).

When $\mathcal{C}$ is essentially a poset, $H^*(B\mathcal{C};G)$ can be described directly by normalizing the chain complex associated with $N\mathcal{C}$, as in [14, p5] and [13, p236]. We let $C_p(\mathcal{C})$ be the free abelian group on the set of all chains $\sigma : \sigma_0 \to \ldots \to \sigma_p$ of non-identity arrows and define $\partial_i : C_p(\mathcal{C}) \to C_{p-1}(\mathcal{C})$ as follows: $\partial_0(\sigma)$ is $\sigma_1 \to \ldots \to \sigma_p$, $\partial_i(\sigma)$ is $\sigma_0 \to \ldots \to \sigma_{i-1} \to \sigma_{i+1} \to \ldots \to \sigma_p$ for $0 < i < p$ and $\partial_p(\sigma)$ is $\sigma_0 \to \ldots \to \sigma_{p-1}$. With $\delta_p = \Sigma(-1)^i \partial_i$ we have a chain complex $C_*(\mathcal{C})$, and $H^*(\mathcal{C};G)$ is the homology of $\mathrm{Hom}(C_*(\mathcal{C}),G)$. It is immediate that $H^q(\mathcal{C};G) = 0$ whenever q is larger than the length of every chain of arrows in $\mathcal{C}$. $H^*(-;G)$ is a functor from small categories to graded abelian groups, taking coproducts (disjoint unions) to products.

From [18, prop 2, p84] we see that if $\mathcal{C}$ has either an initial or terminal object then $B\mathcal{C}$ is contractible and $H^*(\mathcal{C})$ is the cohomology of a point. If $C\mathcal{C}$ is the category obtained from $\mathcal{C}$ by formally adjoining a terminal object, then $BC\mathcal{C}$ is topologically the (contractible) cone of $B\mathcal{C}$; we define the reduced group $\tilde{H}^*(\mathcal{C},G)$ to be the cokernel of $H^*(C\mathcal{C};G) \to H^*(\mathcal{C};G)$.

For an object α of a category $\mathcal{C}$, $\mathcal{C}' \subseteq \mathcal{C}$ we write $\mathcal{C}' \downarrow \alpha$ for the comma category whose objects are all morphisms $\beta \to \alpha$, $\beta \in \mathcal{C}'$. When $\mathcal{C}$ is essentially a finite poset, let $\alpha_1, \ldots, \alpha_n$ be maximal elements of $\mathcal{C}$ and $\mathcal{B}$ the full subcategory on $\mathcal{C} - \{\alpha_1, \ldots, \alpha_n\}$. The following square is a pushout in the category of small categories:

$$(0.1)\qquad \begin{array}{ccc} \coprod_i \mathcal{B}\downarrow\alpha_i & \longrightarrow & \coprod_i \mathcal{G}\downarrow\alpha_i \\ \downarrow & & \downarrow \\ \mathcal{B} & \longrightarrow & \mathcal{G} \end{array}$$

The geometric realization of (0.1) is a pushout in the category of topological spaces.

Since the comma categories $\mathcal{G}\downarrow\alpha$ have terminal objects the $B(\mathcal{G}\downarrow\alpha_i)$ are contractible. In fact $\mathcal{G}\downarrow\alpha_i \cong C(\mathcal{B}\downarrow\alpha_i)$. The standard construction of the Mayer-Vietoris sequence in cohomology (q.v. [10]) yields the exact sequence

$$(0.2)\qquad \to \prod_i \tilde{H}^{q-1}(\mathcal{B}\downarrow\alpha_i) \to \tilde{H}^q(\mathcal{G}) \to \tilde{H}^q(\mathcal{B}) \to \prod_i \tilde{H}^q(\mathcal{B}\downarrow\alpha) \to \ldots$$

<u>Example 0.3</u>: Let X be a simplicial complex and let $\mathcal{G}$ denote the associated poset of simplices in X. The space $B\mathcal{G}$ is the barycentric subdivision of X and the maximal elements of $\mathcal{G}$ correspond to the top cells of X. If α is a top cell $B(\mathcal{B}\downarrow\alpha)$ is homeomorphic to the boundary of this cell, i.e. to a sphere.

<u>Example 0.4</u>: If $\{\alpha_i\}$ is a set of hyperplanes in general position in $\mathbb{A}^{n+1}$, let $\mathcal{G}$ be the poset of intersections of the α_i. This is the opposite of the poset $\mathcal{C}(\{\alpha_i\})$ of [6], where it was shown that $B\mathcal{G}$ is a bouquet of n-spheres, and each $B(\mathcal{B}\downarrow\alpha)$ is a bouquet of (n-1)-spheres.

<u>Example 0.5:</u> When $\mathcal{C}$ is one dimensional (ie. there is no pair of nontrivial composable arrows) $B\mathcal{C}$ is a bipartite graph whose cohomology is discussed in [4]. The objects of $\mathcal{C}$ are the vertices of $B\mathcal{C}$ and the morphisms are edges of $B\mathcal{C}$. For α maximal in $\mathcal{C}$, $\mathcal{B}\downarrow\alpha$ is the discrete set of edges incident to the vertex α.

Motivated by these examples we introduce some definitions. Let $\mathcal{C}$ be essentially a poset. If $\tilde{H}^p(\mathcal{C};G)=0$ for $p\neq n$ we call $\mathcal{C}$ an <u>n-bouquet</u>. We call $\mathcal{C}$ <u>graded</u> if there is a functor $\dim : \mathcal{C} \to \mathbb{N}$ sending distinct objects $\beta < \alpha$ to distinct integers (i.e. $\dim(\beta)<\dim(\alpha)$). By $\mathcal{C}^n$ we will mean the full subcategory of all α with $\dim(\alpha) \leq n$ and we will call α an n-cell if $\dim(\alpha) = n$. Note that if the set of objects is finite, $\mathcal{C}$ can be graded in at least one way with minimal objects having dimension 0. Finally we call a graded category <u>cellular</u> if a) for every α there is an 0-cell v and a map $v \to \alpha$, and b) for every n the maps $H^p(\mathcal{C}^{n+1};G) \to H^p(\mathcal{C}^n;G)$ are isomorphisms when $p < n$ and injections when $p = n$.

<u>Proposition 0.6</u>: The following are equivalent for a graded category $\mathcal{C}$:

a) $\mathcal{C}$ is cellular

b) For every p-cell α in $\mathcal{C}$ with $p \neq 0$ the category $\mathcal{C}^{p-1}\downarrow\alpha$ is a non-empty (p-1)-bouquet.

Furthermore, if $\mathcal{C}$ is cellular then so are the categories $\mathcal{C}^p$, $\mathcal{C}^p\downarrow\alpha$ for every $p \geq 0$ and every α in $\mathcal{C}$.

<u>Proof</u>: The cellularity of the $\mathcal{C}^p$ is immediate from the definition of cellularity. The equivalence of a),b) is now immediate from (0.2). To see that $\mathcal{C}^p\downarrow\alpha$ is cellular when $\mathcal{C}$ is, observe that for every object $\beta \to \alpha$ of $\mathcal{C}^p\downarrow\alpha$ the categories $\mathcal{C}\downarrow\beta$ and $(\mathcal{C}^p\downarrow\alpha)\downarrow(\beta\to\alpha)$ are naturally isomorphic.

<u>Examples 0.7.</u> The following posets are not cellular:

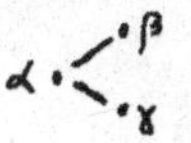

2 1

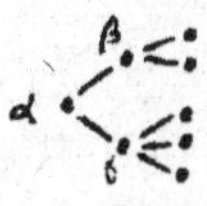

2 1 0

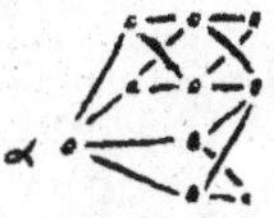

3 2 1 0

The first has no 0-cells, the second has $\mathcal{C}^1\downarrow\alpha$ disconnected, and the third has $B(\mathcal{C}^2\downarrow\alpha) = S^2 \vee S^1$. It is easy to see for any graded category $\mathcal{C}$ that: $\mathcal{C}^0$ is always cellular; $\mathcal{C}^1$ is cellular iff for every 1-cell α there is a map $v \to \alpha$ from a 0-cell v; and $\mathcal{C}^2$ is cellular iff $\mathcal{C}^1$ is cellular and $\mathcal{C}^1\downarrow\alpha$ is connected for every 2-cell α. We can "make" the above three posets cellular as follows:

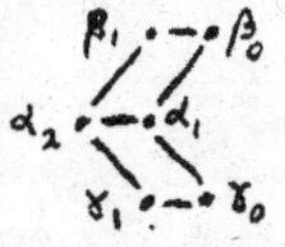

2 1 0

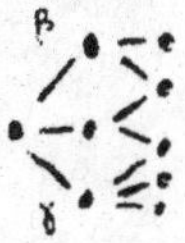

2 1 0

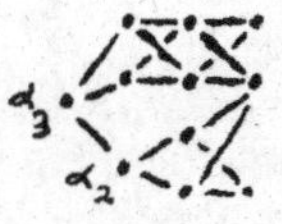

3 2 1 0

In the first example, all cells have been duplicated in one lower dimension to give a new poset. In the second example, a dummy one-cell is added to make $\mathcal{C}^1\downarrow\alpha$ connected.

In the third example, we have added a dummy 2-cell in order to get $B(\mathfrak{a}^2 \downarrow a_3) = S^2$. Unfortunately, these methods of making posets cellular seem completely ad hoc.

In §1 we will apply ring-valued functors A to $\mathfrak{a}$ and study the rings lim A. These ad hoc cellularization techniques do not change the ring lim A.

§1 Limits of Rings

Let $\mathcal{C}$ be a finite category which is essentially a poset and $A : \mathcal{C} \to$ (commutative rings) a contravariant functor. We are interested in the "inverse limit" of A on $\mathcal{C}$, variously denoted $A(\mathcal{C}) = \lim A = \lim A(\mathcal{C})$. Note that if $\{\alpha_1,\ldots,\alpha_n\}$ are the maximal elements of $\mathcal{C}$ then $\lim A = \{(a_1,\ldots,a_n) \in \prod_i A(\alpha_i) : A(f)(a_i) = A(g)(a_j)$ whenever $f : \gamma \to \alpha_i$, $g : \gamma \to \alpha_j$ are morphisms in $\mathcal{C}\}$. Thus $\lim A$ can be thought of as obtained by glueing the rings $A(\alpha_i)$ together along the rings $A(\gamma)$ for γ mediate.

We will refer to a ring as being <u>glued</u> if it arises as $\lim A$ for some $A : \mathcal{C} \to$ (commutative rings). This terminology arose in [23] in the following way. Let $A \subseteq B$ be a finite integral noetherian extension, α a prime of A and $v_1,\ldots,v_n$ the primes of B above α. The ring A' "obtained from B by glueing over α" is the pullback in the diagram

$$\begin{array}{ccc} A' & \longrightarrow & B \\ \downarrow & & \downarrow \\ A/\alpha & \longrightarrow & \prod B/v_i \end{array}$$

Traverso showed that the seminormalization of A in B could be obtained by a sequence of such glueings. The pull back above is a limit and more generally a sequence of such glueings can be obtained as a limit of a more complicated category. The concept of glueing was elaborated in [17] [19], [20] from a more geometric standpoint.

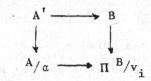

We now address the recognition problem: which affine rings/schemes are glued? If we are given a ring A and a poset of ideals $\mathfrak{a}$ (which we will view as ordered by reverse inclusion) we get a coherent set of maps $\varphi(\alpha): A \to A(\alpha) = A/\alpha$ and hence a map $A \to \lim A(\mathfrak{a})$. We would like to know when this map is an isomorphism. It is injective precisely when $\cap\{\alpha : \alpha \in \mathfrak{a}\} = 0$. Frequently it is easier to write A as a quotient ring of a nicer ring R (eg. if Spec $A \subseteq \mathbb{A}^n_k$ we take $R = k[x_1,\ldots,x_n]$). We then have coherent maps $\varphi(\alpha) = f^*\varphi(\alpha)$ and would like to know when the induced map $R \to \lim A(\mathfrak{a})$ is a surjection. More generally, we may be given a contravariant functor $A : \mathfrak{a} \to$ (commutative rings) and a coherent set of maps $\varphi(\alpha) : R \to A(\alpha)$, and would like to determine whether the induced map $\varphi : R \to \lim A(\mathfrak{a})$ is surjective.

We can somewhat iternalize the problem of when the induced map $\varphi : R \to \lim A(\mathfrak{a})$ is surjective. Let $B(\alpha) = \operatorname{Im}\varphi(\alpha)$, so that B is a functor on $\mathfrak{a}$ factoring through the poset $\tilde{\mathfrak{a}} = \mathfrak{a}/$(parallel arrows). If we define $I(\alpha) = \ker \varphi(\alpha)$ then $B(\alpha) = R/I(\alpha)$ and we can think of $\tilde{\mathfrak{a}}$ as the poset of ideals $I(\alpha)$ of R ordered by reverse inclusion. In this case we have

$$\lim B = \{(\bar{a}_1,\ldots,\bar{a}_n)\in\Pi B(\alpha_i) : a_i \equiv a_j \bmod I(\gamma)$$

$$\text{whenever } \alpha_i, \alpha_j \geq \gamma \text{ in } \mathcal{G}\}.$$

It is clear that lim B is a subring of lim A and that the two are equal if each $\varphi(\alpha)$ is surjective. Note however the pecularity that $I(\alpha) \subseteq I(\gamma)$ may occur without $\mathcal{G}$ possessing a map $\gamma \to \alpha$.

Note that if $\mathcal{G}$ has parallel arrows f,g for which $A(f) \neq A(g)$, then there cannot exist a coherent set of surjections $\varphi(\alpha) : R \to A(\alpha)$. In particular the maps $\lim A \to A(\alpha)$ cannot all be onto. In this case, we can restrict the functor A to $B(\alpha) = \mathrm{Im}(\lim A \to A(\alpha))$ and proceed as before. Therefore for the rest of this section we will make the following

<u>Assumption 1.1</u>: $\mathcal{G} = \{I(\alpha) : \alpha\in\mathcal{G}\}$ is a finite set of ideals of R indexed in reverse order by the poset $\mathcal{G}$. A is the contravariant functor on $\mathcal{G}$ given by $A(\alpha) = R/I(\alpha)$.

<u>Definition</u>: Let $\mathcal{G}$ be as in (1.1) and $\ell \geq 2$. We say the condition $(\mathrm{CRT})_\ell$ holds if

$$I(\alpha) + \bigcap_{\kappa=2}^{\ell} I(\beta_\kappa) = \bigcap\{I(\gamma) : \gamma\leq\alpha, \gamma\leq\beta_\kappa \text{ some } \kappa\}$$

for all ℓ-tuples $(\alpha,\beta_2,\ldots,\beta_\ell)$ of maximal elements of $\mathcal{G}$.

We remark that $(CRT)_2$ says there are enough mediate elements γ so that for maximal elements α, β $I(\alpha) + I(\beta) = \cap I(\gamma)$. Given $(CRT)_2$ it easily follows that $(CRT)_\ell$ is equivalent to the distributivity of + over $\cap$ for the ideals $I(\alpha)$, α maximal, i.e.

$$I(\alpha) + \bigcap_k I(\beta_k) = \bigcap_k (I(\alpha)+I(\beta_k))$$

In [26, Vol I, p280] it is remarked that the Chinese Remainder Theorem follows from this distributivity, hence the name CRT.

If $\mathfrak{a}$,A are as in (1.1) and $\mathfrak{B} \subseteq \mathfrak{a}$, we write $A(\mathfrak{B}) = \lim A(\mathfrak{B})$ for the limit of the functor A restricted to $\mathfrak{B}$.

Theorem 1.2: Fix $k \geq 2$. The following conditions are equivalent under the standing assumptions (1.1):

(a) $(CRT)_\ell$ holds for $2 \leq \ell \leq k$

(b) For every subposet $\mathfrak{B}$ of $\mathfrak{a}$ generated by ℓ maximal elements of $\mathfrak{a}$ $(2 \leq \ell \leq k)$, $R \to \lim A(\mathfrak{B})$ is surjective

If $(CRT)_\ell$ holds for all ℓ, then $R \to \lim A(\mathfrak{a})$ is surjective. Conversely, if $R \to \lim A(\mathfrak{a})$ is surjective and if in addition for each (not necessarily maximal) $\alpha \in \mathfrak{a}$ the map $R \to \lim A(\{\gamma | \gamma < \alpha\})$ is surjective, then $(CRT)_\ell$ holds for all ℓ.

Proof: To show a) => b) let $\mathcal{B}$ be generated by the maximal elements $\beta_1,\ldots,\beta_\ell$ of $\mathcal{C}$. We will do induction on ℓ. When $\ell = 1$, $\lim A(\mathcal{B}) = A(\beta_1) = R/I(\beta_1)$ so b) holds. Now in general $\lim A(\mathcal{B}) = \{(\bar{r}_1,\ldots,\bar{r}_\ell) \in \prod_{i=1}^{\ell} A(\beta_i):$ $r_i \equiv r_j \bmod I(\gamma)$ whenever $\gamma \leq \beta_i,\beta_j\}$. Let $(\bar{r}_1,\ldots,\bar{r}_\ell) \in \lim A(\mathcal{B})$. Then $(\bar{r}_1,\ldots,\bar{r}_{\ell-1}) \in \lim A(\mathcal{B}')$, where $\mathcal{B}'$ is generated by $\beta_1,\ldots,\beta_{\ell-1}$. By induction $(\bar{r}_1,\ldots,\bar{r}_{\ell-1}) = (\bar{r},\ldots,\bar{r})$ for some $r \in R$, i.e., $r_i \equiv r \bmod I(\beta_i)$ for $i=1,\ldots,\ell-1$. Then $r - r_\ell \in \cap\{I(\gamma) \mid \gamma \leq \beta_i, \beta_\ell$ some $i < \ell\} = I(\beta_\ell) + \bigcap_{j<\ell} I(\beta_j)$ so $r - r_\ell = s + t$ for $s \in I(\beta_\ell)$, $t \in \bigcap_{j<\ell} I(\beta_j)$. It follows that $(\bar{r}_1,\ldots,\bar{r}_\ell) = (\bar{r}',\ldots,\bar{r}')$ where $r' = r - t = r_\ell + s$. Thus $R \to \lim A(\mathcal{B})$ is surjective.

Now assume $R \to \lim A(\mathcal{B})$ is surjective whenever $\mathcal{B}$ is generated by the maximal elements $\beta_1,\ldots,\beta_\ell$ of $\mathcal{C}$. It is always true that $I(\beta_\ell) + \bigcap_{j<\ell} I(\beta_j) \subseteq \cap\{I(\gamma): \gamma \leq \beta_\ell$ and $\gamma \leq \beta_j$ for some $j < \ell\}$. For the reverse inclusion let $r \in \cap I(\gamma)$. Then $(\bar{r},\ldots,\bar{r},0) \in \lim A(\mathcal{B})$ so $\exists\, s \in R$ with $(\bar{s},\ldots,\bar{s}) = (\bar{r},\ldots,\bar{r},0)$. Hence $s \equiv r \bmod I(\beta_i)$ for $i < \ell$, $s \equiv 0 \bmod I(\beta_\ell)$ and so $r = s + (r-s)$ where $s \in I(\beta_\ell)$ $r - s \in \bigcap_{j<\ell} I(\beta_j)$. This shows b) implies a).

Finally, the last sentence of Theorem 1.2 is a simple consequence of the following Lemma:

Lemma 1.3: Under our assumption (1.1) assume for each $\alpha \in \mathcal{C}$ that $R \to \lim A(\{\gamma : \gamma < \alpha\})$ is surjective. Then for every $\mathcal{B} \subseteq \mathcal{C}$ generated by its maximal elements, $\lim A(\mathcal{C}) \to \lim A(\mathcal{B})$ is surjective.

<u>Proof:</u> Let $\mathcal{B}'$ be a subposet of $\mathcal{G}$ containing $\mathcal{B}$, maximal with respect to being generated by its maximal elements and having $\lim A(\mathcal{B}') \to \lim A(\mathcal{B})$ be onto. Such a $\mathcal{B}'$ exists as $\mathcal{G}$ is finite. Let $\beta_1,\ldots,\beta_\ell$ be the maximal elts in $\mathcal{B}'$. Assume $\mathcal{B}' \neq \mathcal{G}$, then we can choose $\beta_{\ell+1}$ minimal in $\mathcal{G} - \mathcal{B}'$ and let $\mathcal{B}'' = \mathcal{B}' \cup \{\beta_{\ell+1}\}$. Note $\{\gamma : \gamma<\beta_{\ell+1}\} \subseteq \mathcal{B}'$, $\mathcal{B}''$ is generated by its maximal elements and the set $\{\beta_1,\ldots,\beta_{\ell+1}\}$ includes all maximal elements of $\mathcal{B}''$ (it is possible that $\beta_i \leq \beta_{\ell+1}$ for some $i \leq \ell$). Given $(\bar{a}_1,\ldots,\bar{a}_\ell) \in \prod_{i=1}^{\ell} A(\beta_i)$ in $\lim A(\mathcal{B}')$, its image in $\lim A(\{\gamma:\gamma<\beta_{\ell+1}\})$ is the image of an element $a_{\ell+1} \in R$ by hypothesis, and thus $(\bar{a}_1,\ldots,\bar{a}_{\ell+1}) \in \prod_{i=1}^{\ell+1} A(\beta_i)$ is an element of $\lim A(\mathcal{B}')$ mapping to $(\bar{a}_1,\ldots,\bar{a}_\ell)$. Thus $\lim A(\mathcal{B}'') \to \lim A(\mathcal{B})$ is surjective, contradicting maximality of $\mathcal{B}'$.

<u>Remark:</u> Under the standing assumption (1.1) if we let $\bar{R} = \lim A(\mathcal{G})$ then $R \to A(\alpha_i)$ factors through $\bar{R}$. Letting $\bar{I}(\alpha) = \ker(\bar{R}\to A(\alpha))$, $\bar{A}(\alpha) = \bar{R}/\bar{I}(\alpha)$ then $\bar{A}(\alpha) = A(\alpha)$ and $\lim \bar{A}(\mathcal{G}) = \lim A(\mathcal{G})$. Thus we can replace R by $\lim A(\mathcal{G})$ preserving the assumption (1.1).

<u>Proposition 1.4</u>: If $R = \lim A(\mathcal{G})$ under the standing assumption, $\alpha_1,\ldots,\alpha_n$ the maximal elements of $\mathcal{G}$, then $\lim A(\mathcal{G}) \subseteq \prod A(\alpha_i)$ is an integral extension with conductor

$$c = \sum_{i=1}^{\ell} \bigcap_{j\neq i} I(\alpha_j) = \bigcap\{I(\gamma):\gamma\in\mathfrak{a} \text{ is mediate}\} =$$

$$\bigcap_i (I(\alpha_i) + \bigcap_{j\neq i} I(\alpha_j))$$

Proof: We have each $\lim A(\mathfrak{a}) \to A(\alpha_i)$ surjective and it is well known that this implies $\pi A(\alpha_i)$ is an integral extension of $\lim A(\mathfrak{a})$. Now if $(a_1,\ldots,a_n)$ is in the conductor, multiplication by idempotents of the form $(0,\ldots,1,0,\ldots,0)$ leaves it in the conductor, so each element of the form $(0,\ldots,a_i,0,\ldots,0)$ is in $c \subseteq \lim A(\mathfrak{a})$. Conversely the set of all sums of elements $(0,\ldots,a,0,\ldots,0) \in \lim A(\mathfrak{a})$ forms an $\pi A(\alpha_i)$ ideal and so must be contained in the conductor. It is easy to see from the above description of c that $c = \sum_i \bigcap_{j\neq i} I(\alpha_j) = \bigcap\{I(\gamma):\gamma\in\mathfrak{a}$ is mediate$\}$. Since already $R = \lim A(\mathfrak{a})$, $(CRT)_n$ holds and this immediately gives the last equality.

Remark [6, Prop. 1.12] is seen to be a special case of Proposition 1.4 above, as is [16, Lemma 2.2].

Corollary 1.5: Under the standing assumptions (1.1), assume also that A is a functor to $\underset{\sim}{k}$-algebras, $\underset{\sim}{k}$ noetherian, and that each $A(\alpha)$ is finitely generated as a $\underset{\sim}{k}$-algebra. Then $\lim A(\mathfrak{a})$ is a finitely generated $\underset{\sim}{k}$ algebra. In particular $\lim A(\mathfrak{a})$ is noetherian .

Proof: This is Proposition 7.8 of [2].

The category of $\underset{\sim}{k}$-algebras is dual to the category of affine schemes over Spec $\underset{\sim}{k}$, the correspondence given by $R \to \text{Spec } R$. Given a contravariant functor $A: \mathcal{A} \to \underset{\sim}{k}$-algebras we get a functor $\text{Spec } A: \mathcal{A} \to (\text{Schemes over } \underset{\sim}{k})$ by $(\text{Spec } A)(\alpha) = \text{Spec } A(\alpha)$. From the duality we obtain $\text{Spec}(\lim A(\mathcal{A})) = \text{colim}(\text{Spec } A)(\mathcal{A})$ provided this latter colimit is taken in the category of affine schemes over $\underset{\sim}{k}$. Our next result says that this actually works at the prime ideal level.

<u>Proposition 1.6</u>: Under the standing assumptions (1.1), assume also $\{(I(\alpha):\alpha\in\mathcal{A}\}$ satisfies $(\text{CRT})_2$. Then $\text{colim}(\text{Spec } A)(\mathcal{A}) = \text{Spec}(\lim A(\mathcal{A}))$ as topological spaces.

<u>Proof</u>: Let $\alpha_1,\ldots,\alpha_n$ be the maximal elements of $\mathcal{A}$. From the universal properties of coproducts and colimits we obtain a commutative diagram

$$\begin{array}{ccc}
\coprod\limits_i \text{Spec } A(\alpha_i) & = & \text{Spec} \prod\limits_i A(\alpha_i) \\
\downarrow & & \downarrow \\
\text{colim}(\text{Spec } A)(\mathcal{A}) & \longrightarrow & \text{Spec}(\lim A(\mathcal{A}))
\end{array}$$

where the right hand vertical map is induced by $\lim A \subseteq \prod\limits_i A(\alpha_i)$. Since $\prod\limits_i A(\alpha_i)$ is integral over

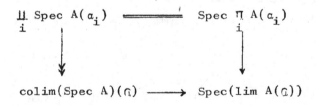

$\lim A(\mathfrak{a})$ this last map is a closed surjection (see [1, pp. 13, 67]. Thus $\operatorname{colim}(\operatorname{Spec} A)(\mathfrak{a}) \to \operatorname{Spec}(\lim A(\mathfrak{a}))$ is a closed surjection so we need only show it is injective. For this it is sufficient to consider points $x_i \in \operatorname{Spec} A(\alpha_i)$, $x_j \in \operatorname{Spec} A(\alpha_j)$ which map to the same point $\mathfrak{p}$ of $\operatorname{Spec}(\lim A(\mathfrak{a}))$. If $i = j$ since $\operatorname{Spec} A(\alpha_i) \to \operatorname{Spec}(\lim A(\mathfrak{a}))$ is injective $x_i = x_j$. Otherwise $I(\alpha_i)$, $I(\alpha_j)$ are both contained in $\mathfrak{p}$ and by $(CRT)_2$ $\cap\{I(\gamma):\gamma\leq\alpha_i,\alpha_j\} = I(\alpha_i) + I(\alpha_j) \subseteq \mathfrak{p}$ so some $I(\gamma) \subseteq \mathfrak{p}$. Hence $\mathfrak{p}$ is the image of $\mathfrak{p}' \in \operatorname{Spec} A(\gamma)$. However $\operatorname{Spec} A(\gamma) \to \operatorname{Spec}(\lim A(\mathfrak{a}))$ factors through the injections $\operatorname{Spec} A(\alpha_k) \to \operatorname{Spec}(\lim A(\mathfrak{a}))$ $k = i,j$. Hence $\mathfrak{p}'$ goes to $x_i \in \operatorname{Spec} A(\alpha_i)$ and $x_j \in \operatorname{Spec} A(\alpha_j)$ so $x_i = x_j$ in $\operatorname{colim}(\operatorname{Spec} A)(\mathfrak{a})$.

We next show that the ring $\lim A(\mathfrak{a})$ is seminormal, under fairly modest hypotheses. We continue with our standing assumptions (1.1), so that in particular the maps $\varphi(\alpha) : R \to A(\alpha)$ are onto. This implies $\lim A(\mathfrak{a}) \to A(\alpha)$ is onto as well.

We let $\alpha_1,\ldots,\alpha_n$ be the maximal elements of $\mathfrak{a}$ and use pr_i to denote the projections $\Pi A(\alpha_i) \to A(\alpha_i)$

<u>Lemma 1.7</u>: The prime ideals of $\Pi A(\alpha_i)$ lying over a prime $\mathfrak{p}$ of $\lim A(\mathfrak{a})$ are the $pr_i^{-1}(\mathfrak{p}/I(\alpha_i))$, where i ranges over all the indices with $I(\alpha_i) \subseteq \mathfrak{p}$.

Proof: The prime ideals of $\Pi A(\alpha_i)$ are the ideals $pr_i^{-1}(P)$ where P is a prime ideal of $A(\alpha_i)$. This prime ideal lies over $\wp$ if and only if $\wp = \varphi(\alpha_i)^{-1}(P)$, and this occurs if and only if both $I(\alpha_i) \subseteq \wp$ and $P = \wp/I(\alpha_i)$.

Theorem 1.8: We make the assumptions of (1.1), i.e., that R is a commutative ring, $\{I(\alpha)\}$ is a finite set of radical ideals of R indexed in reverse by the poset $\mathfrak{a}$, and set $A = \lim R/I(\alpha)$. Let $\alpha_1, \ldots, \alpha_n$ be the maximal elements of $\mathfrak{a}$. Then

a) A is seminormal in $B = \Pi\, R/I(\alpha_i)$

b) If $(CRT)_2$ holds, A is the seminormalization of the image of R in B.

Comment: The proof given here is essentially F. Orecchia's proof of the special case $\{I(\alpha)\} = \{\text{minimal primes } \wp_i\} \cup \{\wp_i + \wp_j\}$, which is [16, Theorem 1.3]. We wish to thank him for pointing out to us the applicability of his result in this situation.

Proof: We can replace R by $R/\cap I(\alpha_i)$ to assume that R is a subring of both A and B. We first show that A contains the seminormalization ${}^+_BR$ of R in B and if $(CRT)_2$ holds that $A \subseteq {}^+_BR$. This proves b), and setting $R = A$ proves a) as the hypotheses remain valid.

Recall from [23] that ${}^+_BR = \{b = (\bar{a}_1, \ldots, \bar{a}_n) \in B :$ $b \in R_P + \text{rad}(B_P)$ for every prime P of $R\}$. In order to show that $b = (\bar{a}_1,\ldots,\bar{a}_n) \in {}^+_B R$ is in A we have to show that $a_i \equiv a_j \bmod I(\gamma)$ whenever $\gamma \leq \alpha_i, \alpha_j$ in $\mathfrak{a}$. Given such a γ, α_i, α_j, pick a prime ideal P of R containing $I(\gamma)$. Since $b \in {}^+_B R$ there are elements $r \in R$, $c \in B \cap \mathrm{rad}(B_P)$, $s \in R - P$ with $sb = r + c$. In the kth coordinate this says that $s\bar{a}_k = \bar{r} + pr_k(c) \in R/I(\alpha_k)$. By Lemma 1.7, $B \cap \mathrm{rad}(B_P)$ is contained in $pr_i^{-1}(P/I(\alpha_i)) \cap pr_j^{-1}(P/I(\alpha_j))$, so $pr_k(c) \subset P/I(\alpha_k)$ for $k = i,j$. Lifting back to R, we have $sa_i \equiv r \equiv sa_j \bmod P$, whence $a_i \equiv a_j \bmod P$. As $I(\gamma)$ is a radical ideal we can quantify over P to obtain $a_i \equiv a_j \bmod I(\gamma)$, as desired.

In order to show that $A \subseteq {}^+_B R$, it is sufficient by [23] to show that

a) For every $P \in \mathrm{Spec}\, R$ there is a unique $\mathfrak{P} \in \mathrm{Spec}\, A$ above P

b) The cannonical homomorphism $k(P) \to k(\mathfrak{P})$ is an isomorphism.

We note b) follows immediately from Lemma 1.7: if $\mathfrak{p} \in \mathrm{Spec}\, A$ lies over $P \in \mathrm{Spec}\, R$, $\mathfrak{p} \subseteq pr_i^{-1}(P/I(\alpha_i)) = P_i$. But $R/P \subseteq A/\mathfrak{p} \subseteq B/P_i = R/P$, and hence all three have the same quotient field.

Before showing a) we note if $b = (a_1,\ldots,a_n) \in A$ then $a_i \equiv a_j \bmod (I(\alpha_i)+I(\alpha_j))$. For by $(CRT)_2$ $I(\alpha_i) + I(\alpha_j) = \cap I(\gamma)$ where γ runs over all $\gamma \leq \alpha_i, \alpha_j$. But for γ $a_i - a_j \in I(\gamma)$ by definition of $A = \lim {}^R/I(\alpha)$.

Now suppose $\mathfrak{P}_1, \mathfrak{P}_2$ are distinct primes of A over $P \in \operatorname{Spec} R$. Then by Lemma 1.7 $\mathfrak{P}_1, \mathfrak{P}_2$ are contained in distinct primes $pr_i^{-1}(P/I(\alpha_i))$, $pr_j^{-1}(P/I(\alpha_j))$ of B. Thus there must exist $b = (\bar{a}_1, \ldots, \bar{a}_n) \in A$ with $a_i \in P$, $a_j \notin P$ But $a_i - a_j \in I(\alpha_i) + I(\alpha_j) \subseteq P$ so this is impossible.

Corollary 1.9: In addition to the standing assumptions (1.1), assume $A(\alpha)$ is seminormal for maximal α, reduced for all α, and that $B = \pi A(\alpha_i)$ is contained in the total quotient ring $Q(\lim A(\mathfrak{a}))$. Then $\lim A(\mathfrak{a})$ is seminormal.

Proof: The hypotheses insure B is seminormal and $\overline{B} = \overline{\lim A(\mathfrak{a})}$ (here $\overline{B}$ = integral closure of B in $Q(B)$). The seminormality of $\lim A(\mathfrak{a})$ then follows from Theorem 1.8 and transitivity of seminormality [23, Lemma 1.2].

Remark: The hypothesis $B \subseteq Q(\lim A(\mathfrak{a}))$ is satisfied, for instance, if the $I(\alpha_i)$'s (α_i maximal) are intersections of minimal primes, no one minimal prime being associated to distinct $I(\alpha_i)$.

Combining Theorems (1.2) and (1.8) we have:

Corollary 1.10: With the hypotheses and notation of (1.8) consider the conditions

i) $(CRT)_\ell$ holds for all ℓ

ii) $R/\cap I(\alpha_i) = \lim A(\mathfrak{a})$

iii) $R/\cap I(\alpha_i)$ is seminormal in B

Then i) => ii) => iii). If $(CRT)_2$ holds, ii) <=> iii). If $R \to \lim A(\{\gamma | \gamma < \alpha\})$ is surjective for each $\alpha \in \mathfrak{a}$ then i) <=> ii).

Example 1.11: Let $R = \underset{\sim}{k}[X,Y,Z]/(XYZ(X+Y+Z))$ where $\underset{\sim}{k}$ is a normal domain. As R is the coordinate ring of 4 planes through the origin, no three passing through a common line, it follows from [16] that R is seminormal, and from [5] that R is not K_0-regular.

We can give a direct proof of the seminormality of R using corollaries 1.9, 1.10. Let $\mathfrak{a}$ be the poset of ideals $\{(X),(Y),(Z),(X+Y+Z),(X,Y),(X,Z),(X,X+Y+Z),(Y,Z)$ $(Y,X+Y+Z),(Z,X+Y+Z),(X,Y,Z)\}$ ordered by reverse inclusion. Now the maximal elements of $\mathfrak{a}$ are the minimal primes of R and the quotient rings are polynomial rings over $\underset{\sim}{k}$, hence seminormal. It thus sufficient to verify the conditions $(CRT)_\ell$ $\ell = 2,3,4$. $(CRT)_2$ holds as the set of ideals is closed under sums. $(CRT)_3$ is satisfied by [6; (1.9)] as any 3 of the polynomials $X,Y,Z,X+Y+Z$ form an admissible set of hyperplanes in $\mathbb{A}^3_{\underset{\sim}{k}}$.

For $(CRT)_4$ it is enough to show (by symmetry) that $(F) + (XYZ) = (F,X)\cap(F,Y)\cap(F,Z)$ where $F = X + Y + Z$. One inclusion is obvious so assume $f \in (F,X)\cap(F,Y)\cap(F,Z)$. Then $f = Fh + Zg$ $g,h \in \underset{\sim}{k}[X,Y,Z]$. The homomorphism $\varphi : \underset{\sim}{k}[X,Y,Z] \to \underset{\sim}{k}[Z]$ given by $\varphi(X) = 0$, $\varphi(Y) = -Z$, $\varphi(Z) = Z$ has kernel (F,X) so $\varphi(f) = \varphi(Fh) = 0$ and hence $0 = \varphi(Zg) = Z\varphi(g)$ so that $\varphi(g) = 0$. Then $g \in (F,X)$ so $f = Fh' + XZg'$. A similar argument shows $g' \in (F,Y)$ so that $f = Fh'' + XYZg''$ as desired.

We remark that the condition $R \to \lim A(\{\gamma:\gamma<\alpha\})$ onto is not satisfied. Letting $\alpha = (Z)$ we see the ring $\underset{\sim}{k}[X,Y,Z]/[(X,Z)\cap(Y,Z)\cap(F,Z)] \approx \underset{\sim}{k}[X,Y]/XY(X+Y)$ is not seminormal so $(CRT)_3$ cannot hold for $\{(X,Z),(Y,Z),(F,Z),(X,Y,Z)\}$. In fact $(X,Z)+(Y,Z)\cap(F,Z)\neq(X,Y,Z)$.

We end this section with a discussion of the limit of a ring-valued functor on a graded category. Given a contravariant functor $A : \mathfrak{C} \to$ (commutative rings), $\mathfrak{C}$ graded, we can also think of A as a functor on the comma categories $\mathfrak{C}^p\downarrow\alpha$ by composing A with the forgetful functor $\mathfrak{C}^p\downarrow\alpha \to \mathfrak{C}$ given by $(\beta\to\alpha) \to \beta$. The fundamental fact is

<u>Lemma 1.12</u>: Let $\mathfrak{C}$ be a graded category, $A : \mathfrak{C} \to$ (commutative rings) a contravariant functor. Then

$$\begin{array}{ccc} \lim A(\mathcal{G}^p) & \longrightarrow & \prod_{\dim\alpha=p} \lim A(\mathcal{G}^p\downarrow\alpha) \\ \downarrow & & \downarrow \\ \lim A(\mathcal{G}^{p-1}) & \longrightarrow & \prod_{\dim\alpha=p} \lim A(\mathcal{G}^{p-1}\downarrow\alpha) \end{array}$$

is cartesian and

$$\begin{array}{ccc} \coprod_{\dim\alpha=p} \operatorname{colim} \operatorname{Spec} A(\mathcal{G}^{p-1}\downarrow\alpha) & \longrightarrow & \coprod_{\dim\alpha=p} \operatorname{colim} \operatorname{Spec} A(\mathcal{G}^p\downarrow\alpha) \\ \downarrow & & \downarrow \\ \operatorname{colim} \operatorname{Spec} A(\mathcal{G}^{p-1}) & \longrightarrow & \operatorname{colim} \operatorname{Spec} A(\mathcal{G}^p) \end{array}$$

is co cartesian.

The proof of the above is strictly categorical and is based on (0.1).

Note for $\dim\alpha=p$ that $\mathcal{G}^p\downarrow\alpha$ has terminal object $\alpha\to\alpha$, and so $\lim A(\mathcal{G}^p\downarrow\alpha) = A(\alpha)$. We will want the right hand vertical maps of the first square of (1.12) to be surjective, i.e. we want each $A(\alpha)\to\lim A(\mathcal{G}^{p-1}\downarrow\alpha)$ to be surjective for $\dim\alpha = p$. In case $\mathcal{G}^{p-1}\downarrow\alpha$ is a poset we have from Theorem (1.2)

<u>Proposition 1.13</u>: Let A be a contravariant commutative ring-valued functor on a graded category $\mathcal{G}$ and suppose $A(\alpha)\to A(\beta)$ is surjective for each $\beta\to\alpha$ in $\mathcal{G}$. Suppose $\mathcal{G}^{p-1}\downarrow\alpha$ is a poset. Then $A(\alpha)\to\lim A(\mathcal{G}^{p-1}\downarrow\alpha)$ is surjective if $\{\ker(A(\alpha)\to A(\beta)) : \beta\to\alpha\in\mathcal{G}^{p-1}\downarrow\alpha\}$ satisfies $(\mathrm{CRT})_\ell$ for each ℓ.

One should note that $\mathfrak{G}^{p-1}\downarrow\alpha$ may be a poset even if $\mathfrak{G}$ is not (see example 4.7 below). If $\mathfrak{G}^p$ is a poset and each $A(\alpha) \to \lim(\mathfrak{G}^{p-1}\downarrow\alpha)$ is surjective, $\dim\alpha=p$, then by (1.12) so is $\lim A(\mathfrak{G}^p) \to \lim A(\mathfrak{G}^{p-1})$. Applying 1.6, colim Spec $A(\mathfrak{G}^{p-1}) \to$ colim Spec $A(\mathfrak{G}^p)$ is injective. Thinking of colim Spec $A(\mathfrak{G}^p)$ as the p-skeleton of colim Spec $A(\mathfrak{G})$, this says the (p-1)-skeleton is a subspace of the p-skeleton.

If $\mathfrak{G}$ is a graded poset and $\dim\alpha=p$, then $\mathfrak{G}^{p-1}\downarrow\alpha$ can be identified with the subposet $\{\beta\in\mathfrak{G} : \beta<\alpha\}$. Thus from Theorem (1.2) we have the following partial converse to 1.13 .

<u>Proposition 1.14:</u> In addition to the standing hypotheses (1.1), assume $\mathfrak{G}$ is graded and $A(\alpha) \to \lim A(\mathfrak{G}^{p-1}\downarrow\alpha)$ is surjective for $\dim\alpha=p$. Then $R \to \lim A(\mathfrak{G})$ is surjective iff $\{I(\alpha) : \alpha\in\mathfrak{G}\}$ satisfies $(CRT)_\ell$ all ℓ .

Finally we consider a special case of the above, which is, however, typical of our applications. We start with our standing hypothesis, $\mathfrak{G}$ graded, and assume in addition for $\dim\alpha = p$ that $A(\alpha) \to \lim A(\mathfrak{G}^{p-1}\downarrow\alpha)$ is surjective and $A(\alpha)$ is a normal domain. Assume also $\beta\in\mathfrak{G}^p$ is maximal iff $\dim\beta = p$ and mediate in $\mathfrak{G}^p$ otherwise. Then the $I(\alpha)$ $\dim\alpha=p$ map to the minimal primes of $\lim A(\mathfrak{G}^p)$ and so $\prod_{\dim\alpha=p} A(\alpha) \subseteq Q(\lim A(\mathfrak{G}^p))$ is the integral closure of $\lim A(\mathfrak{G}^p)$. It is not difficult to see

$\ker(\lim A(\mathcal{G}^p) \to \lim A(\mathcal{G}^{p-1})) = \ker(\prod_{\dim\alpha=p} A(\alpha) \to \prod_{\dim\alpha=p} \lim A(\mathcal{G}^{p-1}\downarrow\alpha))$

$= \bigcap_{\dim\gamma<p} I(\gamma)$ which by (1.4) is the conductor of $\lim A(\mathcal{G}^p)$ in $\prod_{\dim\alpha=p} A(\alpha)$. Thus the first square of (1.12) is simply the "conductor square" of the introduction.

§2 K_1-regularity

In this section we consider the problem of when $\lim A(\mathcal{G})$ is K_1-regular. Although we are unable to give a general solution, we do give some conditions ensuring K_1-regularity, and provide some examples to show when K_1-regularity cannot occur.

The application of this problem is to the spectral sequence developed in §3 in order to compute the groups $KV_*(\lim A)$. In order to apply our spectral sequence we will need to know that the onto maps $A(\alpha) \to \lim A(\mathcal{G}^{p-1}\downarrow\alpha)$ are Gl-fibrations for all p-cells α. In our applications $A(\alpha)$ will be K_1-regular, so these maps are GL-fibrations exactly when the rings $\lim A(\mathcal{G}^{p-1}\downarrow\alpha)$ are K_1-regular (See [5, 9] for this fact).

It was proven in [25] and [6] that a K_1-regular ring is also K_0-regular, hence Pic-regular and hence (if it is noetherian with finite normalization) seminormal by [23]. The rings $\lim A(\mathcal{G})$ that we will be considering will generally be seminormal by the results of §1. However in example (1.11) we have already seen a seminormal ring of the form $\lim A(\mathcal{G})$ which is not K_0-regular. We now give another example which also shows that the concept K_0-regularity is different from that of K_1-regularity even for our rings. In fact this example could be extended to show the concepts of K_n-regularity $1 \geq n > -\infty$ are distinct for different n.

<u>Example 2.1</u>: Let $R = \underset{\sim}{k}[x,y,z,w]$, $\underset{\sim}{k}$ a field, and let $\mathfrak{a}$ be the poset of the following ideals of R (ordered by reverse inclusion): $\alpha_3 = (w)$, $\beta_3 = (w-z(z-xy))$ $\alpha_2 = (z,w)$, $\beta_2 = (w,z-xy)$, $\alpha_1 = (y,z,w)$, $\beta_1 = (x,z,w)$. $v_0 = (x,y,z,w)$. This category is graded by its subscripts. It is evident that each $A(\gamma) = R/\gamma$ is a polynomial ring over $\underset{\sim}{k}$ in $\dim(\gamma)$ variables. The following equalities are straightforward:

i) $\alpha_1 + \beta_1 = v_0$

ii) $\alpha_2 + \beta_2 = \alpha_1 \cap \beta_1 = (xy,z,w)$

iii) $\alpha_3 + \beta_3 = \alpha_2 \cap \beta_2 = (w,z(z-xy))$

Since these are $(CRT)_2$ conditions (for $\mathfrak{a} = \mathfrak{a}^3, \mathfrak{a}^2, \mathfrak{a}^1$) it follows by Theorem 1.2 that R maps onto $A_p = \lim A(\mathfrak{a}^p)$ for $p = 1,2,3$. Hence $A_1 = \underset{\sim}{k}[x,y]/(xy)$, $A_2 = \underset{\sim}{k}[x,y,z]/(z(z-xy))$ $A_3 = {}^R/\alpha_3 \cap \beta_3 = \underset{\sim}{k}[x,y,z,w]/(w(w-z(z-xy)))$. We have the cartesian squares

$$\begin{array}{ccc} A_3 & \longrightarrow & {}^R/\beta_3 \\ \downarrow & & \downarrow \\ {}^R/\alpha_3 & \longrightarrow & A_2 \end{array} \qquad\qquad \begin{array}{ccc} A_2 & \longrightarrow & {}^R/\beta_2 \\ \downarrow & & \downarrow \\ {}^R/\alpha_2 & \longrightarrow & A_1 \end{array} .$$

Now from [5], [7] we know that A_1 is K_1-regular but not K_2-regular. From [5] it follows that A_2 is K_0-regular but not K_1-regular, and that A_3 is K_{-1}-regular but neither K_0 nor K_1-regular. On the other hand, Corollary 1.9 applies and states that A_p is seminormal for $p = 1,2,3$.

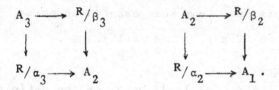

The most effective tool we know for proving rings K_1-regular is Vorst's Theorem [24], which states that K_1-regularity is a local property. In order to localize the ring $\lim A(\mathcal{G})$ one need only consider the relevant part of the indexing category. This reflects the geometric fact that the local structure of a scheme at a point only depends on the components of the scheme (and of its singular set) passing through the point.

In order to formalize this concept we introduce the following definition. If a finite category is essentially a poset, $A : \mathcal{G} \to$(commutative rings) a functor with each $A(\alpha)$ a domain, S a multiplicatively closed set of a ring R, and $\varphi(\alpha) : R \to A(\alpha)$ a coherent set of maps, then we define $\mathcal{G}_S$ to be the full subcategory of $\mathcal{G}$ whose objects α are such that $S \cap \ker \varphi(\alpha) = \emptyset$. Note that we have assumed each $I(\alpha) = \ker \varphi(\alpha)$ to be a prime ideal. We remark that if S is disjoint from $\cap I(\alpha)$ then $\mathcal{G}_S$ is nonempty. This is because S is disjoint from some associated prime of $\cap I(\alpha)$, i.e. from some $I(\alpha)$, α maximal in $\mathcal{G}$.

<u>Proposition 2.2</u>: Let $\mathcal{G}, A, R, \varphi(\alpha), I(\alpha)$ be as in the definition above with $R = \lim A(\mathcal{G})$ and suppose $S \subseteq R$ is a multiplicatively closed set satisfying $S \cap I(\alpha) = \emptyset$ for all maximal $\alpha \in \mathcal{G}$. Then

$$S^{-1}R = S^{-1}[\lim A(\mathcal{G}_S)] = \lim S^{-1}A(\mathcal{G}_S)$$

where $S^{-1}A$ is the contravariant functor on $\mathfrak{a}_S$ given by $S^{-1}A(\alpha) = [\varphi(\alpha)(S)]^{-1}A(\alpha)$.

Proof: Let $\alpha_1,\ldots,\alpha_n$ be the maximal elements of $\mathfrak{a}$. The hypotheses insure $\alpha_i \in \mathfrak{a}_S$ for each i and $A(\alpha_i) \subseteq S^{-1}A(\alpha_i)$ so that $R = \lim A(\mathfrak{a}) \subseteq \prod A(\alpha_i) \subseteq \prod S^{-1}A(\alpha_i)$. Further no $s \in S$ is a zero divisor of R so $R \subseteq S^{-1}R \subseteq S^{-1}\lim A(\mathfrak{a}_S) \subseteq \lim S^{-1}A(\mathfrak{a}_S)$. It suffices to show that given $b \in \lim S^{-1}A(\mathfrak{a}_S)$ $\exists\, s \in S$ so that $sb \in R$. So let $b = (\frac{a_1}{s_1},\ldots,\frac{a_n}{s_n}) \in \lim S^{-1}A(\mathfrak{a}_S) \subseteq \prod S^{-1}A(\alpha_i)$. We can first find $s' \in S$ so that $s'b = (b_1,\ldots,b_n) \in \prod A(\alpha_i)$. The condition that $s'b \in \lim S^{-1}A(\mathfrak{a}_S)$ is that whenever $\gamma \xrightarrow{f} \alpha_i$, $\gamma \xrightarrow{g} \alpha_j$ are morphisms in $\mathfrak{a}_S$ $A(f)b_i = A(g)b_j$ in $S^{-1}A(\gamma)$; but $A(\gamma)$ is a domain, so already $A(f)b_i = A(g)b_j$ in $A(\gamma)$. On the other hand if $\gamma \notin \mathfrak{a}_S$ there exists $r_\gamma \in I(\gamma)\cap S$ and so for $\gamma \xrightarrow{f} \alpha_j$, $A(f)r_\gamma s' b_j = 0$. Letting $s'' = \prod_{\gamma\notin\mathfrak{a}_S} r_\gamma$ we find $s''s'b \in \lim A(\mathfrak{a}) = R$ as desired.

Although $\mathfrak{a}$ is quite general above, the hypothesis on S is too strong. We have however

Theorem 2.3: Let $\mathfrak{a}$, A be as in the standing assumption (1.1) and assume $R = \lim A(\mathfrak{a})$, each $A(\alpha)$ is a domain and $\{I(\alpha):\alpha\in\mathfrak{a}\}$ satisfies $(\mathrm{CRT})_\ell$ for all ℓ. Then for every multiplicatively closed set $S \subseteq R$

$$S^{-1}R = S^{-1}[\lim A(\mathfrak{a}_S)] = \lim S^{-1}A(\mathfrak{a}_S).$$

<u>Proof</u> As noted earlier the maximal elements of $\mathcal{C}_S$ are maximal elements of $\mathcal{C}$; let $\mathcal{B}$ be the subposet of $\mathcal{C}$ generated by these elements. By Theorem 1.2 $\theta : R \to \lim A(\mathcal{B})$ is surjective and $\ker \theta = \{(\bar{a}_1,\ldots,\bar{a}_n) \in \prod A(\alpha_i) : a_i \in I(\alpha_i)$ whenever $\alpha_i \in \mathcal{C}_S\}$ (again $\alpha_1,\ldots,\alpha_n$ are the maximal elements of $\mathcal{C}$). However $(\bar{a}_1,\ldots,\bar{a}_n) \in \ker(R \to S^{-1}R)$ precisely when there exist $s_i \in S$ so that $s_i a_i \in I(\alpha_i)$ for each $i = 1,\ldots,n$. For $\alpha_i \notin \mathcal{C}_S$ there is always such an s_i; for $\alpha_i \in \mathcal{C}_S$ as $A(\alpha_i)$ is a domain such an s_i exists precisely if $a_i \in I(\alpha_i)$. Hence $\ker \theta = \ker(R \to S^{-1}R)$ and so $R \to S^{-1}R$ factors injectively through $\lim A(\mathcal{B})$, i.e. $[\theta(S)]^{-1}[\lim A(\mathcal{B})] = S^{-1}R$. But for each maximal $\alpha_i \in \mathcal{B}$, $\alpha_i \in \mathcal{C}_S$ so $S \cap I(\alpha_i) = \emptyset$ and as $\lim A(\mathcal{C}) \to A(\alpha_i)$ factors through $\lim A(\mathcal{B})$ $\theta(S) \cap \ker[\lim A(\mathcal{B}) \to A(\alpha_i)] = \emptyset$. Thus Proposition 2.2 applies to this situation giving (as $\mathcal{C}_S = \mathcal{B}_S$) $S^{-1}R = S^{-1}[\lim A(\mathcal{B})] = S^{-1}[\lim A(\mathcal{C}_S)] = \lim S^{-1}A(\mathcal{C}_S)$ where, of course, $S^{-1}[\lim A(\mathcal{C}_S)]$ means $[\theta(S)]^{-1}[\lim A(\mathcal{C}_S)]$.

<u>Example 2.4</u>: As an application of Vorst's Theorem, we show that any union V of a linear subspaces (of differing dimensions) of $\mathbb{A}^{n+1}_k$ is K_1-regular, if it is in "general enough" position. By "general enough position" we mean that there is a set $\mathcal{H}$ of rational hyperplanes of $\mathbb{A}^{n+1}_k$ for which every linear subspace in V is an intersection of hyperplanes in $\mathcal{H}$ and that $\mathcal{H}$ is "admissible" in the sense of [6]. "Admissible" means that every subset $\{H_i\}$ of $\mathcal{H}$ maximal relative to $\cap H_i \neq \emptyset$ has exactly $n+1$ elements intersecting in exactly

one point.

Formally, we let $\mathcal{C}$ be the poset of intersections of hyperplanes in $\mathcal{H}$ and define $A : \mathcal{C} \to$ ($\underset{\sim}{k}$-algebras), $\underset{\sim}{k}$ a K_1-regular domain, by letting $A(L)$ be the coordinate ring of the linear subspace L of $\mathbb{A}^{n+1}_{\underset{\sim}{k}}$. If V is the union of $\{L_i\}$ we set $\mathcal{G}$ equal to the subposet of $\mathcal{C}$ generated by the $\{L_i\}$. We will show that the coordinate ring $A(V)$ of V is the ring $\lim A(\mathcal{G})$, and that $A(V)$ is K_1-regular.

In order to correspond to the notation of [6], we choose a linear equation $f_H \in \underset{\sim}{k}[x_0,\ldots,x_n]$ for each H in $\mathcal{H}$ and set $\mathcal{F} = \{f_H\}$. The poset $\mathcal{C}(\mathcal{F})$ of [6] is the present poset $\mathcal{C}$ with the reverse ordering.

We first do the special case in which every irreducible component of V passes through the origin. By the "general enough" position hypothesis, we can assume $\mathcal{H}$ is the coordinate hyperplanes, i.e., that $\mathcal{F} = \{x_0,\ldots,x_n\}$. In this case, $I(\alpha)$ is generated by the set $\mathcal{E}(\alpha)$ of x_i in $I(\alpha)$ for each α in $\mathcal{G}$. Checking the CRT conditions is now easy: if $\alpha, \beta_2, \ldots, \beta_\ell$ are elements of $\mathcal{G}$, then $I(\alpha\cap\beta_k)$ is generated $\mathcal{E}(\alpha) \cup \mathcal{E}(\beta_k)$. $\cap I(\alpha \cap \beta_k)$ is generated by monomials of the form $y = \prod \{x_i \in \mathcal{M}\}$ for $\mathcal{M}$ a subset of $\mathcal{F}$ meeting each $\mathcal{E}(\alpha) \cup \mathcal{E}(\beta_k)$. Either $\mathcal{M} \cap \mathcal{E}(\alpha) \neq \emptyset$ and $y \in I(\alpha)$ or else $\mathcal{M} \cap \mathcal{E}(\beta_k) \neq \emptyset$ for each k and $y \in \cap I(\beta_k)$. This shows that $(CRT)_\ell$ holds, so by Theorem (1.2) we have $A(V) = \lim A(\mathcal{G})$.

To see that $A(V)$ is K_1-regular, we argue by

induction: assume $\mathcal{A} \subseteq \mathcal{C}$ is generated by its maximal elements and minimal among such posets which have not yet been proven to be K_1-regular. Pick σ maximal in $\mathcal{A}$ and let $\mathcal{A}{\downarrow}\sigma = \{\alpha \in \mathcal{A}: \alpha \leq \sigma\}$, $\mathcal{B} = \{\alpha \in \mathcal{A}: \alpha < \sigma\}$. Then $\mathcal{A} = (\mathcal{A}-\{\sigma\}) \cup (\mathcal{A}{\downarrow}\sigma)$, $\mathcal{B} = (\mathcal{A}-\{\sigma\}) \cap (\mathcal{A}{\downarrow}\sigma)$, both $\mathcal{A} - \{\sigma\}$ and $\mathcal{B}$ are generated by their maximal elements. By [6, (1.7)]

$$\begin{array}{ccc} \lim A(\mathcal{A}) & \longrightarrow & \lim(\mathcal{A}{\downarrow}\sigma) = A(\sigma) \\ \downarrow & & \downarrow \\ \lim A(\mathcal{A}-\{\sigma\}) & \longrightarrow & \lim A(\mathcal{B}) \end{array}$$

is cartesian and $\lim A(\mathcal{A}{\downarrow}\sigma) \to \lim A(\mathcal{B})$ is surjective by [6, (1.4) and (1.11)]. $A(\sigma)$ is K_1-regular and so is $\lim A(\mathcal{A}-\{\sigma\})$ by induction. We claim the bottom map of the above square splits and so $\lim A(\mathcal{A})$ is K_1-regular by [5, (1. 3)].

To see this splitting we produce coherent maps $\varphi(\alpha)$: $\lim A(\mathcal{B}) \to A(\alpha \cap \sigma) \to A(\alpha)$ for each $\alpha \in \mathcal{A} - \{\sigma\}$. The first map is defined since $\alpha \cap \sigma \in \mathcal{B}$: it is a proper subspace of σ (σ is maximal in $\mathcal{A}$) containing the origin. The second map is the composite $A(\alpha \cap \sigma) \cong \underset{\sim}{k}[\mathcal{F} - \mathcal{E}(\alpha \cap \sigma)] \subseteq k[\mathcal{F} - \mathcal{E}(\alpha)] \cong A(\alpha)$. Since both composites of $\varphi(\alpha)$ are natural in α, the $\varphi(\alpha)$ are coherent. Since the maps $A(\alpha \cap \sigma) \to A(\alpha)$ split the natural maps $A(\alpha) \to A(\alpha \cap \sigma)$, the $\varphi(\alpha)$ induce the desired splitting $\lim A(\mathcal{B}) \to \lim A(\mathcal{A}-\{\sigma\})$.

Now we consider the general case when $\mathfrak{A} \subseteq \mathcal{C}$ is generated by its maximal elements and $\mathcal{F}$ is an arbitrary admissible set of hyperplanes. We will show that the natural map $\varphi: A(V) \to \lim A(\mathfrak{A})$ satisfies the CRT conditions. By Theorem 1.2 it will follow that φ is an isomorphism (it is injective since both rings are contained in $\pi A(\alpha_i)$). By Theorem 2.3 it will follow that $S^{-1}A(V) = S^{-1}\lim A(\mathfrak{A}_S)$ for every multiplicatively closed set S, and we can apply Vorst's theorem as in [6].

The CRT conditions and K_1-regularity are both local conditions, so it suffices to consider a maximal ideal $\underset{\sim}{m}$ of A(V) and $S = A(V) - \underset{\sim}{m}$. Now $\underset{\sim}{m} \cap \mathcal{F}$ is an element of $\mathcal{C}(\mathcal{F})$ and hence contained in a coordinate system which we may assume to be $\{x_0, \ldots, x_n\}$. Now $\mathfrak{A}_S = \{\sigma \in \mathfrak{A} : I(\sigma) \cap S = \phi\} = \{\sigma \in \mathfrak{A} : I(\sigma) \subseteq \underset{\sim}{m}\}$ so we see for $\sigma \in \mathfrak{A}$ that $\sigma \in \mathfrak{A}_S$ precisely if $I(\sigma)$ is generated by some subset of the x_i's . Thus $\mathfrak{A}_S \subseteq \mathcal{C}'$ where $\mathcal{C}'$ is the poset associated with the admissible set $\{x_0, \ldots, x_n\}$ Let V_o denote the part of V passing through the origin, and note that the surjection $A(V) \to A(V_o)$ induces an isomorphism $A(V)_{\underset{\sim}{m}} \to S^{-1}A(V_o)$. By the special case above, $A(V_o) = \lim A(\mathfrak{A}_S)$, the $(CRT)_\ell$ hold for $A(V_o)$, and $A(V_o)$ is K_1-regular. Hence the localization $S^{-1}A(V_o) = A(V)_{\underset{\sim}{m}}$ satisfies the $(CRT)_\ell$ and is K_1-regular. As this is true for every maximal ideal, A(V) satisfies the $(CRT)_\ell$, is K_1-regular, and $A(V) = \lim A(\mathfrak{A})$.

We conclude this section by showing that $\lim A(\mathfrak{a})$ is K_1-regular on certain one-dimensional graded categories. $\mathfrak{a}$ one-dimensional means that $\mathfrak{a}$ is essentially a poset and there are no composable (non identity) arrows. Then $\dim(\alpha) = 1$ only if there is a map $\alpha \to \beta$ in $\mathfrak{a}$ with $\alpha \neq \beta$, and otherwise $\dim(\alpha) = 0$.

It is often convenient to view seminormal curves with rational singularities in the form $\mathrm{Spec}(\lim A(\alpha))$ for a one-dimensional category $\mathfrak{a}$; here the 0-cells would correspond to the singular points and the 1-cells to the components. Thus the following is an analog of Vorst's Theorem A, [24], which says seminormality implies K_1-regularity for curves with rational (or separable) singularities.

Proposition 2.5: Let $\underset{\sim}{k}$ be a K_2 regular domain, $\mathfrak{a}$ a one-dimensional graded category, and $A : \mathfrak{a} \to (\underset{\sim}{k}\text{-algebras})$. Suppose that $A(\gamma) = \underset{\sim}{k}$ for every 0-cell γ, $A(\gamma)$ is K_1-regular for every 1-cell α, and for each 1-cell α the kernels of the maps $A(\alpha) \to A(\gamma)$ are pairwise comaximal. Then the ring $A = \lim A(\mathfrak{a})$ is K_1-regular.

Proof: This is implicit in [8], [22]. There it is shown [8, Theorem 2] that $\Omega_{B/A} = 0$ where $B = \prod_{\dim \alpha = 1} A(\alpha)$ and thus that excision holds. From the cartesian square

$$\begin{array}{ccc} A & \longrightarrow & B \\ \downarrow & & \downarrow \\ \prod_{\dim \nu=0} \underset{\sim}{k} = \lim A(\mathcal{C}^0) & \longrightarrow & \prod_{\dim \alpha=1} \lim A(\mathcal{C}^0 \downarrow \alpha) \end{array}$$

we extract the exact sequences

$$\prod_{\dim \alpha=1} N^i K_2(\lim A(\mathcal{C}^0 \downarrow \alpha)) \to N^i K_1(A) \to \prod_{\dim \nu=0} N^i K_1(\underset{\sim}{k}) \oplus \prod_{\dim \alpha=1} N^i K_1(A(\alpha)).$$

Now $\lim A(\mathcal{C}^0 \downarrow \alpha)$ is a product of copies of $\underset{\sim}{k}$, one for each map from α in $\mathcal{C}$ so it is K_2-regular. Thus for $i \geq 1$ the outside terms vanish, forcing $N^i K_1(A) = 0$, i.e. A to be K_1-regular.

§3 The Spectral Sequence

In this section we prove the following theorem.

<u>Theorem 3.1</u> Let $\mathcal{G}$ be a finite, graded, cellular category, $\underset{\sim}{k}$ a commutative K_0-regular ring and $A : \mathcal{G} \to (\underset{\sim}{k}\text{-algebras})$ a contravariant functor. Assume for every p-cell α that

i) $A(\alpha) \approx \underset{\sim}{k}[x_1,\ldots,x_p]$

ii) $A(\alpha) \to \lim A(\mathcal{G}^{p-1}\downarrow\alpha)$ is a GL-fibration (p>0).

Then there is a spectral sequence

$$E_2^{pq} = H^p(\mathcal{G};KV_{-q}(\underset{\sim}{k})) \Rightarrow KV_{-p-q}(\lim A(\mathcal{G}))$$

Moreover, $H^0(\mathcal{G};KV_q(\underset{\sim}{k})) = E_2^{0,-q} = E_\infty^{0,-q}$ is a summand of $KV_q(\lim A(\mathcal{G}))$.

It seems appropriate to begin by explaining the need for the hypotheses. To simplify notation let us write A^p, $A^p(\alpha)$ for $\lim A(\mathcal{G}^p)$ and $\lim A(\mathcal{G}^p\downarrow\alpha)$, respectively. The cartesian squares of (1.12) may be written as

$$(3.2)\qquad \begin{array}{ccc} A^p & \longrightarrow & \prod\limits_{\dim(\alpha)=p} A(\alpha) \\ \downarrow & & \downarrow \\ A^{p-1} & \longrightarrow & \prod\limits_{\dim(\alpha)=p} A^{p-1}(\alpha) \end{array}$$

Our hypothesis (ii) guarantees that the right-hand map is a GL-fibration, so there are exact Mayer-Vietoris sequences [12]

$$(3.3) \quad \to KV_q(A^p) \to KV_q(A^{p-1}) \oplus KV_q(\Pi A^p(\alpha)) \to KV_q(\Pi A^{p-1}(\alpha)) \to \cdots$$

Now KV-theory commutes with finite products and agrees with classical K-theory for $q = 0,-1,-2,\ldots$. As in [6] the K_0-regularity of $\underset{\sim}{k}$ and hypotheses (i) ensure that for all q

$$KV_q(A(\alpha)) = KV_q(\underset{\sim}{k}[x_1,\ldots,x_p]) = KV_q(\underset{\sim}{k}).$$

Finally, this collection of interlocking sequences will give a spectral sequence

$$E_1^{pq} = KV_{-p-q}(A^p, A^{p-1}) \Rightarrow KV_{-p-q}(A^{\infty})$$

The problem is that in general it is difficult to identify the E_2-terms. In order to overcome this obstacle, we borrow an idea from the Atiyah-Hirzebruch spectral sequence for simplicial complexes [1], thinking of KV-theory as a cohomology theory on the associated affine schemes. The topological trick is that for complexes the space X^p/X^{p-1} is homotopy equivalent to a bouquet of p-spheres.

We make the assumption that $\mathcal{C}$ is cellular, guaranteeing that the categories $\mathcal{C}^{p-1}\downarrow\alpha$ have realizations which are homotopy equivalent to a bouquet of (p-1)-spheres by Proposition 0.6 .

For the remainder of this section we will further enhance the analogy with cohomology by raising indices and writing KV^q for KV_{-q}. In addition we will use the convention $KV^q(A^p) = 0$ for $p < 0$.

We now begin the proof of Theorem 3.1 . For any $\underset{\sim}{k}$ algebra R it is customary to define the reduced groups $\widetilde{KV}^q(R) = \mathrm{coker}(KV^q(\underset{\sim}{k}) \to KV^q(R))$. Since each cell of $\mathcal{C}$ contains a point all our $\underset{\sim}{k}$-algebras A^p, $A^p(\alpha)$ are augmented so there is a splitting $KV^q(R) = KV^q(\underset{\sim}{k}) \oplus \widetilde{KV}^q(R)$. In the sequence (3.3) each $KV^q(A(\alpha)) = KV^q(\underset{\sim}{k})$ maps therefore injectively to $KV^q(A^{p-1}(\alpha))$ and so (3.3) gives the exact sequence

$$(3.4)\quad \to KV^q(A^p)\to KV^q(A^{p-1})\to \prod_{\dim\alpha=p} \widetilde{KV}^q(A^{p-1}(\alpha))\to KV^{q+1}(A^p)\to\dots$$

It is then reasonable to write $KV^q(A^p,A^{p-1}) = \prod_{\dim\alpha=p} \widetilde{KV}^{q-1}(A^{p-1}(\alpha))$, when $p>0$. When $p=0$ we set $KV^q(A^0,A^{-1}) = KV^q(A^0)$. Thus the sequences

$$(3.5)\quad \to KV^q(A^p)\to KV^q(A^{p-1})\to KV^{q+1}(A^p,A^{p-1})\to KV^{q+1}(A^p)\to\dots$$

are exact for all p,q . Interlocking these sequences,

setting $D_1^{pq} = KV^{p+q-1}(A^{p-1})$, $E_1^{pq} = KV^{p+q}(A^p, A^{p-1})$,
we have a cohomology exact couple [13, p. 336]. Since G
is finite it is seen that the spectral sequence of this
couple converges to $KV^{p+q}(A)$, i.e. if $\dim G=n$ there is
a filtration

$$0 = F^{n+1} \subseteq F^n \subseteq \ldots \subseteq F^1 \subseteq F^0 = KV^{p+q}(A) = KV_{-p-q}(A)$$

where $F^p/F^{p+1} \approx E_\infty^{pq}$. Here $F^j = \ker(KV^{p+q}(A) \to KV^{p+q}(A^{j-1}))$
for fixed $p+q$.

It remains to show $E_2^{pq} = H^p(G; KV^q(\underline{k}))$. We first
explore the relationship of the pth cohomology group of G^p
and the KV-theory of A^p.

To facilitate this computation we adjoin a
terminal object to G to obtain CG as in §0. We will
call this terminal object ζ and if $\dim G=n$ we let
$\dim(\zeta) = n+1$. The forgetful functor $(\alpha \to \zeta) \to \alpha$ induces
an isomorphism $CG \downarrow \zeta \to CG$, making this identification
$CG^n \downarrow \zeta = CG^n = G$, $CG^p \downarrow \alpha = G^p \downarrow \alpha$ for $\alpha \in G$ (note CG^p
means $(CG)^p$). Finally we extend A to CG by $A(\zeta) = \underline{k}$
and for $\alpha \in G$ $A(\zeta) \to A(\alpha)$ is just the inclusion
$\underline{k} \subseteq \underline{k}[x_1, \ldots, x_p] = A(\alpha)$. Note $A^{n+1} = \lim A(CG) = \underline{k}$.

Each morphism $\sigma \to \tau \in CG$ induces a functor
$CG^p \downarrow \sigma \to CG^p \downarrow \tau$ by $(\alpha \to \sigma) \to (\alpha \to \sigma \to \tau)$ and consequently a map
$A^p(\tau) \to A^p(\sigma)$ for each p. When $\tau = \zeta$ these are
just our usual maps $G^p \downarrow \sigma \to G^p$, $A^p \to A^p(\sigma)$ because

of our identification $\mathcal{C} = C\mathcal{C}^n \downarrow \zeta$.

Given p,q integers, $\sigma \in C\mathcal{C}$ we will construct maps $\eta_{pq}(\sigma) : \tilde{H}^p(C\mathcal{C}^p \downarrow \sigma; KV^{q-p}(\underset{\sim}{k})) \to \tilde{K}V^q(A^p(\sigma))$ such that each morphism $\sigma \to \tau$ induces a commutative diagram

$$
\begin{array}{ccc}
\tilde{H}^p(C\mathcal{C}^p \downarrow \tau; KV^{q-p}(\underset{\sim}{k})) & \xrightarrow{\eta_{pq}} & \tilde{K}V^q(A^p(\tau)) \\
\downarrow & & \downarrow \\
\tilde{H}^p(C\mathcal{C}^p \downarrow \sigma; KV^{q-p}(\underset{\sim}{k})) & \xrightarrow{\eta_{pq}} & \tilde{K}V^q(A^p(\sigma))
\end{array} \qquad (3.6)
$$

We use induction on p. (When $p<0$ both groups are trivial.) When $p=0$ there are cannonical isomorphisms

$$\tilde{H}^0(C\mathcal{C}^0 \downarrow \sigma; KV^q(\underset{\sim}{k})) \approx \operatorname{coker}\left[KV^q(\underset{\sim}{k}) \xrightarrow{\text{diag}} \prod_{\substack{\alpha\to\sigma \\ \dim \alpha = 0}} KV^q(\underset{\sim}{k})\right] \approx \tilde{K}V^q(A^0(\sigma)).$$

Setting η_{0q} to be this composition, we clearly have the desired naturality.

Now assume $p>0$ and the $\eta_{p-1,q}$ are defined. Reducing sequence (3.4) and using (0.2) we obtain (noting $H^p(C\mathcal{C}^{p-1} \downarrow \sigma)=0$) Figure (3.7). The naturality of $\eta_{p-1,q}$ gives commutativity of the left hand square and thus induces $\eta_{pq}(\sigma)$ (dotted arrow). The naturality of η_{pq} follows from the fact that $\sigma \to \tau$ induces a diagram of squares

$$\begin{array}{ccccccc}
\coprod\limits_{\substack{\alpha\to\sigma \\ \dim\alpha=p}} C\mathcal{C}^{p-1}\downarrow\alpha & \longrightarrow & \coprod\limits_{\substack{\alpha\to\sigma \\ \dim\alpha=p}} C\mathcal{C}^{p}\downarrow\alpha & & \coprod\limits_{\substack{\alpha\to\tau \\ \dim\alpha=p}} C\mathcal{C}^{p-1}\downarrow\alpha & \longrightarrow & \coprod\limits_{\substack{\alpha\to\tau \\ \dim\alpha=p}} C\mathcal{C}^{p}\downarrow\alpha \\
\downarrow & & \downarrow & \longrightarrow & \downarrow & & \downarrow \\
C\mathcal{C}^{p-1}\downarrow\sigma & \longrightarrow & C\mathcal{C}^{p}\downarrow\sigma & & C\mathcal{C}^{p-1}\downarrow\tau & \longrightarrow & C\mathcal{C}^{p}\downarrow\tau
\end{array} \qquad (3.8)$$

In general $\eta_{pq}(\zeta)$ will not be an isomorphism. The purpose of our cellularity hypothesis, however, is to insure that $\eta_{p-1,q}(\sigma)$ is an isomorphism for $\sigma \neq \zeta$, $\dim\sigma=p$.

<u>Lemma 3.9</u>: If $C\mathcal{C}^k\downarrow\sigma$ is a cellular bouquet then $\eta_{pq}(\sigma)$ is an isomorphism for $p\leq k$. In particular this is true for all $\sigma\neq\zeta$.

<u>Proof</u>: We first remark that if $C\mathcal{C}^k\downarrow\sigma$ is a k-bouquet then $C\mathcal{C}^p\downarrow\sigma$ is a p-bouquet for $p\leq k$. To see this we use downward induction on p. By cellularity $H^q(C\mathcal{C}^{p+1}\downarrow\sigma;G) \to H^q(C\mathcal{C}^p\downarrow\sigma;G)$ is an isomorphism for $q<p$ (any abelian group G) so if $\mathcal{C}^{p+1}\downarrow\sigma$ is a $p+1$ bouquet $\tilde{H}^q(C\mathcal{C}^{p+1}\downarrow\sigma;G) = 0$ $q \leq p$ so $H^q(C\mathcal{C}^p\downarrow\sigma;G) = 0$ for $q \neq p$ i.e. $C\mathcal{C}^p\downarrow\sigma$ is a p-bouquet.

We now show $\eta_{pq}(\sigma)$ is an isomorphism by upward induction on p. We saw η_{0q} is always an isomorphism so suppose $\eta_{p-1,q}(\alpha)$ is an isomorphism for all $\alpha \neq \zeta$. Since $C\mathcal{C}^p\downarrow\sigma$ is a bouquet

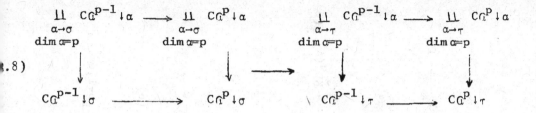

Figure (3.7)

$$\begin{array}{ccccccc}
\to \tilde{H}^{p-1}(C\mathfrak{a}^{p-1}\downarrow\sigma;KV^{q-p}(\underset{\sim}{k})) & \to & \prod\limits_{\substack{\alpha\to\sigma\\ \dim\alpha=p}} \tilde{H}^{p-1}(C\mathfrak{a}^{p-1}\downarrow\alpha;KV^{q-p}(\underset{\sim}{k})) & \to & \tilde{H}^{p}(C\mathfrak{a}^{p}\downarrow\sigma;KV^{q-p}(\underset{\sim}{k})) & \longrightarrow & 0 \\
\downarrow{\scriptstyle \eta_{p-1,q-1}(\sigma)} & & \downarrow{\scriptstyle \Pi\eta_{p-1,q-1}(\alpha)} & & \downarrow{\scriptstyle \eta_{p,q}(\sigma)} & & \\
\longrightarrow \tilde{K}V^{q-1}(A^{p-1}(\sigma)) & \longrightarrow & \prod\limits_{\substack{\alpha\to\sigma\\ \dim\alpha=p}} \tilde{K}V^{q-1}(A^{p-1}(\alpha)) & \longrightarrow & \tilde{K}V^{q}(A^{p}(\sigma)) & \longrightarrow & \tilde{K}V^{q}(A^{p-1}(\sigma))
\end{array}$$

Figure (3.11)

$$\begin{array}{ccccc}
\prod\limits_{\dim\alpha=p} \tilde{H}^{p-1}(\mathfrak{a}^{p-1}\downarrow\alpha;KV^{q}(\underset{\sim}{k})) & \longrightarrow & \tilde{H}^{p}(\mathfrak{a}^{p};KV^{q}(\underset{\sim}{k})) & \longrightarrow & \prod\limits_{\dim\alpha=p+1} \tilde{H}^{p}(\mathfrak{a}^{p}\downarrow\alpha;KV^{q}(\underset{\sim}{k})) \\
\downarrow{\scriptstyle \Pi\eta_{p-1,p+q-1}(\alpha)} & & \downarrow{\scriptstyle \eta_{p,p+q}(\zeta)} & & \downarrow{\scriptstyle \Pi\eta_{p,p+q}(\alpha)} \\
E_1^{pq} = \prod\limits_{\dim\alpha=p} \tilde{K}V^{p+q-1}(A^{p-1}(\alpha)) & \longrightarrow & \tilde{K}V^{p+q}(A^{p}) & \longrightarrow & \prod\limits_{\dim\alpha=p+1} \tilde{K}V^{p+q}(A^{p}(\alpha)) = E_1^{p+1,q}
\end{array}$$

$\tilde{H}^{p-1}(C\mathfrak{a}^p\downarrow\sigma;KV^{q-p}(\underset{\sim}{k})) = 0$ so the upper left hand map of (3.7) is injective, and hence so is the lower left. But this holds for all q so $\tilde{K}V^q(A^{p-1}(\sigma)) \to \Pi\tilde{K}V^q(A^{p-1}(\alpha))$ is injective implying that $\Pi\tilde{K}V^{q-1}(A^{p-1}(\alpha)) \to \tilde{K}V^q(A^p(\sigma))$ is surjective. The three lemma applied to (3.7) shows $\eta_{pq}(\sigma)$ is an isomorphism.

As an aside we remark that if $\mathfrak{a} = C\mathfrak{a}^n(\zeta)$ is a bouquet then (3.9) is actually a sharper statement then (3.1) i.e.

<u>Corollary 3.10</u>: Suppose in addition to the hypotheses of Theorem 3.1 $\mathfrak{a}$ is an n-bouquet, $n = \dim\mathfrak{a}$. Then there are isomorphisms

$$\eta_{n,-i}(\zeta) : \tilde{H}^n(\mathfrak{a};KV_{n+i}(\underset{\sim}{k})) \to \tilde{K}V_i(A^n)$$

$$KV_i(A^n) \approx KV_i(\underset{\sim}{k}) \oplus H^n(\mathfrak{a};KV_{n+i}(\underset{\sim}{k}))$$

We may now complete the computation of the E_2-term. Letting $\sigma = \zeta$ in (3.7) we obtain Figure (3.11) for each p,q where the outside vertical maps are isomorphisms. Inspection of our exact couple shows the composite of the bottom map is just d_1: $KV^{p+q}(A^p)=D^{p+1,q}\to E^{p+1,q}$ factors through $\tilde{K}V^{p+q}(A^p)$ as the map $KV^{p+q}(\underset{\sim}{k}) \to KV^{p+q}(A^p)$ factors through $KV^{p+q}(A^{p+1}) = D^{p+2,q-1}$. From (0.2) we may interpret the groups $\prod_{\dim\alpha=p} \tilde{H}^{p-1}(\mathfrak{a}^{p-1}\downarrow\alpha;KV^q(\underset{\sim}{k}))$ as relative groups

$H^p(\mathcal{G}^p, \mathcal{G}^{p-1}; KV^q(\underset{\sim}{k}))$ and so (3.11) says that for fixed q the chain complexes E_1^{*q}, $H^*(\mathcal{G}^*, \mathcal{G}^{*-1}; KV^q(\underset{\sim}{k}))$ are equivalent. Thanks to the cellularity of $\mathcal{G}$ a standard diagram chase shows that the homology of this complex is $H^*(\mathcal{G}; KV^q(\underset{\sim}{k}))$ (See also the discussion in §5).

It remains to prove the assertion about the E_r^{0q} terms. As $D_1^{pq} = KV^{p+q-1}(A^{p-1})$ vanishes for $p \leq 0$ and $E_1^{0q} = D_1^{1q} = KV^q(A^0)$, it follows that $E_r^{0q} = \mathrm{Im}(D_1^{r,q-r+1} \to D_1^{1q})$ $= \mathrm{Im}(KV^q(A^{r-1}) \to KV^q(A^0))$ and $E_\infty^{0q} = E_{n+1}^{0q}$, where $n = \dim \mathcal{G}$. It thus suffices to show $\mathrm{Im}(KV^q(A^n) \to KV^q(A^0))$ $= \mathrm{Im}(KV^q(A^1) \to KV^q(A^0))$ and this image is isomorphic to a summand of $KV^q(A^n)$.

Now $\mathcal{G}$ is the disjoint union of its connected components, $\lim A$ takes disjoint unions to products and KV commutes with products, so it is enough to check this on each component of $\mathcal{G}$ separately, i.e. we may assume $\mathcal{G}$ to be connected. Furthermore cohomology takes disjoint unions to products, so each component of a cellular graded category is cellular, and we may still assume $\mathcal{G}$ is cellular. Then $H^0(\mathcal{G}^n) \to H^0(\mathcal{G}^i)$ is an isomorphism, so $\mathcal{G}^1$ is also connected.

But now the standard argument using connectedness and KV regularity of the $A(\lambda) = \underset{\sim}{k}[x]$, $\dim \lambda = 1$ (see for instance [19]) shows $\mathrm{Im}(KV^q(\underset{\sim}{k}) \to KV^q(A^0))$ $\subseteq \mathrm{Im}(KV^q(A^n) \to KV^q(A^0)) \subseteq \mathrm{Im}(KV^q(A^1) \to KV^q(A^0)) =$ $\mathrm{Im}(KV^q(\underset{\sim}{k}) \to KV^q(A^0))$. But this image is isomorphic to

$KV^q(\underset{\sim}{k})$ which is a summand of $KV^q(A^n)$.

That completes the proof of Theorem 3.1. We note that the last part of the theorem implies that the differentials $d_r : E_r^{0q} \to E_r^{r,q-r+1}$ vanish. If $\dim \mathcal{C} \leq 2$ these are the only possible non zero differentials so $E_2^{pq} = E_\infty^{pq}$ for all p,q. Hence

<u>Corollary 3.12</u>: If (in addition to the hypotheses of Theorem 3.1) $\mathcal{C}$ is a curve (i.e. $\dim \mathcal{C} = 1$), then

$$KV_q(\lim A(\mathcal{C})) = H^0(\mathcal{C};KV_q(\underset{\sim}{k})) \oplus H^1(\mathcal{C};KV_{q+1}(\underset{\sim}{k})).$$

<u>Corollary 3.13</u> If (in addition to the hypotheses of Theorem 3.1) $\dim \mathcal{C} = 2$, then

$$KV_q(\lim A(\mathcal{C})) = H^0(\mathcal{C};KV_q(\underset{\sim}{k})) \oplus F$$

where

$$0 \to H^2(\mathcal{C};KV_{q+2}(\underset{\sim}{k})) \to F \to H^1(\mathcal{C};KV_{q+1}(\underset{\sim}{k})) \to 0$$

is exact.

Theorem 3.1 also implies that if $H^p(\mathcal{C}) = 0$, $0 < p < n = \dim \mathcal{C}$ then $KV_q(\lim A(\mathcal{C})) = H^0(\mathcal{C};KV_q(\underset{\sim}{k})) \oplus H^n(\mathcal{C};KV_{q+n}(\underset{\sim}{k}))$ as again all differentials d_r, $r \geq 2$ must vanish. But this is simply 3.10 applied to each component of $\mathcal{C}$.

§4 Examples

To avoid repetition we will make the following standing assumption for this section

Assumption 4.1: $\mathfrak{G}$ is a graded poset, each cell of $\mathfrak{G}$ has a point, $\underset{\sim}{k}$ a commutative ring, $A : \mathfrak{G} \to (\underset{\sim}{k}\text{-algebras})$ is a contravariant functor such that $A(\alpha) \simeq \underset{\sim}{k}[x_1,\ldots,x_p]$ when $\dim \alpha = p$.

We first look at curves, i.e. $\dim \mathfrak{G} = 1$. We make the following additional assumptions

Assumption 4.2: (4.1) is satisfied and

i) $\underset{\sim}{k}$ is K_1-regular

ii) For each 1-cell α, $A(\alpha) \to \lim A(\mathfrak{G}^0 \downarrow \alpha)$ is surjective.

We note that $A(\mathfrak{G}^0 \downarrow \alpha) = \prod_{\substack{v \to \alpha \\ \dim v = 0}} \underset{\sim}{k}$, i.e. a product of copies of $\underset{\sim}{k}$, one for each map $v \to \alpha$ where v is a 0-cell. Then ii) of (4.2) is satisfied if and only if for each 1-cell α the ideals $\{\ker(A(\alpha) \to A(v)) : v \to \alpha, \dim v = 0\}$ are pairwise co-maximal.

We now note that the hypotheses of Theorem 3.1 are satisfied. Cellularity of $\mathfrak{G}$ follows from (4.1) and Proposition (0.6) as $\mathfrak{G}^0 \downarrow \alpha$ is a non-empty collection of 0-cells for each $\alpha \in \mathfrak{G}$, $\dim \alpha = 1$. $A(\alpha) \to \lim A(\mathfrak{G}^0 \downarrow \alpha)$ is a GL-fibration from conditions i), ii) of (4.2).

Thus we may apply Corollary 3.12 to give for $A^1 = \lim A(\mathcal{G})$

$$KV_q(A^1) = H^0(\mathcal{G};KV_q(\underset{\sim}{k})) \oplus H^1(\mathcal{G};KV_{q+1}(\underset{\sim}{k}))$$

If in addition we assume $\mathcal{G}$ is connected, $H^0(\mathcal{G};KV_q(\underset{\sim}{k})) = KV_q(\underset{\sim}{k})$. Now because $\mathcal{G}$ is connected and $\dim \mathcal{G} = 1$ it follows from the description of $H^*(\mathcal{G};G)$ given in §0 that $H^1(\mathcal{G};\mathbb{Z})$ is a free group on $M-m-n+1$ generators where m is the number of 0-cells, n the number of 1-cells and M the number of arrows $v \to \alpha$, v a 0-cell and α a 1-cell. From the universal coefficient theorem we obtain $H^1(\mathcal{G};KV_{q+1}(\underset{\sim}{k})) = [KV_{q+1}(\underset{\sim}{k})]^{M-n-m+1}$. Hence

<u>Proposition 4.3</u> In addition to the assumptions (4.2) assume $\mathcal{G}$ is connected. Then for M,n,m as above

$$KV_q(\lim A(\mathcal{G})) = KV_q(\underset{\sim}{k}) \oplus [KV_{q+1}(\underset{\sim}{k})]^{M-n-m+1}$$

The above result is essentially that of [20].

We now turn our attention to surfaces. We will assume

<u>Assumption 4.4</u>: The conditions of (4.1) are satisfied, $\dim \mathcal{G} = 2$ and

i) $\underset{\sim}{k}$ is a K_2-regular domain.

ii) $A(\alpha) \to \lim A(\mathcal{G}^{p-1}\downarrow\alpha)$ is surjective for $\dim \alpha = p = 1,2$

iii) For each 2-cell α, $\mathcal{C}^1 \downarrow \alpha$ is connected.

We claim that under the assumptions of (4.4) the hypotheses of Theorem 3.1 are satisfied. Condition iii) of (4.4) implies $\mathcal{C}^1 \downarrow \alpha$ is a bouquet for each 2-cell α, which together with the fact that each cell has a point implies by (0.6) that $\mathcal{C}$ is cellular. As in the case of curves, $A(\alpha) \to \lim A(\mathcal{C}^0 \downarrow \alpha)$ is a GL-fibration for $\dim \alpha = 1$. $A(\alpha) \to \lim A(\mathcal{C}^1 \downarrow \alpha)$ is a GL fibration for 2-cells α from i), ii) of (4.4) and (2.5). This completes the proof of the claim and so the KV-theory of $\mathcal{C}$ is given by (3.13).

<u>Example 4.5</u>: In example (2.1) consider the functor A restricted to the category $\mathcal{C}^2$. From the discussion given there it is clear that all the hypotheses of (4.4) are satisfied. $\mathcal{C}^2$ has an initial object (the one 0-cell) and so $\mathcal{C}^2$ has the cohomology of a point. Hence from (3.13) we obtain $KV_q(\underset{\sim}{k}[X,Y,Z]/(Z(Z-XY)) = KV_q(\lim A(\mathcal{C}^2))$ $= KV_q(\underset{\sim}{k})$. We remark that for $q \geq 1$ this result follows from the "homotopy axiom" of KV-theory, a "null homotopy" being given by $X \to tX$, $Y \to tY$, $Z \to t^2Z$. For $q \leq 0$ one would need to know in addition that the ring is K_0-regular. This will follow from a generalization of our spectral sequence given in §5.

Example 4.6: Consider the ring $R = \underset{\sim}{k}[X,Y,Z]/XYZ(X+Y+Z)$ $= \lim A(\mathcal{C})$ where $\mathcal{C}$ is the poset of Example (1.11), graded in the obvious way, and $A(\alpha) = k[X,Y,Z]/\alpha$. The conditions (4.4) are not satisfied: if $\alpha = (Z)$ ($\dim \alpha = 2$) it was shown $A(\alpha) \to \lim A(\mathcal{C}^1 \downarrow \alpha)$ is not surjective. $\mathcal{C}$ again has the cohomology of a point so the spectral sequence would have predicted $KV_q(R) = KV_q(\underset{\sim}{k})$. Again by homotopy this is true for $q \geq 1$. However this is false for $q = 0$ since R is not K_0-regular (see [5], p. 136).

We next give an example where the spectral sequence is non-trivial. We call this example the "topological projective plane".

Example 4.7: Let e_2 be the subspace of $\mathbf{R}^2$ given by $e_2 = \{(x,y) \in \mathbf{R}^2 : 0 \leq x \leq 1, 0 \leq y \leq x - x^2\}$, e_1 the closed unit interval $[0,1]$, $e_0 = \{0\}$. Define $\pi_i : e_0 \to e_1$ by $\pi_i(0) = i$ $i=0,1$ $\rho_0 : e_1 \to e_2$ by $\rho_0(t) = (t,0)$ and $\rho_1 : e_1 \to e_2$ by $\rho_1(t) = (1-t, t-t^2)$. We then have a small subcategory $\mathcal{C}$ of the category of topological spaces with objects e_0, e_1, e_2 generated by the maps π_i, ρ_i. $\mathcal{C}$ is graded in the obvious way. Let $X = \operatorname{colim} \mathcal{C}$ i.e. the colimit of the identity functor on $\mathcal{C}$. X is obtained from e_2 by identifying points $\rho_0(t)$, $\rho_1(t)$ $t \in e_1$; e_2 is topologically a disk and we are identifying "antipodal" points, so X is homeomorphic to $\mathbf{RP}^2$, the real projective plane.

In view of the finite category theory developed in this paper (see also §5) one would anticipate that the realization of the category $\mathcal{G}$, $B\mathcal{G}$ would be $\mathbb{RP}^2$. This is in fact true as can be seen from the picture

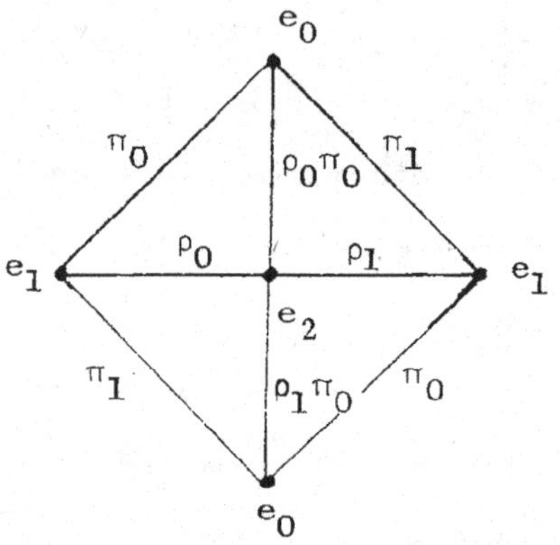

where similarly labeled edges are to be identified. The boundary of this diagram (edges not identified) corresponds to $\mathcal{G}^1 \downarrow e_2$ and is, in particular, connected.

Let $\underset{\sim}{k}$ be a K_2-regular domain and define $A : \mathcal{G} \to (\underset{\sim}{k}\text{-algebras})$ by $A(e_0) = \underset{\sim}{k}$, $A(e_1) = \underset{\sim}{k}[t]$ $A(e_2) = \underset{\sim}{k}[x,y]$, $A(\pi_i)(t) = i$, $i = 0,1$, $A(\rho_0)(x) = t$ $A(\rho_0)(y) = 0$, $A(\rho_1)(x) = 1-t$, $A(\rho_1)(y) = t-t^2$. Write $R = A^2 = \lim A(\mathcal{G})$, $A^1 = \lim A^1(\mathcal{G})$, $A^1(e_2) = \lim A(\mathcal{G}^1 \downarrow e_2)$, etc.,as in the proof of (3.1). We note

$R = A^2 = \{f(x,y) \in \underset{\sim}{k}[x,y] : f(t,0) = f(1-t,t-t^2)\}$,

$A^1 = \{f(t) \in \underset{\sim}{k}[t] : f(0) = f(1)\}$ (the coordinate ring of a node) and $A^1(e_2) = \{(f_0,f_1) \in \underset{\sim}{k}[t] \times \underset{\sim}{k}[t] :$ $f_0(0) = f_1(1)\ ,\ f_0(1) = f_1(0)\}$.

Although (1.4), (1.5), (1.6), (1.8) do not apply to this situation ($\mathcal{G}$ is not a poset, $R \to A(e_2)$ is not surjective), we remark $R = \underset{\sim}{k}[x-x^2, g, h, hx, hxy]$ where $g = (1-2x)y + x(x-x^2)$, $h = y(y-x+x^2)$. $\underset{\sim}{k}[x,y]$ is the integral closure of R with conductor $h\underset{\sim}{k}[x,y]$; R is obtained from $\underset{\sim}{k}[x,y]$ by glueing over the prime $hk[x,y] \subseteq R$. Hence R is seminormal by [23] (see also [17]). In addition $\operatorname{Spec} R = \operatorname{colim}(\operatorname{Spec} A)(\mathcal{G})$. As we will not need any of these results we omit the proofs.

We next show the conditions of (4.4) are satisfied. Certainly the conditions of (4.1) are satisfied as are i) and iii) of (4.4). Now $A(e_1) \to A^0(e_1)$ is surjective since $\ker A(\Pi_0) = t\underset{\sim}{k}[t]$, $\ker A(\Pi_1) = (1-t)\underset{\sim}{k}[t]$ and these are comaximal. Let $(f_0, f_1) \in A^1(e_2)$. Then $f_0(0) = f_1(1)$, $f_0(1) = f_1(0)$ so $f_0(1-t) - f_1(t)$ vanishes at $0,1$ and hence $f_1(t) - f_0(1-t) = (t-t^2)g(1-t)$ for some $g \in k[t]$. As $A(e_2) \to A^1(e^2)$ is given by $f(x,y) \to (f(t,0), f(1-t, t-t^2))$ we see $f_0(x) + yg(x)$ maps to (f_0, f_1) showing $A(e_2) \to A^1(e_2)$ is surjective.

Thus (3.13) applies and we obtain $KV_q(R) = H^0(\mathcal{G}; KV_q(\underset{\sim}{k})) \oplus F$ where

$$0 \to H^2(\mathcal{G}; KV_{q+2}(\underset{\sim}{k})) \to F \to H^1(\mathcal{G}; KV_{q+1}(\underset{\sim}{k})) \to 0$$

is exact. As $\mathcal{G}$ has the cohomology of $\mathbb{RP}^2$ it follows (eg [10, (19.23)] and the universal coefficient theorem)

that $H^0(\mathcal{G};KV_q(\underset{\sim}{k})) = KV_q(\underset{\sim}{k})$, $H^1(\mathcal{G};KV_{q+1}(\underset{\sim}{k})) = \ker(KV_{q+1}(\underset{\sim}{k}) \xrightarrow{2} KV_{q+1}(\underset{\sim}{k}))$ (i.e. the 2-torsion of $KV_{q+1}(\underset{\sim}{k})$) and $H^2(\mathcal{G};KV_{q+2}(\underset{\sim}{k})) = KV_{q+2}(\underset{\sim}{k})/2KV_{q+2}(\underset{\sim}{k})$.
If $\underset{\sim}{k} = \mathbb{Z}$ or a field of characteristic $\neq 2$ the two torsion of $KV_1(\underset{\sim}{k}) = K_1(\underset{\sim}{k})$ is $\{\pm 1\}$ so $K_0(R) = \mathbb{Z} \oplus \tilde{K}_0(R)$ and $0 \to K_2(\underset{\sim}{k})/2K_2(\underset{\sim}{k}) \to \tilde{K}_0(R) \to \mathbb{Z}/2\mathbb{Z} \to 0$ is exact. If $\underset{\sim}{k} = \mathbb{Z}$ or $\mathbb{R}$ we will see in §5 that the above sequence does not split and $K_2(\underset{\sim}{k})/2K_2(\underset{\sim}{k}) = \mathbb{Z}/2\mathbb{Z}$, so $\tilde{K}_0(R) = \mathbb{Z}/4\mathbb{Z}$. On the other hand if $\underset{\sim}{k} = \mathbb{C}$, $K_2(\mathbb{C})$ is divisible so $K_2(\underset{\sim}{k})/2K_2(\underset{\sim}{k}) = 0$. Hence $\tilde{K}_0(R) = \mathbb{Z}/2\mathbb{Z}$.

As an application of Theorem 3.1 in higher dimensions we consider the posets of Example 2.4. Let $\mathcal{H}$ be a collection of hyperplanes in $\mathbb{A}^n_{\underset{\sim}{k}}$, $\mathcal{C}$ the set of linear subspaces of $\mathbb{A}^n_{\underset{\sim}{k}}$ formed by intersecting the hyperplanes in $\mathcal{H}$. $\mathcal{C}$ is then graded as each linear subspace has a well defined dimension. Now suppose $\mathcal{G} \subseteq \mathcal{C}$ is a subposet generated by its maximal elements. We claim $\mathcal{G}$ is cellular. By Proposition (0.6) it suffices to show $\mathcal{G}^{p-1}\downarrow L$ is a non-empty $(p-1)$ bouquet for each $L \in \mathcal{G}$ $\dim L = p$. To this end let $L \in \mathcal{G}$, $\dim L = p$. Then L is isomorphic to $\mathbb{A}^p_{\underset{\sim}{k}}$. If $H \in \mathcal{H}$ is a hyperplane of $\mathbb{A}^n_{\underset{\sim}{k}}$ such that H does not contain L and is not disjoint from L, then $H \cap L$ corresponds to a hyperplane of $\mathbb{A}^p_{\underset{\sim}{k}}$. Further, $\mathcal{G}$ is generated by its maximal elements and $H \cap L \subseteq L$, so $H \cap L \in \mathcal{G}$. It is then clear that $\mathcal{G}^{p-1}\downarrow L$ is generated by such linear subspaces so $\mathcal{G}^{p-1}\downarrow L$ is isomorphic to $\mathcal{C}'$, where $\mathcal{C}'$ is the poset of linear subspaces of $\mathbb{A}^p_k$ corresponding to the set of hyperplanes $\mathcal{H}' = \{H\cap L \subseteq L \approx \mathbb{A}^p_k : H \in \mathcal{H}, H\cap L \neq L, H\cap L \neq \emptyset\}$. By [6, Theorem 3.10] $\mathcal{C}'$ is a bouquet, proving our assertion.

Now letting $A : \mathcal{G} \to$ ($\underset{\sim}{k}$-algebras) be as in Example 2.4 it is clear all the hypotheses of Theorem 3.1 hold except possibly ii). But this holds by [6] as by the above $\mathcal{G}^{p-1}\downarrow L$ can be identified with the poset of an admissible set of hyperplanes of $\mathbb{A}^p_k$ and so $A(L) \to \lim A(\mathcal{G}^{p-1}\downarrow L)$ is onto with K_1-regular target. Summarizing we have

<u>Theorem 4.8</u> Let $\mathcal{H}$ be an admissible set of hyperplanes of $\mathbb{A}^n_k$, $\mathcal{C}$ the poset of linear subspaces formed as intersections of members of $\mathcal{H}$. Let $\underset{\sim}{k}$ be a K_1-regular domain, $A : \mathcal{C} \to$ ($\underset{\sim}{k}$-algebras) the functor assigning to $L \in \mathcal{C}$ its coordinate ring $A(L)$. Then for every subposet $\mathcal{G} \subseteq \mathcal{C}$ generated by its maximal elements (if $\sigma \in \mathcal{G}$ and $\beta < \sigma$ then $\beta \in \mathcal{G}$) there is a spectral sequence

$$E_2^{pq} = H^p(\mathcal{G}; KV_{-q}(\underset{\sim}{k})) \Rightarrow KV_{-p-q}(\lim A(\mathcal{G}))$$

Furthermore $K_1(\lim A(\mathcal{G})) = KV_1(\lim A(\mathcal{G}))$.

In the special case of (4.8) where $\mathcal{G} = \mathcal{C}$ we simply recover, by (3.10), the results of [6] modulo the computation in [6] of the rank of $H^{n-1}(\mathcal{G};\mathbb{Z})$. This rank was calculated as $g = (-1)^n + \Sigma(-1)^{\dim(L)}$ where the sum runs over all $L \in \mathcal{C}$.

Example 4.9: Let $\mathcal{S}$ be an ordinary simplicial complex, i.e. $\mathcal{S}$ is a set of non-empty subsets of a set $\{v_0,\ldots,v_n\}$ such that $\{v_i\} \in \mathcal{S}$ all $i = 1,\ldots,n$ and $\alpha \in \mathcal{S}$, $\beta \subseteq \alpha \Rightarrow \beta \in \mathcal{S}$. Let $\underset{\sim}{k}$ be a K_1-regular domain, $\mathcal{H}$ the set of hyperplanes of $\mathbb{A}^{n+1}_{\underset{\sim}{k}} = \operatorname{Spec} \underset{\sim}{k}[x_0,\ldots,x_n]$ $\{H_0,\ldots,H_{n+1}\}$ where H_i is given by the equation $x_i = 0$ $i = 0,\ldots,n$ and H_{n+1} is given by the equation $x_0 + x_1 +\ldots+ x_n = 1$. For $\alpha \in \mathcal{S}$ let $L_\alpha = H_{n+1} \cap (\cap\{H_i : v_i \notin \alpha$ $i=0,\ldots,n\})$ Let $\mathcal{G} = \{L_\alpha : \alpha \in \mathcal{S}\}$. Then $\mathcal{G}$ is a subposet of $\mathcal{C}$ generated by its maximal elements, where $\mathcal{C}$ is the poset corresponding to the admissible set $\mathcal{H}$.
Theorem 4.8 applies and we write $\underset{\sim}{k}[\mathcal{S}] = \lim A(\mathcal{G})$ and call $\operatorname{Spec} \underset{\sim}{k}[\mathcal{S}]$ the $\underset{\sim}{k}$-realization of $\mathcal{S}$ and $\underset{\sim}{k}[\mathcal{S}]$ the $\underset{\sim}{k}$-coordinate ring of $\mathcal{S}$. Writing $KV^q(\operatorname{Spec} \underset{\sim}{k}[\mathcal{S}]) = KV_{-q}(\underset{\sim}{k}[\mathcal{S}])$ our spectral sequence becomes

$$E_2^{pq} = H^p(\mathcal{S};KV^q(\underset{\sim}{k})) \Rightarrow KV^{p+q}(\operatorname{Spec} \underset{\sim}{k}[\mathcal{S}])$$

where here H^* is just simplicial cohomology (see 0.3). This may be interpreted as a Karoubi-Villamayor K-theory analog of the Atiyah-Hirzebruch spectral sequence [1].

§5 A Generalization

In the proof of 3.1 we used only some very general facts about Karoubi-Villamayor K-theory. In this section we state a more general result and give some applications.

A category $\mathcal{K}$ of graded categories will be called <u>ample</u> if whenever $\mathcal{C} \in \mathcal{K}$ so are $\mathcal{C}^p$, $\mathcal{C}\downarrow\alpha$ for $p \geq 0$, $\alpha \in \mathcal{C}$, and, given $f : \mathcal{C} \to \mathcal{B}$ a morphism of $\mathcal{K}$, so are $\mathcal{C}^p \to \mathcal{C}$, $\mathcal{C}\downarrow\alpha \to \mathcal{C}$, $\mathcal{C}^p \to \mathcal{B}^p$ and $\mathcal{C}\downarrow\alpha \to \mathcal{B}\downarrow f(\alpha)$.

If $\mathcal{C}$ is a graded category there is a smallest ample category $\mathcal{K}(\mathcal{C})$ containing $\mathcal{C}$, namely the category formed by repeatedly taking skeleta and forming comma categories. By (0.6) if $\mathcal{C}$ is cellular so is each member of $\mathcal{K}(\mathcal{C})$.

We remark that if $\mathcal{C} \in \mathcal{K}$, $\mathcal{K}$ an ample category, there is a functor $\Psi_{\mathcal{C}} : \mathcal{C} \to \mathcal{K}$ defined on objects by $\Psi_{\mathcal{C}}(\sigma) = \mathcal{C}\downarrow\sigma$ and on morphisms in the obvious way.

If $G : \mathcal{C} \to$ (Abelian groups) is a contravariant functor then Quillen [18] defines cohomology groups $H^*(\mathcal{C};G)$. If G is constant (and hence morphism inverting) then these groups are the ones described in §0, where here we interpret G as the value of G on an object of $\mathcal{C}$.

<u>Definition (5.1)</u>: Let $\mathcal{K}$ be an ample category of graded categories, $F = \bigoplus F^p : \mathcal{K} \to$ (graded abelian groups) a contravariant functor. F will be called a M.V. (Mayer-Vietoris) theory on $\mathcal{K}$ if

i) F takes finite coproducts to products

ii) $F\Psi_{\mathcal{G}}$ is constant for each $\mathcal{G} \in \mathcal{K}$

iii) For each square

$$\begin{array}{ccc} \coprod_{\dim\alpha=p} \mathcal{G}^{p-1}\downarrow\alpha & \longrightarrow & \coprod_{\dim\alpha=p} \mathcal{G}^{p}\downarrow\alpha \\ \downarrow & & \downarrow \\ \mathcal{G}^{p-1} & \longrightarrow & \mathcal{G}^{p} \end{array}$$

in $\mathcal{K}$ there is a long exact sequence, natural with respect to maps of squares,

$$\to \prod_{\dim\alpha=p} F^{q-1}(\mathcal{G}^{p-1}\downarrow\alpha) \to F^{q}(\mathcal{G}^{p}) \to F^{q}(\mathcal{G}^{p-1}) \oplus \prod_{\dim\alpha=p} F^{q}(\mathcal{G}^{p}\downarrow\alpha) \to \cdots$$

Our theorem is

Theorem 5.2 Let $F = \oplus F^{p}$ be a M.V. theory on an ample category $\mathcal{K}$ of graded categories. Then for every finite cellular category $\mathcal{G} \in \mathcal{K}$ there is a spectral sequence

$$E_2^{pq}(\mathcal{G};F) \approx H^{p}(\mathcal{G};F^{q}\Psi_{\mathcal{G}}) \Rightarrow F^{p+q}(\mathcal{G})$$

These spectral sequences are functorial both for maps $\mathcal{G} \to \mathcal{B}$ in $\mathcal{K}$ and natural transformations $F \to F'$ of M.V. theories on $\mathcal{K}$.

The M.V. theory considered in §3 was the theory $F^{q}(\mathcal{G}^{p}\downarrow\sigma) = KV_{-q}(\lim A(\mathcal{G}^{p}\downarrow\sigma))$ where $\mathcal{K} = \mathcal{K}(C\mathcal{G})$, (see §0 for $C\mathcal{G}$), $\mathcal{G}$ a graded cellular category and $A : \mathcal{G} \to$ (k-algebras) satisfies the hypotheses of (3.1). The proof given in §3 of this special case is essentially the same as the proof of this mor general result, hence we will omit the proof of Theorem 5.2 .

Another example is where $\mathcal{K}$ is any ample category of finite graded categories and $F^q(\mathcal{C}) = H^q(\mathcal{C}) = H^q(\mathcal{C};\mathbb{Z})$. Note here that $\Psi_{\mathcal{C}}(\alpha) = \mathcal{C}\downarrow\alpha$ which has terminal object α hence is contractible. Thus $F^q\Psi_{\mathcal{C}}(\alpha) = 0$ if $q \neq 0$, $F^0\Psi_{\mathcal{C}}(\alpha) = \mathbb{Z}$, so $F^q\Psi_{\mathcal{C}}$ is constant. If instead of assuming $\mathcal{C}$ is cellular we assume only that each cell of $\mathcal{C}$ has a point, we still obtain a spectral sequence

$$E_1^{pq}(\mathcal{C};F) \approx H^{p+q}(\mathcal{C}^p,\mathcal{C}^{p-1}) \Rightarrow H^{p+q}(\mathcal{C})$$

as in the proof of 3.1 . This spectral sequence degenerates and it is easy to see that cellularity, or some similar condition on $\mathcal{C}$, is necessary in order that the correct E_2 term is obtained.

As another example let $X : \mathcal{C} \to$ (topological spaces) be a functor, $\mathcal{C}$ a cellular category. Let $F = \bigoplus F^q$: $\mathcal{K}(C\mathcal{C}) \to$ (graded abelian groups) be given by $F^q(\mathcal{C}^p\downarrow\sigma) = H^q_{sing}(\text{colim } X(\mathcal{C}^p\downarrow\sigma);G)$ where $H^*_{sing}(\cdot;G)$ is singular cohomology. F will take finite coproducts to products, and if each $X(\alpha)$ is contractible then $F\Psi_{\mathcal{C}}$ is constant. Thus theorem (5.2) applies as long as the cocartesian squares of $\mathcal{C}$ give rise to Mayer-Vietoris sequences. By the "dimension axiom" of singular homology $F^q\Psi_{\mathcal{C}} = 0$ for $q \neq 0$, so the spectral sequence degenerates giving $H^q_{sing}(\text{colim } X(\mathcal{C});G) = H^q(\mathcal{C};G)$. From this point of view it was not surprising that in Example (4.7) if X is the identity functor on $\mathcal{C}$ we saw that $\operatorname{colim} X(\mathcal{C}) = \mathbf{RP}^2 \approx B\mathcal{C}$.

If $\mathcal{C}$ is a simplicial complex, which, as we have seen, can be viewed as a cellular category, and $X : \mathcal{C} \to$ (Topological spaces) is the functor assigning to $\alpha \in \mathcal{C}$ the corresponding cell in the realization $B\mathcal{C}$ of $\mathcal{C}$, then $\operatorname{colim} A(\mathcal{C}) = B\mathcal{C}$. If we let F be defined on $\mathcal{K}(C\mathcal{C})$ by $F^q(\mathcal{C}^p \downarrow \sigma) = K^q_{Top}(\operatorname{colim} X(\mathcal{C}^p \downarrow \sigma))$ then theorem 5.2 applies and gives the classical Atiyah-Hirzebruch spectral sequence [1].

Returning to algebraic K-theory, we can now show that the hypotheses of Theorem 3.1 (the spectral sequence for KV-theory) imply that $R=\lim A$ is K_o-regular. Note that Example (4.5) shows that K_1-regularity of R is not necessary in order for Theorem 3.1 to apply. Let $\underset{\sim}{k}$ be a K_o-regular ring, and let $A:\mathcal{C}\to(\underset{\sim}{k}\text{-algebras})$ be a functor satisfying the hypotheses of Theorem 3.1. Let $A[t]: \mathcal{C}\to(\underset{\sim}{k}\text{-algebras})$ be given by $A[t](\alpha)=A(\alpha)[t]$. There are associated functors $F, F[t]$ on $\mathcal{K}(C\mathcal{C})$ defined by $F^q=KV_{-q}(\lim A)$, $F[t]^q=KV_{-q}(\lim A[t])$. By (5.2), the evident natural transformation $F\to F[t]$ induces a natural transformation of spectral sequences $E_r^{pq}(\mathcal{C};F)\to E_r^{pq}(\mathcal{C};F[t])$. But this is an isomorphism at the E_2 level as KV is a homotopy functor and $\underset{\sim}{k}$ is K_o-regular. Thus $KV_i(\lim A(\mathcal{C}))\cong KV_i(\lim A[t](\mathcal{C}))=KV_i((\lim A(\mathcal{C}))[t])$ for all i. This holds in particular for i=o, from which it follows that $R=\lim A(\mathcal{C})$ is K_o-regular.

We can derive this result in another way which applies in a slightly more general situation. Let $N_rKV_q(R) = \ker(KV_q(R[t_1,\ldots,t_r]) \to KV_q(R))$, where the map sends each t_i to 0. Of course $N_rKV_q(R) = 0$ for $q \geq 1$ by the homotopy property of KV-theory and $N_rKV_q(R) = N_rK_q(R)$ for $q \leq 0$. Thus R is K_0-regular iff $N_rKV_q(R) = 0$ for all q,r. Now let $\mathcal{C}$ be a finite cellular category, $A : \mathcal{C} \to$ (commutative rings) a contravariant functor such that $A(\alpha)$ is K_0-regular for each $\alpha \in \mathcal{C}$. Let $F_r = \bigoplus F_r^q : \mathcal{K}(C\mathcal{C}) \to$ (graded abelian groups) be given by $F_r^q(\mathcal{C}^p{\downarrow}\sigma) = N_rKV_{-q}(\lim A(\mathcal{C}^p{\downarrow}\sigma))$. Then F_r takes finite products to coproducts and $F_r\Psi_{\mathcal{C}}$ is constant , in fact constantly 0. The Mayer-Vietoris sequence for F_r comes from the Karoubi-Villamayor Mayer-Vietoris sequence provided for each p-cell σ of $\mathcal{C}$, $p>0$, $A(\sigma) = \lim A(\mathcal{C}^{p-1}{\downarrow}\sigma)$ is a GL-fibration. Theorem (4.2) applies and as the E_2-term of the spectral sequence vanishes so does $N_rKV_q(\lim A(\mathcal{C}))$ for each q,r. Thus

<u>Theorem 5.3:</u> Let $\mathcal{C}$ be a finite cellular category, $A : \mathcal{C} \to$ (commutative rings) a contravariant functor such that

i) $A(\alpha)$ is K_0-regular for each $\alpha \in \mathcal{C}$

ii) For each p-cell σ, $A(\sigma) \to \lim A(\mathcal{C}^{p-1}{\downarrow}\sigma)$ is a GL-fibration.

Then $\lim A(\mathcal{C})$ is K_0-regular.

We remark that the rings of Examples (2.1) and (4.5) show that the theorem is not valid if one replaces K_0 by K_1 or if one weakens ii) to simply require $A(\sigma) \to \lim A(\mathcal{G}^{p-1}\downarrow\sigma)$ to be a surjection.

We now return to example 4.7 . As we will be changing our base $\underset{\sim}{k}$ we will denote the functor A by $A_{\underset{\sim}{k}}$ and $A^2 = \lim A(\mathcal{G}^2)$ by $A^2_{\underset{\sim}{k}}$ to indicate our ground ring. For $\alpha \in \mathcal{G}$ we let $\mathbb{R}(\alpha)$ be the ring of continuous real valued functions on α and for morphisms $\alpha \xrightarrow{f} \beta$ in $\mathcal{G}$ we let $\mathbb{R}(f)$ be the homomorphism induced by the continuous function f. If $\underset{\sim}{k}$ is a subring of $\mathbb{R}$ we may identify $A_{\underset{\sim}{k}}(\alpha)$ as a subring of $\mathbb{R}(\alpha)$ in such a way that the maps $A_{\underset{\sim}{k}}(f)$ are restrictions of $\mathbb{R}(f)$. Thus we have a natural transformation $A_{\underset{\sim}{k}} \to \mathbb{R}$ which induces a homomorphism $A^2_{\underset{\sim}{k}} = \lim A_{\underset{\sim}{k}}(\mathcal{G}) \to \lim \mathbb{R}(\mathcal{G}) \simeq \mathbb{R}(X_{Top})$ where $X_{top} = \text{colim } \mathcal{G} \simeq \mathbb{RP}^2$. As the rings $\mathbb{R}(\alpha)$, $\mathbb{R}(X_{top})$ are Banach algebras, in addition to the K-theory of $B(\alpha)$ as a discrete ring, there is the KV-theory of the Banach algebra $\mathbb{R}(\alpha)$ denoted KV^*_{top}. There is a natural transformation $KV_*(\mathbb{R}(\alpha)) \to KV^{-*}_{top}(\mathbb{R}(\alpha))$ [12]. As each $\alpha \in \mathcal{G}$ is compact we have an isomorphism $KV^{-*}_{top}\mathbb{R}(\alpha) \approx KO^{-*}(\alpha)$ Thus we have a natural transformation of functors on $\mathcal{G}$: $F^q = KV_{-q}(\lim A_{\underset{\sim}{k}}(\cdot)) \to G^q = KO^q$ This in turn gives a natural transformation of spectral sequences which, as in (4.7) degenerate to give (for $\underset{\sim}{k} \subseteq \mathbb{R}$)

$$\begin{array}{ccccccccc} 0 \to & H^2(G;KV_{q+2}(\underset{\sim}{k})) & \to & \widetilde{KV}_q(A^2_{\underset{\sim}{k}}) & \to & H^1(G;KV_{q+1}(\underset{\sim}{k})) & \to 0 \\ & \downarrow & & \downarrow & & \downarrow & \\ 0 \to & H^2(G;KO^{-q-2}(pt)) & \to & KO^{-q}(\mathbf{RP}^2) & \to & H^1(G;KO^{-q-1}(pt)) & \to 0 \end{array}$$

where the maps between cohomology groups are induced by the maps between coefficient groups $KV_q(\underset{\sim}{k}) \to KV_q(\mathbf{R}) \to KO^{-q}(pt)$.

We now restrict ourselves, for the moment, to the case where $q = 0$, $k = \mathbf{Z}$. Now $H^1(G;KV_1(\mathbf{Z})) = KV_1(\mathbf{Z}) = \{\pm 1\}$ and $H^1(G;KO^{-1}(pt)) = KO^{-1}(pt) = \{\pm 1\}$. The map $KV_1(Z) \to KO^{-1}(pt)$ is evidentally an isomorphism (see [15, §7]). Also $H^2(G;KV_2(\mathbf{Z})) = K_2(\mathbf{Z})$, as this latter group is cyclic of order 2 generated by the symbol $\{-1,-1\}$. On the other hand $H^2(G;KO^{-2}(pt)) = \pi_1(SL(\mathbf{R})) = \{\pm 1\}$ and, by [15, 8.4], $\{-1,-1\}$ maps to the generator of $KO^{-2}(pt)$. Thus the outside maps of the above diagram are isomorphisms and hence $\widetilde{KV}_0(A^2_{\mathbf{Z}}) \approx KO(\mathbf{RP}^2)$. By [11, p. 223] $KO(RP^2) \approx Z/4\mathbf{Z}$ and hence the filtration for $\widetilde{KV}_0(A^2_{\mathbf{Z}})$ of (3.1) does not split. We can extend this result as follows:

<u>Proposition 5.4</u>: Let $\underset{\sim}{k}$ be a field of characteristic $\neq 2$. If the symbol $\{-1,-1\} \in K_2(\underset{\sim}{k})$ is in $2K_2(\underset{\sim}{k})$ then $\widetilde{K}_0(A^2_{\underset{\sim}{k}}) \approx K_2(\underset{\sim}{k})/2K_2(\underset{\sim}{k}) \oplus Z/2\mathbf{Z}$. Otherwise the exact sequence

$$0 \to K_2(\underset{\sim}{k})/2K_2(\underset{\sim}{k}) \to \widetilde{K}_0(A_{\underset{\sim}{k}}) \to \mathbf{Z}/2\mathbf{Z} \to 0$$

does not split.

Proof: The map $\mathbb{Z} \to \underset{\sim}{k}$ induces a natural transformation $A_{\mathbb{Z}}$ to $A_{\underset{\sim}{k}}$ inducing by (5.2) a transformation of the spectral sequences of (3.1). Looking at the resulting filtrations we obtain a diagram

$$\begin{array}{ccccccc}
0 \to & K_2(\mathbb{Z}) & \to & \tilde{K}_0(A^2_{\mathbb{Z}}) & \to & \mathbb{Z}/2\mathbb{Z} & \to 0 \\
& \downarrow & & \downarrow & & \downarrow & \\
0 \to & \mathbb{Z}/2\mathbb{Z}\otimes K_2(\underset{\sim}{k}) & \to & \tilde{K}_0(A^2_{\underset{\sim}{k}}) & \to & \mathbb{Z}/2\mathbb{Z} & \to 0
\end{array}$$

The generator ξ of $\tilde{K}_0(A^2_{\mathbb{Z}})$ maps to the generator -1 of $\mathbb{Z}/2\mathbb{Z} = K_1(\mathbb{Z})$ and 2ξ is the image of $\{-1,-1\}$. Thus if $\bar{\xi}$ is the image of ξ in $\tilde{K}_0(A^2_{\underset{\sim}{k}})$, $\bar{\xi}$ maps to -1 in $\mathbb{Z}/2\mathbb{Z}$ and $2\bar{\xi}$ is the image of the class of $\{-1,-1\}$ in $\mathbb{Z}/2\mathbb{Z} \otimes K_2(\underset{\sim}{k})$. Since both $\mathbb{Z}/2\mathbb{Z}$ and $\mathbb{Z}/2\mathbb{Z} \otimes K_2(\underset{\sim}{k})$ have exponent 2 we see the lower sequence splits precisely if $2\bar{\xi} = 0$, i.e. $\{-1,-1\} = 0$ in $\mathbb{Z}/2\mathbb{Z} \otimes K_2(\underset{\sim}{k})$.

We end by giving the following amusing extension of some of the ideas of the preceeding example. Let $\mathcal{S}$ be a two dimensional ordinary simplicial complex with vertices $\{v_0,\ldots,v_n\}$. Let $\underset{\sim}{k} = \mathbb{Z}$ and $\mathbb{Z}[\mathcal{S}]$ be the integral coordinate ring of $\mathcal{S}$ as in example (4.9). Let $\mathcal{G}$ be the category associated to $\mathcal{S}$ as in (4.9), $A : \mathcal{G} \to$ (Z-algebras) the functor of Theorem 4.8 and suppose $F = \oplus F^q$ given by $F^q(\mathcal{G}^p\downarrow\sigma) = KV_{-q}\lim A(\mathcal{G}^p\downarrow\sigma))$ as above. On the other hand there is a functor $X : \mathcal{G} \to$ (Topological spaces) given by $X(\alpha) = \{(x_0,\ldots,x_n)\in\mathbb{R}^n \mid x_i\geq 0,\ x_0+\ldots+x_n=1$ and $x_i=0, i\notin\alpha\}$, and $|\mathcal{S}| = \operatorname{colim} X(\mathcal{G}) = \bigcup_{\alpha\in\mathcal{G}} X(\alpha)$ is the ordinary realization of $\mathcal{S}$. Comparing this to the construction of 4.9 we see there is a natural inclusion $A(\alpha) \subseteq \mathbb{R}^{X(\alpha)}$ where $\mathbb{R}^{X(\alpha)}$ is the ring of continuous real valued functions on $X(\alpha)$. This gives, as above, a natural transformation $F \to G$ of functors on $\mathcal{M}(\mathcal{C}\mathfrak{a})$ where $G^q(\mathfrak{a}^p \downarrow \sigma) = KO^q(\operatorname{colim} X(\mathfrak{a}^p \downarrow \sigma))$. This gives by (5.2) a natural transformation of spectral sequences. Now $KO^0(pt) = Z = KV_0(Z)$ and by the remarks of the preceeding example this transformation induces isomorphisms $K_1(Z) \to KO^{-1}(pt)$ and $K_2(Z) \to KO^{-2}(pt)$. We thus obtain

Proposition 5.5: Let $\mathcal{S}$ be a 2-dimensional simplicial complex and $Z[\mathcal{S}]$ its integral coordinate ring. Then $KO(|\mathcal{S}|) \approx K_0(Z[\mathcal{S}])$.

References

1. M.F. Atiyah and F. Hirzebruch, Vector Bundles and Homogeneous Spaces, Proc. of Symp. in Pure Math., Amer. Math. Soc. 3 (1961), pp 7-35.

2. M.F. Atiyah and I.G. Macdonald, Introduction to Commutative Algebra, Addison-Wesley, Reading, Mass, 1969.

3. H. Bass, Algebraic K-theory, Benjamin, New York, 1968.

4. N. Biggs, Algebraic Graph Theory, Cambridge University Press, 1974.

5. B.H. Dayton, K-Theory of Tetrahedra, J. Alg. 56 (1979), 129-144.

6. B.H. Dayton and C.A. Weibel, K-theory of Hyperplanes, Trans. AMS 257 (1980), 119-141.

7. R.K. Dennis and M. Krusemeyer, $K_2(A[X,Y]/XY)$, a problem of Swan, and related computations, J. Pure App. Alg. 15 (1979), 125-148.

8. S.C. Geller and L.G. Roberts, Further Results on Excision for K_1 of Algebraic Curves, preprint (1978).

9. S. Gersten, Higher K-theory of Rings, Lecture Notes in Math. 341, Springer Verlag, New York, 1973.

10. M.J. Greenburg, Lectures on Algebraic Topology, Benjamin, New York, 1967.

11. D. Husemoller, Fibre Bundles, McGraw Hill, New York, 1966.

12. M. Karoubi and O. Villamayor, K-theorie algebrique et K-theory topologique, Math. Scand. 28 (1971), 265-307.

13. S. MacLane, Homology, Springer-Verlag, New York, 1967.

14. J.P. May, Simplicial Objects in Algebraic Topology, Van Nostrand, Princeton, 1967.

15. J. Milnor, Introduction to Algebraic K-Theory, Princeton University Press, Princeton, 1971.

16. F. Orecchia, Sulla seminormalita di certe varieta affini riducibili, Boll. Un. Mat. Ital. (2) 13 B (1976) pp. 588-600.

17. C. Pedrini, Incollamenti di ideali primi e gruppi di Picard, Rend. Sem. Mat. Univ. Padova, 48 (1973), 39-66.

18. D. Quillen, Higher Algebraic K-theory I, Lecture Notes in Math 341 Springer-Verlag, New York, 1973.

19. L.G. Roberts, The K-theory of some reducible affine varieties, J. Alg. 35 (1975), 516-527.

20. L.G. Roberts, The K-theory of some reducible affine curves, a combinatorial approach, Lecture Notes in Math 551, Springer-Verlag, New York, 1976.

21. L.G. Roberts, SK_1 of n-lines in the plane, Trans. AMS 222 (1976).

22. S.C. Geller and L.G. Roberts, Excision and K_1-regularity for curves with normal crossings, J. Pure App. Alg. 15 (1979) 11-21.

23. C. Traverso, Seminormality and Picard Group, Annali della Scoula Norm. Sup. Pisa 24 (1970), 585-595.

24. T. Vorst, Polynomial Extensions and Excision for K_1, Math. Ann. 244 (1979), 193-204.

25. T. Vorst, Localisation of the K-Theory of Polynomial Math. Ann. 244 (1979), 33-53.

26. O. Zariski and P. Samuel, Commutative Algebra Vol. I, Van Nostrand, Princeton, 1958.

Seminormality of Unions of Planes

by

Barry H. Dayton
Northeastern Illinois University
Chicago, Illinois, U.S.A.

and

Leslie G. Roberts
Queen's University
Kingston, Ontario

In this paper we try to decide when the union of planes in affine space is seminormal. The corresponding problem with lines has a trivial solution: a union of lines is seminormal if and only if at each intersection point the directions of the lines are linearly independent. With hyperplanes there is also a simple characterization of seminormality: a union of hyperplanes is seminormal if and only if no three of them contain the same affine subspace of codimension two (Theorem 3 below). However, for planes (even in affine four space) the situation is more complicated. But we do reduce the problem to one in finite dimensional linear algebra. We give several interesting examples, including a union of planes that is seminormal, but if any one plane is removed the result is no longer seminormal (Example 16).

First recall the definition: a commutative ring R is seminormal if it is reduced and whenever b, c ϵ R satisfy

$b^3 = c^2$, there is an $a \varepsilon R$ with $a^2 = b$, $a^3 = c$ [S] . This is a local condition (Proposition 3.7 of [S]). Thus there is no loss in generality if we discuss only planes passing through the origin.

Our main tool for discussing seminormality is the relation between seminormality and the Chinese Remainder Theorem. Let A be a commutative ring and $\mathfrak{a} = \{I_1,\ldots,I_n\}$ a set of ideals of A . Then $\mathfrak{a}$ satisfies the Chinese Remainder Theorem (CRT) if given $a_1,\ldots,a_n \varepsilon A$ such that $a_i \equiv a_j \bmod (I_i+I_j)$ for all i,j there exists $x \varepsilon A$ such that $x \equiv a_i \bmod I_i$ $(1 \leq i \leq n)$. Assume also that $\cap I_i = 0$. Then A is a subring of $B = \prod_{i=1}^{n} (A/I_i)$.

<u>Lemma 1</u> Let C be the conductor of A in B . Then $C = \Sigma_i(\cap_{j\neq i} I_j) = \cap_i(I_i + \cap_{j\neq i} I_j)$.

<u>Proof.</u> Both these formulas are known, but the discussions that we have seen ([B], [O1]) make the unnecessary assumption that $\mathfrak{a}$ is an irredundant collection of minimal primes. Hence we sketch a proof. Since C is a B ideal, $C = \prod_{i=1}^{n} C_i$ where C_i is an ideal of A/I_i . Clearly $C_i = \{a \varepsilon A/I_i | (0,\ldots,a,0\ldots0) \varepsilon A\}$. Let $\tilde{C}_i$ be the subgroup of C canonically isomorphic to C_i (i.e. $\tilde{C}_i = \{(0,\ldots0,a,0\ldots0) | a \varepsilon C_i\}$). Then $\tilde{C}_i = \cap_{j\neq i} I_j$. Thus $C = \Sigma_i \tilde{C}_i = \Sigma_i(\cap_{j\neq i} I_j)$ which proves the first formula for C . Let $\pi_i : A \to A/I_i$ be the projection. Then

$C = \cap_i(\pi_i^{-1}C_i) = \cap_i(I_i + \cap_{j\neq i} I_j)$ which proves the second formula.

Done.

The relationship between seminormality and the CRT is now given by

Theorem 2. (CRT) Let A be a commutative ring and $\mathfrak{a} = \{I_1, \ldots, I_n\}$ be a set of ideals of A with $\cap I_i = 0$. Let $B = \prod_{i=1}^{n}(A/I_i)$, C be the conductor of A in B and C_i be the projection of C in A/I_i. Assume also that A/I_i is seminormal for each i and that $I_j + I_k$ is radical all j,k. Then conditions (1) and (2) are equivalent.

(1) A is seminormal

(2) CRT holds for $\mathfrak{a} = \{I_1, \ldots I_n\}$.

These imply the next 5 conditions which are equivalent.

(3) For each i $\cap_{j\neq i}(I_i + I_j) = I_i + \cap_{j\neq i} I_j$

(4) For each i $I_i + \cap_{j\neq i} I_j$ is radical in A

(5) For each i $C_i = \cap_{j\neq i}((I_i + I_j)/I_i)$

(6) For each i C_i is radical in A/I_i

(7) C is radical in B

The above imply the next 2 conditions which are equivalent

(8) C is radical in A

(9) $C = \cap_{\substack{j,k \\ j\neq k}} (I_j + I_k)$

If, in addition $A/(\cap_{j\neq i} I_j)$ is seminormal for any i then

(7) $\Rightarrow$ (1).

We will show by Examples 14, 15 below that in general (7) does not imply (1). If we remove the hypotheses on the I_i's then A seminormal in B implies (2), (2) $\Rightarrow$ (3), and (1) $\Rightarrow$ (7) $\Leftrightarrow$ (6) $\Rightarrow$ (8). We don't know if (8) $\Rightarrow$ (7). Whenever we speak of the conductor being radical, we mean radical in B.

<u>Proof</u>. (2) $\Rightarrow$ (1) is easy and (1) $\Rightarrow$ (2) follows by applying [S] 2.2 and 3.4 to obtain Traverso's formulation of seminormality, and then applying the argument of [O2] as modified in [DW2] Theorem 1.8. For a direct proof see [D].

For 2)$\Rightarrow$3) note first that always $I_i + \cap_{j\neq i} I_j \subseteq \cap_{j\neq i}(I_i+I_j)$. Conversely suppose $b \in \cap_{j\neq i}(I_i+I_j)$. Set $a_j = 0$ $j \neq i$ and $a_i = b$; then by CRT there is an $x \in A$ with $x \equiv a_j$ mod I_j . Then $b = (b-x) + x \in I_i + \cap_{j\neq i} I_j$. Now 3)$\Rightarrow$4) is immediate as I_i+I_j is radical for all i, j and 4)$\Rightarrow$3) comes from Lemma 2.5 below. 3)$\Leftrightarrow$5) follows from the fact that $C_i = \pi_i\,(I_i+\cap_{j\neq i} I_j)$. 5)$\Rightarrow$6) follows as $(I_i+I_j)/I_i$ is radical in A/I_i while 6)$\Rightarrow$5) follows from Lemma 2.5. 6)$\Leftrightarrow$7) is immediate. 8$\Leftrightarrow$9 by Lemma 2.5 and 7)$\Rightarrow$8) is clear.

Finally if $A/\cap_{j\neq i} I_i$ is seminormal and 4) holds, we may apply the NU, NPic exact sequence to the cartesian square

$$\begin{array}{ccc} A & \longrightarrow & A/\cap_{j\neq i} I_j \\ \downarrow & & \downarrow \\ A/I_i & \longrightarrow & A/(I_i+\cap_{j\neq i} I_j) \end{array}$$

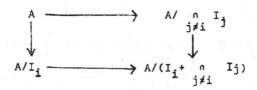

to obtain NPic $A = 0$ as NPic $(A/I_i) = $ NPic $(A/\cap_{j\neq i} I_j) = 0$ by [S] Theorem 1 and NU $(A/(I_i + \cap_{j\neq i} I_j)) = 0$ by 4). Then by [S] Theorem 1, A is seminormal.

<u>Lemma 2.5</u>: With hypotheses as above $\sqrt[A]{C} = \cap_{j\neq k}(I_j+I_k)$ and $\sqrt[A/I_i]{C_i} = \cap_{j\neq i}((I_i+I_j)/I_i)$ for each i.

<u>Proof</u>. For both statements it is enough to show for fixed i $\sqrt[A]{I_i + \cap_{j\neq i} I_j} = \cap_{j\neq i}(I_i + I_j)$. Now $I_i + \cap_{j\neq i} I_j \subseteq \cap_{j\neq i}(I_i + I_j)$ for fixed i and this latter ideal is radical so we must show (for fixed i) $\cap_{j\neq i}(I_i + I_j) \subseteq \sqrt[A]{I_i + \cap_{j\neq i} I_j}$. Let $a \in \cap_{j\neq i}(I_i + I_j)$. For each $j \neq i$ $a = b_j + c_j$ where $b_j \in I_i$, $c_j \in I_j$ so $a^{n-1} = \prod_{j\neq i}(b_j + c_j) = (\prod_{j\neq i} c_i) + b$ where $b \in I_i$. But $\prod_{j\neq i} c_j \in \cap_{j\neq i} I_j$ so $a^{n-1} \in I_i + \cap_{j\neq i} I_j$ as desired.

As an illustration of the CRT we derive the known condition for a union of hyperplanes in $\mathbb{A}^p$ to be seminormal. Let k be a field $R = k[X_1,\ldots,X_p]$, $f_i = a_{i1}X_1 + \ldots + a_{ip}X_p + b_i$, not all $a_{ij} = 0$, and we assume no two f_i are scalar multiples of each other, so that the $H_i = V(f_i) = \{P \in \text{Spec } R \mid f_i \in P\}$ are distinct hyperplanes.

Let $A = R/f_1 \cdots f_N$, I_i the image of (f_i) in A. Since R is a UFD $\cap I_i = 0$, further $A/I_i \cong k[t_1, \ldots, t_{p-1}]$, $A/(I_i+I_j) \cong k[t_1,\ldots,t_{p-2}]$ so the hypotheses of Theorem 2 are satisfied. Assume $H_i \cap H_j$ are all distinct. Now $(I_i+I_j)/I_i$ is generated by the image of f_j in A/I_i so $\bigcap_{j \neq i} (I_i+I_j) = (\bar{f}_1 \ldots \hat{f}_i \ldots \bar{f}_n)$ where $\bar{f}_j$ is the image of f_j in A/I_i and $\hat{f}_i$ means that $\bar{f}_i$ has been omitted. (Here we use that the $\bar{f}_j \in A/I_i$ are not scalar multiples of each other, ie. the $H_i \cap H_j$ are distinct). Now $F_i = f_1 \ldots \hat{f}_i \ldots f_n$ vanishes in B except for the ith co-ordinate, so $F_i \in \tilde{C}_i$. Thus $C_i = \bigcap_{j \neq i} (I_i+I_j)/I_i$ so by 5) of CRT the conductor of A in B is radical. By induction, using the last sentence of Theorem 2, A is seminormal.

If we had not assumed the $H_i \cap H_j$ were distinct then $\bigcap_{j \neq i} (I_i+I_j)/I_i$ is generated by $\bar{f}_{i_1} \ldots \bar{f}_{i_k}$ where $\bar{f}_{ij}$ correspond to the distinct varieties $H_i \cap H_j$. But $\bar{f}_{i_1} \ldots \bar{f}_{i_k}$ is not in C_i unless all $f_j (j \neq i)$ are present so 5) of Theorem 2 would not hold. Thus C is not radical in B and A would not be seminormal. Summarizing we have

<u>Theorem 3</u> Let $H_1, \ldots, H_N$ be distinct hyperplanes in $\mathbb{A}^p$, $H_i = V(f_i)$. Then $\cup H_i$ (i.e. $k[x_1,\ldots,x_p]/(f_1 \ldots f_N)$) is seminormal iff the codimension 2 subvarieties $H_i \cap H_j$ are distinct.

Now let A be the co-ordinate ring of N distinct planes Π_i through the origin in p-space over a field k. That is, $R = k[X_1, X_2, \ldots, X_p]$ $p \geq 3$, $\tilde{I}_i = \{f \in R \mid f \text{ vanishes on } \Pi_i\}$

$1 \leq i \leq N$ is an ideal generated by $p-2$ k-independent linear forms, and $A = R/(\cap_i \tilde{I}_i)$. Let I_i be the image of $\tilde{I}_i$ in A. Then $R/\tilde{I}_i = A/I_i = k[t_i,u_i]$. If $i \neq j$ Π_i and Π_j either meet in a line or in the origin. In the first case $R/(\tilde{I}_i+\tilde{I}_j)=A/(I_i+I_j) \cong k[t]$, and in the second case $R/(\tilde{I}_i+\tilde{I}_j) = A/(I_i+I_j) \cong k$. The rings A/I_i are all semi-normal and each $I_i + I_j$ is radical, so the hypotheses of theorem 2 (CRT) are satisfied. In particular A semi-normal implies that I is radical in B.

By lemma 2.5 the radical of C_i in A/I_i is $\bigcap_{j \neq i} (I_i+I_j)/I_i$. Now if Π_i & Π_j meet in a point $I_i+I_j = M$, the maximal ideal at the origin so that $(I_i+I_j)/I_i$ is the maximal ideal at the origin of $A/I_i \approx k[t_i,u_i]$. Otherwise Π_i meets Π_j in a line and so $(I_i+I_j)/I_i$ is a principal ideal generated by an equation of this line in $\mathbb{A}^2 = \operatorname{Spec} A/I_i$. This equation is a form of degree 1. Then the intersection $\cap (I_i+I_j)/I_i$ is either M/I_i (if Π_i intersects the other planes in points only) or a principal ideal whose generator is a product of equations of distinct lines $\Pi_i \cap \Pi_j$. So if Π_i meets the Π_j in d distinct lines $d > 0$ $\bigcap_{j \neq i} (I_i+I_j)/I_i$ is a principal ideal generated by a form of degree d. Consequently any form of degree d in $\bigcap_{j \neq i} (I_i+I_j)/I_i$ is a scalar multiple of the generator and so, if non-zero, is a generator. Now $C_i = (I_i + \cap_{j \neq i} I_j)/I_i \subseteq \bigcap_{j \neq i} (I_i+I_j)/I_i$ is the set of images of elements of A which vanish on all Π_j $(j \neq i)$. We conclude that C_i is radical iff there is a form of degree d in A (or equivalently, R)

which vanishes on Π_j $j\neq i$ but not on Π_i.

There is a classical correspondance between planes through the origin in $\mathbb{A}^p$ and **lines** in $\mathbb{P}^{p-1}$. Planes intersecting only at the origin correspond to skew lines, planes intersecting in lines correspond to intersecting lines. A form f in R vanishes on the plane Π_i iff it vanishes on the corresponding line in $\mathbb{P}^{p-1}$. Thus the condition that C_i be radical can be translated to a condition on the corresponding lines in $\mathbb{P}^{p-1}$ - namely that C_i is radical iff whenever the line corresponding to Π_i has d intersection points (with other lines) $d > 0$ then there is a form of degree d in R which vanishes on all lines but this. (i.e. a hyper-surface of degree d containing all but this one line). (If $d = 0$ there must be two forms of degree 1 which are independent on this line but vanish on all others). In what follows we will freely change between planes in $\mathbb{A}^p$ and lines in $\mathbb{P}^{p-1}$.

Some of the simplest cases are:

Example 3.5 (a) Suppose all the lines in $\mathbb{P}^{p-1}$ pass through one point P. Then the conductor $\mathcal{C}$ is radical if and only if the directions of the lines are linearly independent.

(b) A necessary condition for $\mathcal{C}$ to be radical is that the lines in $\mathbb{P}^{p-1}$ through any intersection point be linearly independent.

(c) If p=4 and some line ℓ does not meet any of the others, then $\mathcal{C}$ is radical if and only if we have exactly two skew lines in $\mathbb{P}^3$.

(d) If p=4 and some line ℓ meets the other lines in exactly one point, then $\mathcal{C}$ is radical if and only if the remaining lines are coplanar with no three through a point.

Proof In (a), d=1 for all lines. Thus C is radical if and only if for each line there is a hyperplane containing all the others, but not it. This is exactly what linearly independent means. Part (b) follows from (a) since the conductor being radical is a local property, and (c), (d) are easy.

One interesting set of lines in $\mathbb{P}^3_k$ is given by the rulings on the non-singular quadric XY-ZW (where X,Y,Z,W are homogeneous co-ordinates on $\mathbb{P}^3_k$ - where $\mathbb{P}^3_k$ = Proj k[X,Y,Z,W]). This quadric contains two families of rulings $\{\ell\}$ and $\{m\}$, each family parametrized by $\mathbb{P}^1_k$. Distinct ℓ's are skew, and each ℓ meets each m in one point. We wish to determine when the set of planes through the origin in $\mathbb{A}^4_k$ given by some set of rulings on our quadric surface is **seminormal**.

We first explore some consequences of the conductor being radical in B . Motivated by the rulings example we divide our planes into two subsets $\Pi_i = V(\tilde{I}_i)$ $(1 \leq i \leq n)$ and $\Gamma_j = V(\tilde{J}_j)$ $(1 \leq j \leq m)$ m + n = N. (Except for this partition we keep the notation introduced after the proof of the CRT.) Note that our ideals $\tilde{I}_i$, $\tilde{J}_i$ are homogeneous, so A and B are graded rings, and I_i, J_i, C, C_i are all homogeneous ideals. For a homogeneous ideal J in any graded k-algebra let $F_d(J)$ be the dth graded piece of J . Let $M = (X_1,\ldots X_p)R$ and d_i be the number of distinct ideals unequal to M of the form $\tilde{I}_i + \tilde{I}_j$ $(i \neq j)$ or $\tilde{I}_i + \tilde{J}_j$ (geometrically this is the number of distinct lines in Π_i of the form $\Pi_i \cap \Pi_j$ $(i \neq j)$ or $\Pi_i \cap \Gamma_j$) .

Let $\sqrt{\ } \subseteq (\cap \tilde{I}_i) \cap (\cap \tilde{J}_i)$ (ie. $(\cup T_i) \cup (\cup \Pi_i) \subseteq V(\sqrt{\ })$).

<u>Lemma 4</u> If the conductor C of A in B is radical in B then

$$n + \Sigma_{i=1}^{n} \{r-d_i\} \leq \dim_k F_r(\cap_{j=1}^{m} \tilde{J}_j) - \dim F_r(\sqrt{\ })$$

where $\{a\}=a$ if $a \geq 0$ and $\{a\} = -1$ if $a < 0$ (r any integer).

<u>Proof</u> If C is radical, then by condition (5) of the CRT, $C_i \subset A/I_i \cong k[t,u]$ is generated by a polynomial of degree d_i. Then $F_r(C_i) \cong F_{(r-d_i)}(R/\tilde{I}_i)$ and the latter has dimension $r - d_i + 1$ if $r \geq d_i$ and 0 if $r < d_i$. If $d_i = 0$ then $C_i = (t,u)$ (still under our assumption that C is radical) so the formula holds here also. The conductor C is generated by those elements of B which lie in A and have all but one coordinate 0. Thus there is a vector space $V_i \subset F_r(R)$ such that V_i maps isomorphically onto $F_r(C_i)$, and projects to 0 in $R/\tilde{I}_j$ $(j \neq i)$ and $R/(\tilde{J}_i)$ $(1 \leq i \leq m)$.

Let $V = \oplus_{i=1}^{n} V_i$. Then V maps isomorphically onto $\Pi_{i=1}^{n} F_r(C_i)$ and projects to zero in $\Pi_{i=1}^{m} R/(\tilde{J}_i)$. Thus $V \cap \ker(R \to B) = 0$ and $V \subseteq F_r(\cap \tilde{J}_i)$. As $\sqrt{\ } \subseteq \ker(R \to B)$, $V \cap F_r(\sqrt{\ }) = 0$. Now $F_r(\sqrt{\ }) \subseteq F_r(\cap \tilde{J}_i)$ so $V + F_r(\sqrt{\ }) \subseteq F_r(\cap \tilde{J}_i)$ and $\dim(V + F_r(\sqrt{\ })) = \dim V + \dim F_r(\sqrt{\ }) \leq \dim F_r(\cap \tilde{J}_i)$. Lemma 4 now follows. Done.

Note that in lemma 4 we do allow more then two planes to pass through a line. The corresponding linesin $\mathbb{P}^{p-1}$ pass through one point. We have seen that the conductor cannot be radial unless these directions are independent.

In order to use lemma 4 we have to know $\dim_k F_r(\cap \tilde{J}_i)$. We calculate this in the next lemma.

Lemma 5 If the planes Γ_j, Γ_k $(j \neq k)$ meet only in the origin then

$$\dim F_d(\cap_{j=1}^{m} \tilde{J}_j) = \binom{d+p-1}{p-1} - (d+1)m \quad d \geq m-1$$

$$\dim F_d(\cap_{j=1}^{m} \tilde{J}_j) \leq \binom{d+p-1}{p-1} - (d+1)^2 \quad d \leq m-1$$

Proof Let $B' = \Pi_{j=1}^{m} R/\tilde{J}_j$. We have an exact sequence

$$0 \to F_d(\cap \tilde{J}_j) \to F_d(R) \to F_d(B').$$

We claim that if $d \geq m-1$ then $F_d(R) \to F_d(B')$ is onto. If $j \neq \ell$, $F_1(\tilde{J}_j) \to F_1(R/\tilde{J}_\ell)$ is onto because $\Gamma_j \cap \Gamma_\ell$ is the origin. For each ℓ fix an isomorphism $R/\tilde{J}_\ell \cong k[t,u]$ and for each $j \neq \ell$ pick $\beta_{\ell j}, \gamma_{\ell j} \in \tilde{J}_j$ such that $\beta_{\ell j}$ has image t and $\gamma_{\ell j}$ has image u in $R/\tilde{J}_\ell$. Both $\beta_{\ell j}$ and $\gamma_{\ell j}$ vanish in $R/\tilde{J}_j$. If $d \geq m-1$ and f is any monomial in $F_d(R/\tilde{J}_\ell)$ then we can find a monomial in the $\beta_{\ell j}, \gamma_{\ell j}$ which maps to f in $R/\tilde{J}_\ell$ and 0 in all other co-ordinates. Thus $F_d(R) \to F_d(B')$ is onto if $d \geq m-1$. Now $\dim F_d(R) = \binom{d+p-1}{p-1}$ and $\dim F_d(B') = (d+1)m$, so the first formula follows. To prove the second formula note that $F_d(\cap_{j=1}^{m} \tilde{J}_j) \subseteq F_d(\cap_{j=1}^{d+1} \tilde{J}_j)$ $(d \leq m-1)$. By the first formula $\dim F_d(\cap_{j=1}^{d+1} \tilde{J}_j) = \binom{d+p-1}{p-1} - (d+1)^2$. The second now follows.

Remark 6 If $m = 0$ then lemma 4 holds with $\cap_{j=1}^{m} \tilde{J}_j = R$. In particular this says that if the number of intersection lines on a plane is bounded and the conductor is radical in B then the number of planes is bounded.

Example 7 Let Π_i $(1 \leq i \leq n)$ be planes in $\mathbb{A}_k^4$ arising from lines in one ruling system on the quadric, XY-ZW and Γ_j $(1 \leq j \leq m)$ arise from lines in the other ruling system. We claim that the following are equivalent:

(1) A is seminormal.

(2) The conductor of A in B is radical in B.

(3) $|n-m| \leq 1$.

Proof (1)$\Rightarrow$(2) is the implication (1)$\Rightarrow$(7) of theorem 2. Suppose $m \leq n$. Then each $d_i = m$. Let $r = m$. Lemma 4 says that $n \leq \dim F_m(\cap_{j=1}^{m} \tilde{J}_j) - \dim F_r(Q)$ where $(Q) = (XY-ZW)R$. By lemma 5 $\dim F_m(\cap_{j=1}^{m} \tilde{J}_j) = \binom{m+3}{3} - (m+1)m$. Also $\dim F_m(Q) = \binom{m+1}{3}$ (so long as we interpret $\binom{m+1}{3}$ as being 0 if $m = 0$ or 1). This yields $n \leq \frac{(m+3)(m+2)(m+1)}{6} - (m+1)m - \frac{(m+1)(m)(m-1)}{6} = m+1$. If $n \leq m$ we similarly get $m \leq n + 1$ which proves (3). We prove that (3)$\Rightarrow$(2) by exhibiting explicit elements in the conductor. It suffices to find a non-zero element of degree m in C_i (= component of the conductor in $R/\tilde{I}_i$). Renumbering if necessary we can assume

$i = n$. Let F_{ij} be the hyperplane in $\mathbb{A}_k^4$ which contains Π_i and Γ_j . Then $F_{1,1}\ F_{2,2}\ \ldots.\ F_{m,m}$ is the desired element of C_n if $n = m + 1$. If $n = m$ or $m - 1$ we can take $F_{1,1}\ F_{2,2}\ \ldots.\ F_{m-1,m-1}\ F_{m-1,m}$ or $F_{11}\ F_{22} \ldots.\ F_{m-2,m-2}\ F_{m-2,m-1}\ F_{m-2,m}$ respectively. (1) now follows from the last sentence of Theorem 2. We build up A one plane at a time (starting from m=1, n=1) in any way so long as the difference $|m-n|$ never exceeds 1.

Remark 8 Example 7 shows that a non-seminormal union of planes in $\mathbb{A}^4$ can sometimes be made seminormal by adding additional planes. This is contrary to the experience with planes in 3 space or curves anywhere.

Remark 9 Here is an alternative argument to show that the difference n-m cannot exceed one if the conductor is radical in B . We illustrate it with m=2 n=4 .

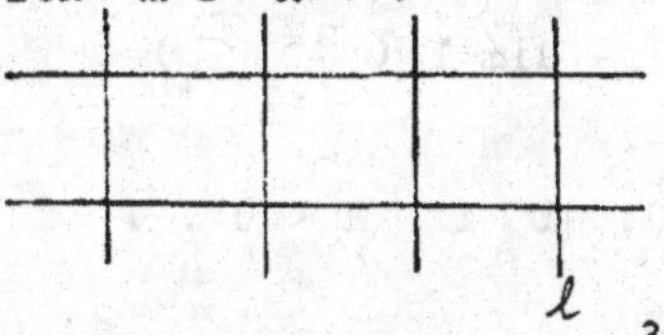

The diagram represents the lines in $\mathbb{P}_k^3$ and lines with no intersection are skew. Suppose there exists $f \in A$ of degree 2 that vanishes on all lines but ℓ . We can draw in a third line m of the horizontal ruling system, ie.

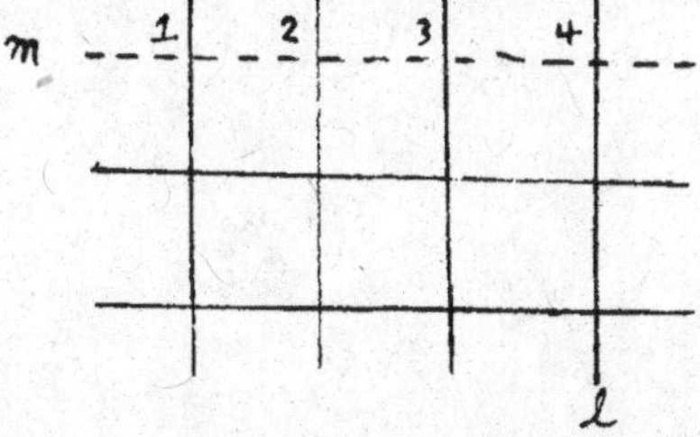

Then f vanishes at the points 1,2,3, of m. Since f is of degree 2, f vanishes on m. Hence f vanishes at 4. Since f is of degree 2 and vanishes at three points of ℓ, f must also vanish on ℓ. Thus the conductor is not radical.

One might try to build up our variety plane by plane. Let $\tilde{I}_1$ be the ideal of the new plane and $\{\tilde{J}_j\}$ the ideals of the planes already there. Set $\mathcal{J} = \tilde{I}_1 \cap (\cap_j \tilde{J}_j)$. Lemma 4 yields

$1 + \{r-d_1\} \leq \dim F_r(\cap_j \tilde{J}_j) - \dim F_r(\tilde{I}_1 \cap (\cap_j \tilde{J}_j))$. Iterating this process, we get

<u>Theorem 10</u> Suppose $V(\tilde{I}_\ell)$ meets $\cup_{i<\ell} V(\tilde{I}_i)$ in d'_ℓ distinct lines. If $\cup_{i=1}^{\ell} V(\tilde{I}_i)$ has radical conductor for each $\ell = 1,\ldots,n$ then for all integers r

$$(n-1) + \Sigma_{\ell=2}^{n} \{r-d'_\ell\} \leq \dim F_r(\tilde{I}_1) - \dim F_r(\cap_{i=1}^{n} \tilde{I}_i) .$$

(As in lemma 4, $\{a\} = a$ if $a \geq 0$, -1 if $a < 0$.)

<u>Example 11</u> Let $p = 4$, ie. $R = k[X,Y,Z,W]$. If we start with one plane, and add others keeping the conductor radical, so that each new plane has 2 (or fewer) lines of intersection with the previous planes, then one can have at most 6 planes. The union is K_0-regular.

The first assertion follows from theorem 10. First of all, dim $F_2(\tilde{I}_1) = 7$. If $d_2' = 0$ we see already that n can be at most 6. If $d_2' = 1$, and $d_3' = 2$ then the ^{7} ℓ_i in $\mathbb{P}^3$ corresponding to Π_1, Π_2, Π_3 must lie in a $d/d\alpha)\Psi_L$ so $d_4' = 1$ (if ℓ_4 meets two of ℓ_1, ℓ_2, ℓ_3, it must lie in this plane, hence meets all of ℓ_1, ℓ_2, ℓ_3). In all cases we get $(2-d_2') + (2-d_3') + (2-d_4') \geq 2$ so n can be at most 6. The assertion about K_0 follows by induction using the Mayer-Vietoris sequences of the cartesian squares

$$\begin{array}{ccc}
R/\cap_{i=1}^{\ell} (\tilde{I}_i) & \longrightarrow & R/\tilde{I}_\ell \\
\downarrow & & \downarrow \\
R \cap_{i<\ell} (\tilde{I}_i) & \longrightarrow & R/(\tilde{I}_\ell + \cap_{i<\ell} \tilde{I}_i)
\end{array}$$

since on assumptions imply that $R/(\tilde{I}_\ell + \cap_{i<\ell} \tilde{I}_i) \cong k, k[t]$, or $k[t,u]/(tu)$.

One allowable configuration of 6 lines in $\mathbb{P}^3_k$ is the edges of a tetrahedron. (Thm. 4.8 of DW2) Another possibility is the configuration

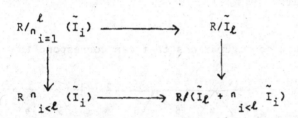

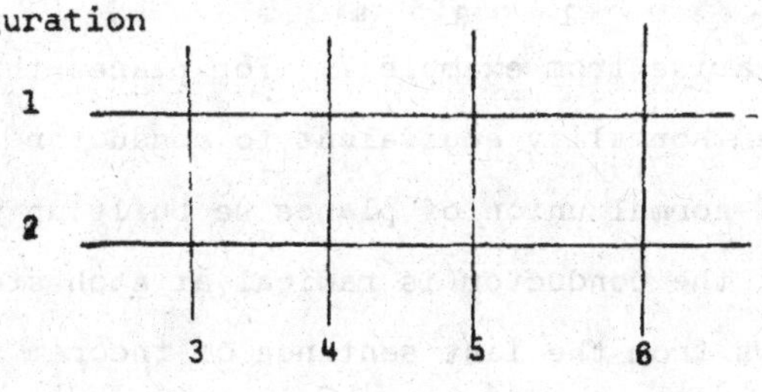

There is a unique quadric containing the skew lines 3,4,5. This quadric is ruled non-singular, and 3,4,5 belong to one of the systems of rulings. Lines 1 and 2 must lie in the quadric since they meet it in three points. Lines can be added in the order 1,2,3,4,5 always keeping the conductor radical. If 6 lies in the quadric we have already seen that the conductor is not radical. If 6 does not lie in the quadric the conductor is radical. For example the quadric through 3,4,5 is the desired element of degree 2 that vanishes on all lines but 6. That C_1 and C_2 contain an element of degree 4 is trivial.

Other such configurations that can correspond to seminormal A are

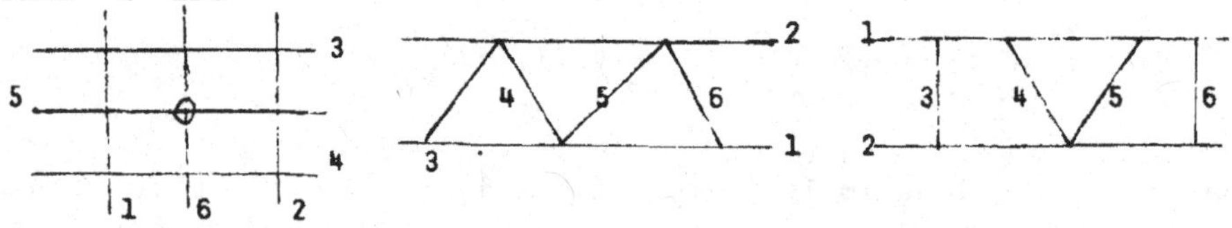

Some configurations of 6 lines for which the A cannot be semi-normal are

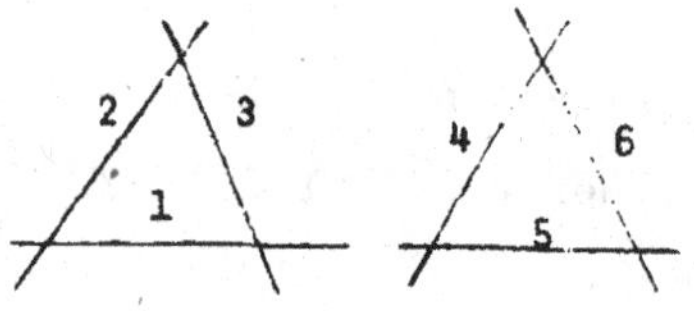

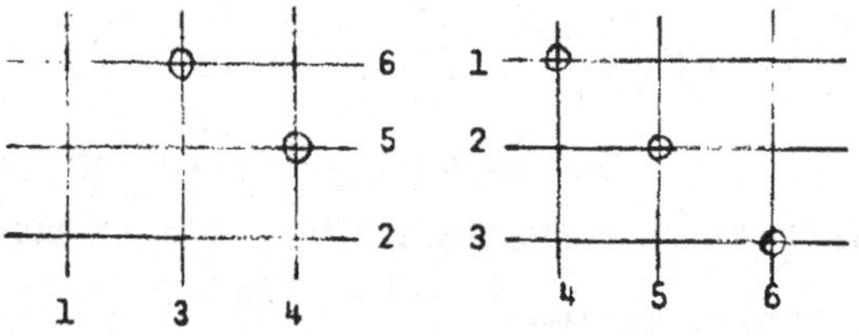

Two questions arise from example 7. For planes through the origin in p-space is seminormality equivalent to conductor radical, and can every semi-normal union of planes be built up plane by plane in such a way that the conductor is radical at each step? (Then semi-normality follows from the last sentence of theorem 2).

The answer to both these questions is no, as we will show by considering configurations related to the double six.

Any non-singular cubic F in $\mathbb{P}^3$ (k algebraically closed) contains 27 lines. ([M] 8.20, [H]) A subconfiguration is the double 6

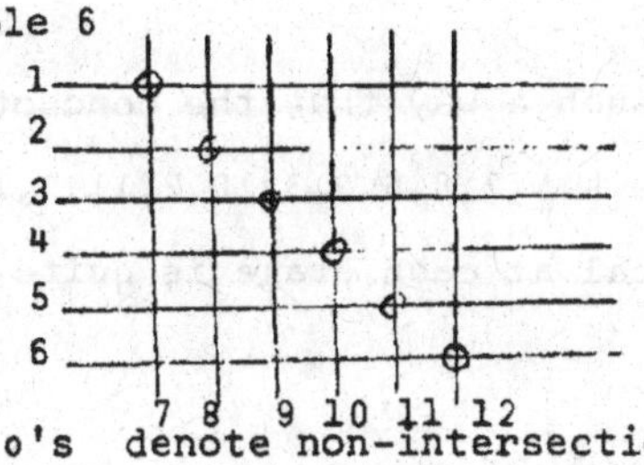

where the o's denote non-intersection. We first note that F is the unique cubic (up to a constant) that vanishes on all 12 lines. This is undoubtedly well known, but we could not find a clear reference, hence we give a proof. By lemma 5 the vector space V of cubics vanishing on the 4 skew lines 1,2,3,4 has dimension $\binom{3+3}{3} - 4.4 = 20 - 16 = 4$. Consider the homomorphism $g: V \longrightarrow k^3$ obtained by evaluation at the points $5\cap 7$, $5\cap 8$, $5\cap 9$. (Here, and later, whenever we evaluate a form at a point P of $\mathbb{P}^{p-1}$ we first choose a set of homogeneous co-ordinates for P.) Let Q_{234} be the quadric vanishing on lines 2,3,4. This quadric does not vanish on line 5, however it does equal zero on lines 7 and 12. It therefore vanishes at the points $5\cap 7$ and $5\cap 12$. It can equal zero at no other points on line 5. Let H_{68} be the hyperplane containing lines 6 and 8. Then H_{68} equals zero at $5\cap 8$ but no other point on line 5, so $Q_{234}\ H_{68}$ has image $(0,0,\lambda)$ $\lambda \neq 0$ under the homomorphism $V \rightarrow k^3$. Similarly for the other co-ordinates.

Thus g is onto, and the vector space of cubics vanishing on the entire double 6 must have dimension 1 (spanned by F).

Example 12 The co-ordinate ring of planes through the origin in $\mathbb{A}^4_k$ corresponding to the lines of the double 6 is semi-normal.

Proof The union can be built up in such a way that the conductor is radical at each stage, for example 4,5,7,8,6,9,3,10,2,11,1,12. The proof that the conductor is radical at each stage is quite easy. For example

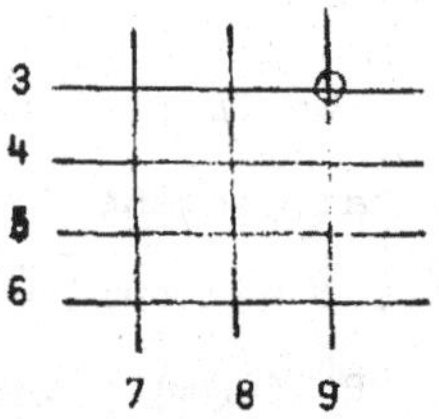

A quadric vanishing on all lines except 3 is Q_{456}, a cubic vanishing on all but 9 is $Q_{345}\ H_{67}$, a quartic vanishing on all but 7 is $H_{38}\ H_{49}\ H_{58}\ H_{68}$, a cubic vanishing on all but 6 is $H_{57}\ H_{38}\ H_{49}$. Or for the entire double 6, $Q_{789}\ H_{3,10}\ H_{2,11}\ H_{1,11}$ is of degree 5 vanishing on all lines but 12. Done.

Now consider the double 5

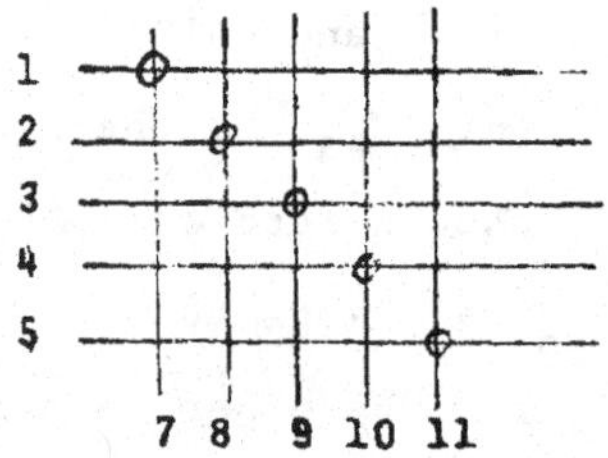

The conductor is radical. For example $Q_{789}\, H_{2,10}\, H_{1,10}$ is of degree 4 and equals 0 on all lines except 11. However, we claim that the double 5 is not seminormal. We will show this by developing an additional theoretical tool.

First recall Swan's definition of relative seminormality. Given an inclusion of commutative rings $A \subset B$, we say that A is seminormal in B if whenever, $b \in B$, $b^2, b^3 \in A$ we have $b \in A$ ([S] theorem 2.5).

<u>Theorem 13</u> Given a cartesian square

$$\begin{array}{ccc} A & \longrightarrow & B \\ \downarrow & & \downarrow \\ A/I & \longrightarrow & B/I \end{array}$$

Then A is seminormal in B if and only if A/I is seminormal in B/I.

<u>Proof</u> Let $b \in B$, with $b^2, b^3 \in A$. Let $\bar{b}$ be the image of b in B/I. Then $\bar{b}^2, \bar{b}^3 \in A/I$ so $\bar{b} \in A/I$ if we assume A/I is seminormal in B/I. The fact that the square is cartesian then implies $b \in A$, so A is seminormal in B. Conversely, suppose A is seminormal in B and suppose $\bar{b} \in B/I$, with $\bar{b}^2, \bar{b}^3 \in A/I$. Then lift $\bar{b}$ to $b \in B$. Since the square is cartesian we must have

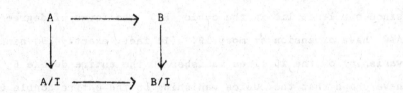

b^2, $b^3 \in A$, so $b \in A$, and $\bar{b} \in A/I$. Done.

We now apply theorem 12 with $A =$ co-ordinate ring of planes in $\mathbb{A}^4$ corresponding to the lines of the double 5, $B =$ normalization of $A = \Pi_{i=1}^{10} k[t,u]$, and C the conductor of A in B. Since I is radical, A/C is the co-ordinate ring of 20 lines through the origin in $\mathbb{A}^4$, and B/C is the product of 10 plane curves each consisting of 4 lines through the origin. The conductor of B/C in $\overline{B/C}$ is $\Pi_{i=1}^{10} (t,u)^3 = \Pi_{i=1}^{40} (t^3)$ by [01] . (The $^-$ denotes normalization. $\overline{B/C}$ consists of 40 disjoint lines). Now consider the conductor of A/C in $\overline{A/C}$ (= disjoint union of 20 lines). The space of forms of degree 3 in 4 variables has dimension 20. But since our lines lie in the cubic $F=0$ the forms of degree 3 in A/C have dimension at most 19. (In fact, exactly 19, since a cubic vanishing on the 20 lines vanishes on the entire double 6, and we have shown that the cubics vanishing on the entire double 6 have dimension 1.) The cubic forms cannot span the 20 dimensional vector space $F_3(\Pi_{i=1}^{20} k[t])$ so some component of the conductor of A/C in $\overline{A/C}$ (say $1 \cap 11$) must have exponent greater than 3. This exponent must be 4 since there exists a quartic vanishing on all lines but one. Hence let $\bar{b} \in \overline{B/C}$ have component t^3 in the two lines corresponding to the line $1 \cap 11$. Then $\bar{b} \in B/C$, $\bar{b}^{-2}$, $\bar{b}^{-3} \in A/C$, but $\bar{b} \notin A/C$. Thus A/C is not semi-normal in B/C, so A is not semi-normal in B by theorem 12. Thus A is not semi-normal. This proves

Example 14 Let A be the coordinate ring of planes in $\mathbb{A}_k^4$ corresponding to the lines of a double 5 that arises as above from a non-singular cubic. Then A has radical conductor, but is not seminormal.

If any line is removed from the double 5, the conductor is not radical (by the method of remark 9).

Example 15 Let A be the co-ordinate ring of planes through the origin in $\mathbb{A}_k^4$ corresponding to the lines of a double 4. If the double 4 is contained in a cubic the conductor is not radical. If the double 4 is not contained in a cubic the conductor is radical but A is not seminormal.

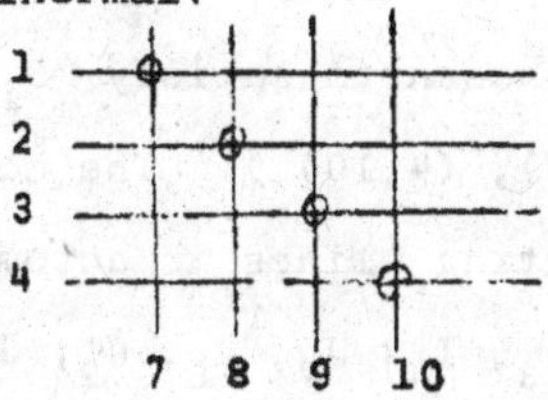

Proof Suppose the double 4 is contained in a cubic $G = 0$. We apply lemma 4 with the $\tilde{I}_i$ corresponding to 1,2,3,4, the $\tilde{J}_j$ corresponding to 7,8,9,10, $r = 3$, and $\mathfrak{I} = (G)$. The inequality yields $4 \leq 4 - 1$ which is not true, so the conductor is not radical. If the double 4 is not contained in a cubic consider the 19 points consisting of the 12 intersection points, one additional point on each of lines 1,2,3,4, and one additional point on each of lines 7,8,10. One can find a non-zero cubic which equals 0 at these 19 points. This cubic vanishes on all lines but 9, but does not vanish on 9 since the double 4 is assumed to not lie in a cubic. Similarly for the other lines. Thus the conductor is radical. We can show that such a double 4 is not seminormal by appealing to the implication (1) ⇔ (2) of Theorem 2. Let f_i be a homogeneous degree 2 element in $R/\tilde{I}_i$ $(1 \leq i \leq 4)$ or $R/\tilde{J}_j$ $(7 \leq j \leq 10)$. Set $f_3 = f_4 = f_7 = f_8 = f_9 = 0$, f_{10} any quadric in $R/\tilde{J}_{10}$ that equals 0 at $3 \cap 10$. Then f_1 and f_2 can be chosen to equal 0 at $1 \cap 8$, $1 \cap 9$, $2 \cap 7$, $2 \cap 9$, and to agree with f_{10} on the lines in $\mathbb{A}^4$ corresponding to $1 \cap 10$, $2 \cap 10$. Suppose there is an element of f of A such that $f \equiv f_i \mod \tilde{I}_i(\tilde{J}_j)$. Since $f_7 = f_8 = f_9 = 0$ f must be a multiple of Q_{789}. Since this quadric vanishes also at $3 \cap 10$ this multiple must be 0 (otherwise line 3 would be contained in the quadric, which is impossible since $3 \cap 9 = \emptyset$). The (f_i) are not all 0, and have been chosen to satisfy $f_i \equiv f_j \mod \tilde{I}_i + \tilde{J}_j$ $((i,j) \neq (1,7), (2,8), (3,9), (4,10))$. The congruence condition at a pair of non-intersecting lines is automatic. Thus CRT fails for the ideals $I_1, I_2, I_3, I_4, J_7, J_8, J_9, J_{10}$ and A is not seminormal. (We remark that double 4's not contained in a cubic do exist, an explicit one is given in Example 16.)

Suppose we try to construct the same type of counter-example to CRT on a double 5 that is not contained in a cubic.

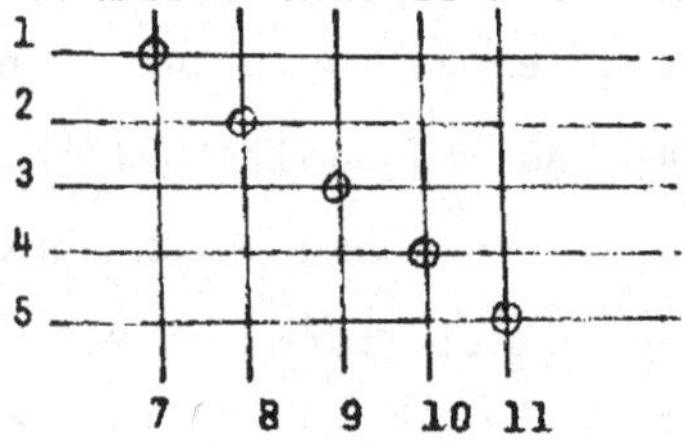

Again let f_i be homogeneous of degree 2 in $R/\tilde{I}_i$ ($1 \le i \le 11$, $i \ne 6$), and suppose $(f_1, f_2, f_3, f_4, f_5, f_7, f_8, f_9, f_{10}, f_{11})$ is compatible along the 20 intersection lines. Choose a set of homogeneous co-ordinates for each of the 20 intersection points, so that we can speak of the value of a homogenous quadric at each of these points in $\mathbb{P}^3_k$. There exists a quadric which agrees with the f_i at the following 9 points:

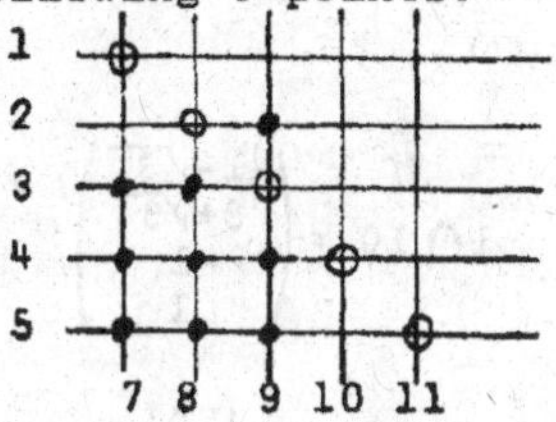

By subtracting this quadric from each co-ordinate we can assume $f_7 = f_8 = f_9 = f_5 = f_4 = 0$. The lifting to A, if it exists, must be a multiple of $Q_{7,8,9}$. As in example 15 $Q_{7,8,9}$ does not vanish at $3 \cap 10$. By subtracting a multiple of $Q_{7,8,9}$ we can assume our compatible quadrics vanish at $3 \cap 10$ also, ie. $f_3 = 0$. This leaves only f_1, f_2, f_{10}, f_{11}, which might be non-zero. Since $f_1(1 \cap 8) = f_1(1 \cap 9) = 0$ we must have $f_1 = c_1 Q_{7,8,9}$. (More precisely f_1 is the image of $c_1 Q_{7,8,9}$ in R/I_1). Similarly $f_2 = c_2 Q_{7,8,9}$, $f_{10} = c_3 Q_{3,4,5}$, and $f_{11} = c_4 Q_{3,4,5}$. Compatibility at $1 \cap 10$, $1 \cap 11$, $2 \cap 10$, $2 \cap 11$, yields the system of equations

$$c_1 \; Q_{7,8,9}(1 \cap 10) = c_3 \; Q_{3,4,5}(1 \cap 10)$$
$$c_1 \; Q_{7,8,9}(1 \cap 11) = c_4 \; Q_{3,4,5}(1 \cap 11)$$
$$c_2 \; Q_{7,8,9}(2 \cap 10) = c_3 \; Q_{3,4,5}(2 \cap 10)$$
$$c_2 \; Q_{7,8,9}(2 \cap 11) = c_4 \; Q_{3,4,5}(2 \cap 11)$$

This system has a nontrivial solution for c_1, c_2, c_3, c_4 if and only if

$$(1)\quad Q_{7,8,9}(1\cap 10)\,Q_{3,4,5}(1\cap 11)\,Q_{3,4,5}(2\cap 10)\,Q_{7,8,9}(2\cap 11)$$
$$= Q_{3,4,5}(1\cap 10)\,Q_{7,8,9}(1\cap 11)\,Q_{7,8,9}(2\cap 10)\,Q_{3,4,5}(2\cap 11).$$

We expect that this will not be satisfied "generically". In any case here is an explicit example of a double 5 not contained in a cubic, where the above equality fails to hold. We will describe it by giving the intersection points ($k=\mathbb{R}$) .

$$1\cap 8 = \begin{pmatrix} 0 \\ 4+2\sqrt{3} \\ 0 \\ 2 \end{pmatrix} \quad 1\cap 9 = \begin{pmatrix} 1+\sqrt{3} \\ 1+\sqrt{3} \\ 1 \\ 1 \end{pmatrix} \quad 1\cap 10 = \begin{pmatrix} -1-\sqrt{3} \\ 3+\sqrt{3} \\ -1 \\ 1 \end{pmatrix} \quad 1\cap 11 = \begin{pmatrix} 2 \\ 4 \\ -1+\sqrt{3} \\ 3-\sqrt{3} \end{pmatrix}$$

$$2\cap 7 = \begin{pmatrix} 1 \\ 0 \\ 1 \\ 0 \end{pmatrix} \quad 2\cap 9 = \begin{pmatrix} 2 \\ 2 \\ 1 \\ 1 \end{pmatrix} \quad 2\cap 10 = \begin{pmatrix} 0 \\ -2 \\ 1 \\ -1 \end{pmatrix} \quad 2\cap 11 = \begin{pmatrix} 1 \\ 2 \\ 0 \\ 1 \end{pmatrix}$$

$$3\cap 7 = \begin{pmatrix} 1 \\ 0 \\ -1 \\ 0 \end{pmatrix} \quad 3\cap 8 = \begin{pmatrix} 0 \\ 1 \\ 0 \\ 1 \end{pmatrix} \quad 3\cap 10 = \begin{pmatrix} 1 \\ 1 \\ -1 \\ 1 \end{pmatrix} \quad 3\cap 11 = \begin{pmatrix} 1 \\ 2 \\ -1 \\ 2 \end{pmatrix}$$

$$4\cap 7 = \begin{pmatrix} 0 \\ 0 \\ 1 \\ 0 \end{pmatrix} \quad 4\cap 8 = \begin{pmatrix} 0 \\ 0 \\ 0 \\ 1 \end{pmatrix} \quad 4\cap 9 = \begin{pmatrix} 0 \\ 0 \\ 1 \\ 1 \end{pmatrix} \quad 4\cap 11 = \begin{pmatrix} 0 \\ 0 \\ 1 \\ -1 \end{pmatrix}$$

$$5\cap 7 = \begin{pmatrix} 1 \\ 0 \\ 0 \\ 0 \end{pmatrix} \quad 5\cap 8 = \begin{pmatrix} 0 \\ 1 \\ 0 \\ 0 \end{pmatrix} \quad 5\cap 9 = \begin{pmatrix} 1 \\ 1 \\ 0 \\ 0 \end{pmatrix} \quad 5\cap 10 = \begin{pmatrix} 1 \\ -1 \\ 0 \\ 0 \end{pmatrix}$$

This was constructed by adding lines in the order 5,7,8,4,9,3,10,2,11,1. The only delicate computation is finding line 1. This was done in the manner described on page 164 of [HC]. Then $Q_{789} = XW - YZ$ and, $Q_{345} = XW + YZ$. The equation (1) becomes $-8 -8\sqrt{3} = -8 + 8\sqrt{3}$

which is false, so non-trivial c_1, c_2, c_3, c_4 do not exist. Now we claim that this double 5 is not contained in any cubic. First of all

$$Q_{8,9,10} = Z^2 + YZ - ZW - XW$$
$$Q_{8,9,11} = X^2 - XY - 3XZ + XW + 2ZY$$
$$Q_{8,10,11} = -X^2 + XZ + 3XW + 2Z^2 + 2ZW - XY$$
$$Q_{9,10,11} = -Z^2 + W^2 - X^2 + Y^2 + 2XZ + 4XW - 2YZ - 4YW.$$

Four cubics vanishing on 8,9,10,11 are $f_1 = Q_{8,9,10}(2X-Y)$, $f_2 = Q_{8,9,11}(Z+W)$, $f_3 = Q_{8,10,11}(X-Y)$, $f_4 = Q_{9,10,11}(Z)$. If we evaluate these cubics at the four intersection points on line 7 we get a matrix of rank 4. Thus the $\{f_i\}$ are linearly independent, and so form a basis of the four dimensional space V of cubics vanishing on 8,9,10,11. No non-trivial element of V vanishes on 7 so no non-zero cubic vanishes on the entire double 5.

<u>Example 16</u> Let A be the co-ordinate ring of planes through the origin in $\mathbb{A}_k^4$ corresponding to the above double 5. (Or more generally, any double 5 for which equality (1) is false, and which is not contained in a cubic.) Then A is seminormal. If any plane is removed the union is no longer seminormal, so A cannot be built up one plane at a time in such a way that the conductor is radical at each stage.

We will prove this by first giving a brief theoretical digression. Let A be the co-ordinate ring of a union of planes Π_i through the origin in p-space. As before $\tilde{I}_i$ = ideal of Π_i in $R = k[X_1 \ldots X_p]$ and I_i = image of $\tilde{I}_i$ in A. Let $B = \Pi A/I_i$

be the normalization of A , and assume that the conductor C of A in B is radical. As before let $\overline{\quad}$ denote normalization. Then we have a square

$$(2) \qquad \begin{array}{ccc} A & \longrightarrow & B \\ \downarrow & & \downarrow \\ \overline{A/C} & \xrightarrow{\text{diagonal}} & \overline{B/C} \end{array}$$

By definition CRT holds for $\{I_i\}$ in A if and only if this square is cartesian. We claim

<u>Lemma 17</u> The square

$$(3) \qquad \begin{array}{ccc} A/C & \longrightarrow & B/C \\ \downarrow & & \downarrow \\ \overline{A/C} & \xrightarrow{\text{diagonal}} & \overline{B/C} \end{array}$$

is cartesian (ie. $A/C = \overline{A/C} \cap B/C$) if and only if CRT holds for $\{I_i\}$.

<u>Proof of Lemma 17</u> Suppose (3) is cartesian. The square

$$(4) \qquad \begin{array}{ccc} A & \longrightarrow & B \\ \downarrow & & \downarrow \\ A/C & \longrightarrow & B/C \end{array}$$

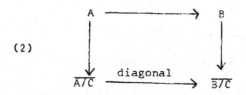

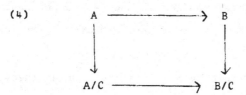

is always cartesian, and putting (3) and (4) together we conclude that (2) is cartesian. Conversely, suppose (2) is cartesian. It suffices to show that $(B/C) \cap \overline{A/C} \subset A/C$. Let $\bar{b} \in (B/C) \cap \overline{A/C}$. Lift $\bar{b}$ to $b \in B$. Since (2) is cartesian $b \in A$ so $\bar{b} \in A/C$ as required. Done.

To prove (3) cartesian we must show that $A/C = (B/C) \cap \overline{A/C}$. Since all rings are graded it suffices to show that $F_r(A/C) = F_r(B/C) \cap F_r(\overline{A/C})$ for all $r \geq 0$. Since A/C is a curve we have $F_r(A/C) = F_r(\overline{A/C})$ for r sufficiently large. (Determining whether A is seminormal is now a problem in finite dimensional linear algebra.) Lower values of r can often be handled conveniently by the following lemma:

<u>Lemma 18</u> Let r be fixed. Let W_r be a subset of the intersection lines $\Pi_i \cap \Pi_j$ with the following property: if $f = (f_i) \in F_r(B)$ $(f_i \in A/I_i)$ satisfies the hypotheses of the CRT and f vanishes on all the lines of W_r, then $f = 0$. Then

(a) $\dim F_r(A) = \dim F_r(A/C)$

(b) $\dim (F_r(\overline{A/C}) \cap F_r(B/C)) \leq \# W_r$.

<u>Proof of Lemma 18</u> Let L_{ij} be the ideal in R of $\Pi_i \cap \Pi_j$ (whenever the latter intersection is a line). Consider the diagram

$$\begin{array}{ccccccccc} 0 & \longrightarrow & F_r(\cap I_i) & \longrightarrow & F_r(R) & \longrightarrow & Fr(A) & \longrightarrow & 0 \\ & & \downarrow & & \| & & \downarrow & & \\ 0 & \longrightarrow & F_r(\cap L_{ij}) & \longrightarrow & F_r(R) & \longrightarrow & F_r(A/C) & \longrightarrow & 0 \\ & & \downarrow & & & & & & \\ & & F_r(\underset{W_r}{\cap} L_{ij}) & & & & & & \end{array}$$

The assumptions say that the inclusion $F_r(\cap I_i) \longrightarrow F_r(\bigcap_{W_r} L_{ij})$ is an isomorphism, hence so also $F_r(\cap I_i) = F_r(\cap L_{ij})$. Thus $F_r(A) \cong F_r(A/C)$. The projection $\overline{A/C} = \Pi(R/L_{ij}) \longrightarrow \prod_{W_r}(R/L_{ij})$ induces a homomorphism

$$\phi : F_r(\overline{A/C}) \longrightarrow \prod_{W_r} F_r(R/L_{ij}) \cong k^{\#W_r}$$

Our hypotheses imply that ϕ restricted to $F_r(\overline{A/C}) \cap F_r(B/C)$ is an inclusion. (b) follows. Done.

<u>Proof of example 16</u> Here $F_r(A/C) = F_r(\overline{A/C})$ for $r \geq 3$. (The case $r = 3$ is true because our double 5 is not contained in a cubic). By lemma 18 (a) dim $F_r(A) = \dim F_r(A/C)$ for $r \leq 2$ (take W_r to be all 20 intersection lines). Dim $F_0(A) = 1$, dim $F_1(A) = 4$, dim $F_2(A) = 10$ so we seek sets W_r of these cardinalities:

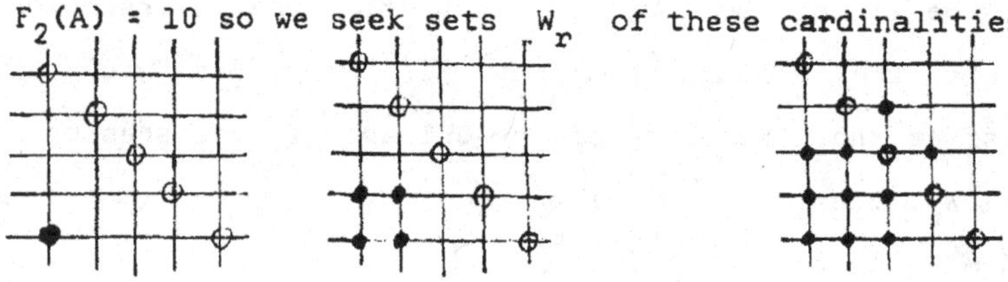

W_0 W_1 W_2

Clearly W_0 and W_1 are satisfactory. W_2 is satisfactory because equation (1) is not satisfied for our double 5.

To show that A (of example 16) cannot be build up plane by plane it suffices to show that if any plane is removed the result is not seminormal. The removal of any plane results in a configuration that looks like

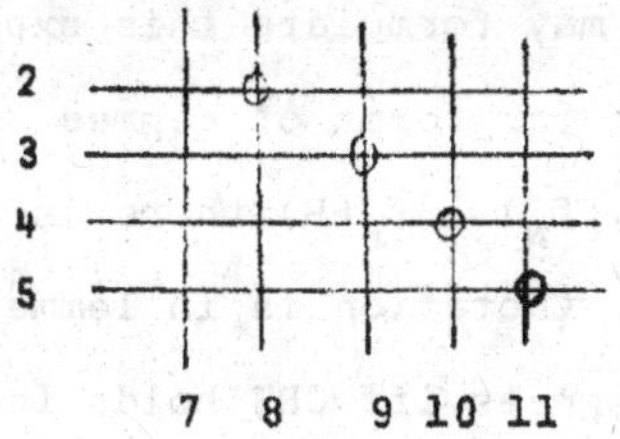

Let $r = 2$. Set $f_3 = f_4 = f_5 = f_7 = f_8 = f_9 = 0$. Let $f_2 = Q_{7,8,9}$ (restricted to Π_2), $f_{10} = \lambda Q_{3,4,5}$, $f_{11} = \mu Q_{3,4,5}$. Choose λ and μ so that $f_2(2\cap 10) = \lambda Q_{3,4,5}(2\cap 10)$ and $f_2(2\cap 11) = \mu Q_{3,4,5}(2\cap 11)$. Then the f_i satisfy the hypotheses of CRT but do not lift to R. (Since $f_7 = f_8 = f_9 = 0$ the lifting if it exists must be $\gamma Q_{7,8,9}$. But $Q_{7,8,9}(3\cap 10) \neq 0$ so γ must be 0.) Thus the new configuration does not have seminormal co-ordinate ring.

<u>Remark 19</u> The seminormality of the double 6 can be proved similarly using, W_o, W_1 as above and

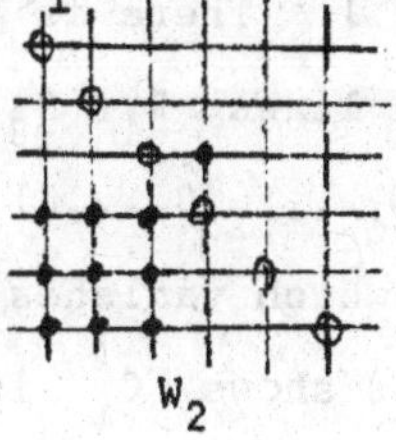

W_2

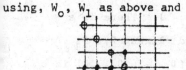

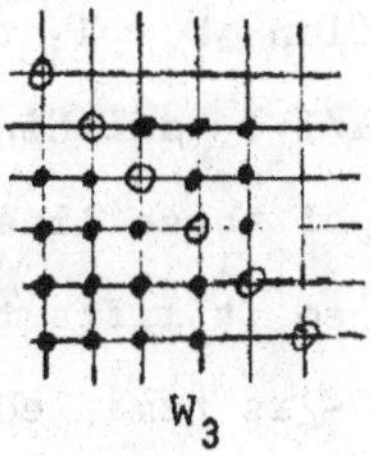

W_3

($\#W_3 = 19$).

We have not given a totally satisfactory necessary and sufficient criterion for a union of planes in $\mathbb{A}^p$ to be seminormal. However the methods we have used in our examples will, at least in principle, give a definite answer after a finite amount of finite dimensional linear algebra. One may formulate this explicitely as follows. We say the CRT holds for forms of degree r if for every compatible collection $(f_1,..f_N) \in F_r(B)$ there is a form f in $F_r(A)$ with $f \equiv f_i \mod I_i$. (Notation is in lemma 17). Clearly the CRT holds for forms of all degrees iff CRT holds for $\mathfrak{a} = \{I_1 \ldots I_N\}$. We actually have.

<u>Theorem 20</u>: With notation as above (N planes in $\mathbb{A}^p$) A is seminormal iff CRT holds for forms of degree $\leq N - 1$.

<u>Proof</u>: If A is seminormal CRT holds (by Theorem 2) and thus it holds for forms of degree $\leq N - 1$.

Suppose CRT holds for forms of degree $\leq N - 1$. We first note that this implies the conductor C is radical (or equivalently that each C_i is radical). For if Π_i contains d distinct lines of intersection $\Pi_i \cap \Pi_j$ then $d \leq N - 1$. There is a non-zero form $f_i \in F_d(A/I_i)$ vanishing on the d lines $\Pi_i \cap \Pi_j$. (i.e. the equation in Π_i of these lines). Then $(0,..,f_i,0...,0)$ is a compatible collection so it lifts to $f \in F_d(A)$ which vanishes on all planes Π_j but Π_i - as remarked earlier this shows C_i is radical. (a similar argument may be given when $d = 0$).

Next an argument similar to that of lemma 5 shows that $F_r(R) \longrightarrow F_r(\overline{A/C}) = F_r(\Pi_{i<j} R/(I_i+I_j))$ is onto if $r \geq N - 2$. The homomorphism $R \longrightarrow \overline{A/C}$ factors through A/C so $F_r(A/C) \longrightarrow F_r(\overline{A/C})$ is also onto. Then square (3) of lemma 17 is cartesian at the F_r

level, so (2) is cartesian at the F_r level by lemma 17. That is, CRT holds, and A is seminormal.

References

[B] H. Bass, Algebraic K-Theory, Benjamin, New York, 1968.

[D] B. Dayton, Seminormality implies the Chinese Remainder Theorem, these proceedings.

[DW1] B. Dayton and C. Weibel, K-Theory of Hyperplanes, Trans. A.M.S. 257 (1980), 119-141

[DW2] B. Dayton and C. Weibel, A spectral sequence for the K-theory of affine glued schemes, these proceedings.

[H] R. Hartshorne, Algebraic Geometry, Springer-Verlag, New York, 1977.

[HC] D. Hilbert and S. Cohn-Vossen, Geometry and the Imagination, translated by P. Nemenyi, Chelsea, New York 1952.

[M] D. Mumford, Algebraic Geometry I Complex Projective Varieties, Springer-Verlag, Berlin Heidelberg 1976.

[O1] F. Orecchia, Sul gruppo di Picard di certe algebre finite non integre. Ann. Univ. Ferrara Sez. VII, 21 (1975), 25-36.

[O2] F. Orecchia, Sulla seminormalita di certe varieta affini riducibili, Boll. Un. Math. Ital (2) B (1976), 588-600.

[S] R. G. Swan, On Seminormality, to appear.

SEMINORMALITY IMPLIES THE CHINESE REMAINDER THEOREM

Barry H. Dayton

The purpose of this note is to provide a simple, self-contained proof of the following theorem.

<u>Theorem A</u>: Let A be a commutative ring, $I_1,\ldots,I_n$ be ideals of A such that $I_i + I_j$ is radical for $i \neq j$. Let $B = \prod A/I_i$. Then the Chinese Remainder Theorem holds for $\{I_1,\ldots,I_n\}$ if and only if $A/(\cap I_i)$ is seminormal in B.

As in [S] R is seminormal in B if $R \subseteq B$ and whenever $b \in B$ satisfies $b^2, b^3 \in R$ then $b \in R$. As in [DR] we say the Chinese Remainder Theorem (CRT) holds for $\{I_1,\ldots,I_n\}$ if given $a_1,\ldots,a_n$ such that $a_i \equiv a_j \bmod (I_i+I_j)$, $i \neq j$, then there exists $x \in A$ with $x \equiv a_i \bmod I_i$ for all i.

Theorem A implies (1) $\Leftrightarrow$ (2) of Theorem 2 of [DR], for if, in addition the A/I_i are assumed to be seminormal, then B is seminormal, so $A/(\cap I_i)$ is seminormal in B if and only if $A/(\cap I_i)$ is seminormal.

This theorem is originally due to F. Orecchia ([O] Theorem 1.3) in a slightly different form. Another version of this theorem is presented as [DW2] Corollary 1.10 where we interpreted CRT as saying that the ring $A/(\cap I_i)$ is the limit of the inverse system consisting of the rings A/I_i, $A/(I_i+I_j)$ and the cannonical projections between them. In these two versions of Theorem A, Traverso's notion of seminormality was used and additional restrictions were needed on the I_i's, eg in [O] the I_i's were the minimal primes of A.

In [DR] the discussion of Theorem 2 is based on Swan's

notion of seminormality. Thinking of CRT as saying that $A/(\cap I_i)$ is a limit, the direction (2) $\Rightarrow$ (1) is easy, being motivated by [S] Corollary 3.3. However, the proof of the converse involved applying the non-trivial argument of Swan ([S] 2.5 and 3.4) to recover Traverso's definition of seminormality and then adapting the argument of [DW2]. I can now give a simple direct proof of this converse, in fact it is implied by the following theorem.

Theorem B: Let $I_1,\dots,I_n$ be ideals of A, $B = \prod A/I_i$. If $A/(\cap I_i)$ is seminormal in B then CRT holds.

For the proof of Theorems A and B we note, passing from A to $A/(\cap I_i)$, that there is no loss of generality in assuming $\cap I_i = 0$. Then we may identify A as a subring of B by $a \leftrightarrow (\bar{a},\dots,\bar{a})$. In B we will suppress the bars over the a_i's as no confusion will arise.

We first prove that CRT implies that A is seminormal in B in Theorem A. Suppose CRT holds for $\{I_1,\dots,I_n\}$ and $I_i + I_j$ are radical, $i \neq j$. Let $b = (b_1,\dots,b_n) \in B$ be such that $b^2, b^3 \in A$. Then $b_i^2 \equiv b_j^2 \bmod (I_i+I_j)$ and $b_i^3 \equiv b_j^3 \bmod (I_i+I_j)$ for $i \neq j$. As in [S] Lemma 3.1 $(b_i - b_j)^3 = b_i^3 - 3b_i^2 b_j + 3b_i b_j^2 - b_j^3 \equiv b_i^3 - 3b_j^2 b_j + 3b_i b_i^2 - b_i^3 = -3b_j^3 + 3b_i^3 \equiv 0 \bmod (I_i+I_j)$. But as I_i+I_j is radical, this implies $b_i \equiv b_j \bmod (I_i+I_j)$, $i \neq j$ and so, by CRT, $b \in A$.

The converse implication of Theorem A follows from Theorem B. To prove this, we first make two remarks.

Remark 1: For fixed $j \geq 2$, if $a \in \bigcap_{i<j} (I_i+I_j)$ then $a^r \in I_j + \bigcap_{i<j} I_i$ for $r \geq j-1$.

Remark 2: A seminormal in B implies that if $b \in B$ with $b^r \in A$ for all large r then $b \in A$.

The first remark follows as in [DR] Lemma 2.5. For the second, note that if r were the largest positive integer with $b^r \notin A$ then $(b^r)^2, (b^r)^3 \in A$ so b^r would be in A.

Now let $C = \{(a_1,\dots,a_n) \in B : a_i \equiv a_j \bmod (I_i+I_j)\}$, C is a subring of B containing A; to prove Theorem B it suffices to show $C \subseteq A$. We first prove, by downward induction on j, $2 \leqslant j \leqslant n+1$, that if $\alpha = (a_1,\dots,a_n) \in C$ is such that $a_i = 0$ for $i < j$ then $\alpha \in A$. When $j = n+1$ $\alpha = (0,\dots,0) \in A$, so suppose the assertion is true for $j+1$, $2 \leqslant j \leqslant n$, and $\alpha = (0,\dots,0,a_j,\dots,a_n) \in C$. Then $a_j \equiv 0$ mod (I_i+I_j) for $i < j$ so $a_j \in \bigcap_{i<j} (I_i+I_j)$. By Remark 1 $a_j^r \in I_j + \bigcap_{i<j} I_i$ for $r \geqslant j-1$ so for each such r $a_j^r = b_r + c_r$ where $b_r \in I_j$ and $c_r \in \bigcap_{i<j} I_i$. Then $\alpha^r - c_r = (0,\dots,0,a_{j+1}^r - c_r,\dots,a_n^r - c_r) \in A$ by the induction hypothesis, so $\alpha^r \in A$ for $r \geqslant j-1$. By Remark 2 $\alpha \in A$ since A is seminormal in B, and this completes the induction.

Thus if $\alpha = (a_1,\dots,a_n) \in C$, $\alpha - a_1 = (0,a_2-a_1,\dots,a_n-a_1) \in A$ and thus $\alpha \in A$. Done.

REFERENCES

[DR] B.H. Dayton and L.G. Roberts, Seminormality of Unions of Planes, these proceedings.

[DW2] B.H. Dayton and C.A. Weibel, A Spectral Sequence for the K-theory of Affine Glued Schemes, these proceedings.

[O] F. Orecchia, Sulla seminormalita di certe varieta' affini riducibili, Boll. Un. Mat. Ital. (2) B (1976), pp. 588-600.

[S] R.G. Swan, On Seminormality, to appear.

Northeastern Illinois University
Chicago, Illinois 60625

Etale Cohomology of Reductive Groups

by

Eric M. Friedlander and Brian Parshall*

A central theorem in much of the first author's work is the following.

Theorem 1. Let $G_{\mathbb{C}}$ be a reductive complex algebraic group and let $G_{\mathbb{Z}}$ denote the Chevalley reductive integral group scheme of the same type. Let k be a separably closed field, let R denote the Witt vectors of k, and let $R \to K \leftarrow \mathbb{C}$ be embeddings of R and the complex numbers $\mathbb{C}$ into an algebraically closed field K. Then the base change maps associated to $G_{\mathbb{Z}}$,

$$G_k \to G_R \leftarrow G_K \to G_{\mathbb{C}},$$

induce isomorphisms in etale cohomology

$$H^*(G_k, \mathbb{Z}/m) \overset{\sim}{\leftarrow} H^*(G_R, \mathbb{Z}/m) \overset{\sim}{\to} H^*(G_K, \mathbb{Z}/m) \overset{\sim}{\leftarrow} H^*(G_{\mathbb{C}}, \mathbb{Z}/m)$$

for any integer m invertible in k.

The purpose of this note is to provide a complete proof of this result. Because the "classical comparison theorem" of [1;XVI.4.] implies that $H^*(G_{\mathbb{C}}, \mathbb{Z}/m)$ is naturally isomorphic to the mod-m singular cohomology of the associated complex Lie group $G(\mathbb{C})$, Theorem 1 provides a determination of $H^*(G_k, \mathbb{Z}/m)$.

If G_A is a group scheme over Spec A, the bar construction yields a simplicial scheme BG_A such that $(BG_A)_n$ is the n-fold fibre product of G_A with itself over Spec A (cf. [5; Defn. 2.1] where BG is denoted $\overline{W}\{G\}$). Most applications of Theorem 1 have been based on the following corollary, an inadequate

*Both authors were partially supported by the N.S.F.

proof of which is sketched in [5]. This corollary is an immediate consequence of Theorem 1 together with the existence of a natural spectral sequence in etale cohomology

$$E_1^{p,q} = H^q(X_p, \mathbb{Z}/m) \Rightarrow H^{p+q}(X., \mathbb{Z}/m)$$

associated with any simplicial scheme X. [3; 5.2.3.2].

Corollary 2. Assume the hypotheses and notation of Theorem 1. Then the base change maps associated to $BG_{\mathbb{Z}}$

$$BG_k \to BG_R \leftarrow BG_K \to BG_{\mathbb{C}}$$

induce isomorphisms in etale cohomology

$$H^*(BG_k, \mathbb{Z}/m) \overset{\sim}{\leftarrow} H^*(BG_R, \mathbb{Z}/m) \overset{\sim}{\to} H^*(BG_K, \mathbb{Z}/m) \overset{\sim}{\leftarrow} H^*(BG_{\mathbb{C}}, \mathbb{Z}/m)$$

for any integer m invertible in k.

Theorem 1 is proved in two steps. First, we prove in Proposition 5 that $G_{\mathbb{Z}}/B_{\mathbb{Z}}$ exists (i.e., the quotient sheaf is representable as a scheme) and is projective and smooth over Spec $\mathbb{Z}$. We have attempted to provide a more elementary proof of this result than that of [4;XXII.5.8.3]. Our proof uses the existence of a closed immersion $G_{\mathbb{Z}} \subseteq GL_{n,\mathbb{Z}}$ for general $G_{\mathbb{Z}}$. Second, we prove in Proposition 7 that the quotient map $\pi: G_{\mathbb{Z}[1/m]} \to G_{\mathbb{Z}[1/m]}/B_{\mathbb{Z}[1/m]}$ is well-behaved cohomologically. This is a consequence of the fact that π is locally a product projection and that $B_{\mathbb{Z}[1/m]} \to \mathrm{Spec}\ \mathbb{Z}[1/m]$ is well-behaved cohomologically.

We begin the proof of Theorem 1 with the following proposition asserting the existence of a very well-behaved quotient map $GL_{n,\mathbb{Z}} \to GL_{n,\mathbb{Z}}/P_{\mu,\mathbb{Z}}$ for any parabolic subgroup scheme $P_{\mu,\mathbb{Z}}$ of $GL_{n,\mathbb{Z}}$. The proof of Propositon 3 below is a straightforward generalization of the arguments of S. Kleiman presented in [7].

Proposition 3. Let $\mu = (i_1, \ldots, i_t)$ be an ordered partition of n and let $P_{\mu,\mathbb{Z}}$ be a parabolic subgroup scheme of $GL_{n,\mathbb{Z}}$ of type μ. Let $D(\mu)$ denote

The Σ_n-set of ordered t-tuples of disjoint subsets $(S_1,\ldots,S_t)$ of $\{1,\ldots,n\}$ with $\#S_j = i_j$.

(a) The sheaf-theoretic quotient of $GL_{n,\mathbb{Z}}$ by $P_{\mu,\mathbb{Z}}$ is isomorphic to the functor F_μ sending a scheme X to the set $F_\mu(X)$ of isomorphism classes of "quotient flags" $\underset{\rightarrow}{E}: \mathcal{O}_X^n \to E_{t-1} \to \ldots \to E_1$ where E_j is a locally free, rank $e_j = i_1 + \ldots + i_j$ $\mathcal{O}_X$-module and each map is surjective (two such quotient flags are isomorphic if they determine the same (t-1)-tuple of kernel subsheaves of $\mathcal{O}_X^n$).

(b) F_μ is covered by open subsheaves F_μ^α, $\alpha \in D(\mu)$, each of which is representable by a scheme $D^\alpha \approx \mathbb{A}_{\mathbb{Z}}^N$, where $N = \Sigma_{j=1}^t i_j e_{j-1}$.

(c) The restriction of the quotient map $GL_{n,\mathbb{Z}} \to F_\mu$ above each F_μ^α is representable by a product projection $P_{\mu,\mathbb{Z}} \times D^\alpha \to D^\alpha$.

(d) The quotient map $GL_{n,\mathbb{Z}} \to F_\mu$ is representable by a smooth morphism of schemes $\pi: GL_{n,\mathbb{Z}} \to GL_{n,\mathbb{Z}}/P_{\mu,\mathbb{Z}}$ with $GL_{n,\mathbb{Z}}/P_{\mu,\mathbb{Z}}$ projective and smooth over Spec $\mathbb{Z}$.

Proof: We consider the natural action

$$GL_{n,\mathbb{Z}} \times F_\mu \to F_\mu$$

defined for any scheme X by sending $g \in GL_{n,\mathbb{Z}}(X)$ and $\underset{\rightarrow}{E} \in F_\mu(X)$ to $\underset{\rightarrow}{E} \circ g^{-1}$. A choice of a "standard" quotient flag $\underset{\rightarrow}{Q} \in F_\mu(\text{Spec } \mathbb{Z})$ thus determines a $GL_{n,\mathbb{Z}}$-equivariant morphism $GL_{n,\mathbb{Z}} \to F_\mu$ sending $g \in GL_{n,\mathbb{Z}}(X)$ to $\underset{\rightarrow}{Q} \circ g^{-1} \in F_\mu(X)$ for any scheme X (where we are also denoting the pull-back of $\underset{\rightarrow}{Q}$ to X by $\underset{\rightarrow}{Q}$). The stabilizer of the element $\underset{\rightarrow}{Q} \in F_\mu(X)$ is clearly $P_{\mu,\mathbb{Z}}(X)$, where $P_{\mu,\mathbb{Z}}$ is the parabolic subgroup scheme of $GL_{n,\mathbb{Z}}$ of type μ determined by $\underset{\rightarrow}{Q}$.

Because F_μ is clearly a sheaf, to prove (a) it suffices to show that any $\underset{\rightarrow}{E} \in F_\mu(X)$ when restricted to $F_\mu(U_\beta)$ for each U_β of some Zariski open covering $\{U_\beta\}$ of X is in the image of $GL_{n,\mathbb{Z}}(U_\beta)$. This follows from the observation that the restriction of $\underset{\rightarrow}{E}$ to $F_\mu(U)$ is in the image of $GL_{n,\mathbb{Z}}(U)$ whenever U is sufficiently small so that $\ker(\mathcal{O}_U^n \to E_j|_U)$ is a free summand of $\ker(\mathcal{O}_U^n \to E_{j-1}|_U)$ for each j.

For $S \subseteq \{1, \ldots, n\}$, let $\mathcal{O}_X^S$ be the free summand of $\mathcal{O}_X^n$ spanned by the basis elements indexed by S. To prove (b), for $\alpha = (S_1, \ldots, S_t)$, let $F_\mu^\alpha(X) \subseteq F_\mu(X)$ consist of those quotient flags $\underset{\rightarrow}{E}$ such that $\mathcal{O}_X^{S_1 \cup \ldots \cup S_j} \to E_j$ is surjective (and hence bijective) for each j. Clearly, such an $\underset{\rightarrow}{E}$ is obtained uniquely by specifying for each j, $e_{j-1}i_j$ global sections of $\mathcal{O}_X$ corresponding to a choice of basis for E_j compatible with a basis already selected by E_{j-1}. Hence, $F_\mu^\alpha \approx \mathbb{A}^N$. To prove that the F_μ^α constitute an open covering of F_μ, we must verify for every map $\underset{\rightarrow}{E}: X \to F_\mu$ that $\{F_\mu^\alpha \times_{F_\mu} X\}$ represents an open covering of X. But $F_\mu^\alpha \times_{F_\mu} X$ is represented by the open $X^\alpha \subseteq X$ consisting of those points at whose stalks each of the maps $\mathcal{O}_X^{S_1 \cup \ldots \cup S_j} \to E_j$ is surjective.

To prove (c), we consider $\pi_\alpha : GL_{n,\mathbb{Z}}^\alpha \to D^\alpha$ representing the restriction of $Gl_{n,\mathbb{Z}} \to F_\mu$ above F_μ^α. Observe that π_α is principal homogeneous for $P_{\mu,\mathbb{Z}}$. Consequently, π_α is isomorphic to a product projection if it admits a section. Such a section is defined for $\alpha_0 = (S_1, \ldots, S_t)$ with $S_j = \{e_{j-1}+1, \ldots, e_j\}$ by sending D^{α_0} to the "opposite" of the unipotent radical of $P_{\mu,\mathbb{Z}}$. More generally, π_α is the translate of π_{α_0} by any $\sigma \in \Sigma_n$ sending α_0 to α, so that a section for π_α is obtained by translating a section for π_{α_0}.

The representability of the sheaf F_μ follows from (b): $GL_{n,\mathbb{Z}}/P_{\mu,\mathbb{Z}}$ is obtained by patching together the open subschemes D^α. The map π is smooth because it is locally the pull-back of the smooth map $P_{\mu,\mathbb{Z}} \to \mathrm{Spec}\,\mathbb{Z}$. Also, $GL_{n,\mathbb{Z}}/P_{\mu,\mathbb{Z}}$ is smooth by (b).

Now we show that F_μ is projective. Since the Grassmannian scheme $\mathrm{Grass}_{n,e,\mathbb{Z}}$ is projective [7], it is enough to show that F_μ is a closed subfunctor of

$$T = \prod_{j=1}^{t-1} \mathrm{Grass}_{n,e_j,\mathbb{Z}} .$$

More precisely, given a commutative ring A and an element $x \in T(A)$ (= T(Spec A)), we must show that there exists an ideal I of A such that for any homomorphism $f: A \to B$ of commutative rings, $T(f)(x) \in F_\mu(B)$ iff $f(I) = 0$. By considering the kernels of the various surjections $\mathcal{O}_A^n \to E_j$, this is an easy consequence of the following fact. Let M,N be direct factors of the free A-module A^n, and let J be the ideal of A generated by the entries of the matrix representing the composite of the projection of A^n onto M (along some complement) with the projection along N onto some complement of N. Then $M \otimes_A B \subseteq N \otimes_A B$ iff $f(J) = 0$. This completes the proof of Proposition 3.

Let Φ be an abstract (reduced) root system in euclidean space $\mathbb{R}^\ell$. Let $Q(\Phi)$ and $P(\Phi)$ denote respectively the root lattice of Φ and the full weight lattice (generated by the fundamental dominant weights). Fix an intermediate lattice Γ, $Q(\Phi) \subseteq \Gamma \subseteq P(\Phi)$. Associated to the pair (Φ,Γ) there is an integral group scheme $G_{\mathbb{Z}}$ with the following property: given any algebraically closed field k, $G_k = G_{\mathbb{Z}} \otimes k$ is the semisimple algebraic group over k with root system Φ and weight lattice Γ. We will describe below the construction of $G_{\mathbb{Z}}$, together with some of its basic properties. The reader may consult Borel's article [2] for more details.

Let $\mathfrak{g}$ be a complex semisimple Lie algebra with root system Φ relative to a fixed Cartan subalgebra $\mathfrak{h}$. Let $\pi = \{\alpha_1, \ldots, \alpha_\ell\}$ be a set of simple roots of Φ, and Φ^+ the corresponding set of positive roots. Choose a Chevalley basis $\{X_\alpha, h_i \mid \alpha \in \Phi, 1 \leq i \leq \ell\}$ for $\mathfrak{g}$, where X_α belongs to the α-root space and $h_i = h_{\alpha_i}$ is the coroot associated to α_i in $\mathfrak{h}$. Now let $\mathcal{U}_{\mathbb{Z}}$ be the associated Kostant $\mathbb{Z}$-form of the universal enveloping algebra $\mathcal{U}$ of $\mathfrak{g}$. Thus, $\mathcal{U}_{\mathbb{Z}}$ is the $\mathbb{Z}$-subalgebra of $\mathcal{U}$ generated by the $X_\alpha^n/n!$, $\alpha \in \Phi$, $n \in \mathbb{Z}^+$. Fix a (faithful) finite dimensional representation $\rho: \mathfrak{g} \to \mathfrak{gl}(V)$ such that the weights of $\mathfrak{h}$ in V generate the lattice Γ. Inside V we choose a $\mathcal{U}_{\mathbb{Z}}$-stable lattice $V_{\mathbb{Z}}$, and we identify GL(V) with $GL_n(\mathbb{C})$ by selecting a basis for $V_{\mathbb{Z}}$ consisting of weight vectors, ordered compatibly with the usual partial ordering on $P(\Phi)$

(i.e., $\psi \geq \chi$ if and only if $\psi - \chi$ is a sum of positive roots). Let $B_n(\mathbb{C})$ be the Borel subgroup of $GL_n(\mathbb{C})$ consisting of upper triangular matrices, let $T_n(\mathbb{C})$ be the maximal torus consisting of diagonal matrices in $GL_n(\mathbb{C})$, let $U_n(\mathbb{C})$ be the unipotent radical of $B_n(\mathbb{C})$, and let $U_n^-(\mathbb{C})$ denote the subgroup of lower triangular unipotent matrices. The algebraic group $GL_{n,\mathbb{C}}$ associated to $GL_n(\mathbb{C})$ admits a natural $\mathbb{Z}$-form $GL_{n,\mathbb{Z}}$ represented by the ring $A_{n,\mathbb{Z}} = \mathbb{Z}[x_{11}, \ldots, x_{nn}, t]/I$, where I is the ideal $(t\det(x_{ij}) - 1)$. Also, we have closed subgroup schemes $U^{\pm}_{n,\mathbb{Z}}$, $T_{n,\mathbb{Z}}$ of $GL_{n,\mathbb{Z}}$ corresponding to $U^{\pm}(\mathbb{C})$, $T_n(\mathbb{C})$ and defined in the obvious way.

We define $G(\mathbb{C})$ to be the subgroup of $GL_n(\mathbb{C})$ generated by the $x_\alpha(t) = \exp(t\rho(X_\alpha))$, $\alpha \in \Phi$, $t \in \mathbb{C}$. Thus, $G(\mathbb{C})$ is the complex Lie group associated to the complex semisimple algebraic group $G_{\mathbb{C}}$ with root system Φ and weight lattice Γ. Using elementary properties of Borel subgroups, we conclude that our choice of basis for V guarantees that $B(\mathbb{C}) = B_n(\mathbb{C}) \cap G(\mathbb{C})$ is a Borel subgroup of $G(\mathbb{C})$, with maximal torus $T(\mathbb{C}) = T_n(\mathbb{C}) \cap G(\mathbb{C})$ and unipotent radical $U(\mathbb{C}) = U_n(\mathbb{C}) \cap G(\mathbb{C})$. The integral group scheme $G_{\mathbb{Z}}$ is represented by $A_{n,\mathbb{Z}}/J$, where J is the ideal of functions in $A_{n,\mathbb{Z}}$ which vanish on $G(\mathbb{C})$. Then $G_{\mathbb{Z}}$ is smooth over Spec $\mathbb{Z}$; and for any algebraically closed field k, $G_k = G_{\mathbb{Z}} \otimes k$ is semisimple with root system Φ and weight lattice Γ.

We call the associated closed immersion of integral group schemes, $G_{\mathbb{Z}} \subseteq GL_{n,\mathbb{Z}}$, a <u>Chevalley closed immersion</u> associated to the root system Φ and weight lattice Γ.

After fixing an ordering $(\beta_1, \ldots, \beta_N)$ of Φ^+, multiplication defines an isomorphism

$$\mathbb{A}^N \simeq U_{\mathbb{C}}, \quad (t_1, \ldots, t_N) \to \Pi x_{\beta_i}(t_i).$$

Let $\zeta_\alpha \in \mathbb{C}[U(\mathbb{C})]$, $\alpha \in \Phi^+$, define the α-component via this isomorphism. Because $V_{\mathbb{Z}}$ is $\mathcal{U}_{\mathbb{Z}}$-stable, it follows that $\mathbb{Z}[\zeta_\alpha | \alpha \in \Phi^+]$ defines a $\mathbb{Z}$-form $U_{\mathbb{Z}}$ for the group $U(\mathbb{C})$ (or, strictly speaking, the associated algebraic group $U_{\mathbb{C}}$).

Similarly, $U_{\mathbb{Z}}^-$ is defined. A $\mathbb{Z}$-form $T_{\mathbb{Z}}$ of $T(\mathbb{C})$ is represented by $\mathbb{Z}[h_i'^{\pm 1}\ 1 \leq i \leq \ell]$, where the h_i' are a basis for Γ. It follows that the inclusion morphisms $U_{\mathbb{Z}}^{\pm} \to G_{\mathbb{Z}}$, $T_{\mathbb{Z}} \to G_{\mathbb{Z}}$ are closed immersions, since they clearly become so when composed with the closed immersion $G_{\mathbb{Z}} \subseteq GL_{n,\mathbb{Z}}$. We let $B_{\mathbb{Z}} = U_{\mathbb{Z}}.T_{\mathbb{Z}}$, the semidirect product of $U_{\mathbb{Z}}$ and $T_{\mathbb{Z}}$, and we set $\Omega_{\mathbb{Z}} = U_{\mathbb{Z}}^-.B_{\mathbb{Z}} \stackrel{\sim}{\approx} U_{\mathbb{Z}}^- \times B_{\mathbb{Z}}$.

Lemma 4. Let $G_{\mathbb{Z}} \subseteq GL_{n,\mathbb{Z}}$ be a Chevalley closed immersion as above. Then we have (scheme-theoretic) intersections

$$U_{\mathbb{Z}}^{\pm} = G_{\mathbb{Z}} \cap U_{n,\mathbb{Z}}^{\pm}, \quad T_{\mathbb{Z}} = G_{\mathbb{Z}} \cap T_{n,\mathbb{Z}}, \quad B_{\mathbb{Z}} = G_{\mathbb{Z}} \cap B_{n,\mathbb{Z}},$$

and $\Omega_{\mathbb{Z}} = G_{\mathbb{Z}} \cap \Omega_{n,\mathbb{Z}}$ (where $\Omega_{n,\mathbb{Z}} = U_{n,\mathbb{Z}}^-.B_{n,\mathbb{Z}}$).

Proof. We recall from [2] that $\Omega_{n,\mathbb{Z}}$ and $\Omega_{\mathbb{Z}}$ are open affine subschemes of $GL_{n,\mathbb{Z}}$ and $G_{\mathbb{Z}}$ respectively. Since $G_{\mathbb{Z}}$ is irreducible and $\Omega_{\mathbb{Z}}$ is closed in $\Omega_{n,\mathbb{Z}}$, the following square is cartesian:

$$\begin{array}{ccc} \Omega_{\mathbb{Z}} & \longrightarrow & \Omega_{n,\mathbb{Z}} \\ \downarrow & & \downarrow \\ G_{\mathbb{Z}} & \longrightarrow & GL_{n,\mathbb{Z}} \end{array}$$

(because $\Omega_{\mathbb{Z}}$ is open in $G_{\mathbb{Z}}$ and thus open in the pull-back of $\Omega_{n,\mathbb{Z}}$ by $G_{\mathbb{Z}}$, and $\Omega_{\mathbb{Z}}$ is closed in $\Omega_{n,\mathbb{Z}}$ and thus closed in the pull-back). In other words, $\Omega_{\mathbb{Z}} = G_{\mathbb{Z}} \cap \Omega_{n,\mathbb{Z}}$. On the other hand, we also have the cartesian squares

$$\begin{array}{ccc} B_{\mathbb{Z}} & \longrightarrow & B_{n,\mathbb{Z}} \\ \downarrow a & & \downarrow b \\ \Omega_{\mathbb{Z}} & \longrightarrow & \Omega_{n,\mathbb{Z}} \end{array} \qquad \begin{array}{ccc} T_{\mathbb{Z}} & \longrightarrow & T_{n,\mathbb{Z}} \\ \downarrow i & & \downarrow i \\ B_{\mathbb{Z}} & \longrightarrow & B_{n,\mathbb{Z}} \end{array} \qquad \begin{array}{ccc} U_{\mathbb{Z}}^{\pm} & \longrightarrow & U_{n,\mathbb{Z}}^{\pm} \\ \downarrow i & & \downarrow i \\ B_{\mathbb{Z}} & \longrightarrow & B_{n,\mathbb{Z}} \end{array}$$

where $b(x) = (e,x)$ and a is defined similarly. Consequently, we may immediately conclude the other asserted equalities.

With the aid of Proposition 3 and Lemma 4, we can now construct the quotient of $G_{\mathbb{Z}}$ by $B_{\mathbb{Z}}$ and prove the properties we require.

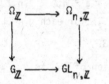

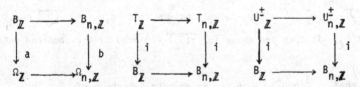

Proposition 5. Let $G_{\mathbb{Z}} \subseteq GL_{n,\mathbb{Z}}$ be a Chevalley closed immersion associated to some root system Φ and weight lattice Γ as constructed above. Let $B_{\mathbb{Z}} = G_{\mathbb{Z}} \cap B_{n,\mathbb{Z}}$ as in Lemma 4. Then the sheaf-theoretic quotient of $G_{\mathbb{Z}}$ by $B_{\mathbb{Z}}$ is represented by a smooth map of schemes

$$\tau: G_{\mathbb{Z}} \to G_{\mathbb{Z}}/B_{\mathbb{Z}}$$

which is locally for the Zariski topology a product projection. Moreoever, $G_{\mathbb{Z}}/B_{\mathbb{Z}}$ is projective and smooth.

Proof. It follows easily from the Bruhat decomposition that $G_{\mathbb{Z}}(k) = G_{\mathbb{Z}}(\mathbb{Z}).\Omega_{\mathbb{Z}}(k)$ for any algebraically closed field k. Thus, we have $G_{\mathbb{Z}} = G_{\mathbb{Z}}(\mathbb{Z}).\Omega_{\mathbb{Z}}$, and so we can choose $g_1, \ldots, g_r \in G_{\mathbb{Z}}(\mathbb{Z})$ such that $G_{\mathbb{Z}} = g_1\Omega_{\mathbb{Z}} \cup \ldots \cup g_r\Omega_{\mathbb{Z}}$. We apply Proposition 3 in the special case in which $\mu = (1, \ldots, 1)$ and $B_{n,\mathbb{Z}} = P_{\mu,\mathbb{Z}}$ to conclude the existence of $\pi: GL_{n,\mathbb{Z}} \to GL_{n,\mathbb{Z}}/B_{n,\mathbb{Z}}$. We see by inspection that the restriction of π to $g\Omega_{n,\mathbb{Z}}$ is the product projection map $\pi_g: g\Omega_{n,\mathbb{Z}} \to gU^-_{n,\mathbb{Z}}$ for any $g \in GL_{n,\mathbb{Z}}(\mathbb{Z})$. Moreoever, $gU^-_{n,\mathbb{Z}}$ is open in $GL_{n,\mathbb{Z}}/B_{n,\mathbb{Z}}$ for each $g \in GL_{n,\mathbb{Z}}(\mathbb{Z})$ because π is locally a product projection by Proposition 3.

Let $\tau = \pi \circ i: G_{\mathbb{Z}} \to GL_{n,\mathbb{Z}} \to GL_{n,\mathbb{Z}}/B_{n,\mathbb{Z}}$ be the restriction of π to $G_{\mathbb{Z}}$, and let $\overline{\tau(G_{\mathbb{Z}})}$ denote the closure of the image of $G_{\mathbb{Z}}$ in $GL_{n,\mathbb{Z}}/B_{n,\mathbb{Z}}$ (provided with the unique structure of a reduced closed subscheme of $GL_{n,\mathbb{Z}}/B_{n,\mathbb{Z}}$). Then $\overline{\tau(G_{\mathbb{Z}})} \cap g_iU^-_{n,\mathbb{Z}}$ is a Zariski open of $\overline{\tau(G_{\mathbb{Z}})}$ given as a reduced closed subscheme of $g_iU^-_{n,\mathbb{Z}}$ (where $g_iU^-_{n,\mathbb{Z}}$ is open in $GL_{n,\mathbb{Z}}/B_{n,\mathbb{Z}}$). Because $g_iU^-_{\mathbb{Z}}$ is closed in $g_iU^-_{n,\mathbb{Z}}$ and dense in $\overline{\tau(G_{\mathbb{Z}})}$ $(g_iU^-_{\mathbb{Z}} = \tau(g_i\Omega_{\mathbb{Z}}))$, we conclude that $g_iU^-_{\mathbb{Z}} = \overline{\tau(G_{\mathbb{Z}})} \cap g_iU^-_{n,\mathbb{Z}}$. In particular, $\tau(G_{\mathbb{Z}})$ is open and dense in $\overline{\tau(G_{\mathbb{Z}})}$ (covered by the open and dense $g_iU^-_{\mathbb{Z}}$).

We observe that $\tau^{-1}(g_iU^-_{\mathbb{Z}}) = g_i\Omega_{\mathbb{Z}}$ as an open subscheme of $G_{\mathbb{Z}}$: by construction, $\tau(g_i\Omega_{\mathbb{Z}}) \subseteq g_iU^-_{\mathbb{Z}}$; on the other hand, $\tau^{-1}(g_iU^-_{\mathbb{Z}}) \subseteq \pi^{-1}(g_iU^-_{n,\mathbb{Z}}) \cap G_{\mathbb{Z}}$ $= g_i\Omega_{n,\mathbb{Z}} \cap G_{\mathbb{Z}} = g_i\Omega_{\mathbb{Z}}$. Consequently, the facts that each $\pi_{g_i}: \pi^{-1}(g_iU^-_{n,\mathbb{Z}}) \to g_iU^-_{n,\mathbb{Z}}$ is a product projection and that $\Omega_{\mathbb{Z}} \cong U^-_{\mathbb{Z}} \times B_{\mathbb{Z}} \subseteq U^-_{n,\mathbb{Z}} \times B_{n,\mathbb{Z}} \cong \Omega_{n,\mathbb{Z}}$ is a product map imply that $\tau: G_{\mathbb{Z}} \to \tau(G_{\mathbb{Z}})$ is locally a product projection with each $\tau_{g_i}: \tau^{-1}(g_iU^-_{\mathbb{Z}}) = g_i\Omega_{\mathbb{Z}} \to g_iU^-_{\mathbb{Z}}$ a product projection.

We next verify that $\tau: G_{\mathbb{Z}} \to \tau(G_{\mathbb{Z}})$ is the sheaf-theoretic quotient of $G_{\mathbb{Z}}$ by $B_{\mathbb{Z}}$ (defined to be the sheaf associated to the presheaf $X \to G_{\mathbb{Z}}(X)/B_{\mathbb{Z}}(X)$). The injectivity of $G_{\mathbb{Z}}(X)/B_{\mathbb{Z}}(X) \to \mathrm{Hom}(X, \tau(G_{\mathbb{Z}}))$ follows from the injectivity of each of the maps

$$G_{\mathbb{Z}}(X)/B_{\mathbb{Z}}(X) \to GL_{n,\mathbb{Z}}(X)/B_{n,\mathbb{Z}}(X) \to \mathrm{Hom}(X, GL_{n,\mathbb{Z}}/B_{n,\mathbb{Z}})$$

whose composition factors through $\mathrm{Hom}(X, \tau(G_{\mathbb{Z}}))$. Local surjectivity of $G_{\mathbb{Z}}(\ )/B_{\mathbb{Z}}(\ ) \to \mathrm{Hom}(\ ,\tau(G_{\mathbb{Z}}))$ follows from the observation that the restriction of any $f \in \mathrm{Hom}(X, \tau(G_{\mathbb{Z}}))$ to each $f_i \in \mathrm{Hom}(f^{-1}(g_i U_{\mathbb{Z}}^-), \tau(G_{\mathbb{Z}}))$ is in the image of $G_{\mathbb{Z}}(f^{-1}(g_i U_{\mathbb{Z}}^-))$ and $\{f^{-1}(g_i U_{\mathbb{Z}}^-) | i = 1, \ldots, r\}$ is an open covering of X.

Consequently, for any geometric point $\mathrm{Spec}\, k \to \mathrm{Spec}\, \mathbb{Z}$, the base change of τ, $\tau_k: G_k \to \tau(G_{\mathbb{Z}})_k$, is the sheaf-theoretic quotient of G_k by B_k, $\tau_k: G_k \to G_k/B_k$. Because each G_k/B_k is a connected, projective algebraic variety and because $\tau(G_{\mathbb{Z}}) \to \mathrm{Spec}\, \mathbb{Z}$ admits a section, [6;15.7.10] implies that $\tau(G_{\mathbb{Z}}) \to \mathrm{Spec}\, \mathbb{Z}$ is proper. Hence, $\tau(G_{\mathbb{Z}}) \to GL_{n,\mathbb{Z}}/B_{n,\mathbb{Z}}$ is also proper, so that $\tau(G_{\mathbb{Z}}) = \overline{\tau(G_{\mathbb{Z}})}$. We conclude that $\tau(G_{\mathbb{Z}})$ is projective.

We now denote $\tau: G_{\mathbb{Z}} \to \tau(G_{\mathbb{Z}})$ by $\tau: G_{\mathbb{Z}} \to G_{\mathbb{Z}}/B_{\mathbb{Z}}$. Because $B_{\mathbb{Z}}$ is smooth over $\mathrm{Spec}\, \mathbb{Z}$ and because τ is locally a pull-back of $B_{\mathbb{Z}} \to \mathrm{Spec}\, \mathbb{Z}$, τ is smooth. Because each $g_i U_{\mathbb{Z}}^-$ is smooth, $G_{\mathbb{Z}}/B_{\mathbb{Z}}$ is smooth as well as proper. This completes the proof.

We next consider the etale cohomology of Borel subgroups. As seen in the following lemma, such group schemes are particularly well-behaved with respect to cohomological base change.

<u>Lemma 6</u>. Let A be a commutative ring and m a positive integer such that $1/m \in A$ and $\mu_m(A) \simeq \mathbb{Z}/m$. For a positive integer s, let $T_{s,A} = \mathrm{Spec}\, A[y_1^{\pm 1}, \ldots, y_s^{\pm 1}]$, the s-dimensional torus over Spec A. Consider

$$\pi: B_A = \mathbb{A}_A^r \times_{\mathrm{Spec}\, A} T_{s,A} \to \mathrm{Spec}\, A,$$

so that $B_A = \mathrm{Spec}\, A[x_1, \ldots, x_r, y_1^{\pm 1}, \ldots, y_s^{\pm 1}]$. Then $R^q \pi_* \mathbb{Z}/m$ is constant on

Spec A, and $(R^q\pi_* \mathbb{Z}/m)_y \simeq H^q(B_k, \mathbb{Z}/m)$, $B_k = B_A \times_{\text{Spec } A} k$, for all $q \geq 0$ and all geometric points $y:\text{Spec } k \to \text{Spec } A$.

<u>Proof</u>. We apply the smooth base change theorem [8;VI.4.1] to the cartesian square

$$\begin{array}{ccc} B_A & \longrightarrow & \mathbb{A}^r_A \\ \Big\downarrow \pi_2 & & \Big\downarrow p \\ T_{s,A} & \xrightarrow[\pi_1]{} & \text{Spec } A \end{array}$$

(where p, π_1 are the structure morphisms and π_2 is the projection pr_2) to obtain that $R^q\pi_{2*}\mathbb{Z}/m \simeq \pi_1^* R^q p_* \mathbb{Z}/m$ on $T_{s,A}$. Because $R^q p_* \mathbb{Z}/m = 0$ for $q > 0$ and $R^0 p_* \mathbb{Z}/m = \mathbb{Z}/m$, we conclude that $R^q\pi_{2*}\mathbb{Z}/m = 0$ for $q > 0$ and $R^0\pi_{2*}\mathbb{Z}/m = \mathbb{Z}/m$. Thus, $R^q\pi_{1*}\mathbb{Z}/m \xrightarrow{\sim} R^q\pi_*\mathbb{Z}/m$ and $\pi_2 : B_k \to T_{s,k}$ induces an isomorphism $H^*(T_{s,k}, \mathbb{Z}/m) \xrightarrow{\sim} H^*(B_k, \mathbb{Z}/m)$ for each geometric point $y:\text{Spec } k \to \text{Spec } A$.

We recall that, in the special case $s = 1$, $R^q\pi_{1*}\mathbb{Z}/m = 0$ for $q > 1$, $R^1\pi_{1*}\mathbb{Z}/m \simeq R^1\pi_{1*}\mu_m = \mathbb{Z}/m$, and $R^0\pi_{1*}\mathbb{Z}/m = \mathbb{Z}/m$; moreover, $(R^q\pi_{1*}\mathbb{Z}/m)_y \simeq H^q(T_{1,k}, \mathbb{Z}/m)$ for all $q > 0$ and all geometric points $y:\text{Spec } k \to \text{Spec } A$. (This can be proved using the proper, smooth base change theorem [8;VI.4.2] and the Gysin sequence [8;VI.5.3]). Because the pull-back of $T_{1,A} \to \text{Spec } A$ by $T_{s-1,A} \to \text{Spec } A$ is the projection map $T_{s,A} \to T_{s-1,A}$, we may employ induction and the Leray spectral sequence for $\pi_1 : T_{s,A} \to T_{s-1,A} \to \text{Spec } A$ to conclude that $R^q\pi_{1*}\mathbb{Z}/m \simeq R^q\pi_*\mathbb{Z}/m$ is constant on Spec A for all $q \geq 0$. Moreover, comparing this Leray spectral sequence to that for $\pi_1 : T_{s,k} \to T_{s-1,k} \to \text{Spec } k$, we conclude that

$$(R^q\pi_*\mathbb{Z}/m)_y \simeq (R^q\pi_{1*}\mathbb{Z}/m)_y \simeq H^q(T_{s,k}, \mathbb{Z}/m) \simeq H^q(B_k, \mathbb{Z}/m)$$

for all $q \geq 0$ and all geometric points $y:\text{Spec } k \to \text{Spec } A$. This completes the proof.

We now consider an arbitray complex reductive algebraic group $G_{\mathbb{C}}$. The radical $R_{\mathbb{C}}$ of $G_{\mathbb{C}}$ is a torus and the commutator subgroup $G'_{\mathbb{C}}$ is semisimple. Moreover,

$G'_{\mathbb{C}} \cap R_{\mathbb{C}}$ is a finite central subgroup (which we denote by H) and the multiplication map $G'_{\mathbb{C}} \times R_{\mathbb{C}} \to G_{\mathbb{C}}$ is a principal covering space for H. If $G'_{\mathbb{Z}} \subseteq GL_{n,\mathbb{Z}}$ is a Chevalley closed immersion of integral group schemes associated to the root system and weight lattice of $G'(\mathbb{C})$ (as constructed prior to Lemma 4) and if $R_{\mathbb{Z}}$ is a $\mathbb{Z}$-form for $R_{\mathbb{C}}$ suitably chosen with respect to H, then $G'_{\mathbb{Z}} \times R_{\mathbb{Z}}$ contains a central subgroup scheme $H_{\mathbb{Z}} = \coprod_H \operatorname{Spec} \mathbb{Z}$ etale over Spec $\mathbb{Z}$. The sheaf-theoretic quotient of $G'_{\mathbb{Z}} \times R_{\mathbb{Z}}$ by $H_{\mathbb{Z}}$ is represented by a group scheme $G_{\mathbb{Z}}$ which we shall call a Chevalley $\mathbb{Z}$-form for $G_{\mathbb{C}}$.

We now verify the second important step in our proof of Theorem 1.

Proposition 7. Let $G_{\mathbb{C}}$ be a complex reductive algebraic group and let $G_{\mathbb{Z}}$ be a Chevalley $\mathbb{Z}$-form for $G_{\mathbb{C}}$ (as discussed above). Let A be a commutative ring and m a positive integer with $1/m \in A$. Then $R^q\pi_* \mathbb{Z}/m$ is locally constant on Spec A (for the etale topology) and the base change map $(R^q\pi_* \mathbb{Z}/m)_y \to H^q(G_k, \mathbb{Z}/m)$ is an isomorphism for all $q \geq 0$ and all geometric points y:Spec k $\to$ Spec A, where $\pi: G_A \to \operatorname{Spec} A$ is obtained by base change from $G_{\mathbb{Z}} \to \operatorname{Spec} \mathbb{Z}$.

Proof. We first assume that $G_{\mathbb{C}}$ is semisimple, admitting a $\mathbb{Z}$-form $G_{\mathbb{Z}}$ provided with a Chevalley closed immersion $G_{\mathbb{Z}} \subseteq GL_{n,\mathbb{Z}}$ as in Proposition 5, and that A contains a primitive m^{th} root of unity (so that the sheaf μ_m on any scheme over Spec A is isomorphic to $\mathbb{Z}/m$). Then $B_{\mathbb{Z}} = G_{\mathbb{Z}} \cap B_{n,\mathbb{Z}} \simeq U_{\mathbb{Z}} \times T_{\mathbb{Z}}$, so that Lemma 6 applies to $p:B_A \to \operatorname{Spec} A$. Because $\pi_1:G_A \to G_A/B_A$ is locally (on G_A/B_A for the Zariski topology) the pull-back of $p:B_A \to \operatorname{Spec} A$, the smooth base change theorem and Lemma 6 imply that $R^t\pi_{1*} \mathbb{Z}/m$ is locally constant and that $(R^t\pi_{1*} \mathbb{Z}/m)_y \xrightarrow{\sim} H^t(B_k, \mathbb{Z}/m)$ for all $t \geq 0$ and all geometric points y:Spec k $\to$ G_A/B_A. In fact, because $R^t\pi_{1*} \mathbb{Z}/m$ is locally constant for the Zariski topology, it is actually constant.

By the proper, smooth base change theorem [8; VI.4.2] applied to $\pi_2: G_A/B_A \to \operatorname{Spec} A$, we conclude that $R^s\pi_{2*}(R^t\pi_{1*} \mathbb{Z}/m)$ is locally constant on Spec A and that $(R^s\pi_{2*}(R^t\pi_{1*} \mathbb{Z}/m))_y \xrightarrow{\sim} H^s(G_k/B_k, H^t(B_k, \mathbb{Z}/m))$ for all $s,t \geq 0$

and all geometric points $y:\mathrm{Spec}\ k \to \mathrm{Spec}\ A$. By the naturality of the Leray spectral sequence, we can compare spectral sequences for $\pi = \pi_2 \circ \pi_1$ when restricted to various etale opens of Spec A to conclude that $R^q\pi_* \mathbb{Z}/m$ is locally constant for all $q \geq 0$. Similarly, comparing the limit of these Leray spectral sequences with that for the composition $G_k \to G_k/B_k \to \mathrm{Spec}\ k$, we conclude that $(R^q\pi_* \mathbb{Z}/m)_y \xrightarrow{\sim} H^q(G_k, \mathbb{Z}/m)$ for all $q \geq 0$ and all geometric points $y:\mathrm{Spec}\ k \to \mathrm{Spec}\ A$.

More generally, we consider $G'_{\mathbb{Z}} \times R_{\mathbb{Z}}$ where $G'_{\mathbb{Z}}$ is semisimple and $R_{\mathbb{Z}}$ is a torus. Continuing to assume that A contains a primitive m^{th} root of unity, Lemma 6 and the smooth base change theorem imply that $R^t pr_{1*} \mathbb{Z}/m$ is constant on G'_A and $(R^t pr_{1*} \mathbb{Z}/m)_y \xrightarrow{\sim} H^q(R_k, \mathbb{Z}/m)$ for all geometric points $y:\mathrm{Spec}\ k \to G'_A$, where $pr_1: G'_A \times R_A \to R_A$. Consequently, we conclude with the aid of the Leray spectral sequence and the special case proved above that $R^q\tilde{\pi}_* \mathbb{Z}/m$ is locally constant on Spec A and $(R^q\tilde{\pi}_* \mathbb{Z}/m)_y \xrightarrow{\sim} H^q(G'_k \times R_k, \mathbb{Z}/m)$ for all geometric points $y:\mathrm{Spec}\ k \to G'_A \times_{\mathrm{Spec}\ A} R_A$, where $\tilde{\pi}: G'_A \times R_A \to \mathrm{Spec}\ A$. If $G'_{\mathbb{Z}} \times R_{\mathbb{Z}} \to G_{\mathbb{Z}}$ is a galois covering with group H, we consider the functor sending an etale map $A \to B$ to the spectral sequence

$$E_2^{p,q} = H^p(H, H^q(G'_B \times_{\mathrm{Spec}\ B} R_B, \mathbb{Z}/m) \Longrightarrow H^{p+q}(G_B, \mathbb{Z}/m).$$

Because the sheaf associated to $B \to H^q(G'_B \times_{\mathrm{Spec}\ B} R_B, \mathbb{Z}/m)$ (namely, $R^q\tilde{\pi}_* \mathbb{Z}/m$) is locally constant on Spec A we conclude that $R^{p+q}\pi_* \mathbb{Z}/m$ is also locally constant on Spec A, for $\pi: G_A \to \mathrm{Spec}\ A$. Using the base change map to compare the above spectral sequence to

$$E_2^{p,q} = H^p(H, H^q(G'_k \times_{\mathrm{Spec}\ k} R_k, \mathbb{Z}/m) \Longrightarrow H^{p+q}(G_k, \mathbb{Z}/m)$$

we conclude that $R^{p+q}\pi_* Z/m \xrightarrow{\sim} H^{p+q}(G_k, Z/m)$.

Finally, we no longer assume that A contains a primitive m^{th} root of unity. Let $A \to A'$ be the finite etale extension obtained by adjoining to A a primitive m^{th} root of unity. Applying the smooth base change theorem to $\mathrm{Spec}\ A' \to \mathrm{Spec}\ A$, we conclude for $\pi: G_A \to \mathrm{Spec}\ A$ that $R^q\pi_* Z/m$ on Spec A is locally constant when restricted to Spec A' (and thus locally constant on Spec A as well) and that $(R^q\pi_* \mathbb{Z}/m)_y \xrightarrow{\sim} H^q(G_k, \mathbb{Z}/m)$ for all $q \geq 0$ and all geometric points $y:\mathrm{Spec}\ k \to \mathrm{Spec}\ A$.

Proof of Theorem 1. We apply Proposition 7 with A equal to R, the Witt vectors of k. Because R is a strict hensel local ring, $(R^q\pi_* \mathbb{Z}/m)_y = H^q(G_R, \mathbb{Z}/m)$ for all $q \geq 0$ and all geometric points $y:\mathrm{Spec}\ k \to \mathrm{Spec}\ R$. Hence, Proposition 7 implies that $G_k \to G_R \leftarrow G_K$ induces isomorphisms $H^*(G_k, \mathbb{Z}/m) \xleftarrow{\sim} H^*(G_R, \mathbb{Z}/m) \xrightarrow{\sim} H^*(G_K, \mathbb{Z}/m)$. Moreoever, if we apply Proposition 7 with A equal to $\mathbb{C}$ (also, of course, a strict hensel local ring) and view $\mathrm{Spec}\ K \to \mathrm{Spec}\ \mathbb{C}$ as a geometric point of Spec $\mathbb{C}$, we conclude that $G_K \to G_{\mathbb{C}}$ induces isomorphisms $H^*(G_K, \mathbb{Z}/m) \xleftarrow{\sim} H^*(G_{\mathbb{C}}, \mathbb{Z}/m)$.

References

[1] M. Artin, A. Grothendieck, and J. Verdier, Théorie des Topos et Cohomologie Etale des Schemas, III, Lecture notes in mathematics 305 (1973), Springer-Verlag, Berlin.

[2] A. Borel, Properties and linear representations of Chevalley groups. In: Seminar on Algebraic Groups and Related Finite Groups, Lecture notes in mathematics 131 (1970), 1-50, Springer-Verlag, Berlin.

[3] P. Deligne, Théorie de Hodge, III, Publ. Math. I.H.E.S. no. 44 (1974), 5-77.

[4] M. Demazure and A. Grothendieck, Schemas en Groupes, III, Lecture notes in mathematics 153 (1970), Springer-Verlag, Berlin.

[5] E. Friedlander, Computations of K-theories of finite fields, Topology 15 (1976), 87-109.

[6] A. Grothendieck, EGA IV: Étude locale des schemas et des morphismes de schemas (troisieme partie), Publ. Math. I.H.E.S. no. 28 (1966).

[7] S. Kleiman, Geometry on grassmannians and applications to splitting bundles and smoothing cycles, Publ. Math. I.H.E.S. no. 36 (1969), 281-297.

[8] J.S. Milne, Etale Cohomology, Princeton University Press, Princeton (1980).

Department of Mathematics
Northwestern University
Evanston, Ill. 60201

Department of Mathematics
University of Virginia
Charlottesville, Va. 22903

Comparison of K-theory Spectral Sequences, with Applications

Henri Gillet

Contents

Introduction

In [2], Bloch and Ogus proved Washnitzer's conjecture that the filtration on algebraic de Rham cohomology coming from the hypercohomology spectral sequence coincides with the filtration by coniveau. In this paper we prove in §2 a K-theoretic analogue of their result; that from E_2 onward the coniveau spectral sequence of Quillen ([13]) for the K-theory of a regular scheme X satisfying Gersten's conjecture coincides with a 'local to global' spectral sequence. The local to global spectral sequence is the direct limit of the Bousfield-Kan spectral sequences associated to open covers of X, rather than the 'Postnikov tower' spectral sequence of [4], and its construction (in §1) was suggested by Thomason's use of similar spectral sequences for the étale topology. The advantage of the local to global spectral sequence is that it is compatible with the product structure in K-theory (Thm. (1.7)). This compatibility provides a very direct proof that the cup product in K-theory may be used to define the intersection product for algebraic cycles, extending this result to all regular varieties over a field rather than just smooth varieties as in [11], [9]. The proof consists of the remark that Serre's intersection product ([14]) is already defined using products in K-theory, so the product defined using the Bloch-Quillen map (3.1) should be essentially the same as Serre's.

Finally in §5 we show how the compatibility of the product in K-theory with the topological filtration may be used to prove the Grothendieck Riemann-Roch theorem for proper, non projective, morphisms of smooth varieties over a field, generalizing [8], [9]. In order to set up this result we need to prove in §4 that the K-theory of coherent sheaves is covariant for all proper morphisms, not just projective morphisms as proved in [13].

Note; All schemes are supposed noetherian separated.

§1. K-Theory Spectral Sequences

Preliminaries

If X is a noetherian scheme Quillen has defined ([13]):

$$K'_i(X) = \pi_{i+1} BQ\underline{\underline{M}}(X)$$

$$K_i(X) = \pi_{i+1} BQ\underline{\underline{P}}(X)$$

where $\underline{\underline{M}}(X)$ and $\underline{\underline{P}}(X)$ are the exact categories of coherent and of locally free $\mathcal{O}_X$ modules respectively. If $\underline{\underline{E}}$ is any exact category, then $BQ\underline{\underline{E}}$ has a natural infinite loop space structure ([16]), so we may define a C-W Ω-spectrum ([1]) $\mathcal{K}(\underline{\underline{E}})$:

$$\Omega^i \mathcal{K}(\underline{\underline{E}}) = \Omega^{i+1} BQ\underline{\underline{E}} \quad i \in \mathbb{Z}.$$

We write $\mathcal{K}'(X)$ and $\mathcal{K}(X)$ for the spectra corresponding to $\underline{\underline{M}}(X)$ and $\underline{\underline{P}}(X)$ respectively. Note that $\mathcal{K}(X)$ is a contravariant functor of X and that $\mathcal{K}'(X)$ is contravariant witn respect to flat morphisms and covariant with respect to projective morphisms ([13]). Given any biexact functor

$$\underline{\underline{E}} \times \underline{\underline{F}} \longrightarrow \underline{\underline{G}}$$

Waldhausen has defined a product ([16])

$$BQ\underline{\underline{E}} \quad BQ\underline{\underline{F}} \longrightarrow BQ\underline{\underline{G}}$$

which may be extended to a pairing (see [9] for details)

$$\mathcal{K}(\underline{\underline{E}}) \quad \mathcal{K}(\underline{\underline{F}}) \longrightarrow \mathcal{K}(\underline{\underline{G}}) .$$

If $f\colon Y \longrightarrow X$ is a morphism of noetherian schemes there is a pairing

$$\otimes_{\mathcal{O}_Y} : \underline{\underline{P}}(X) \times \underline{\underline{M}}(Y) \longrightarrow \underline{\underline{M}}(Y)$$

and hence a product:

$$\mathcal{K}(X) \quad \mathcal{K}'(Y) \longrightarrow \mathcal{K}'(Y).$$

Relative to this product, if f is a projective morphism of schemes defined over a fixed base S, then

$$f_*\colon \mathcal{K}'(X) \longrightarrow \mathcal{K}'(Y)$$

is a map of $\mathcal{K}(S)$ module spectra (where $\mathcal{K}(S)$ is a ring spectrum via $\otimes_{\mathcal{O}_S} : \underline{\underline{P}}(S) \times \underline{\underline{P}}(S) \longrightarrow \underline{\underline{P}}(S)$). Quillen's localization theorem ([13]) may also be expressed in terms of spectra: let Y be a closed subscheme of X, with complement U then (where i: $Y \longrightarrow X$ and α: $U \longrightarrow X$ are the natural maps):

$$K'(Y) \xrightarrow{i_*} K'(X) \xrightarrow{\alpha^*} K'(U)$$

is a cofibration sequence. If X is an S-scheme then i_* and α^* are both maps of $K(S)$ module spectra. We may think of $K'(Y)$ as being the relative K-theory $K'(X,U)$.

The Quillen Spectral Sequence

The category $\underline{\underline{M}}(X)$ has a filtration:

$$\underline{\underline{M}}(X) = \underline{\underline{M}}^{0}(X) \supset \ldots \supset \underline{\underline{M}}^{i}(X) \supset \ldots$$

where $\underline{\underline{M}}^{i}(X)$ is the full exact subcategory of coherent sheaves on X with support of codimension at least i. The associated filtration $\{K'^{(i)}(X)\}$ of $K'(X)$ will be called the coniveau filtration and gives rise to a spectral sequence ([13]§7):

$$(1.1) \qquad E_1^{p,q} = \pi_{-p-q}(K'^{(p)}(X)/K'^{(p+1)}(X)) \Rightarrow K'_{-p-q}(X).$$

ecall that (where $X^{(p)}$ is the set of points of codimension p in X):

$$(1.2) \qquad E_1^{p,q} \simeq \bigoplus_{x \in X^{(p)}} K_{-p-q}(\mathbb{k}(x)) .$$

This spectral sequence is contravariant with respect to flat morphisms, and hence the $E_1^{p,q}$ term (1.2) defines a family R_q^* $q \geq 0$ of flasque sheaves of abelian groups on X:

$$R_q^p(U) = E_1^{p,-q}(U).$$

There are natural augmentations of sheaves $K'_q \longrightarrow R_q^*$ and we shall say that X satisfies <u>Gersten's condition</u> if for all $q \geq 0$ these augmentations are resolutions, so that:

$$E_2^{p,q}(X) = H^p(X,K'_{-q}).$$

Quillen has proved Gersten's conjecture, that this condition holds for all regular schemes, when X is of finite type over a field. Note that $Y \subset X$ is a closed codimension d subscheme of a scheme satisfying Gersten's condition, then for $U \subset X$:

$$\Gamma_Y(U,R_q^i) \simeq R_{q-d}^{i-d}(U \cap Y) .$$

Hence the Quillen spectral sequence for Y may be interpreted (after a shift in degree $E_2^{p,q} \longmapsto E_2^{p+d,q-d}$) as a spectral sequence:

(1.3) $$E_2^{p,q} = H_Y^p(X,K'_q) \Rightarrow K'_{q-p}(X,X-Y).$$

The Local to Global Spectral Sequence

First we recall Lubkin's approach to the cohomology of sheaves in the Zariski topology ([12]).

Definition 1.4. a) A Lubkin cover of a topological space X is a full subcategory $\underline{\underline{U}}$ of the category of open sets in X, such that

(i) Each $x \in X$ is contained in a finite number of $U \in \underline{\underline{U}}$ and is contained in at least one such U.

(ii) For each $x \in X$, there is a $U_x \in \underline{\underline{U}}$ such that $V \in \underline{\underline{U}}$ and $x \in \underline{\underline{V}}$ implies $U_x \subset V$.

b) A Lubkin cover $\underline{\underline{V}}$ is said to refine a Lubkin cover $\underline{\underline{U}}$, if for all $x \in X$, $V_x \subset U_x$.

c) Let F be a presheaf of abelian groups on X, and $\underline{\underline{U}}$ a Lubkin cover of X. Then we set:

$$H^p(X,\underline{\underline{U}},F) = \underset{\longleftarrow\ \underline{\underline{U}}^{op}}{\mathrm{Lim}}{}^p (F)$$

where F defines in a natural way a $\underline{\underline{U}}^{op}$ diagram of abelian groups. Note that $H^p(X,\underline{\underline{U}},F)$ is computed as the cohomology of the complex:

$$p \longrightarrow \prod_{U_{\alpha_0} \supset \dots U_{\alpha_p}} F(U_{\alpha_p}) = C^p(X,\underline{\underline{U}},F).$$

If $Y \subset X$ is a closed subscheme and $\underline{\underline{U}}$ is a Lubkin refining the cover $\{X,X-Y\}$ we define

$$C_Y^p(X,\underline{\underline{U}},F) = \mathrm{Ker}[C^p(X,\underline{\underline{U}},F) \longrightarrow C^p(U,\underline{\underline{U}}|_U,F)]$$

where $\underline{\underline{U}}|_U = \{V \in \underline{\underline{U}} \mid V \subset U\}$, and set

$$H_Y^p(X,\underline{\underline{U}},F) = H^p(C_Y^*(X,\underline{\underline{U}},F)).$$

Theorem 1.5. Let F be a presheaf of abelian groups on the Zariski site of a noetherian scheme X. Then

$$H^p_Y(X,\tilde{F}) = \varinjlim_{\underline{\underline{U}}} H^p_Y(X,\underline{\underline{U}},F)$$

where $\tilde{F}$ is the sheaf associated to F.

Proof. [12].

One may associate to a Lubkin cover $\underline{\underline{U}}$ of X a simplicial sheaf $B\cdot\underline{\underline{U}}$, such that for a Zariski open $V \subset X$,

$$B.U(V) = B.\underline{\underline{U}}(V)^{op}$$

where $\underline{\underline{U}}(V)$ is the full subcategory of $\underline{\underline{U}}$ consisting of those $U \in \underline{\underline{U}}$ with $V \subset U$. By a)ii) of (1.4) the stalks of $B.\underline{\underline{U}}$ are contractible, hence the natural map $B.\underline{\underline{U}} \longrightarrow *$ (the punctual sheaf) is a weak equivalence in the sense of [4]. Since $K'(X)$ is contravariant functor of X with respect to flat morphisms, we have a natural sheaf of infinite loop spaces K' on X, associated to the presheaf $U \longrightarrow K'(U)$. We may replace K' and all its deloopings by flasque resolutions, and since $U \longrightarrow K'(U)$ is a pseudo flasque presheaf ([4]§3) for each $U \subset X$ we have:

$$R\Gamma(U,K') = K'(U)$$

so without confusion we may write K' for the flasque resolution of the 'old' K'. Given a Lubkin cover $\underline{\underline{U}}$ of X, K' defines a $\underline{\underline{U}}^{op}$ diagram of infinite loop spaces, fibrant (in each degree) in the sense of ([3]Ch XI).

Lemma 1.6. Let $\underline{\underline{U}}$, X be as above. Then:

$$\underset{\longleftarrow\, \underline{\underline{U}}^{op}}{\mathrm{Holim}}\, K' \simeq K'(X) .$$

Proof. By definition ([3]) $\underset{\longleftarrow\, \underline{\underline{U}}^{op}}{\mathrm{Holim}}\, K'$ is $\mathrm{Hom}_{\underline{\underline{S}}(\underline{\underline{U}}^{op})}(B.\underline{\underline{U}},K')$, the function space of maps between diagrams of $\underline{\underline{U}}^{op}$ spaces, which has a natural infinite loop structure. Since for all $x \in X$, $B.\underline{\underline{U}}$ restricted to U_x is a constant sheaf of simplicial sets we know that

$$\mathrm{Hom}_{\underline{\underline{S}}(\underline{\underline{U}}^{op})}(B.\underline{\underline{U}},K') = \mathrm{Hom}_{\underline{\underline{S}}(\underline{\underline{X}})}(B.\underline{\underline{U}},K')$$

where $\underline{\underline{S}}(\underline{\underline{X}})$ is the category of simplicial sheaves on X. Since K' is flasque, and

$B.\underline{\underline{U}}$ is weak equivalent to $*$,

$$\mathrm{Hom}_{\underline{\underline{S}}(X)}(B.\underline{\underline{U}},K') \simeq K'(X)\ .$$

Proposition 1.7. For each closed subscheme Y of a scheme X there is a spectral sequence

$$E_2^{p,q}(X,X-Y) = H_Y^p(X,K'_{-q}) \Rightarrow K'_{-p-q}(X,X-Y) \tag{1.8}$$

and if X is regular the product structure on the K-theory of X induces for every pair of closed subsets Y,Z of X a pairing of spectral sequences

$$E_r^{p,q}(X,X-Y) \otimes E_r^{p',q'}(X,X-Z) \longrightarrow E_r^{p+p',q+q'}(X,X-(Y \cap Z)).$$

Proof. If $S \subset T$ is a pair of simplicial sheaves on X, we define $K'(T,S)$ as the fibre of the restriction map:

$$\mathrm{Hom}_{\underline{\underline{S}}(X)}(T,K') \longrightarrow \mathrm{Hom}_{\underline{\underline{S}}(X)}(S,K')$$

and set $K_{-q}(T,S) = \pi_{-q}K(T,S)$. If $\underline{U}$ is a Lubkin cover refining (X,X-Y), by (1.6) we know

$$K'(B.\underline{\underline{U}},B.\underline{U}|_{X-Y}) = K'(X,X-Y).$$

Now we filter $B.\underline{U}$ by $\{B^s.\underline{\underline{U}} \cup B.\underline{\underline{U}}|_{X-Y}\}$ where $B^s.\underline{U}$ is the sheaf of s-skeleta of $B.\underline{U}$. The associated tower of fibrations $K'(B^s.\underline{\underline{U}} \cup B.\underline{\underline{U}}_{X-Y}, B.\underline{\underline{U}}_{X-Y})$ has limit $K'(X,X-Y)$ and layers:

$$\prod_{\substack{U_{\alpha_0} \supset \ldots U_{\alpha_s} \\ U_{\alpha_0} \cap Y \neq \emptyset}} \Omega^s K'(U_{\alpha_s})\ .$$

We get a spectral sequence ([3]IX)

$$E_r^{p,q}(X,X-Y,\underline{\underline{U}}) \Rightarrow K'_{-p-q}(X,X-Y)$$

with

$$E_1^{p,q} = \prod_{\substack{U_{\alpha_0} \supset \ldots \supset U_{\alpha_p} \\ U_{\alpha_0} \cap V \neq 0}} \pi_{-q-p}(\Omega^p K'(U_{\alpha_p})) = \prod_{\substack{U_{\alpha_0} \supset \ldots \supset U_{\alpha_p} \\ U_{\alpha_0} \cap V \neq \emptyset}} K_{-q}(U_{\alpha_p})$$

and

$$E_2^{p,q} = H_Y^p(X,U,K_{-q}).$$

Taking the direct limit over all covers $\underline{U}$ of X we get the desired spectral sequence (1.8).

Now pick a cover $\underline{U}$ of X (which we suppose regular) refining $\{X, X-Y, X-Z, X-(Y \cup Z)\}$ and consider the pairing (where $\underline{V}' = \underline{U}|_{X-Y}$, $\underline{V}'' = \underline{U}|_{X-Z}$, $\underline{V} = \underline{U}|_{X-(Y \cap Z)}$) :

(1.9) $$K(B.\underline{U},B.\underline{V}') \quad K(B.\underline{U},B\underline{V}'') \longrightarrow K(B.\underline{U},B.\underline{V}).$$

which is constructed as follows. The product in K-theory induces a pairing of sheaves of infinite loop spaces on X, $K \;\; K \longrightarrow K$, and hence a pairing

(1.10) $$K(B.\underline{U},B.\underline{V}') \quad K(B.\underline{U},B.\underline{V}'') \longrightarrow K(B.\underline{U}\times B.\underline{U},B.\underline{U}\times B.\underline{V}''\cup B.\underline{V}'\times B.\underline{U}).$$

The pairing (1.10) comes from (1.9) by composing with the map induced by the diagonal

$$\Delta: \quad (B.U,B.V) \longrightarrow (B.U\times B.U,B.U\times B.V''\cup B.V'\times B.U).$$

There is a natural pairing from $E_*^{**}(X,X-Y) \otimes E_*^{**}(X,X-Z)$ to the spectral sequence $E_r^{pq}((\underline{U},\underline{V}') \times (\underline{U},\underline{V}''))$ corresponding to the product filtration of the codomain of Δ:

$$\{ \bigcup_{i+j=s} B^i.\underline{U}\times B^j.\underline{U}\cup B.\underline{U}\times B.\underline{V}''\cup B.\underline{V}'\cup B\underline{U}, B.\underline{U}\times B.\underline{V}''\cup B.\underline{V}'\times B.\underline{U}\}.$$

We must identify the E_2 term of this spectral sequence. It is the cohomology of the complex $E_1^{*,q}$:

$$E_1^{p,q} = \bigoplus_{i+j=p} \prod_{\substack{U_{\alpha_0}\supset\ldots\supset U_{\alpha_i} \\ U_{\alpha_0}\cap Y\neq\emptyset}} \prod_{\substack{U_{\beta_0}\supset\ldots\supset U_{\beta_j} \\ U_{\beta_0}\cap Z\neq\emptyset}} K_{-q}(U_{\alpha_i} \cap U_{\beta_j}) .$$

If we compare this with the spectral sequence obtained by replacing K by the presheaf of Eilenberg MacJane spaces $U \longrightarrow K(K_{-q}(U),-q)$ $(q \leq 0)$, we see that $E_2^{p,q}$ is

is

(1.11) $$\underset{\longleftarrow}{\mathrm{Lim}}^p{}_{(\underline{U}^{op}\times\underline{U}^{op},\ \underline{U}^{op}\times\underline{V}''^{op}\cup\underline{V}'^{op}\times\underline{U}^{op})}((U_\alpha\times U_\beta) \longrightarrow K_{-q}(U_\alpha\cap U_\beta))$$

where the 'relative' $\underset{\longleftarrow}{\mathrm{Lim}}^*$ is defined so as to fit into the right long exact sequence.

Now (1.11) is the E_2 term of the spectral sequence $E_r^{p,q}(\underline{\underline{U}}\times\underline{\underline{U}}, \underline{\underline{U}}\times\underline{\underline{V}}''\cup\underline{\underline{V}}'\times\underline{\underline{U}})$ converging to:

$$K_{-p-q}(\mathcal{B}.(\underline{\underline{U}}\times\underline{\underline{U}}),\mathcal{B}.(\underline{\underline{U}}\times\underline{\underline{V}}'\cup\underline{\underline{V}}'\times\underline{\underline{U}}))$$

coming from the skeletal filtration of $\mathcal{B}.(\underline{\underline{U}}\times\underline{\underline{U}})$.

Note that $\mathcal{B}.(\underline{\underline{U}}\times\underline{\underline{U}}) = \mathcal{B}.\underline{\underline{U}} \times \mathcal{B}.\underline{\underline{U}}$ and that

$$\bigcup_{i+j=p} \mathcal{B}^i\underline{\underline{U}}\times\mathcal{B}^j\underline{\underline{U}} \subset \mathcal{B}^p(\underline{\underline{U}} \times \underline{\underline{U}}) \ .$$

So there is a map of spectral sequences

$$E_r^{pq}(\underline{\underline{U}}\times\underline{\underline{U}},\underline{\underline{U}}\times\underline{\underline{V}}''\cup\underline{\underline{V}}'\times\underline{\underline{U}}) \longrightarrow E_r^{pq}((\underline{\underline{U}},\underline{\underline{V}}') \times (\underline{\underline{U}},\underline{\underline{V}}'))$$

which is an isomorphism for $r \geq 2$. Composing with the map

$$\Delta^*: \ E_r^{pq}(\underline{\underline{U}}\times\underline{\underline{U}},\underline{\underline{U}}\times\underline{\underline{V}}''\cup\underline{\underline{V}}'\times\underline{\underline{U}}) \longrightarrow E_r^{pq}(\underline{\underline{U}},\underline{\underline{V}})$$

and taking the direct limit over all covers $\underline{\underline{U}}$ gives the desired product.

Note following ([15] ch. 9) that the pairing on $E_2^{p,-q}$ terms is $(-1)^{qp'}$ times the cup product pairing:

$$H_Y^p(X,\underline{\underline{K}}_q) \otimes H_Z^{p'}(X,\underline{\underline{K}}_{q'}) \longrightarrow H_{Y\cap Z}^{p+p'}(X,\underline{\underline{K}}_{q+q'}).$$

2. The Comparison Theorem

Theorem 2.1. Let X be a regular scheme on which satisfies Gersten's condition and $Y \subset X$ a closed subscheme. Then the spectral sequences (1.3) and (1.8) coincide from E_2 onward.

Proof. We shall start with a reinterpretation of the Quillen spectral sequence. Associated to any subscheme $Y \subset X$, we have a differential spectrum:

$$R_Y^0 \longrightarrow R_Y^1 \longrightarrow \dots \longrightarrow R_Y^i \longrightarrow \dots$$

where $R_Y^i = \Omega^{-i}\,(K'^{(i-d)}(Y)/K'^{(i-d+1)}(Y))$ $(d = \operatorname{codim}_X(Y))$. Following [5], we can replace R_Y^* with a cosimplicial spectrum $\Pi^\cdot(R_Y^*)$ as follows:

$$\Pi^n(R_Y^*) = \bigvee_{0 \le p \le n} \ \bigvee_{f:\,[n] \twoheadrightarrow [p]} R_Y^{p,f}$$

where $R_Y^{p,f} = R_Y^p$, and $[n]$, $[p]$ represent the typical elements $\{0 < \dots < n\}$, $\{0 < \dots < p\}$ of the category Δ of finite totally ordered sets. If $u \in \operatorname{Hom}_\Delta([m],[n])$ we define $\Pi^\cdot(R_Y^*)(u)$ as follows. There is precisely one square for each g:

$$\begin{array}{ccc} [m] & \xrightarrow{u} & [n] \\ {\scriptstyle g}\big\downarrow & & \big\downarrow{\scriptstyle f} \\ [q] & \xrightarrow{j} & [p] \end{array}$$

If $q = p$, we map $R_Y^{q,g}$ to $R_Y^{p,f}$ by the identity. If $q = p-1$ and $j = d^0$, we map $R_Y^{q,g}$ to $R_Y^{p,f}$ by the differential in R_Y^*. In all other cases we map $R_Y^{q,g}$ to $R_Y^{p,f}$ by the trivial map to the basepoint.

Lemma 2.2. There is a weak equivalence

$$\operatorname{Tot}(\Pi^\cdot(R_Y^*)) \simeq K(X, X-Y)$$

(See [3] for the definition of Tot)

Proof. Recall ([3] X §6) that the Bousfield-Kan spectral sequence of $\Pi^\cdot(R_Y^*)$ may be constructed as follows.

Define

$$M^n\Pi^\cdot(R_Y^*) = \{(x^0,\dots,x^n) \in (\Pi^n(R_Y^*))^{n+1} \,\big|\, s^i x^j = s^{j-1} x^i \text{ for all } i,j\}$$

also define a map $\Pi^{n+1}(R_Y^*) \longrightarrow M^n\Pi^{\cdot}(R_Y^*)$ by sending x to $(s^o x,\ldots,s^m x)$. A simple calculation shows that the kernel $N^n R^*$ of this map is R_Y^n. Following (op.cit) we know that the fibre of

$$\mathrm{Tot}_n \Pi^{\cdot}(R_Y^*) \longrightarrow \mathrm{Tot}_{n-1}\Pi^{\cdot}(R_Y^*)$$

is $\Omega^n N^n R_Y^* = \Omega^n R_Y^n$ and another easy calculation shows that the differential

$$\Omega^n N^n R_Y^* \longrightarrow \Omega^n N^{n+1} R_Y^*$$

is just the n-fold loop of the differential in the complex R_Y^*. Hence $\mathrm{Tot}(\Pi^{\cdot} R_Y^*)$ and $K(X,X-Y)$ are weak equivalent since the limit of a tower of the fibratious of infinite loop spaces is determined by the corresponding complex of infinite loop spaces, and the Bousfield-Kan spectral sequence for $\Pi^{\cdot}(R_Y^*)$ is the same as the Quillen spectral sequence.

Returning to the theorem, we observe that the cosimplicial spectrum $\Pi^{\cdot}(R_Y^*)$ is contravariant, with respect to flat maps; similarly for $\Pi^{\cdot}(R_X^*)$ and each can therefore be viewed as a presheaf. Hence for each Lubkin cover $\underline{\underline{U}}$ of X refining $\{X,(X-Y)\}$ we can form the bicosimplicial space $C_Y^*(X,\underline{\underline{U}},\Pi^*(R_X^*))$:

$$C_Y^p(X,U,\Pi^q(R_X^*)) = \prod_{\substack{U_{\alpha_o} \supset \ldots \subset U_{\alpha_p} \\ U_{\alpha_o} \cap Y \neq \emptyset}} \Pi^q(R_X^*(U_{\alpha_p})).$$

The natural maps

$$K \xrightarrow{e} R_X^* \xleftarrow{f} R_Y^*$$

corresponding to the maps $K(X) \longrightarrow K^0(X)/K^1(X)$ and $K^{(i-d)}(Y)/K^{(i-d+1)}(Y) \longrightarrow K^{(i)}(X)/K^{(i)}(X)$ induce a pair of augmentations:

$$\begin{array}{ccc} C_Y^*(X,\underline{\underline{U}},K) & \xrightarrow{e} & C_Y^*(X,\underline{\underline{U}},\Pi^*(R_X^*)) \\ & & \uparrow f \\ & & \Pi^*(R_Y^*) \end{array}$$

(The vertical augmentation is via the natural map $\Pi^*(R_Y^*) \longrightarrow \prod_{U \in \underline{\underline{U}}} (\Pi^*(R_X^*(U))))$.

There is a spectral sequence

(2.3) $$E_2^{p,q} = H^p(C_Y^*(X,\underline{\underline{U}},\pi_q\Pi^*(R_X^*)))$$

$$\Rightarrow \pi_{-p-q}\mathrm{Tot}(C_Y^*(X,\underline{\underline{U}},\Pi^*(R_X^*)))$$

The E_2 term may be written

$$E_2^{p,q} = H_Y^p(X,\underline{\underline{U}},R_{-q}^*) = H_Y^p(X,K_{-q}) .$$

Hence the augmentation f induces an isomorphism on E_2 terms of the corresponding spectral sequences. Similarly, after taking the direct limit over all $\underline{\underline{U}}$, e induces an isomorphism between the local to global spectral sequence (1.8) and the spectral sequence (1.1).

Corollary 2.4. Let X be a regular noetherian scheme satisfying Gersten's condition. Then the filtration of $K_*(X)$ by codimension of support is compatible with the product $\mu: K_*(X) \otimes K_*(X) \longrightarrow K_*(X)$.

Proof. The $E_\infty^{p,q}$ term of the Quillen spectral sequence is $F^pK_{-q}(X)/F^{p+1}K_{-q}(X)$ where $F^pK_{-q}(X)$ is the subgroup of $K_{-q}(X)$ of all elements supported on closed subsets of codimension at least p. By the theorem, from E_2 on the Quillen spectral sequence coincides with the local to global sequence (1.8) which is compatible with products (1.7). Hence $\mu(F^iK_p(X) \otimes F^jK_q(X)) \subset F^{i+j}K_{p+q}(X)$.

Remark. For p = q = 0 and X quasi-projective this is in SGA6; I don't know if that method (via the moving lemma) extends to higher K-theory.

§3. Intersection Theory on Regular Schemes

Recall ([13]) that if X is a regular scheme satisfying Gersten's condition then for all codimension d closed subschemes $Y \subset X$,

$$H^p_Y(X,\mathcal{K}_p) \simeq CH^{p-d}(Y) \tag{3.1}$$

where $CH^{p-d}(Y)$ is the Chow group of codimension p-d (i.e. of codimension p on X) cycles on Y modulo rational equivalence (c.f. [7]). If Y is integral it therefore has a cycle class

$$\eta(Y) \in H^p_Y(X,\underline{\mathcal{K}}_p) .$$

Given two integral subschemes Y and Z of X of codimension p and q respectively which intersect properly one can ask what the relationship between $\eta(Y) \cup \eta(Z)$ and $\eta(Y.Z) \in H^{p+q}_{Y\cap Z}(X,\mathcal{K}_{p+q})$ is. Here Y.Z is defined via Serre's formula ([14] ch V,C)

$$Y.Z = \sum_{x \in (Y\cap Z)} \chi^{X,x}(\mathcal{O}_Y,\mathcal{O}_Z)[\bar{x}] \tag{3.2}$$

where the sum runs over the generic points x of $Y \cap Z$.

The cycle class $\eta(Y)$ may be defined as follows. The Quillen spectral sequence (1.3):

$$E_2^{p,q} = H^p_Y(X,\mathcal{K}_{-q}) \Rightarrow K_{-p-q}(X,X-Y) \simeq K'_{-p-q}(Y)$$

has $E_r^{p,q} = 0$ for $p < \operatorname{codim}_X(Y) = d$. Hence there is an edge homomorphism

$$\eta_q:\ K'_q(Y) \longrightarrow H^p_Y(X,\mathcal{K}_{q+p})$$

and the image of $[\mathcal{O}_Y]$ under η_o is $\eta(Y)$. Following (2.1) we know that

$$\eta([\mathcal{O}_Y].[\mathcal{O}_Z]) = (-1)^{pq}\eta(Y) \cup \eta(Z).$$

In order to give content to this equation we must compute $[\mathcal{O}_Y].[\mathcal{O}_Z] \in K_o(X,X-(Y\cap Z)) \simeq K'_o(Y\cap Z)$. Choose a finite resolution $\mathcal{P}_* \longrightarrow \mathcal{O}_Y$ by locally free $\mathcal{O}_X$ modules, then as an element of $K_o(X)$ $[\mathcal{O}_Y]$ is represented by a loop γ_p in BQP(X):

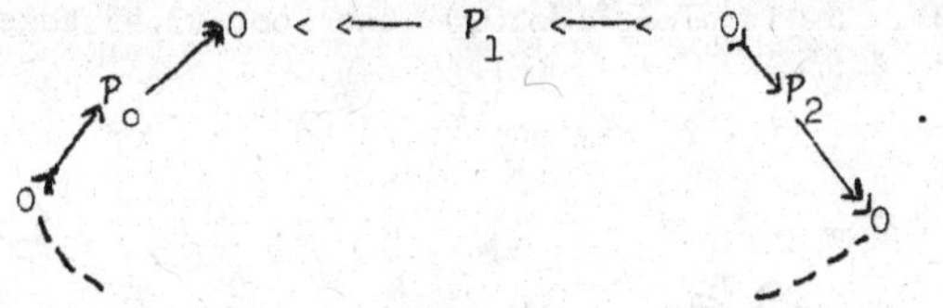

On X-Y the complex P_* is exact and hence there is a nullhomotopy $\eta\colon \gamma_p \simeq *$;

$$\begin{array}{ccc} S^1 & \xrightarrow{\gamma_p} & BQ\underline{P}(X) \\ \downarrow & & \downarrow \\ C(S^1) & \xrightarrow{\eta} & BQ\underline{\underline{P}}(X-Y) \end{array}$$

where $C(S^1)$ is the reduced cone over S^1, and j^* is the map induced by $j\colon (X-Y) \longrightarrow X$. η may be constructed by successfully filling in triangles in $BQ\underline{\underline{P}}(U)$.

(3.3)

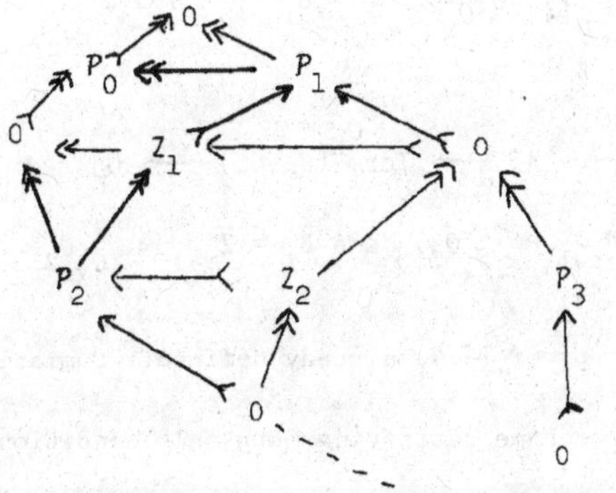

(where $Z_i = \mathrm{Ker}(P_i \longrightarrow P_{i-1}) = \mathrm{Im}(P_{i+1} \longrightarrow P_i)$). The nullhomotopy η defines a loop in the fibre of j^*, which is weak equivalent to $BQ\underline{\underline{M}}(Y)$. The product $[O_Y].[O_Z] \in K_o(X,X-(Y\cap Z)) \simeq K'_o(Z,Z-(Y\cap Z))$ is then represented by the loop:

(3.4)

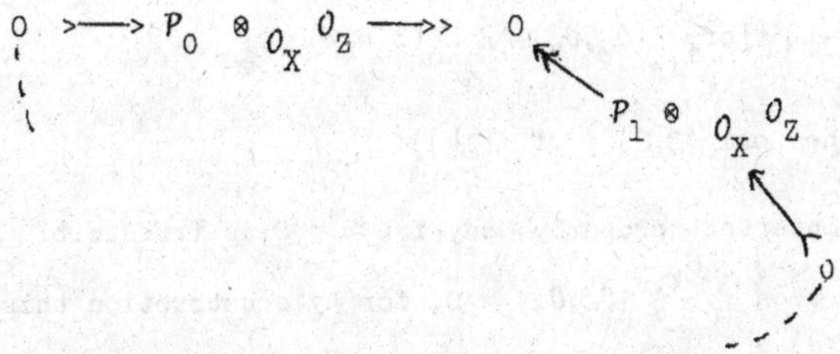

together with the induced null homotopy (note that the Z_i are locally free O_{X-Y} modules so tensoring (3.3) with O_Z defines a diagram in $BQ\underline{\underline{P}}(Z-(Y\cap Z)))$ over $Z-(Y\cap Z)$.

Since the homology of $P_* \otimes_{\mathcal{O}_X} \mathcal{O}_Z$ is $\mathrm{Tor}_*^{\mathcal{O}_X}(\mathcal{O}_Y,\mathcal{O}_Z)$ the loop (3.4) together with the associated null homotopy over Z- (Y ∩ Z) may homotoped to the loop

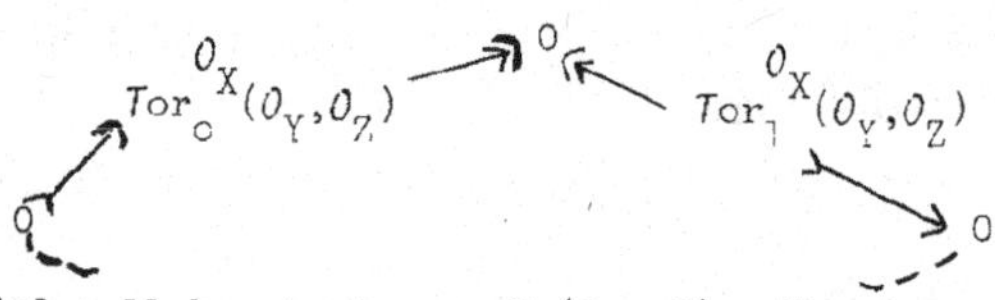

and the trivial null homotopy over Z-(Y ∩ Z). The homotopy is defined by fitting together the singular 2-simplices in $BQ\underline{\underline{M}}(Z)$ corresponding to the short exact sequences

$$0 \rightarrowtail A_i \twoheadrightarrow P_i \otimes_{\mathcal{O}_X} \mathcal{O}_Z \twoheadrightarrow B_i \twoheadrightarrow 0$$

and

$$0 \rightarrowtail B_{i+1} \twoheadrightarrow A_i \twoheadrightarrow \mathrm{Tor}_i^{\mathcal{O}_X}(\mathcal{O}_Y,\mathcal{O}_Z) \twoheadrightarrow 0$$

where $A_i = \mathrm{Ker}(P_i \otimes_{\mathcal{O}_X} \mathcal{O}_Z \longrightarrow P_{i-1} \otimes_{\mathcal{O}_X} \mathcal{O}_Z)$ and $B_i = Z_{i-1} \otimes_{\mathcal{O}_X} \mathcal{O}_Z$. Over Z-Y this restricts to the null homotopy of (3.4) already defined. Summarizing:

<u>Theorem 3.5</u>. Let X be a regular scheme satisfying Gersten's condition. Then for all pairs of properly intersecting subschemes $Y,Z \subset X$ of codimension p and q respectively we have

$$\eta(Y) \cup \eta(Z) = (-1)^{pq}\, \eta(Y.Z) \in H^{p+q}_{Y \cap Z}(X,\mathcal{K}_{p+q})$$

where Y.Z is the cycle corresonding to

$$\sum_{i=0}^{\infty} (-1)^i [\mathrm{Tor}_i^{\mathcal{O}_X}(\mathcal{O}_Y,\mathcal{O}_Z)] \in K_0'(Y \cap Z)$$

which may be written in the form (3.2) (c.f. [14]).

<u>Note</u>. If Y and Z do not intersect properly, say Y ∩ Z = S is irreducible of codimension r less than p+q then $\chi^{\mathcal{O}_{X,S}}(\mathcal{O}_Y,\mathcal{O}_Z) = 0$, for by construction this is the image of the product $[\mathcal{O}_Y].[\mathcal{O}_Z]$ in $H^r_S(X,\mathcal{K}_r) \simeq \mathbb{Z}$, but by (2.4) the product lies in $F^{p+q}K_0(X,X\text{-}S)$ which maps to zero in $H^r_S(X,\mathcal{K}_r)$. This indicates that there is a connection between Gersten's conjecture and Serre's conjectures ([14] V B)§4), a fact which is perhaps not surprising.

4. The Covariance of the K'-theory

Given a proper morphism $f: X \to Y$ between noetherian schemes it is natural to ask whether there is natural homomorphism for all $q \geq 0$,

$$f_*: K'_q(X) \to K'_q(Y) .$$

In [13] Quillen constructed this map for projective morphisms, and for K_o we can define ($\mathfrak{F}$ a coherent sheaf of $\mathcal{O}_X$ modules):

$$f_*[\mathfrak{F}] = \sum_{i\geq 0} (-1)^i [R^i f_* \mathfrak{F}] .$$

Here we show how to pass from the projective situation to the general case. I would like to thank Bob Thomason for pointing out a gap in the proof of an earlier version of theorem 4.1. For generalities on homotopy colimits I refer to [3] and the papers:

[17] R. W. Thomason; Homotopy colimits in the category of small categories. Math. Proc. Comb. Phil. Soc. (1979), 85.

[18] R. W. Thomason; Cat as a closed model category; preprint.

[19] R. W. Thomason; First quadrant spectral sequences in algebraic K-theory via homotopy theory; preprint.

Theorem 4.1. There exists a canonical extension of the functor $\mathcal{K}'$ to the category of all morphisms of noetherian schemes.

Proof. Our primary objective is to construct for each proper morphism $f: X \to Y$ of schemes a map in the stable homotopy category Sho ([1]):

$$f_*: \mathcal{K}'(X) \to \mathcal{K}'(Y)$$

with suitable properties.

Using Chow's lemma we may construct a finite filtration of X by closed subschemes:

$$\phi = X_{-1} \subset X_o \subset \ldots \subset X_i \subset \ldots \subset X_n = X$$

with the following properties.

i) $X_i - X_{i-1}$ is quasi-projective over Y for i=1...n

ii) there exists, for each $i \geq 1$ a commutative diagram ($f_i = f|_{X_i}$):

$$\begin{array}{ccccc} W_i & \hookrightarrow & Z_i & & \\ {\scriptstyle \omega_i}\downarrow & & \downarrow{\scriptstyle \pi_i} & \searrow{\scriptstyle g_i} & \\ X_{i-1} & \hookrightarrow & X_i & & \\ & {\scriptstyle f_{i-1}}\searrow & & \searrow{\scriptstyle f_i} & \\ & & & & Y \end{array}$$

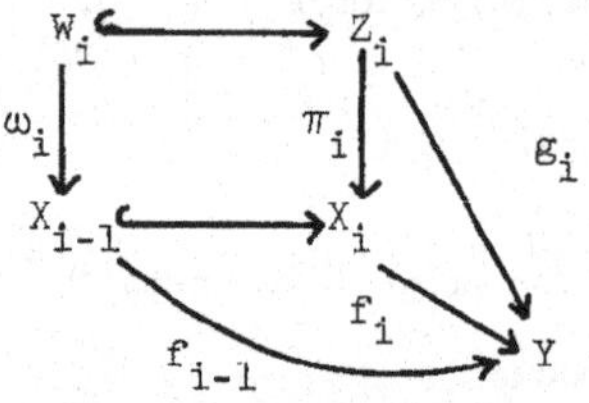

where a) π_i is an isomorphism over $X_i - X_{i-1}$

b) $W_i = Z_i \times_{X_i} X_{i-1}$

c) π_i and g_i are projective morphisms.

Proceeding by induction on i we may define diagrams

$$D_i : \underline{\underline{\Sigma}}_i \to \underline{\underline{Sch}}/Y$$

where each $\underline{\underline{\Sigma}}_i$ is a finite category; as usual we shall frequently identify D_i with a configuration of arrows in $\underline{\underline{Sch}}/Y$. For i=1, D_1 is the diagram:

$$\begin{array}{ccc} W_1 & \to & Z_1 \\ \downarrow & & \\ X_o & & \end{array}$$

Having defined D_{i-1} we have a diagram:

(4.2)
$$\begin{array}{ccccc} & & W_i & \hookrightarrow & Z_i \\ & & \swarrow & & \swarrow \\ D_{i-1} & \searrow & X_{i-1} & \hookrightarrow & X_i \end{array}$$

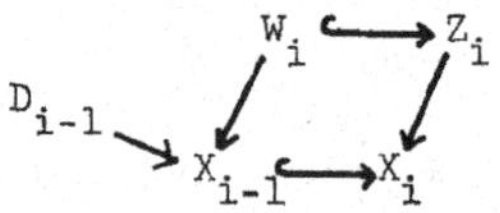

Note that the arrow $D_{i-1} \to X_{i-1}$ actually represents a 'cone' i.e. a compatible family of maps. More generally if D: $\underline{\underline{\Sigma}} \to \underline{\underline{C}}$ and D': $\underline{\underline{\Sigma}}' \to \underline{\underline{C}}$ are diagrams, a morphism f: D $\to$ D' consists of a functor F: $\underline{\Sigma} \to \underline{\Sigma}'$, and a natural transformation $\phi : D \to D' \cdot F$. We can think of a cone under a diagram D as a map from D to a 'punctual' diagram. From (4.2) we form the diagram D_i as:

$$\begin{array}{ccc} & D_{i-1} \times_{X_{i-1}} W_i \to Z_i \\ \swarrow & \\ D_{i-1} & \end{array}$$

Notice that as a scheme X_i is the colimit of D_i and that D_i is a diagram of schemes projective over S.

<u>Definition (4.3)</u>. We shall refer to D_n as a 'projective decomposition' (of length n) of X relative to Y.

<u>Note (4.4)</u>. If $p: P \to X$ is a projective map then $D_n \times_X P$ is a projective decomposition of P relative to X (and also relative to Y).

Consider the general situation of a diagram:

$$D: \underline{\underline{\Sigma}} \to \underline{\underline{Sch}}/Y$$

We wish to define the K'-theory of D. Let us suppose that D is actually a diagram of schemes projective over Y and that D is a subdiagram of a diagram $E: \underline{\underline{\Sigma}}' \to \underline{\underline{Sch}}/Y$. If s is an object of $\underline{\underline{\Sigma}}$ (or s' of $\underline{\underline{\Sigma}}'$) we write X_s for D(s) and $X_{s'}$ for E(s'). For $s \in Ob(\underline{\underline{\Sigma}})$ let $\underline{\underline{A}}_{D,E}(X_s)$ be the full exact subcategory of $\underline{\underline{M}}(X)$ consisting of those coherent $\mathcal{O}_X$ modules $\mathcal{F}$ such that for all morphisms σ in $\underline{\underline{\Sigma}}'$ with domain $s \in Ob(\underline{\underline{\Sigma}}) \subset Ob(\underline{\underline{\Sigma}}')$:

$$R^i D(\sigma)_* \mathcal{F} = 0 \qquad \text{for all } i > 0$$

<u>Lemma 4.5</u>. Let $\sigma: s \to t$ be a morphism of $\underline{\underline{\Sigma}}$ (not $\underline{\underline{\Sigma}}'$!). Then $D(\sigma)_*: \underline{\underline{M}}(X_s) \to \underline{\underline{M}}(X_t)$ induces an exact functor

$$D(\sigma)_*: \underline{\underline{A}}_{D,E}(X_s) \to \underline{\underline{A}}_{D,E}(X_t) .$$

<u>Proof of Lemma</u>. If $\mathcal{F} \in Ob(\underline{\underline{A}}_{D,E}(X_s))$ we must show that

$$D(\sigma)_*(\mathcal{F}) \in Ob(\underline{\underline{A}}_{D,E}(X_t)) .$$

If $\tau: t \to t'$ is a map in $\underline{\underline{\Sigma}}'$ we have the Leray spectral sequence:

$$E_2^{i,j} = R^i D(\tau)_* R^j D(\sigma)_* \mathcal{F} \to R^{i+j} D(\tau.\sigma)_* \mathcal{F}$$

By assumption $E_2^{i,j} = 0$ for $j > 0$; hence

$$R^i D(\tau)_*(D(\sigma)_* \mathcal{F}) = R^i D(\tau\sigma)_* \mathcal{F} = 0 \quad \text{for } i > 0 .$$

Lemma 4.6. For all $s \in \mathrm{Ob}(\underline{\underline{\Sigma}})$, the exact inclusion functor $\underline{\underline{A}}_{D,E}(X_s) \to \underline{\underline{M}}(X_s)$ induces a weak equivalence:

$$BQ\underline{\underline{A}}_{D,E}(X_s) \xrightarrow{\sim} BQ\underline{\underline{M}}(X_s) .$$

Proof of Lemma. First note that any invertible sheaf $\mathcal{L}$ on X_s which is ample relative to Y is ample relative to any $X_{s'}$, for any morphism σ: $s \to s'$ in $\underline{\underline{\Sigma}}'$. Since these morphisms are finite in number given $\mathcal{F} \in \mathrm{Ob}(\underline{\underline{M}}(X_s))$ we can find an integer n_o such that $R^i D(\sigma)_*(\mathcal{F} \otimes_{\mathcal{O}_X} \mathcal{L}^{\otimes n}) = 0$ for all $n \geq n_o$ all σ in $\underline{\underline{\Sigma}}'$ with domain s and so that $\mathcal{F}$ embeds into $(\mathcal{F} \otimes_{\mathcal{O}_X} \mathcal{L}^{\otimes n})^{\otimes rn}$ for some r, as in ([13] §7). The proof finishes as in (op.cit).

Given a pair of diagrams (D, E) as above, we can define a diagram of spectra (if E = D we just write $\underline{\underline{A}}_D$):

$$\mathcal{K}(\underline{\underline{A}}_{D,E}): \underline{\underline{\Sigma}} \to \underline{\underline{\mathrm{Spectra}}}$$

$$s \to \mathcal{K}(\underline{\underline{A}}_{D,E}(X_s))$$

Definition 4.7. The K'-theory of the diagram D is:

$$\mathcal{K}'(D) = \mathrm{Hocolim}(\mathcal{K}(\underline{\underline{A}}_{D,E}))$$

(Note that $\mathcal{K}'(D)$ is a priori only a spectrum, rather than an Ω-spectrum; [19]). We have suppressed the E from $\mathcal{K}'(D)$ since it is clear from the definition that the natural map

$$\mathrm{Hocolim}(\mathcal{K}(\underline{\underline{A}}_{D,E})) \to \mathrm{Hocolim}(\mathcal{K}(\underline{\underline{A}}_D))$$

is a weak equivalence.

If f: D $\to$ E is a map of diagrams of schemes projective over Y there is a corresponding map of diagrams of spectra (we view D as a subdiagram of the composite diagram D $\to$ E)

$$\mathcal{K}'(\underline{\underline{A}}_{D,D\to E}) \to \mathcal{K}'(\underline{\underline{A}}_E)$$

and hence a map

$$\mathcal{K}'(D) \to \mathcal{K}'(E) .$$

One may easily check that $\mathcal{K}'$ becomes a covariant functor from the category of diagrams of schemes projective over Y to $\underline{\underline{\mathrm{Sho}}}$. Let f: X $\to$ Y be a proper morphism, and D $\to$ X a projective decomposition of X relative to Y. Then we have the natural map:

$$\mathcal{K}'(D) = \mathrm{Hocolim}(\mathcal{K}(\underline{\underline{A}}_{D,D\to X})) \to \mathcal{K}(\underline{\underline{M}}(X)) \tag{4.8}$$

<u>Lemma 4.9</u>. The map (4.8) is a weak equivalence.

<u>Proof</u>. By induction on the length n of D $\to$ X. (n=1) D_1 is the diagram:

$$\begin{array}{ccc} W_1 & \to & Z_1 \\ \downarrow & & \\ X_o & & \end{array}$$

We have the spectral sequence for the stable homotopy of $\mathcal{K}'(D_1)$, which in this case is a Meyer-Vietoris sequence ([19]):

$$\to K'_*(W_1) \to K'_*(X_o) \oplus K'_*(Z_1) \to K'_*(D_1) \to \cdots$$

On the other hand, if we write $U = Z_1 - W_1 = X_1 - X_o$ we have a map of localization sequences:

$$\begin{array}{ccccccc} \to K'_*(W_1) & \to & K'_*(Z_1) & \to & K'_*(U) & \to \\ \downarrow & & \downarrow & & \| & \\ \to K'_*(X_o) & \to & K'_*(X_1) & \to & K'_*(U) & \to \end{array}$$

and hence a Meyer-Vietoris sequence:

$$\to K'_*(W_1) \to K'_*(X_o) \oplus K'_*(Z_1) \to K'_*(X_1) \to \cdots$$

Comparing these two sequences we see that the natural map $K'_*(D_1) \to K'_*(X_1)$ is an isomorphism. Now let $m > 1$ and suppose the map $K'_*(D_{m-1}) \to K'_*(X_{m-1})$ is an isomorphism. Then we have the diagram D_m:

$$\begin{array}{ccc} D_{m-1} \times_{X_{m-1}} W_m & \to & Z_m \\ \downarrow & & \\ D_{m-1} & & \end{array}$$

D_m maps to the diagram E_m (with domain $\underline{\underline{\Sigma}}_1$):

(4.10)
$$\begin{array}{ccc} W_m & \rightarrow & Z_m \\ \downarrow & & \\ X_{m-1} & & \end{array}$$

The corresponding functor $\underline{\underline{\Sigma}}_m \rightarrow \underline{\underline{\Sigma}}_1$ makes $\underline{\underline{\Sigma}}_m$ a category cofibered over $\underline{\underline{\Sigma}}_1$. Following the homotopy colimit theorem ([17]) one knows that if F is a diagram of categories:

$$F: \underline{\underline{\Sigma}} \rightarrow \underline{\underline{\mathrm{Cat}}}$$

which corresponds to a category $\underline{\underline{\Sigma}} \int F$ cofibered over $\underline{\underline{\Sigma}}$, then there is a natural homotopy equivalence:

$$\eta: \quad \mathrm{Hocolim}\, B.F \rightarrow B.(\underline{\underline{\Sigma}} \int F)$$

Using the fact that any diagram of simplicial sets can be replaced by a weak equivalent diagram of categories ([18]) and applying the theorem of (op. cit) twice, one arrives at the following:

Generalized Homotopy Colimit Theorem (4.11). Let $F: \underline{\underline{\Sigma}} \rightarrow$ Simplicial Sets be a diagram. Suppose that $\underline{\underline{\Sigma}}$ is cofibered over a category $\underline{\underline{\Delta}}$. Then there is a weak equivalence (d represents the typical object of $\underline{\underline{\Delta}}$ and $\underline{\underline{\Sigma}}_d$ the fibre of $\underline{\underline{\Sigma}}$ over d):

$$\mathrm{Hocolim}(d \rightarrow \mathrm{Hocolim}\, F|_{\underline{\underline{\Sigma}}_d}) \simeq \mathrm{Hocolim}\, F\ .$$

This theorem extends to diagrams of spectra by applying it degreewise. If we view E_m as diagram of schemes over X_m then by the m = 1 case of the lemma:

$$\mathcal{K}'(E_m) \simeq \mathcal{K}'(X_m)\ .$$

Now the natural map $D_m \rightarrow E_m$ (corresponding to a cofibration $\underline{\underline{\Sigma}}_m \rightarrow \underline{\underline{\Sigma}}_1$) induces a map

(4.12)
$$\mathcal{K}'(D_m) \rightarrow \mathcal{K}'(E_m)\ .$$

Hence, following (4.11), if we can prove that the maps

(4.13)
$$\mathcal{K}'(D_{m-1}) \rightarrow \mathcal{K}'(X_{m-1})$$

(4.14) $$\mathcal{K}'(D_{m-1} \times_{X_{m-1}} W_m) \to \mathcal{K}'(W_m)$$

(4.15) $$\mathcal{K}'(Z_m) \to \mathcal{K}'(Z_m)$$

are weak equivalences, and hence that (4.12) is too, we will be done. But (4.13) is a weak equivalence by the induction hypothesis, (4.14) is by the induction hypothesis since $D_{m-1} \times_{W_{m-1}} W_m$ is a projective decomposition of W_m of length m-1, and (4.15) is trivial.

Finally we can define the map

$$f_*: \mathcal{K}'(X) \to \mathcal{K}'(Y)$$

by inverting (in the homotopy category) the map

$$\mathcal{K}'(D) \to \mathcal{K}'(X)$$

and following it by the natural map

$$\mathcal{K}'(D) \to \mathcal{K}'(Y) .$$

We now have to check that:

1) The map $f_*: \mathcal{K}'(X) \to \mathcal{K}'(Y)$ is independent of the choice of projective decomposition $D \to X$.

2) If $f: X \to Y$, $g: Y \to Z$ are proper then $g_* f_* = (gf)_*$.

3) If $f: X \to Y$ is projective our definition coincides with the definition of ([13] §7).

4) f_* is a map of $\mathcal{K}(Y)$ module spectra.

Proof of 1). Let $D_i: \Sigma_i \to \underline{\mathrm{Sch}}/Y$, i = 1, 2 be two projective decompositions of X relative to Y; form the fiber product:

$$D_1 \times_X D_2: \underline{\Sigma}_1 \times \underline{\Sigma}_2 \to \underline{\mathrm{Sch}}/Y$$

We have a commutative diagram:

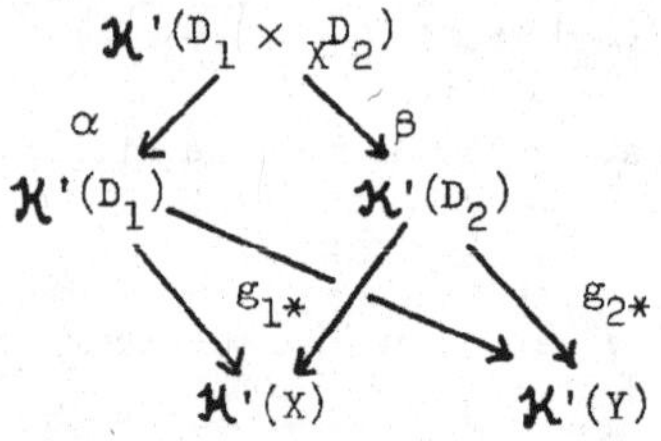

Following (4.11) or ([3] XII) we can compute $\mathcal{H}'(D_1 \times_X D_2)$ as an iterated Hocolim; but by (4.4) and (4.9) for each $s \in \mathrm{Ob}\,(\underline{\underline{\Sigma}}_1)$:

$$\mathcal{H}'(D_1(s) \times_X D_2) \simeq \mathcal{H}'(D_1(s))$$

Hence α (and by symmetry β) is a weak equivalence, and the two maps in $\underline{\underline{\mathrm{Sho}}}$:

$$f_*: \mathcal{H}'(X) \to \mathcal{H}'(Y)$$

induced by g_{1*} and g_{2*} coincide.

<u>Proof of 2)</u>. Consider projective decompositions $D \to X$ of X relative to Z (and hence relative to Y) and $E \to Y$ relative to Z.

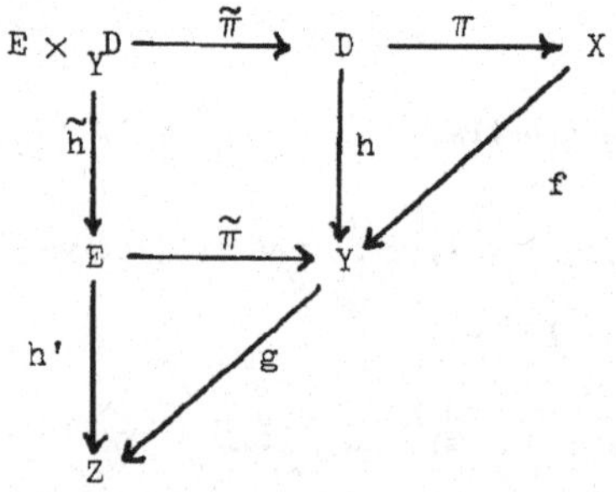

Since D is projective over Y and Z, the maps $f_*: \mathcal{K}'(X) \to \mathcal{K}'(Y)$ and $(gf)_*$: $\mathcal{K}'(X) \to \mathcal{K}'(Z)$ are defined as $f_*\pi_*^{-1}$ and $(gh)_*\pi_*^{-1}$ respectively. As in 1), the map

$$\tilde{\pi}_*: \mathcal{K}'(E \times_Y D) \to \mathcal{K}'(D)$$

is a weak equivalence and applying $\mathcal{K}'$ to diagram (4.16) we get a commutative diagram of spectra in which the horizontal maps are weak equivalences. Then:

$$\begin{aligned} g_*f_* &= h'_*\pi'^{-1}_* h_*\pi_*^{-1} \\ &= h'_*\tilde{h}_*\tilde{\pi}_*^{-1}\pi_*^{-1} \\ &= (gh)_*\pi_*^{-1} \\ &= (gf)_* \end{aligned}$$

<u>Proof of 3)</u>. This is a trivial consequence of 1), regarding $X \xrightarrow{id} X$ as a projective decomposition of X relative to Y.

<u>Proof of 4)</u>. Given a projective decomposition $D \to X$ of X relative to Y, we have for every object Z of D a diagram of functors with biexact rows and exact columns:

$$\begin{array}{ccc} \underline{\underline{P}}(Y) \times \underline{\underline{A}}_D(Z) & \xrightarrow{\otimes_{\mathcal{O}_Y}} & \underline{\underline{A}}_D(Z) \\ \downarrow 1\times g_* & & \downarrow g_* \\ \underline{\underline{P}}(Y) \times \underline{\underline{M}}(Y) & \xrightarrow{\otimes_{\mathcal{O}_Y}} & \underline{\underline{M}}(Y) \end{array}$$

(where $g: Z \to Y$ is the natural map) hence a commutative diagram of maps of diagrams of spectra:

$$\begin{array}{ccc} \mathcal{K}(Y) \wedge \mathcal{K}(\underline{\underline{A}}_D) & \to & \mathcal{K}(\underline{\underline{A}}_D) \\ \downarrow & & \downarrow \\ \mathcal{K}(Y) \wedge \mathcal{K}'(Y) & \to & \mathcal{K}'(Y) \end{array}$$

Now in general if $\mathcal{E}$ is a diagram of spectra, and $\mathcal{F}$ a fixed spectrum (see [19]):

$$\mathrm{Hocolim}(\mathcal{E} \wedge \mathcal{F}) \simeq \mathrm{Hocolim}(\mathcal{E}) \wedge \mathcal{F}$$

Hence there is a commutative diagram

$$\begin{array}{ccccc}
\mathcal{K}(Y) \wedge \mathcal{K}'(X) & \xrightarrow{\sim} & \mathrm{Hocolim}(\mathcal{K}(Y) \wedge \mathcal{K}(\underline{\underline{A}}_D)) & \rightarrow & \mathcal{K}'(D) \\
\downarrow 1\wedge f_* & & \downarrow & & \downarrow f_* \\
\mathcal{K}(Y) \wedge \mathcal{K}'(Y) & \xrightarrow{\sim} & \mathcal{K}(Y) \wedge \mathcal{K}'(Y) & \rightarrow & \mathcal{K}'(Y)
\end{array} \quad .$$

§5. Riemann-Roch theorem for non-projective varieties

Corollary (2.4) permits a very simple proof of the Grothendieck Riemann Roch theorem for proper but not necessarily projective morphisms between smooth varieties over a fixed base field k. For simplicity we consider only etale cohomology, though the result is true for any cohomology theory satisfying the axioms of [9].

Recall that there are Chern classes ($\ell \neq$ char k)

$$C_{i,p}: K_p(X) \longrightarrow H^{2i-p}(X, \mathbb{Z}_\ell(i))$$

and a Chern character:

$$ch_p: K_p(X) \longrightarrow \bigoplus_{2i \geq p} H^{2i-p}(X, \mathbb{Q}_\ell(i))$$

defined on the category of varieties over k. By ([9] §4) we know that if $f: X \longrightarrow Y$ is a projective morphism between non-singular varieties over k that for all $\alpha \in K_p(X)$:

$$(5.1) \qquad f_!(ch(\alpha) \cup Td(X)) = ch(f_*(\alpha)) \cup Td(Y)$$

where $Td(X)$ ($Td(Y)$) is the Todd genus of the tangent bundle to X (Y respectively) and f_*, $f_!$ are the direct image maps in K-theory and etale cohomology respectively.

<u>Theorem 5.2</u>. Formula (5.1) above is true even if f is only proper, rather than projective. (Note that in this case f_* is defined by virtue of (4.1)).

<u>Proof</u>. By Chow's lemma there exists a projective birational morphism $\pi: \tilde{X} \longrightarrow X$ such that $g = f \cdot \pi$ is projective. Then the direct image under π of the unit in $K_o(\tilde{X})$ is:

$$\lambda = \pi_*[\mathcal{O}_{\tilde{X}}] = \sum_{i=0}^{\infty} [R^i f_* \mathcal{O}_{\tilde{X}}](-1)^i.$$

Since π is proper and birational and X is nonsingular $f_*\mathcal{O}_{\tilde{X}} = \mathcal{O}_X$, while for $i > 0$ the $R^i f_* \mathcal{O}_X$ are supported on proper closed subsets of X. Hence $\lambda = [\mathcal{O}_X] + \eta$ where $\eta \in F^1 K_o(X)$. Since the coniveau filtration on the K-theory of X is compatible with the product structure, η is nilpotent and λ is invertible. Given

$\alpha \in K_q(X)$, we have α equal to $\pi_*(\pi^*(\alpha\lambda^{-1}))$ by the projection formula.

Hence

$$
\begin{aligned}
ch_q(f_*(\alpha)) \cup Td(Y) &= ch_q(g_*\pi^*(\alpha\lambda^{-1})) \cup Td(Y) \\
&= g_!(ch_q(\pi^*(\alpha\lambda^{-1})) \cup Td(\tilde{X})) \text{ by } (5.1) \\
&= f_!(\pi_! ch_q(\pi^*(\alpha\lambda^{-1}) \cup Td(\tilde{X})) \\
&= f_!(ch_q(\pi_*\pi^*(\alpha\lambda^{-1})) \cup Td(X)) \text{ by } (5.1) \\
&= f_!(ch_q(\alpha) \cup Td(X)) .
\end{aligned}
$$

References

1. J.F. Adams; Stable homotopy and generalized homology, University of Chicago Lecture Notes, Chicago, 1971.

2. S. Bloch and A. Ogus; Gersten's conjecture and the homology of schemes, Ann. Scient. Ec. Norm. Sup t.7 (1974), 181-202.

3. A.K. Bousfield and D.M. Kan; Homotopy limits, completions and localizations, Lectures Notes in Mathematics 304 (1975), Springer-Verlag, Berlin.

4. K.S. Brown and S.M. Gersten; Algebraic K-theory as generalized cohomology, Lectures Notes in Mathematics 341 (1973), Springer Verlag, Berlin.

5. A. Dold and D. Puppe; Homologie nicht-additiver Functoren anwendungen, Ann. Inst. Fourrier 11 (1961), 201-312.

6. A. Dold; Halbexakte Homotopiefunktoren, Lectures Notes in Mathematics 12 (1966), Springer Verlag, Berlin.

7. W. Fulton; Rational equivalence on singular varieties, Publ. Math. I.H.E.S. No. 45 (1975), 147-167.

8. W. Fulton; A Hirzebruch-Riemann-Roch formula for analytic spaces and non-projective algebraic varieties, Compositio Math. 34 (1977), 279-284.

9. H. Gillet; The Applications of algebraic K-theory to intersection theory, Thesis, Harvard University, (1978).

10. H. Gillet; Riemann Roch theorems for higher algebraic K-theory, to appear in Bull. A.M.S., 1980. (the above is an announcement; a full account is to appear soon).

11. D. Grayson; Products in K-theory and intersecting algebraic cycles, Inv. Math. 47 (1978), 71-84.

12. S. Lubkin; A p-adic proof of Weil's conjectures, Ann. of Math. 87 (1968), 102-255.

13. D. Quillen; Higher algebraic K-theory I, Lectures Notes in Mathematics 341 (1973), Springer Verlag, Berlin.

14. J.P. Serre; Algèbre Local, Multiplicités, Lectures Notes in Mathematics 11 (3rd edition, 1975) Springer Verlag, Berlin.

15. E. Spanier, Algebraic Topology, McGraw-Hill, New York (1966).

16. F. Waldhausen; Algebraic K-theory and generalized free products, Ann. of Math. 108 (1978), 135-256.

Department of Mathematics
Princeton University
Princeton, NJ 08544

Dilogarithm Computations for K_3

Daniel R. Grayson

Columbia University

In this note are presented the results of some machine computations Spencer Bloch suggested I do. They provide convincing evidence for Lichtenbaum's Conjecture for K_3 and K_2 of a quadratic imaginary field F. This is the case where a higher regulator occurs in the formula, and Bloch's work on the dilogarithm gives a way to approximate it (or at least some integer multiple of it).

In addition, Benedict Gross has explained to me how to compute zeta functions for non-totally-real cubic fields, and suggested two fields which have an obvious element in K_3. I present his explanation, and the corresponding computer result. (Since we lack computations of K_2 in this case, there is no confirmation of the conjectures.)

The group $\mathfrak{B}(F)$ is defined by Bloch [Bl] so that

$$0 \longrightarrow \mathfrak{B}(F) \longrightarrow \mathbb{Z}^{(F\setminus 0,1)} \xrightarrow{\ \varphi\ } F^* \otimes_{\mathbb{Z}} F^*$$

is exact, where $\varphi([\lambda]) = \lambda \otimes 1 - \lambda$. Borel's work [Bo] yields a regulator map

$$K_3F \xrightarrow{R} \mathbb{R}$$

(because $K_3\mathcal{O}_F \to K_3F$ is surjective and has finite kernel; Soulé has recently shown the map is injective), whose image is a lattice, the volume R_1 of which figures in Lichtenbaum's conjecture [L] (as modified by Borel)

$$\frac{\#(K_3\mathcal{O}_F)_{tors}}{\#K_2\mathcal{O}_F} = \frac{\pi R_1}{\zeta_F'(-1)}.$$

Let c denote the right hand side of this equation: Borel [Bo 2] has shown it is a rational number.

Bloch [unpublished] has defined a map

$$\mathfrak{B}(F) \longrightarrow K_3F$$

and expects

$$\begin{array}{ccc} \mathfrak{B}(F) & \longrightarrow & K_3F \\ & {\scriptstyle D}\searrow \quad \swarrow{\scriptstyle R} & \\ & \mathbb{R} & \end{array}$$

to commute up to some universal real constant. Here D is Bloch's version of the dilogarithm function, and

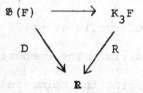

$$D(\sum a_i[\lambda_i]) = \sum a_i D(\lambda_i)$$

shows how it is computed on $\mathfrak{P}(F)$. At this point, we postulate that $D = \pi^2 R$. (We may consider this factor to have been chosen a posteriori to make the answer for one of the fields correct, but then the answer is right for the other four fields as well.)

If S is a finite subset of $\mathfrak{B}(F)$, we let $D(S)$ denote a generator for the lattice in $\mathbb{R}$ spanned by $\{D(s) | s \in S\}$. If the image of S in K_3F spans it mod torsion, then we expect

$$\frac{\#(K_3\mathcal{O}_F)_{tors}}{\#K_2\mathcal{O}_F} = \frac{D(S)}{\pi\zeta_F'(-1)}.$$

Let $c(S)$ denote the right hand side of this equation.

Now suppose F has class number one. Then

$$F^* = \mu_F \oplus \mathbb{Z}^{(\text{primes})}$$

and $F^* \otimes_{\mathbb{Z}} F^*$ is a direct sum of cyclic groups. This means φ can be expressed by some matrix. The computer can factor numbers in F (by first factoring the norm, and then paying attention to which primes split), and thus can compute the entries of the matrix φ.

Let $X \subset (F\setminus 0,1)$ be a finite subset; the machine finds an integral basis L_X for the kernel $\mathfrak{B}_X(F)$ of

$$\mathbb{Z}^X \longrightarrow F^* \underset{\mathbb{Z}}{\otimes} F^*$$

if X is small enough. I was able to get by with X lying in $\mathcal{O}_F$ and having up to 100 elements, in which case the matrix is about 100 by 120. The machine computes the numbers

$$D(v) \qquad \left(v \in L_X\right)$$

up to about 6 digits accuracy, and also finds $c_X = c(L_X)$.

The functional equation yields

$$\zeta_F'(-1) = \frac{|d|^{3/2}}{48\,\pi} L(2,\chi)$$

where d is the discriminant of F, and χ is the quadratic character for F,

$$\chi(p) = \left(\frac{d}{p}\right) \qquad \text{(p odd prime)}.$$

One can compute $L(2,\chi)$ directly, or express it [Bl, p.77] in terms of the dilogarithm:

$$L(2,\chi) = |d|^{-1/2} \left|\sum \chi(a) D(\zeta^a)\right|$$

where ζ is a primitive d^{th} root of 1. In any case,

$c_\chi = 48\ D(L_\chi)/|d|^{3/2}L(2,\chi)$. Now see Table I for the results of the computations.

Table I

d	$\#X$	rank $B_\chi(F)$	D_χ	c_χ
-3	47	20	2.02988	24.00000
	79	31		
-4	47	19	3.66386	24.00000
	119	41		
-7	47	18	5.33349	12.00000
	72	18		
-8	38	14	12.04609	24.00000
	62	18		
-11	42	11	16.59130	24.00000
	79	26		

Now $(K_3\mathcal{O}_F)_{tors}$ is conjectured to be isomorphic to $H^0(\bar{F}/F,\mu^{\otimes 2})$, whose order is 24 for any quadratic imaginary field. Tate [T] has computed $K_2\mathcal{O}_F$ for d = -3, -4, -7, -8, -11, and -15. These results and conjectures appear in Table II.

Table II

d	$\#K_2$	$\#K_3$	$\#K_3/\#K_2$
-3	1	24	24
-4	1	24	24
-7	2	24	12
-8	1	24	24
-11	1	24	24
-15	2	24	12

The agreement between Tables I and II is good.

In Table III are some elements of $\mathcal{B}(F)$, D of which is minimal. Here $\omega = \sqrt{d}$ or $(1 + \sqrt{d})/2$, where $\mathcal{O}_F = \mathbb{Z}[\omega]$.

Table III

$\underline{d}$	
-3	$3[1 + \omega]$
-4	$4[\omega]$
-7	$[-2 + \omega] - 6[\omega]$
-8	$2[1 + 2\omega] - 12[1 - \omega]$
-11	$2[3 + 2\omega] + 12[\omega] - 6[-\omega] - 2[-3 - 2\omega]$

Remark 1: The cokernel of φ is K_2F, where $\lambda \otimes \mu \mapsto \{\lambda,\mu\}$. Two relations in K_2F are $\{\lambda,-\lambda\} = 0$ and $\{\lambda,\mu\} + \{\mu,\lambda\} = 0$. If $F^* \otimes F^*$ is replaced by $F^* \otimes F^*/V$ where V is generated by $\lambda\otimes - \lambda$ and $\lambda \otimes \mu + \mu \otimes \lambda$, and $\mathfrak{B}(F)$ is replaced by $\mathcal{C}(F)$, the kernel of $\mathbb{Z}^{(F\setminus 0,1)} \to F^* \otimes F^*/V$, then the computer still gives the same results as in Table I.

Remark 2: The fact that Table I agrees with Table II is evidence for the map $\mathfrak{B}(F) \to K_3F$ being surjective mod torsion.

Remark 3: The universal constant relating D and the regulator map is unknown. Presumably it is a rational multiple of π^2. Were it known, we could see whether Lichtenbaum's conjecture lacks some rational factor.

Now we present some computations suggested by Benedict Gross. The following analysis is due to him.

Let λ be a root of the equation $x^3 + x - 1 = 0$, and let $F = \mathbb{Q}(\lambda)$. The field F has one real place, and its Galois closure H is an S_3-extension of $\mathbb{Q}$ which is unramified over $\mathbb{Q}(\sqrt{-31})$. (The discriminant of the equation is $d = -31$.) Since the class number of $\mathbb{Q}(\sqrt{-31})$ is 3, H is the Hilbert class field of $\mathbb{Q}(\sqrt{-31})$.

The equation $\lambda^3 = 1 - \lambda$ implies

$\varphi(2[\lambda]) = 2(\lambda \otimes 1 - \lambda) = 2(\lambda \otimes \lambda^3) = 3(\lambda^2 \otimes \lambda) = 6((-\lambda) \otimes \lambda) \in$ the subgroup V defined in remark 1. Thus $2[\lambda] \in \mathcal{C}(F)$; we may hope that its image generates K_3F (mod torsion).

Let W be the permutation representation of $S_3 = \mathrm{Gal}(H/\mathbb{Q})$ or the cosets of $\mathrm{Gal}(H/F)$. Then

$$W \cong 1 \oplus \mathrm{Ind}_{A_3}^{S_3} \varepsilon$$

where ε is a non-trivial cubic character of $A_3 = \mathrm{gal}(H/\mathbb{Q}(\sqrt{-31}))$. Hence

$$\zeta_F(s) = L_H(W,s) = \zeta_{\mathbb{Q}}(s) L_{\mathbb{Q}(\sqrt{-31})}(\varepsilon, s).$$

At $s = -1$ we get

$$\zeta_F'(-1) = -\frac{1}{12} L'(\epsilon,-1).$$

Let C_0, C_1, and C_2 denote the ideal classes of $\mathbb{Q}(\sqrt{-31})$ with C_0 the principal class and $\epsilon(C_1) = e^{2\pi i/3}$. Then for $\mathrm{Re}(s) > 1$

$$L(\epsilon(s) = \sum_{C_0} (N\mathfrak{a})^{-s} + \sum_{C_1} e^{2\pi i/3}(N\mathfrak{a})^{-s} + \sum_{C_2} e^{4\pi i/3}(N\mathfrak{a})^{-s}.$$

Since $\overline{C_1} = C_2$,

$$L(\epsilon,s) = \sum_{C_0} (N\mathfrak{a})^{-s} - \sum_{C_1} (N\mathfrak{a})^{-s}.$$

Since $N(x + y\omega) = x^2 + xy + 8y^2$, where $\omega = \frac{1}{2}(1 + \sqrt{-31})$, we see

$$\sum_{C_0} (N\mathfrak{a})^{-s} = \frac{1}{2} \sum_{(x,y)\neq(0,0)} (x^2 + xy + 8y^2)^{-s}$$

A quadratic form of discriminant -31 inequivalent to the one above is $2x^2 + xy + 4y^2$. Thus

$$\sum_{C_1} (N\mathfrak{a})^{-s} = \frac{1}{2} \qquad (2x^2 + xy + 4y^2)^{-s}.$$

The computer can use these formulas when $s = 2$, and then the functional equation gives the value at -1. (If $\Lambda(s) = (31)^{s/2}(2\pi)^{-s}\Gamma(s)L(\epsilon,s)$ then $\Lambda(s) = \Lambda(1 - s)$.]

The computer found $c(\{2[\lambda]\}) = 12.000019$.

Consider now the alternate equation $\lambda^3 - \lambda + 1 = 0$. Here the discriminant is -23 and $\lambda \otimes 1 - \lambda = \lambda \otimes -\lambda^3 \in H$, so we may repeat the analysis above. In this case the two forms are $x^2 + xy + 6y^2$ and $2x^2 + xy + 3y^2$. The computer found $c(\{[\lambda]\}) = 11.999991$.

For both of these fields w_2F is 24, and we wonder whether $\#K_2\mathcal{O}_F$ is 2.

In fact, since F has a real place, $\{-1, -1\} \neq 0$ and $K_2\mathcal{O}_F$ contains $\mathbb{Z}/2$ as direct summand.

Acknowledgement

I thank both Spencer Bloch and Benedict Gross for the help and suggestions they have given me, and the NSF for partial support.

References

[Bl] S. Bloch, Higher Regulators, Algebraic K-theory, and Zeta Functions of elliptic curves, preprint, 1979.

[Bo 1] A. Borel, Stable Real Cohomology of arithmetic groups, Ann. Sci. ENS, 7 (1974) 235-272.

[Bo 2] A. Borel, Cohomologie de SL_n, et valeurs de fonctions zeta aux pointes entiers, Ann. Sc. N. Sup. - Pisa IV (1977) 613-636.

[L] S. Lichtenbaum, Values of zeta-functions, etale cohomology, and algebraic K-theory; in "Algebraic K-theory II", Lecture Notes in Math. 342, 1973, Springer, Berlin.

[T] J. Tate, appendix to "The Milnor ring of a global field" by H. Bass and J. Tate; in "Algebraic K-theory II", Lecture Notes in Math. 342, 1973, Springer Berlin.

OBSTRUCTION A L'EXCISION EN K-THEORIE ALGEBRIQUE

Dominique GUIN-WALÉRY et Jean-Louis LODAY

L'un des moyens efficaces pour calculer les groupes de K-théorie algébrique est la suite exacte de Mayer-Vietoris [M]. On sait [Sw] que celle-ci n'est valable qu'en basses dimensions, car les homomorphismes d'excision ne sont pas toujours des isomorphismes. Plus précisément, si I et J sont deux idéaux bilatères de l'anneau unitaire Λ tels que $I \cap J = \{0\}$, alors l'homomorphisme d'excision $K_2(\Lambda, I) \rightarrow K_2(\Lambda/J, I+J/J)$ est surjectif, mais il n'est pas, en général, injectif. On est alors amené tout naturellement à étudier le groupe de K-théorie "birelative" $K_2(\Lambda; I, J)$ (défini en fait pour I et J deux idéaux bilatères quelconques) qui s'inscrit dans la suite exacte

$$\cdots \rightarrow K_3(\Lambda, I) \rightarrow K_3(\Lambda/J, I+J/J) \rightarrow K_2(\Lambda; I, J) \rightarrow K_2(\Lambda, I) \rightarrow K_2(\Lambda/J, I+J/J) \rightarrow \cdots$$

Le fait surprenant est que l'on peut calculer très explicitement le groupe $K_2(\Lambda; I, J)$ lorsque $I \cap J = \{0\}$, on trouve

$$K_2(\Lambda; I, J) = I \otimes_{\Lambda^e} J \quad , \qquad \text{avec } \Lambda^e = \Lambda \otimes_{\mathbb{Z}} \Lambda^{op} .$$

Remarquons que $I \otimes_{\Lambda^e} J = I/I^2 \otimes_{\Lambda^e} J/J^2$ et que, si Λ est commutatif, $K_2(\Lambda; I, J) = I \otimes_{\Lambda} J$.

Dans le cas où Λ est commutatif, on généralise ce résultat en donnant une présentation par générateurs et relations du groupe $K_2(\Lambda; I, J)$ lorsque $I \cap J$ est radical, i.e. contenu dans le radical de Jacobson de Λ. Les générateurs sont alors les symboles de Dennis et Stein $\langle a,b\rangle$ définis pour a ou b ε I et a ou b ε J. Les relations sont analogues à celles données par Maazen et Stienstra [M-S] dans le cas d'un seul idéal.

Voici le contenu de cet article. Dans la Section 1, après avoir défini topologiquement les groupes de K-théorie birelative on construit algébriquement le groupe de Steinberg (et le groupe linéaire) birelatif. Le théorème 1 affirme que pour i = 1,2 les définitions algébriques et topologiques de $K_i(\Lambda;I,J)$ coïncident (le cas relatif a été traité dans [L_3]). Ensuite on énonce les résultats concernant $K_2(\Lambda;I,J)$ lorsque $I\cap J$ est nul, puis lorsque Λ est commutatif et $I\cap J$ radical.

La Section 2 contient la démonstration du théorème 1.

La Section 3 est dévolue au calcul de $K_2(\Lambda;I,J)$.

Dans la Section 4 on donne une interprétation homologique des groupes de K-théorie relative et birelative analogue aux isomorphismes $K_1(\Lambda) = H_1(GL(\Lambda))$, $K_2(\Lambda) = H_2(E(\Lambda))$ et $K_3(\Lambda) = H_3(St(\Lambda))$.

La Section 5 donne des applications des théorèmes 2 et 3 au calcul de certains groupes de K-théorie comme $K_2(\mathbb{Z}[x,y]/(xy))$, $K_3(\mathbb{Z}[\mathbb{Z}/2])$, $K_2(\mathbb{Z}[\mathbb{Z}/p][t])$.

Le Département de Mathématiques de l'Université de Northwestern (Evanston) a assuré l'excellente frappe de ce texte, qu'il en soit ici remercié.

1. Définitions et résultats.

Soient Λ un anneau, $GL(\Lambda) = \bigcup_n GL_n(\Lambda)$ le groupe général linéaire de Λ et $BGL(\Lambda)^+$ l'espace construit par Quillen [Q] [L_1]. Ses groupes d'homotopie sont, par définition, les groupes de K-théorie $K_i(\Lambda) = \pi_i(BGL(\Lambda)^+)$, $i\geq 1$. Pour tout idéal bilatère I de Λ, on est amené à définir des groupes de K-théorie relative en posant $K_i(\Lambda,I) = \pi_i K(\Lambda,I)$, $i\geq 1$, où $K(\Lambda,I)$ est la fibre homotopique de l'application $BGL(\Lambda)^+ \to BGL(\Lambda/I)^+$. De même si I et J sont deux idéaux bilatères de Λ, on pose

$K_i(\Lambda;I,J) = \pi_i K(\Lambda;I,J)$, $i\geqslant 1$, où $K(\Lambda;I,J)$ est la fibre homotopique de $K(\Lambda,I) \to K(\Lambda/J,I+J/J)$. On remarque que $K_i(\Lambda;I,J)$ et $K_i(\Lambda;J,I)$ sont canoniquement isomorphes et que $K_i(\Lambda;I,I) = K_i(\Lambda,I)$. Ces groupes de K-théorie relative sont liés entre eux par les suites exactes

$$\cdots \to K_{i+1}(\Lambda) \to K_{i+1}(\Lambda/I) \to K_i(\Lambda,I) \to K_i(\Lambda) \to K_i(\Lambda/I) \to \cdots$$

$$\cdots \to K_{i+1}(\Lambda,I) \to K_{i+1}(\Lambda/J,I+J/J) \to K_i(\Lambda;I,J) \to K_i(\Lambda,I) \to K_i(\Lambda/J,I+J/J) \to \cdots$$

On sait (cf. [Q] [L_1]) que le groupe $K_2(\Lambda)$ (resp. $K_1(\Lambda)$) s'interprète algébriquement comme le noyau (resp. le conoyau) de l'homomorphisme $\phi_\Lambda : St(\Lambda) \to GL(\Lambda)$ où $St(\Lambda)$ est le groupe de Steinberg de Λ.

Dans [S_1] Stein a proposé de définir un groupe de Steinberg relatif de la manière suivante. Soit $D = \Lambda\times_{\Lambda/I}\Lambda$ le produit fibré de Λ par lui-même au-dessus de Λ/I. On pose $St^s(\Lambda,I) = Ker(p_{1*}:St(D) \to St(\Lambda))$ où p_1 est la première projection. Dans [L_3] (cf. aussi [Kn]) on a modifié cette définition en posant $St(\Lambda,I) = St^s(\Lambda,I)/C(I)$ où $C(I) = [Ker\ p_{1*}, Ker\ p_{2*}]$. L'intérêt de cette nouvelle définition est qu'on a une suite exacte

$$1 \to K_2(\Lambda,I) \to St(\Lambda,I) \xrightarrow{\phi_{\Lambda,I}} GL(\Lambda,I) \to K_1(\Lambda,I) \to 1$$

où $GL(\Lambda,I) = Ker(GL(\Lambda) \to GL(\Lambda/I))$.

Soient I et J deux idéaux bilatères de Λ. On se propose de construire le groupe $St(\Lambda;I,J)$ qui permettra d'interpréter algébriquement les groupes $K_i(\Lambda;I,J)$ pour $i = 1$ et 2.

Soit T l'anneau défini par $T = \{(a,b,c,d)\in\Lambda^4 \mid a \equiv b \bmod I,\ c \equiv d \bmod I,\ a \equiv c \bmod J,\ b \equiv d \bmod J\}$. On note p_1 et p_2 (resp. p_1' et p_2') les projections de T sur $D = \Lambda\times_{\Lambda/I}\Lambda$ (resp. $D' = \Lambda\times_{\Lambda/J}\Lambda$) données par $p_1(a,b,c,d) = (a,b)$ et $p_2(a,b,c,d) = (c,d)$ (resp. $p_1'(a,b,c,d) = (a,c)$ et $p_2'(a,b,c,d) = (b,d)$). Les homomorphismes induits sur les groupes de Steinberg sont p_{1*} et $p_{2*}:St(T) \to St(D)$ (resp. p'_{1*} et $p'_{2*}:St(T) \to St(D'))$. On note $C(I,J)$ le sous-groupe normal de $St(T)$ engendré par $[\mathrm{Ker}\, p_{1*}, \mathrm{Ker}\, p_{2*}]$ et $[\mathrm{Ker}\, p'_{1*}, \mathrm{Ker}\, p'_{2*}]$. On constate que $p_{i*}(C(I,J)) = C(I)$ (resp. $p'_{i*}(C(I,J)) = C(J))$ pour $i = 1,2$. On a donc les homomorphismes suivants:

$$\overline{p}_{i*}:St(T)/C(I,J) \to St(D)/C(I)$$

$$\overline{p}'_{i*}:St(T)/C(I,J) \to St(D')/C(J).$$

DEFINITION. *Le groupe de Steinberg* $St(\Lambda;I,J)$ *des idéaux* I *et* J *est* $St(\Lambda;I,J) = \mathrm{Ker}\, \overline{p}_{1*} \cap \mathrm{Ker}\, \overline{p}'_{1*}$.

La transformation de foncteur ϕ induit un homomorphisme de $St(\Lambda;I,J)$ dans $GL(\Lambda;I,J) = \mathrm{Ker}(GL(\Lambda) \to GL(\Lambda/I)) \cap \mathrm{Ker}(GL(\Lambda) \to GL(\Lambda/J))$.

THEOREME 1 *Soient* Λ *un anneau,* I *et* J *deux idéaux bilatères. On a une suite exacte*

$$1 \to K_2(\Lambda;I,J) \to St(\Lambda;I,J) \xrightarrow{\phi_{\Lambda;I,J}} GL(\Lambda;I,J) \to K_1(\Lambda;I,J) \to 1.$$

La démonstration de ce théorème fera l'objet du paragraphe 2. Un cas particulièrement intéressant est celui où $I \cap J = \{0\}$. En effet le carré d'anneaux

$$\begin{array}{ccc} \Lambda & \longrightarrow & \Lambda/I \\ \downarrow & & \downarrow \\ \Lambda/J & \longrightarrow & \Lambda_1 = \Lambda/I+J \end{array}$$

est alors cartésien, ce qui implique $GL(\Lambda;I,J) = K_1(\Lambda;I,J) = 0$. On en déduit une suite exacte de Mayer-Vietoris (cf. [M]):

$$K_2(\Lambda) \to K_2(\Lambda/I) \oplus K_2(\Lambda/J) \to K_2(\Lambda_1) \to K_1(\Lambda) \to K_1(\Lambda/I) \oplus K_1(\Lambda/J) \to K_1(\Lambda_1).$$

Pour pouvoir étendre cette suite vers la gauche, il suffirait que l'homomorphisme d'excision $K_2(\Lambda,I) \to K_2(\Lambda/J,I+J/J)$, qui est surjectif, soit aussi injectif. Donc l'image de $K_2(\Lambda;I,J)$ dans $K_2(\Lambda,I)$ mesure l'obstruction à l'exactitude de la suite de Mayer-Vietoris. Comme l'a montré Swan [Sw], ce groupe est en général différent de 0. On montre, plus précisément, le résultat suivant:

THEOREME 2. *Soient Λ un anneau, I et J deux idéaux bilatères tels que* $I \cap J = \{0\}$. *On a alors les isomorphismes*

$$K_2(\Lambda;I,J) \simeq St(\Lambda;I,J) \simeq I \otimes_{\Lambda^e} J \text{ où } \Lambda^e = \Lambda \otimes_{\mathbb{Z}} \Lambda^{op}.$$

La démonstration sera donnée au paragraphe 3.

COROLLAIRE. *Soient Λ un anneau, I et J deux idéaux bilatères tels que* $I \cap J = \{0\}$. *On a la suite exacte*

$$K_3(\Lambda,I) \to K_3(\Lambda/J,I+J/J) \to I \otimes_{\Lambda^e} J \to K_2(\Lambda,I) \to K_2(\Lambda/J,I+J/J) \to 0.$$

La dernière partie de cette suite exacte, c'est-à-dire l'exactitude en $K_2(\Lambda,I)$ et $K_2(\Lambda/J,I+J/J)$ a été obtenue par Keune [Kn] par une méthode différente (*).

Supposons maintenant que I et J sont deux idéaux bilatères quelconques de l'anneau commutatif Λ. Le groupe abélien $D(\Lambda;I,J)$ est, par définition, engendré par les éléments $\langle a,b\rangle$, a et $b \in \Lambda$ tels que a ou $b \in I$ et a ou $b \in J$ et $(1-ab) \in \Lambda^*$ (cette dernière condition est toujours vérifiée si $I \cap J$ est radical). Les relations de $D(\Lambda;I,J)$ sont

(*) F. Keune a montré l'exactitude complète de la suite dans "Doubly relative K-theory and the relative K_2", preprint, Nijmegen.

(D1) $\langle a,b\rangle\langle b,a\rangle = 1$

(D2) $\langle a,b\rangle\langle a,b'\rangle = \langle a,b+b'-abb'\rangle$

(D3) $\langle a,bc\rangle\langle b,ca\rangle\langle c,ab\rangle = 1$, avec a ou b ou c ε I et a ou b ou c ε J.

En utilisant les symboles de Dennis et Stein [D-S] on construit un homomorphisme de groupes $D(\Lambda;I,J) \to K_2(\Lambda;I,J) \subset St(T)/C(I,J)$ de la manière suivante. Pour tout élément u et v ε T tels que $1-uv \; \varepsilon \; T^*$, on pose $\langle u,v\rangle' = x_{21}^{v(1-uv)^{-1}u} x_{12}^{-v} x_{21}^{-(1-uv)^{-1}u} h_{12}(1-uv)^{-1} \; \varepsilon \; St(T)$ (cf. [M-S], [S_2]). A une permutation près, les générateurs de $D(\Lambda;I,J)$ sont soit de la forme $\langle a,b\rangle$ avec a ε I et b ε J, soit de la forme $\langle \lambda,c\rangle$ avec $\lambda \; \varepsilon \; \Lambda$ et c ε $I \cap J$ (on peut avoir $a = \lambda \; \varepsilon$ I et $b = c \; \varepsilon \; I\cap J$). On définit $D(\Lambda;I,J) \to St(T)/C(I,J)$ en envoyant $\langle a,b\rangle$ sur $\langle(0,a,0,a), (0,0,b,b)\rangle'$ lorsque a ε I, b ε J, et en envoyant $\langle\lambda,c\rangle$ sur $\langle(\lambda,\lambda,\lambda,\lambda), (0,0,0,c)\rangle'$ lorsque $\lambda \; \varepsilon \; \Lambda$, c ε $I\cap J$. Cette application est bien définie et est un homomorphisme de groupes car d'une part $\langle(0,a,0,a),(0,0,c,c)\rangle' = \langle(a,a,a,a), (0,0,0,c)\rangle'$ dans $St(T)/C(I,J)$ (mais pas dans $St(T)$) lorsque a ε I, c ε $I\cap J$, et d'autre part les symboles $\langle -,-\rangle'$ vérifient D1, D2 et D3, cf. [D-S]. L'image de chacun de ces symboles est triviale dans $St(D)/C(I)$, $St(D')/C(J)$ par la première projection et dans $E(T)$, donc l'image de $D(\Lambda;I,J)$ est dans $K_2(\Lambda;I,J)$. On a ainsi défini un homomorphisme $D(\Lambda;I,J) \to K_2(\Lambda,I,J)$.

THÉORÈME 3. <u>Soit</u> Λ <u>un anneau commutatif</u>, I <u>et</u> J <u>deux idéaux bilatères tels que</u> $I\cap J$ <u>soit radical, alors</u> $D(\Lambda;I,J) \simeq K_2(\Lambda;I,J)$.

La démonstration sera donnée au paragraphe 3. Ce résultat généralise le théorème 2, ainsi que le résultat de Maazen-Stienstra (cf. [M-S] et [K]).

2. Démonstration du théorème 1.

Pour tout anneau Λ on désigne par F_Λ la fibre homotopique de l'application $BGL(\Lambda) \to BGL(\Lambda)^+$. Le groupe fondamental de F_Λ s'identifie canoniquement au groupe de Steinberg de Λ, $St(\Lambda) \xrightarrow{\sim} \pi_1 F_\Lambda$.

De même dans le cas relatif $F_{\Lambda,I}$ est la fibre homotopique de $BGL(\Lambda,I) \to K(\Lambda,I)$ et on a l'isomorphisme $St(\Lambda,I) \xrightarrow{\sim} \pi_1 F_{\Lambda,I}$, cf. [$L_3$]. Remarquons que $F_{\Lambda,I}$ est aussi la fibre de $F_\Lambda \to F_{\Lambda/I}$.

Dans le cas birelatif on se propose de montrer:

PROPOSITION 2.1. Soient Λ un anneau, I et J deux idéaux bilatères de Λ. Si $F_{\Lambda;I,J}$ désigne la fibre homotopique de l'application $BGL(\Lambda;I,J) \to K(\Lambda;I,J)$, on a un isomorphisme naturel $St(\Lambda;I,J) \xrightarrow{\sim} \pi_1 F_{\Lambda;I,J}$.

Remarquons que $F_{\Lambda;I,J}$ est aussi la fibre de $F_{\Lambda,I} \to F_{\Lambda/J,I+J/J}$ (voir diagramme (*) dulemme 2.5).

Le théorème 1 est une conséquence immédiate de la proposition 2.1 et de la suite exacte d'homotopie

$$1 \to \pi_2 K(\Lambda;I,J) \to \pi_1 F_{\Lambda;I,J} \to GL(\Lambda;I,J) \to \pi_1 K(\Lambda;I,J) \to 1.$$

Pour démontrer la proposition 2.1 nous utilisons la notion de bicatégorie en groupes et quelques-unes de ses propriétés [L_4].

DEFINITION. Une bicatégorie en groupes $\underline{G}$ (appelée "2-cat-group" dans [L_4]) est la donnée d'un groupe G, de deux sous-groupes N et N' et d'homomorphismes $s,b : G \to N$ et $s',b' : G \to N'$, vérifiant les conditions suivantes:

1) $s|_N = b|_N = id_N$ et $s'|_{N'} = b'|_{N'} = id_{N'}$,

2) $[\mathrm{Ker}\, s, \mathrm{Ker}\, b] = 1 = [\mathrm{Ker}\, s', \mathrm{Ker}\, b']$,

3) ss'=s's, bb'=b'b, sb'=b's et bs'=s'b.

On appelle s et s' les homomorphismes sources, b et b' les homomorphismes buts. Lorsqu'aucune confusion n'est à craindre, on note $\underline{\underline{G}}$ = (G;N,N') cette bicatégorie en groupes.

Un <u>morphisme</u> $\underline{\underline{\psi}}:\underline{\underline{G}}_0 \to \underline{\underline{G}}$ de bicatégories en groupes est un homomorphisme de groupes $\psi:G_0 \to G$ qui commute aux homomorphismes sources et buts. C'est une <u>extension centrale</u> lorsque ψ est surjectif et Ker ψ est dans le centre de G_0. Ce noyau Ker ψ est naturellement muni d'une structure de bicatégorie en groupes.

Dans $[L_4]$ on a montré qu'à toute bicatégorie en groupes $\underline{\underline{G}}$ est associé de manière naturelle un carré commutatif d'espaces dont toutes les flèches sont des fibrations:

$$\underline{\underline{B}}\,\underline{\underline{G}} \qquad \begin{array}{ccc} (\underline{\underline{B}}\,\underline{\underline{G}})^{00} & \xrightarrow{\quad f \quad} & (\underline{\underline{B}}\,\underline{\underline{G}})^{01} \\ g\Big\downarrow & & \Big\downarrow g' \\ (\underline{\underline{B}}\,\underline{\underline{G}})^{10} & \xrightarrow{\quad f' \quad} & (\underline{\underline{B}}\,\underline{\underline{G}})^{11} \end{array}$$

<u>Notations</u>. Soit $h:Y \to X$ une fibration d'espaces connexes dont la fibre est connexe. Soit Ch le "mapping - cone" de h, on note $H_i(h) = H_{i+1}(Ch;\mathbb{Z})$, $i \geq 0$, les groupes d'homologie relative à coefficients dans $\mathbb{Z}$. De même, si on a un carré de fibrations dont toutes les fibres sont connexes, par exemple $\underline{\underline{B}}\,\underline{\underline{G}}$, alors (g,g') induit une application $\alpha:Cf \to Cf'$ et on note $H_i(\underline{\underline{B}}\,\underline{\underline{G}}) = H_{i+2}(C\alpha;\mathbb{Z})$. On utilisera le résultat suivant $[L_4]$:

PROPOSITION 2.2.(*) <u>Soit</u> $\underline{\underline{G}}_0 \to \underline{\underline{G}}$ <u>une extension centrale de bicatégories en groupes</u>

(*) On trouvera aussi une démonstration de ce résultat dans une version préliminaire (Publication I.R.M.A. Strasbourg, Déc. 1979).

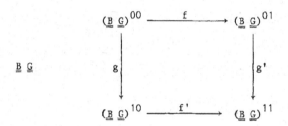

dont le noyau est de la forme (A;1,1). *Si* $H_1(\underline{B}\,\underline{\underline{G}}) = H_2(\underline{B}\,\underline{\underline{G}}) = 0$ *et si le groupe* G_0 *est parfait, alors cette extension est un isomorphisme* (i.e. A = 1). □

La démonstration de la proposition 2.1. consiste à construire une extension centrale de bicatégories en groupes $\underline{\underline{\psi}}:\underline{\underline{G}}^{st} \to \underline{\underline{G}}^{top}$ telle que l'intersection des morphismes sources de $\underline{\underline{G}}^{st}$ (resp. $\underline{\underline{G}}^{top}$) soit $St(\Lambda;I,J)$ (resp. $\pi_1 F_{\Lambda;I,J}$). Puis on démontre que cette extension satisfait aux hypothèses de la proposition 2.2. De l'isomorphisme entre $\underline{\underline{G}}^{st}$ et $\underline{\underline{G}}^{top}$ on déduit l'isomorphisme entre $St(\Lambda;I,J)$ et $\pi_1 F_{\Lambda;I,J}$.

Construction de $\underline{\underline{G}}^{st}$, $\underline{\underline{G}}^{top}$ *et* ψ.

On pose $\underline{\underline{G}}^{st} = (St(T)/C(I,J);\ St(D)/C(I),\ St(D')/C(J))$. Le groupe $St(D)/C(I)$ (resp. $St(D')/C(J)$) s'identifie à un sous-groupe de $St(T)/C(I,J)$ grâce au morphisme induit par la diagonale $(a,b) \longmapsto (a,b,a,b)$ (resp. $(a',b') \longmapsto (a',a',b',b')$). Les homomorphismes source et but à valeurs dans $St(D)/C(I)$ (resp. $St(D')/C(J)$) sont $\bar{p}_{1*}$ et $\bar{p}_{2*}$ (resp. $\bar{p}'_{1*}$ et $\bar{p}'_{2*}$). Les axiomes 1) et 3) des bicatégories en groupes sont immédiats. L'axiome 2) est vérifié car on a quotienté par $C(I,J)$. Par définition $St(\Lambda;I,J)$ est l'intersection des morphismes sources.

En ce qui concerne $\underline{\underline{G}}^{top}$ c'est la bicatégorie en groupes associée au carré commutatif de fibrations (cf. $[L_4]$)

$$\underline{\underline{F}}_{\Lambda;I,J} \qquad \begin{array}{ccc} F_{\Lambda} & \longrightarrow & F_{\Lambda/I} \\ \big\downarrow & & \big\downarrow \\ F_{\Lambda/J} & \longrightarrow & F_{\Lambda/I+J} \end{array} .$$

Rappelons-en la construction. Posons

$$V = F_{\Lambda} \times_{F_{\Lambda/I}} F_{\Lambda},\ V' = F_{\Lambda} \times_{F_{\Lambda/J}} F_{\Lambda},\ V_1 = F_{\Lambda/J} \times_{F_{\Lambda/I+J}} F_{\Lambda/J} \text{ et } V'_1 = F_{\Lambda/I} \times_{F_{\Lambda/I+J}} F_{\Lambda/I}.$$

Clairement on a une application $V \to V_1$ (resp. $V' \to V'_1$). On note alors $W = V \times_{V_1} V$ et on remarque que W est canoniquement homéomorphe à $V' \times_{V'_1} V'$. La bicatégorie en groupes $\underline{\underline{G}}^{top}$ est alors $\underline{\underline{G}}^{top} = (\pi_1 W; \pi_1 V, \pi_1 V')$ où les homomorphismes sources (resp. buts) sont induits par les premières (resp. secondes) projections. On constate que le noyau des morphismes sources est $\pi_1 \, F_{\Lambda;I,J}$.

Puisque $D = \Lambda \times_{\Lambda/I} \Lambda$ (resp. $D_1 = \Lambda/J \times_{\Lambda/I+J} \Lambda/J$) il existe une application canonique $F_D \to V$ (resp. $F_{D_1} \to V_1$). Puisque $T = D \times_{D_1} D$, il existe une application canonique $F_T \to W$ qui induit un homomorphisme $\tilde{\psi}: St(T) = \pi_1 \, F_T \to \pi_1 W$. Puisque $\underline{\underline{G}}^{top}$ est une bicatégorie en groupes (et en particulier satisfait à l'axiome 2), $\tilde{\psi}$ est trivial sur $C(I,J)$ et définit

$$\psi: St(T)/C(I,J) \to \pi_1 W.$$

Il est clair que $\underline{\psi}: \underline{\underline{G}}^{st} \to \underline{\underline{G}}^{top}$ est un morphisme de bicatégories en groupes.

LEMME 2.3. <u>L'homomorphisme ψ est surjectif.</u>

<u>Démonstration</u>. Il suffit de montrer que $\tilde{\psi}: St(T) \to \pi_1 W$ est surjectif. Considérons le diagramme commutatif

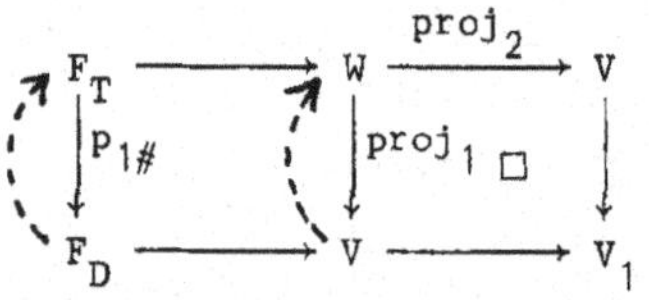

Puisque $\pi_1 F_D = St(D) \to St(D)/C(I) = \pi_1 V$, cf. [$L_3$], est surjectif et que $p_{1\#}$ et $proj_1$ sont scindés, il suffit de démontrer que $Ker\ (p_{1*}) \to Ker\ (proj_{1*})$ est surjectif. Puisque le carré de droite est cartésien il suffit de montrer l'exactitude de la suite

$$\begin{array}{ccccccc}
\mathrm{Ker}(p_{1*}) & \longrightarrow & \pi_1 V & \longrightarrow & \pi_1 V_1 & \longrightarrow & 1. \\
\| & & \| & & \| & & \\
\mathrm{Ker}(\mathrm{St}(T) \xrightarrow{p_{1*}} \mathrm{St}(D)) & & \mathrm{St}(D)/C(I) & & \mathrm{St}(D_1)/C(I+J/J) & &
\end{array}$$

Ceci se dèmontre aisèment en utilisant les présentations de ces groupes comme dans le lemme 6.1 de [M]. □

LEMME 2.4. <u>L'extension</u> $\psi : \mathrm{St}(T)/C(I,J) \to \pi_1 W$ <u>est centrale</u>.

<u>Démonstration</u>. Le carré de fibrations $\underline{\underline{F}}_{\Lambda;I,J}$ s'envoie dans le carré

$$\begin{array}{ccc}
BE(\Lambda) & \longrightarrow & BE(\Lambda/I) \\
\downarrow & & \downarrow \\
BE(\Lambda/J) & \longrightarrow & BE(\Lambda/I+J)
\end{array}$$

Donc l'espace W s'envoie sur son analogue qui est BE(T). L'homomorphisme composé

$$\mathrm{St}(T) \longrightarrow \mathrm{St}(T)/C(I,J) \xrightarrow{\psi} \pi_1 W \longrightarrow E(T)$$

est ϕ_T et Ker ϕ_T est central, donc Ker ψ est central. □

Montrons que $\underline{\underline{G}}^{top}$ satisfait au

LEMME 2.5. <u>Avec les notations précédentes</u>

$H_1(\underline{\underline{B}}\,\underline{\underline{G}}^{top}) = H_2(\underline{\underline{B}}\,\underline{\underline{G}}^{top}) = 0$ et $H_3(\underline{\underline{B}}\,\underline{\underline{G}}^{top}) = \pi_2 F_{\Lambda;I,J} = K_3(\Lambda;I,J)$.

<u>Démonstration</u>. Considérons le diagramme de fibrations

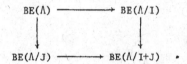

$$(*)\qquad \begin{array}{ccccc}
F_{\Lambda;I,J} & \longrightarrow & F_{\Lambda,J} & \longrightarrow & F_{\Lambda/I;I+J/J} \\
\downarrow & & \downarrow & & \downarrow \\
F_{\Lambda,I} & \longrightarrow & F_{\Lambda} & \longrightarrow & F_{\Lambda/I} \\
 & & & \underline{\underline{F}}_{\Lambda;I,J} & \\
\downarrow & & \downarrow & & \downarrow \\
F_{\Lambda/J,I+J/J} & \longrightarrow & F_{\Lambda/J} & \longrightarrow & F_{\Lambda/I+J}
\end{array}$$

Le Carre´ en haut à gauche sera appelé (par abus de langage) le "carré des fibres" de $\underline{\underline{F}}_{\Lambda;I,J}$. On a un diagramme analogue pour le carré $\underline{\underline{B}}\ \underline{\underline{G}}^{top}$ et d'après [L_4] le carré des fibres de $\underline{\underline{B}}\ \underline{\underline{G}}^{top}$ est

$$\begin{array}{ccc}
B\pi_1 F_{\Lambda;I,J} & \longrightarrow & B\pi_1 F_{\Lambda,J} \\
\downarrow & & \downarrow \\
B\pi_1 F_{\Lambda,I} & \longrightarrow & B\pi_1 F_{\Lambda}
\end{array}$$

Toujours d'après [L_4] il y a un morphisme naturel $\underline{\underline{F}}_{\Lambda;I,J} \to \underline{\underline{B}}\ \underline{\underline{G}}^{top}$ qui induit un isomorphisme sur le π_1, y compris au niveau du carré des fibres. Par conséquent si on note $\tilde{\underline{\underline{F}}}_{\Lambda;I,J}$ la fibre du morphisme de carre´s $\underline{\underline{F}}_{\Lambda;I,J} \to \underline{\underline{B}}\ \underline{\underline{G}}^{top}$, le carré des fibres de $\tilde{\underline{\underline{F}}}_{\Lambda;I,J}$ est formé de revêtements universels, soit

carré des fibres de $\tilde{\underline{\underline{F}}}_{\Lambda;I,J}$:

$$\begin{array}{ccc}
\tilde{F}_{\Lambda;I,J} & \longrightarrow & \tilde{F}_{\Lambda,I} \\
\downarrow & & \downarrow \\
\tilde{F}_{\Lambda,J} & \longrightarrow & \tilde{F}_{\Lambda}
\end{array}$$

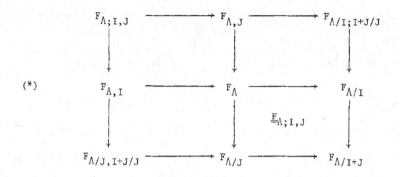

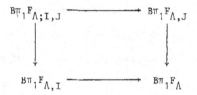

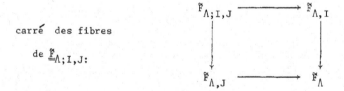

Les groupes d'homologie du carré $\underline{\underline{B}}\ \underline{\underline{G}}^{top}$ sont canoniquement isomorphes à ceux du carré $(\underline{\underline{B}}\ \underline{\underline{G}}^{top})^+$ obtenus en remplaçant chacun des espaces X par X^+. Nous allons tout d'abord calculer les groupes d'homotopie des espaces du carré $(\underline{\underline{B}}\,\underline{\underline{G}}^{top})^+$ à l'aide de $\tilde{\underline{\underline{F}}}_{\Lambda;I,J}$.

Pour tout anneau R l'espace F_R est simple et acyclique (cf. $[L_1]$), par conséquent on a une fibration $\tilde{F}_R \to F_R^+ \to (B\pi_1 F_R)^+$ dans laquelle l'espace F_R^+ est contractile. On a donc une équivalence d'homotopie canonique $\varepsilon_R : \Omega(B\pi_1 F_R)^+ \xrightarrow{\sim} \tilde{F}_R$. Ce résultat, appliqué au morphisme de carré $\underline{\underline{F}}_{\Lambda;I,J} \to \underline{\underline{B}}\,\underline{\underline{G}}^{top}$ donne une équivalence d'homotopie de carrés

$$\varepsilon : \Omega(\underline{\underline{B}}\,\underline{\underline{G}}^{top})^+ \xrightarrow{\sim} \tilde{\underline{\underline{F}}}_{\Lambda;I,J} .$$

Ainsi dans le diagramme de fibrations ci-dessous les neuf espaces sont 2-connexes et

$$\pi_3(Z) = \pi_2(\tilde{F}_{\Lambda;I,J}) = \pi_2(F_{\Lambda;I,J}) = K_3(\Lambda;I,J).$$

(La dernière égalité résulte du fait que $F_{\Lambda;I,J}$ est homotopiquement équivalent à la fibre de $BGL(\Lambda;I,J) \to \mathcal{K}(\Lambda;I,J)$):

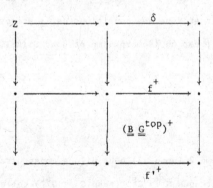

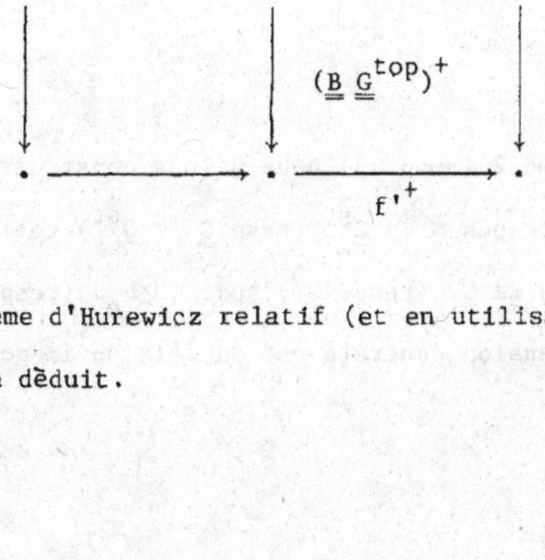

D'après le théorème d'Hurewicz relatif (et en utilisant les notations introduites ci-dessus) on en déduit.

$$H_1(f^+) = H_2(f^+) = 0 \quad , \quad H_3(f^+) = \pi_3(\text{fibre } f^+) \quad ,$$
$$H_1(f'^+) = H_2(f'^+) = 0 \quad , \quad H_3(f'^+) = \pi_3(\text{fibre } f'^+) \quad ,$$
$$H_1(\delta) = H_2(\delta) = 0 \quad \text{et} \quad H_3(\delta) = \pi_3(Z) = K_3(\Lambda;I,J).$$

Appliqués à la suite exacte d'homologie relative

$$H_3(f) \to H_3(f') \to H_2(\underline{\underline{B}}\ \underline{\underline{G}}) \to H_2(f) \to H_2(f') \to H_1(\underline{\underline{B}}\ \underline{\underline{G}}) \to H_1(f)$$

ces résultats montrent que $H_1(\underline{\underline{B}}\ \underline{\underline{G}}) = 0$ et $H_2(\underline{\underline{B}}\ \underline{\underline{G}}) = 0$. En effet la flèche de gauche, qui est $\pi_3(\text{fibre } f^+) \to \pi_3(\text{fibre } f'^+)$ est surjective car Z est 2-connexe.

Le calcul de $H_3(\underline{\underline{B}}\ \underline{\underline{G}})$ résulte des isomorphismes $H_3(\underline{\underline{B}}\ \underline{\underline{G}}) \simeq H_3(\delta)$ et $H_3(\delta) \simeq K_3(\Lambda;I,J)$. □

Fin de la démonstration de la proposition 2.1.

D'après les lemmes 2.3 et 2.4 $\underline{\underline{\psi}}:\underline{\underline{G}}^{st} \to \underline{\underline{G}}^{top}$ est une extension centrale de bicatégories en groupes dont le noyau est de la forme (A;1,1) car $St(D)/C(I) \simeq \pi_1 V$ et $St(D')/C(J) \simeq \pi_1 V'$.

D'aprés le lemme 2.5 on a $H_1(\underline{\underline{B}}\ \underline{\underline{G}}^{top}) = H_2(\underline{\underline{B}}\ \underline{\underline{G}}^{top}) = 0$. D'autre part le groupe $G^{st} = St(T)/C(I,J)$ est parfait et donc les conditions de la proposition 2.2 sont remplies. On en conclut que ψ est un isomorphisme et donc $St(\Lambda;I,J) \simeq \pi_1 F_{\Lambda;I,J}$. □

3. Calcul de $K_2(\Lambda;I,J)$.

Pour démontrer le théorème 2 (resp. 3) nous allons construire une extension centrale de bicatégories en groupes $\underline{\underline{G}}^{\otimes} \to \underline{\underline{G}}^{st}$ (resp. $\underline{\underline{G}}^{D} \to \underline{\underline{G}}^{st}$) telle que l'intersection des morphismes sources de $\underline{\underline{G}}^{\otimes}$ (resp. $\underline{\underline{G}}^{D}$) soit $I \otimes_{\Lambda^e} J$ (resp. $D(\Lambda;I,J)$). Puis nous montrerons que cette extension centrale est en fait un isomorphisme grâce aux propositions 2.1 et 2.2.

Avant d'entamer la démonstration proprement dite nous avons besoin de réinterpréter la notion de bicatégorie en groupes. Soit $\underline{\underline{G}} = (G;N,N')$ une bicatégorie en groupes. Puisque $s|_N = id_N$ le groupe G est isomorphe au produit semi-direct $\mathrm{Ker}\, s \rtimes N$. En utilisant les propriétés de s', à savoir $s'|_{N'} = id_{N'}$ et $ss' = s's$ on exhibe des isomorphismes $\mathrm{Ker}\, s \simeq (\mathrm{Ker}\, s \cap \mathrm{Ker}\, s') \rtimes (\mathrm{Ker}\, s \cap N')$ et $N \simeq (N \cap \mathrm{Ker}\, s') \rtimes (N \cap N')$. Posons $L = \mathrm{Ker}\, s \cap \mathrm{Ker}\, s'$, $M = \mathrm{Ker}\, s' \cap N$, $M' = N' \cap \mathrm{Ker}\, s$ et $P = N \cap N'$, on a alors $G \simeq (L \rtimes M) \rtimes (M' \rtimes P)$ et aussi $G \simeq (L \rtimes M') \rtimes (M \rtimes P)$. L'homomorphisme b (resp. b') envoie L dans M' (resp. L dans M) et M dans P (resp. M' dans P). On rebaptise ces applications λ' (resp. λ) et μ (resp. μ'). Puisque $bb' = b'b$ on a un diagramme commutatif

$$(*) \qquad \begin{array}{ccc} L & \xrightarrow{\lambda} & M \\ {\scriptstyle\lambda'}\downarrow & & \downarrow{\scriptstyle\mu} \\ M' & \xrightarrow{\mu'} & P \end{array}$$

Le groupe P (resp. P, resp. P, resp. M, resp. M') opère sur le groupe M (resp. M', resp. L, resp. L, resp. L) par conjugaison dans le groupe G. D'autre part, si $m \in M$ et $m' \in M'$, alors le commutateur $[m,m']$ appartient à $L = \mathrm{Ker}\, s \cap \mathrm{Ker}\, s'$; on note h l'application $M \times M' \to L$ ainsi définie, $h(m,m') = [m,m']$. On vérifie que toutes ces données satisfont aux propriétés suivantes:

i) Les homomorphismes $\lambda, \lambda', \mu, \mu'$ et $\kappa = \mu\lambda = \mu'\lambda'$ sont des modules croisés (voir ci-dessous) pour les actions précédentes et les morphismes d'applications $(\lambda) \to (\kappa)$, $(\kappa) \to (\mu)$, $(\lambda') \to (\kappa)$ et $(\kappa) \to (\mu')$ sont des homomorphismes de modules croisés,

ii) $\lambda h(m,m') = m\, \mu(m')\cdot m^{-1}$ et $\lambda' h(m,m') = \mu(m)\cdot m'\, m'^{-1}$,

iii) $h(\lambda(\ell),m') = \ell\, m'\cdot\ell^{-1}$ et $h(m,\lambda'(\ell)) = m\cdot\ell\, \ell^{-1}$,

iv) $h(m_1m_2,m') = m_1\cdot h(m_2,m')\, h(m_1,m')$ et $h(m,m_1'm_2') = h(m,m_1')\, m_1'\cdot h(m,m_2')$,

v) $h(n\cdot m, n\cdot m') = n\cdot h(m,m')$,

vi) $m\cdot(m'\cdot\ell)\, h(m,m') = h(m,m')\, m'\cdot(m\cdot\ell)$, pour tout $m,m_1,m_2 \in M$, $m',m'_1,m'_2 \in M'$, $n \in N$ et $\ell \in L$.

On rappelle qu'un *module croisé* [Wh] est la donnée d'un homomorphisme de groupes $\rho:A \to B$ et d'une action de B sur A (notée $(b,a) \longmapsto b\cdot a$) vérifiant

$$\rho(b\cdot a) = b\rho(a)b^{-1} \text{ et } \rho(b)\cdot b' = bb'b^{-1}.$$

Cette notion est équivalente à celle de catégorie en groupes cf. section 4 et [L_4].

Définition. Un *carré croisé* est la donnée d'un carré commutatif de groupes (*), d'actions de P sur M,M' et L, de M sur L et de M' sur L, et d'une application $h:M \times M' \to L$ vérifiant les propriétés i) à vi) énoncées ci-dessus.

On a une notion évidente de morphismes de carrés croisés. On montre qu'on peut reconstruire une bicatégorie en groupes $\underline{\underline{G}}$ à partir d'un carré croisé, plus précisément on a la

PROPOSITION 3.1. [L_4] *La catégorie des bicatégories en groupes est équivalente à celle des carrés croisés.* □

Exemple. A un isomorphisme canonique près, le carré croisé associé à la bicatégorie en groupes $\underline{\underline{G}}^{st} = (St(T)/C(I,J);\ St(D)/C(I), St(D')/C(J))$ est

$$\begin{array}{ccc} St(\Lambda;I,J) & \longrightarrow & St(\Lambda,I) \\ \downarrow & & \downarrow{\scriptstyle \mu'} \\ St(\Lambda,J) & \xrightarrow{\ \mu\ } & St(\Lambda) \end{array}$$

Le groupes $St(\Lambda)$ s'identifie canoniquement à un sous-groupe de $St(T)/C(I,J)$ par Δ'_* où Δ' est l'application diagonale $\Lambda \to T \subset \Lambda^4$. Rappelons la présentation par générateurs et relations du groupe $St(\Lambda,I)$ (cf. [Kn], [L_3]). $St(\Lambda,I)$ est le

$St(\Lambda)$-groupe engendré par les éléments y^a_{ij}, où $i \neq j$, $a \in I$, et soumis aux relations

A1 $\quad y^a_{ij} y^b_{ij} = y^{a+b}_{ij}$

B1 $\quad x^\lambda_{ij} \cdot y^a_{ij} = y^a_{ij}$

B2 $\quad x^\lambda_{ij} \cdot y^a_{k\ell} = y^a_{k\ell} \qquad j \neq k,\ i \neq \ell$

B3 $\quad x^\lambda_{ij} \cdot y^a_{jk} = y^{\lambda a}_{ik} y^a_{jk} \qquad i \neq k$

B3' $\quad x^\lambda_{ij} \cdot y^a_{ki} = y^{-a\lambda}_{kj} y^a_{ki} \qquad j \neq k$

C $\quad x^a_{12} \cdot y^b_{21} = y^a_{12} y^b_{21} y^{-a}_{12}$,

avec $\lambda \in \Lambda$, a et $b \in I$. L'action de $St(\Lambda)$ sur $St(\Lambda, I)$ est celle de $St(\Lambda)$-groupe. L'application μ est donnée par $z \cdot y^a_{ij} \longmapsto z\, x^a_{ij} z^{-1}$. Le produit semi-direct $St(\Lambda, I) \rtimes St(\Lambda)$ est isomorphe à $St(D)/C(I)$, cf. $[L_3]$, l'élément $z \cdot y^a_{ij} \in St(\Lambda, I)$ ayant pour image $\Delta_*(z) x^{(0,a)}_{ij} \Delta_*(z)^{-1}$ par cet isomorphisme. De même $St(\Lambda, I)$ s'envoie injectivement dans $St(T)/C(I,J)$ par $z \cdot y^a_{ij} \longmapsto \Delta'_*(z) x^{(0,a,0,a)}_{ij} \Delta'_*(z)^{-1}$. On a des formules analogues pour $St(\Lambda, J)$.

3.a. Le cas $I \cap J = \{0\}$.

Construisons tout d'abord le carré croisé $\underline{\underline{G}}^{\otimes}$ dont le groupe L est $I \otimes_{\Lambda^e} J$. Le carré

$$(**) \qquad \begin{array}{ccc} I \otimes_{\Lambda^e} J & \xrightarrow{\ \lambda\ } & St(\Lambda, I) \\ {\scriptstyle \lambda'} \downarrow & & \downarrow {\scriptstyle \mu} \\ St(\Lambda, J) & \xrightarrow{\ \mu'\ } & St(\Lambda) \end{array}$$

où $\lambda(a\otimes b) = y_{12}^{a}x_{21}^{b}\cdot y_{12}^{-a}$ et $\lambda'(a\otimes b) = x_{12}^{a}\cdot y_{21}^{b}y_{21}^{-b}$ est commutatif car

$$\mu(y_{12}^{a}x_{21}^{b}\cdot y_{12}^{-a}) = [x_{12}^{a},x_{21}^{b}] = \mu'(x_{12}^{a}\cdot y_{21}^{b}y_{21}^{-b}).$$

Afin de munir ce carré d'une structure de carré croisé on définit une application h grâce au lemme ci-dessous.

LEMME 3.2. Il existe une et une seule application

$$h: St(\Lambda,I) \times St(\Lambda,J) \to I \otimes_{\Lambda^e} J$$

vérifiant:

a) $h(uu',v) = h(u,v)+h(u',v)$,

b) $h(u,vv') = h(u,v)+h(u,v')$,

c) $h(z.u,z.v) = h(u,v)$,

d) $h(y_{ij}^{a},v) = a\otimes\phi(v)_{ji}$ avec $u,u' \in St(\Lambda,I)$, $v,v' \in St(\Lambda,J)$, $z \in St(\Lambda)$.

La notation $\phi(v)_{ji}$ désigne le coefficient de la ligne j et de la colonne i de la matrice $\phi(v) \in GL(\Lambda,J)\subset GL(\Lambda)$.

Démonstration. On se sert des relations a), c) et d) pour définir h sur un élément général du type $(\Pi(z.y_{ij}^{a}),v)$. On vérifie que cette définition de h passe bien aux relations de définition du groupe $St(\Lambda,I)$. Vérifions par exemple la relation (B3):

On calcule $h(x_{ij}^{\lambda}\cdot y_{jk}^{a},v) = a\otimes\phi(x_{ij}^{-\lambda}\cdot v)_{kj} = a\otimes(\phi(v)_{ki}\lambda+\phi(v)_{kj})$. D'autre part. on a $h(y_{ik}^{\lambda a}y_{jk}^{a},v) = \lambda a\otimes\phi(v)_{ki}+a\otimes\phi(v)_{kj}$. Ces deux expressions sont égales.

D'après la construction précédente, il est clair que h vérifie les relations a), c) et d). Montrons que h vérifie la relation b). Pour cela, il suffit de montrer que $\phi(vv')_{ji} = \phi(v)_{ji}+\phi(v')_{ji} \bmod J^2$, soit encore $((1+\alpha)(1+\alpha'))_{ji} = \alpha_{ji}+\alpha'_{ji}$ mod J^2 pour α et α' des matrices à coefficients dans J. Cette égalité est évidente. ⌐

LEMME 3.3. Dans le carré (**) le groupe St(Λ) opère sur St(Λ,I) et St(Λ,J) comme dans le carré $\underline{\underline{G}}^{st}$, et on fait opérer St(Λ), St(Λ,I) et St(Λ,J) trivialement sur $I \otimes_{\Lambda^e} J$. Muni de ces actions et de l'application h définie par le lemme 3.2. le carré (**) devient un carré croisé $\underline{\underline{G}}^{\otimes}$.

Démonstration. La vérification de la propriété i) est immédiate car l'image de $I \otimes_{\Lambda^e} J$ dans St(Λ,I) (resp. St(Λ,J)) est centrale.

La propriété vi) est immédiate compte-tenu des actions triviales et du fait que $I \otimes_{\Lambda^e} J$ est abélien.

La propriété v) (resp. iv)) est conséquence de c) (resp. a) et b)) du lemme 3.2.

Démontrons la propriété iii). Il faut tout d'abord montrer que $h(\lambda(\ell),m') = 0$ pour tout $\ell = a \otimes b$ et tout $m' \in St(\Lambda,J)$. On a

$$\begin{aligned} h(\lambda(\ell),m') &= h(y_{12}^{a} x_{21}^{b} \cdot y_{12}^{-a},m') \\ &= h(y_{12}^{a},m')-h(y_{12}^{a},x_{21}^{-b}\cdot m') \\ &= a \otimes \phi(m')_{21}-a \otimes \phi(x_{21}^{-b}\cdot m')_{21} \\ &= a \otimes (\phi(m')_{21}-\phi(x_{21}^{-b}\cdot m')_{21}). \end{aligned}$$

Or, puisque $b \in J$ et $\phi(m') \in GL(\Lambda,J)$, on a $\phi(m')_{21} \equiv \phi(x_{21}^{-b}\cdot m')_{21} \bmod J^2$. Donc $h(\lambda(\ell),m') = 0$. Pour montrer l'autre relation de iii) il suffit de montrer que $h(m,y_{ij}^{b}) = \phi(m)_{ji} \otimes b$. Or $\phi(m_1 m_2)_{ji} \equiv \phi(m_1)_{ji}+\phi(m_2)_{ji} \bmod I^2$, donc on peut supposer que m est un générateur de St(Λ,I). On a alors $h(z.y_{k\ell}^{a},y_{ij}^{b}) = a \otimes \phi(z^{-1})_{\ell i} b\phi(z)_{jk} = \phi(z)_{jk} a\phi(z)_{\ell i}^{-1} \otimes b = \phi(z\cdot y_{k\ell}^{a})_{ji} \otimes b$.

La propriété ii), à savoir $\lambda h(m,m') = m\, \mu(m')\cdot m^{-1}$ se démontre de la manière suivante. D'après les propriétés précédentes on peut supposer que $m' = y_{ij}^{b}$ et on a alors

$$\lambda h(m,y_{ij}^{b}) = \lambda(a \otimes b) = y_{12}^{a}x_{21}^{b} \cdot y_{12}^{-a} \text{ avec } a = \phi(m)_{ji}.$$ Le même calcul que dans [Kn] lemme 10 montre que $m\, x_{ij}^{b} \cdot m^{-1} = y_{12}^{a}x_{21}^{b} \cdot y_{12}^{-a}$, soit $\lambda h(m,y_{ij}^{b}) = m\, x_{ij}^{b} \cdot m^{-1}$. □

L'homomorphisme de carré croisé $\psi : \underline{\underline{G}}^{\otimes} \to \underline{\underline{G}}^{st}$ est l'identité sur St(Λ,I), St(Λ,J), St(Λ), et défini par

$$a \otimes b \longmapsto [x_{12}^{(0,a,0,a)}, x_{21}^{(0,0,b,b)}] \text{ sur } I \otimes_{\Lambda^e} J.$$

On vérifie sans (trop de) peine que c'est bien un homomorphisme de carrés croisés. On a ainsi construit en fait (cf. 3.1) un morphisme de bicatégories en groupes $(G^{\otimes};St(D)/C(I),St(D')/C(J)) \to (G^{st};St(D)/C(I),St(D')/C(J))$. Son noyau est de la forme (A;1,1) et central car A est inclus dans $I \otimes_{\Lambda^e} J$ qui est central dans $G^{\otimes}$. Montrons que $\psi : G^{\otimes} \to G^{st}$ est surjectif. On note $h^{\otimes}$ (resp. h^{st}) la fonction h du carré croisé $\underline{\underline{G}}^{\otimes}$ (resp. $\underline{\underline{G}}^{st}$). Puisque $I \cap J = \{0\}$, on a GL(Λ;I,J) = 1 et donc (théorème 1) $St(\Lambda;I,J) = K_2(\Lambda;I,J)$. Ainsi, St(Λ;I,J) est central dans G^{st}. Il est facile de voir que tout élément de St(Λ;I,J) peut s'écrire comme une somme d'éléments de la forme $h^{st}(z,z')$, $z \in St(\Lambda,I)$ et $z' \in St(\Lambda,J)$. Cet élément est l'image de $h^{\otimes}(z,z')$ et ainsi ψ est surjectif. D'après la proposition 2.1 et le lemme 2.5 on a $H_1(\underline{\underline{B}}\, \underline{\underline{G}}^{st}) = H_2(\underline{\underline{B}}\, \underline{\underline{G}}^{st}) = 0$. En fin le groupe $G^{\otimes}$ est parfait (en particulier $a \otimes b = [y_{12}^{a}, y_{21}^{b}]$ dans $G^{\otimes}$) et donc toutes les conditions de la proposition 2.2 sont remplies. De l'isomorphisme entre $G^{\otimes}$ et G^{st} on déduit

$$I \otimes_{\Lambda^e} J \simeq St(\Lambda;I,J) = K_2(\Lambda;I,J).$$ □

3.b. <u>Le cas</u> $I \cap J$ <u>radical</u>. Dans cette sous-section l'anneau Λ est commutatif et l'idéal $I \cap J$ est dans le radical de Jacobson de Λ; donc pour tout $c \in I \cap J$ l'élément $1-c$ est inversible. Le principe de démonstration du théorème 3 est le même que celui du théorème 2 (voir 3.a.). Les calculs sont rendus un peu plus compliqués par le fait que l'image de $\phi_{\Lambda;I,J}$, à savoir $E(\Lambda;I,J)$, n'est plus nulle. Par contre on a $E(\Lambda;I,J) = E(\Lambda;I \cap J)$.

<u>Le groupe</u> $S(\Lambda;I,J)$. Pour tout idéal bilatère I de Λ on pose $D(\Lambda,I) = D(\Lambda;I,I)$. L'homomorphisme $t:D(\Lambda,I\cap J) \to D(\Lambda;I,J)$ est induit par l'inclusion naturelle de $I \cap J$ dans I et J.

On définit une application $D(\Lambda,I) \to St(\Lambda,I)$, $\langle a,u\rangle \longmapsto \langle a,u\rangle_*$, pour $a \in I$ et $u \in \Lambda$ de la manière suivante. Considérons $St(\Lambda,I)$ comme un sous-groupe de $St(D)/C(I)$, alors $\langle a,u\rangle_*$ est la classe de $\langle (0,a),(u,u)\rangle' \in St(D)$. Il est clair que cet élément est dans le noyau de la première projection et donc appartient à $St(\Lambda,I)$. L'application $D(\Lambda,I) \to St(\Lambda,I)$ ainsi définie est un homomorphisme de groupes grâce aux propriétés de $\langle -,-\rangle'$ démontrées par Dennis et Stein [D-S]. On utilisera le fait que, si I est radical cet homomorphisme induit un isomorphisme (*) entre $D(\Lambda,I)$ et $K_2(\Lambda,I)$, cf. [M-S] et [Kn].

Par définition le groupe $S(\Lambda;I,J)$ est la somme amalgamée centrale dans le carré

$$\begin{array}{ccc} D(\Lambda,I\cap J) & \xrightarrow{\ t\ } & D(\Lambda;I,J) \\ \downarrow & & \downarrow \\ St(\Lambda,I\cap J) & \longrightarrow & S(\Lambda;I,J) \end{array}$$

(*) Le même type de démonstration que celui utilisé ici pour deux idéaux permet de donner une démonstration relativement simple de l'isomorphisme $D(\Lambda,I) \cong K_2(\Lambda,I)$ lorsque I est radical. Les deux ingrédients principaux sont la surjectivité (première partie de [M-S]) et la théorie des extensions centrales de modules croisés $[L_3]$.

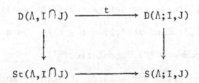

On en déduit immédiatement que $D(\Lambda;I,J)$ est central dans $S(\Lambda;I,J)$ et que le quotient est $E(\Lambda;I\cap J)$. Il nous sera utile par la suite d'avoir une présentation de $S(\Lambda;I,J)$ en tant que $St(\Lambda)$-groupe. Les générateurs sont les y_{ij}^{c} avec $c \in I\cap J$ et les $\langle a,b\rangle$ avec a et $b \in \Lambda$ tels que a ou $b \in I$ et a ou $b \in J$. Les relations sont celles de $St(\Lambda,I\cap J)$, plus celles de $D(\Lambda;I,J)$ plus

$$z\cdot\langle a,b\rangle = \langle a,b\rangle \quad , \quad z \in St(\Lambda),$$

$$[y_{ij}^{c}, \langle a,b\rangle] = 1 \quad ,$$

$$\langle c,u\rangle = \langle c,u\rangle_{*} \quad , \quad c \in I\cap J \text{ et } u \in \Lambda.$$

<u>Le carré croisé</u> $\underline{\underline{G}}^{D}$. Considérons le carré de groupes

$$(***) \qquad \begin{array}{ccc} S(\Lambda;I,J) & \xrightarrow{\lambda} & St(\Lambda,I) \\ \downarrow{\scriptstyle \lambda'} & & \downarrow{\scriptstyle \mu} \\ St(\Lambda,J) & \xrightarrow{\mu'} & St(\Lambda) \end{array}$$

où, ici, λ est défini de la manière suivante. L'inclusion $I\cap J \to I$ induit un homomorphisme $St(\Lambda,I\cap J) \to St(\Lambda,I)$. L'application $\langle a,b\rangle \longmapsto \langle a,b\rangle_{*}$ définit un homomorphisme $D(\Lambda;I,J) \to St(\Lambda,I)$. Ces deux homomorphismes coïncident sur $D(\Lambda,I\cap J)$ et l'image de $D(\Lambda;I,J)$ est centrale dans $St(\Lambda,I)$; il en résulte un unique morphisme de $S(\Lambda;I,J)$ dans $St(\Lambda,I)$ qui est λ. On a une définition analogue pour λ' qui rend le diagramme (***) commutatif.

Les actions du groupe $St(\Lambda)$ sur $St(\Lambda,I)$ et $St(\Lambda,J)$ sont celles du carré croisé $\underline{\underline{G}}^{St}$. Le groupe $St(\Lambda)$ opère sur $St(\Lambda,I\cap J)$ et on le fait opérer trivialement sur $D(\Lambda;I,J)$. Ceci définit une action sur $S(\Lambda;I,J)$. Pour faire opérer un élément de $St(\Lambda,I)$ (resp. $St(\Lambda,J)$) sur $S(\Lambda;I,J)$ on l'envoie tout d'abord dans $St(\Lambda)$ par μ (resp. μ') puis on utilise l'action de $St(\Lambda)$ sur $S(\Lambda;I,J)$. Afin d'avoir un carré croisé il nous reste à construire la fonction h^D, que nous noterons simplement h dans cette sous-section.

LEMME 3.4. Il existe une et une seule application $h: St(\Lambda,I) \times St(\Lambda,J) \to S(\Lambda;I,J)$ vérifiant les axiomes iv) et v) des carrés croisés ainsi que

$$h(y_{ij}^{a},y_{k\ell}^{b}) = 1 \qquad , i \neq \ell,\ j \neq k,$$

$$h(y_{ij}^{a},y_{j\ell}^{b}) = y_{i\ell}^{ab} \qquad , i \neq \ell,$$

$$h(y_{ij}^{a},y_{ki}^{b}) = y_{kj}^{-ba} \qquad , j \neq k,$$

$$h(y_{ij}^{a},y_{ji}^{b}) = \langle b,a\rangle y_{ij}^{ac} y_{ji}^{-ab} (x_{ji}^{1} x_{ij}^{-1}) \cdot (y_{ij}^{c} y_{ji}^{-ab}) y_{ji}^{-bc}$$

avec $a \in I$, $b \in J$ et $c = 1-(1-ab)^{-1} \in I \cap J$.

Remarque. Dans le groupe G^{St} on a $h^{St}(y_{ij}^{a},y_{k\ell}^{b}) = [x_{ij}^{(0,a,0,a)},x_{k\ell}^{(0,0,b,b)}] = 1$ lorsque $i \neq \ell$ et $j \neq k$. De même on a $h^{St}(y_{ij}^{a},y_{j\ell}^{b}) = x_{i\ell}^{(0,0,0,ab)}$ lorsque $i \neq \ell$, $h^{St}(y_{ij}^{a},y_{ki}^{b}) = x_{kj}^{(0,0,0,-ba)}$ lorsque $j \neq k$ et

$$h^{St}(y_{ij}^{a},y_{ji}^{b}) = \langle (0,0,b,b),(0,a,0,a)\rangle' x_{ij}^{(0,0,0,ac)} h_{ji}(1,1,1,1-ab) x_{ji}^{(0,0,0,-bc)} .$$

Ceci explique le choix des formules du lemme ci-dessus.

Démonstration du lemme 3.4. Le problème est de définir h sur tout élément du produit $St(\Lambda,I) \times St(\Lambda,J)$. Comme tout élément de $St(\Lambda,I)$ (resp. $St(\Lambda,J)$) est un produit d'éléments de la forme $z\cdot y_{ij}^{a}$ avec $z \in St(\Lambda)$ et $a \in I$ (resp. J) on se ramène, en utilisant les relations iv) des carrés croisés, à calculer des expressions de la forme $h(z\cdot y_{ij}^{a},z'\cdot y_{k\ell}^{b})$. En utilisant la relation v) des modules croisés on écrit $h(z\cdot y_{ij}^{a},z'\cdot y_{k\ell}^{b}) = z\cdot h(y_{ij}^{a},(z^{-1}z')\cdot y_{k\ell}^{b})$ donc il suffit de connaitre la valeur de $h(y^a_{ij}, z\cdot y^b_{k\ell})$. En fait on peut supposer que z est un produit d'éléments du type $x^u_{k\ell}$ et $x^v_{\ell k}$ car ces éléments $z\cdot y^b_{k\ell}$ engendrent $St(\Lambda,J)$ (cf. [Sw] p.---). Si le couple (ij) est différent de $(k\ell)$ et de (ℓk) on utilise de nouveau la relation v) puis la relation iv) pour se ramener à un produit d'éléments du type $h(y_{:.}, y_{:.})$.

Reste le cas où $(ij) = (k\ell)$ avec toujours z = produit de x^u_{ij} et de x^v_{ji}. Pour calculer $h(y^a_{ij}, z\cdot y^b_{ij})$ on utilise l'égalité $y^b_{ij} = x^1_{ik}\cdot y^b_{kj}(y^b_{kj})^{-1}$ pour remplacer $z\cdot y^b_{ij}$ par $(zx^1_{ik}z^{-1})\cdot(z\cdot y^b_{kj})(z\cdot y^b_{kj})^{-1}$. Or on a $zx^1_{ik}z^{-1} = x^\gamma_{jk}x^\alpha_{ik}$ et $z\cdot y^b_{kj} = y^{\gamma' b}_{ki}y^{\delta' b}_{kj}$ avec $\phi(z)_{ji} = \gamma$, $\phi(z)_{ii} = \alpha$, $\phi(z^{-1})_{ji} = \gamma'$ et $\phi(z^{-1})_{jj} = \delta'$. Il nous faut alors calculer $h(y^a_{ij}, (x^\gamma_{jk}x^\alpha_{ik})\cdot(y^{\gamma' b}_{ki}y^{\delta' b}_{kj}))$ et $h(y^a_{ij}, y^{-\delta' b}_{kj}y^{-\gamma' b}_{ki})$. De nouveau grâce aux relations iv) et v) on se ramène à $h(y_{:.}, y_{:.})$.

Il reste essentiellement à montrer que h passe bien aux relations des groupes de Steinberg relatifs ce qui est une vérification automatique. □

LEMME 3.5. <u>Le carré commutatif de groupes</u> (***) <u>muni des actions définies ci-dessus et de la fonction</u> h <u>du lemme</u> 3.4. <u>est un carré croisé que nous noterons</u> $\underline{\underline{G}}^D$.

<u>Démonstration</u>. Les vérifications des axiomes i) à vi) des carrés croisés sont automatiques et laissées à la vigilance du lecteur. □

<u>Le morphisme</u> $\psi : \underline{\underline{G}}^D \to \underline{\underline{G}}^{St}$. Ce morphisme est l'identité sur $St(\Lambda,I)$, $St(\Lambda,J)$ et $St(\Lambda)$. Pour le définir sur $S(\Lambda;I,J)$ on utilise l'homomorphisme naturel $St(\Lambda,I\cap J) = St(\Lambda;I\cap J, I\cap J) \to St(\Lambda;I,J)$ ainsi que $D(\Lambda;I,J) \to St(\Lambda;I,J) \subset St(T)/C(I,J)$

$$\langle a,b\rangle \longmapsto \langle(0,a,0,a),(0,0,b,b)\rangle'.$$

Puisque ces deux homomorphismes coïncident sur $D(\Lambda,I\cap J)$ (cf. section 1) et que l'image de $D(\Lambda,I\cap J)$ est centrale dans $St(\Lambda;I,J)$ ils définissent un unique homomorphisme $S(\Lambda;I,J) \to St(\Lambda;I,J)$.

On vérifie aisément que l'homomorphisme de carrés ψ ainsi défini est bien un homomorphisme de carrés croisés (cf. la remarque du lemme 3.4.).

<u>La surjectivité de</u> ψ. Puisque $S(\Lambda;I,J)$ est une extension de $E(\Lambda;I\cap J) = E(\Lambda;I,J)$ par $D(\Lambda;I,J)$ il suffit de démontrer que $D(\Lambda;I,J) \to K_2(\Lambda;I,J)$ est surjectif. Pour celà on utilise les deux lemmes suivants.

LEMME 3.6. <u>Soit</u> $\overline{\Lambda} = \Lambda/I\cap J$, $\overline{I} = I/I\cap J$, $\overline{J} = J/I\cap J$, <u>alors la suite</u> $D(\Lambda,I\cap J) \xrightarrow{t} D(\Lambda;I,J) \xrightarrow{p} D(\overline{\Lambda};\overline{I},\overline{J}) \to 1$ <u>est exacte</u>.

<u>Démonstration</u>. Définissons $p:D(\Lambda;I,J) \to D(\overline{\Lambda};\overline{I},\overline{J})$ par $p(\langle a,b\rangle) = \langle\overline{a},\overline{b}\rangle$, où $\overline{a}$ et $\overline{b}$ sont les images dans $\overline{\Lambda}$ de a et b. On vérifie aisément que p est un homomorphisme de groupes, qu'il est surjectif et que $p\circ t = 0$. On a donc un homomorphisme $\overline{p}$ de Coker t dans $D(\overline{\Lambda};\overline{I},\overline{J})$. Contruisons un homomorphisme réciproque $s:D(\overline{\Lambda};\overline{I},\overline{J}) \to$ Coker t. On pose $s(\langle\overline{a},\overline{b}\rangle) = \langle a,b\rangle_0$ = image de $\langle a,b\rangle$ dans Coker t, où a et b sont deux relèvements quelconques de $\overline{a}$ et $\overline{b}$ dans Λ. En fait, cette définition ne dépend pas du représentant choisi, car si c est un élément de $I\cap J$, $\langle a,b+c\rangle = \langle a,b\rangle\langle a,c(1-ab)^{-1}\rangle$, donc, dans Coker t, $\langle a,b+c\rangle_0 = \langle a,b\rangle_0$. L'application s est un homomorphisme de groupes, puisque les relations de $D(\Lambda;I,J)$ sont vérifiées dans Coker t; de plus elle admet $\overline{p}$ comme homomorphisme réciproque. La suite est donc exacte en $D(\Lambda;I,J)$. □

LEMME 3.7. <u>Avec les notations précédentes la suite ci-dessous est exacte</u>

$$K_2(\Lambda,I\cap J) \to K_2(\Lambda;I,J) \to K_2(\overline{\Lambda};\overline{I},\overline{J}) \to 1.$$

<u>Démonstration</u>. Considérons le carré commutatif d'espaces

$$\underline{K}(\Lambda;I,J) \qquad \begin{array}{ccc} K(\Lambda) & \longrightarrow & K(\Lambda/I) \\ \downarrow & & \downarrow \\ K(\Lambda/J) & \longrightarrow & K(\Lambda/I+J) \end{array}$$

associé au triplet $(\Lambda;I,J)$. Par définition (cf. section 2) la fibre homotopique de l'application induite entre les fibres horizontales est $K(\Lambda;I,J)$. De même, pour le triplet $(\bar{\Lambda};\bar{I},\bar{J})$ l'espace correspondant est $K(\bar{\Lambda};\bar{I},\bar{J})$.

La fibre du morphisme de carrés $\underline{\underline{K}}(\Lambda;I,J) \to \underline{\underline{K}}(\bar{\Lambda};\bar{I},\bar{J})$ est le carré

$$\begin{array}{ccc} K(\Lambda,I\cap J) & \longrightarrow & * \\ \downarrow & & \downarrow \\ * & \longrightarrow & * \end{array}$$

dont la fibre homotopique de l'application induite entre les fibres horizontales est bien évidemment $K(\Lambda,I\cap J)$. On en déduit une fibration homotopique

$$K(\Lambda,I\cap J) \to K(\Lambda;I,J) \to K(\bar{\Lambda};\bar{I},\bar{J})$$

La suite exacte du lemme résulte de la suite exacte d'homotopie de cette fibration et de l'isomorphisme $K_1(\Lambda,I\cap J) \simeq K_1(\Lambda;I,J)$. □

PROPOSITION 3.8. <u>Si l'idéal</u> $I\cap J$ <u>est radical</u>, <u>l'homomorphisme</u> $D(\Lambda;I,J) \to K_2(\Lambda;I,J)$ <u>est surjectif</u>.

<u>Démonstration</u>. Dans le diagramme commutatif

$$\begin{array}{ccccccc} D(\Lambda,I\cap J) & \xrightarrow{t} & D(\Lambda;I,J) & \xrightarrow{p} & D(\bar{\Lambda};\bar{I},\bar{J}) & \longrightarrow & 1 \\ \downarrow \simeq & & \downarrow & & \downarrow \simeq & & \\ K_2(\Lambda,I\cap J) & \longrightarrow & K_2(\Lambda;I,J) & \longrightarrow & K_2(\bar{\Lambda};\bar{I},\bar{J}) & \longrightarrow & 1 \end{array}$$

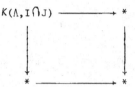

les lignes sont exactes d'après les lemmes précédents. Les flèches verticales extrêmes sont des isomorphismes d'après [M-S], [Kn] (*) et d'après le théorème 2. Il s'ensuit que la flèche verticale du milieu est surjective. □

Comme le noyau de $S(\Lambda;I,J) \to St(\Lambda;I,J)$ est dans $D(\Lambda;I,J)$ qui est central dans

$$G^D = (S(\Lambda;I,J) \rtimes St(\Lambda,I)) \rtimes (St(\Lambda,J) \rtimes St(\Lambda)),$$

on vient de démontrer (cf. prop. 3.1) que $\psi:\underline{\underline{G}}^D \to \underline{\underline{G}}^{St}$ est une extension centrale de bicatégories en groupes. D'après la proposition 2.1 et le lemme 2.5 on a $H_1(\underline{\underline{B}}\,\underline{\underline{G}}^{St}) = H_2(\underline{\underline{B}}\,\underline{\underline{G}}^{St}) = 0$ il nous suffit donc de montrer que G^D est parfait pour compléter les hypothèses de la proposition 2.2. Il est clair que les éléments de $St(\Lambda)$, $St(\Lambda,I)$ et $St(\Lambda,J)$ sont des produits de commutateurs dans G^D. Puisque $S(\Lambda;I,J)$ est une extension de $E(\Lambda;I\cap J)$ par $D(\Lambda;I,J)$ il suffit de montrer que tout élément de $D(\Lambda;I,J)$ est un produit de commutateurs dans G^D. En utilisant la suite exacte du lemme 3.6 et le fait que G^D est parfait on se ramène ā montrer que tout image d'un élément de $D(\Lambda,I\cap J)$ dans $D(\Lambda;I,J)$ est un produit de commutateurs dans G^D. Ce fait résulte de ce que $St(\Lambda,I\cap J) \rtimes St(\Lambda)$ est parfait et $D(\Lambda,I\cap J) = K_2(\Lambda,I\cap J)$.

Ainsi on peut appliquer la proposition 2.2, ce qui nous donne un isomorphisme $S(\Lambda;I,J) \simeq St(\Lambda;I,J)$. L'isomorphisme $D(\Lambda;I,J) \simeq K_2(\Lambda;I,J)$ en résulte, ce qui achève la preuve du théorème 3. □

Remarque. Lorsque $I\cap J=0$ l'isomorphisme entre $D(\Lambda;I,J)$ et $I \otimes_{\Lambda^e} J$ est donné par $\langle a,b\rangle \longmapsto -\, a \otimes b$.

4. Interprétation homologique des groupes $K_i(\Lambda,I)$ et $K_i(\Lambda;I,J)$.

On généralise aux cas relatif et birelatif les isomorphismes $K_1(\Lambda)\simeq H_1(GL(\Lambda))$, $K_2(\Lambda) \simeq H_2(E(\Lambda))$ et $K_3(\Lambda) \simeq H_3(St(\Lambda))$, (cf. [Kv], [G], [$L_1$]).

4.a. Les groupes $K_i(\Lambda,I)$. On utilise la notion de catégorie en groupes introduite dans [L_4], qui est équivalente à celle de module croisé. Une catégorie en groupes $\underline{G}$ = (G,N) est la donnée d'un groupe G, d'un sous-groupe N et de deux homomorphismes s et b:G → N tels que

1) $s|_N = b|_N = id_N$

2) [Ker s, Ker b] = 1.

A toute catégorie en groupes $\underline{G}$ est associé un espace $B\underline{G}$ et une application continue $\underline{B}\ \underline{G}$:B N → B $\underline{G}$ dont la fibre est homotopiquement équivalente à B(Ker s), (cf. [L_4]).

Soient Λ un anneau et I un idéal bilatère. Le produit fibré $D = \Lambda \times_{\Lambda/I} \Lambda$ permet de définir une catégorie en groupes $\underline{G}^{GL}_{\Lambda,I}$ = (GL(D),GL(Λ)), s et b étant in duits par les projections de D sur Λ. En effet l'axiome 2) des catégories en groupes est satisfait car le foncteur GL est exact à gauche. De même si E(Λ) désigne le sous-groupe des matrices élémentaires de GL(Λ), $\underline{G}^{E}_{\Lambda,I}$ = (E(D), E(Λ)) est une catégorie en groupes. Pour le foncteur St, il faut poser $\underline{G}^{St}_{\Lambda,I}$ = (St(D)/C(I), St(Λ)) car sinon, c'est-à-dire si l'on ne quotiente pas par C(I), l'axiome 2) n'est pas en général satisfait. En utilisant les notations introduites au paragraphe 2 concernant l'homologie d'une application, on a le

THEOREME 4. Pour tout idéal bilatère I de Λ, on a les isomorphismes $K_1(\Lambda,I) \simeq H_1(\underline{B}\underline{G}^{GL}_{\Lambda,I},\mathbb{Z})$, $K_2(\Lambda,I)\simeq H_2(\underline{B}\underline{G}^{E}_{\Lambda,I},\mathbb{Z})$ et $K_3(\Lambda,I)\simeq H_3(\underline{B}\underline{G}^{St}_{\Lambda,I},\mathbb{Z})$ avec

$$\underline{B}\,\underline{G}^{GL}_{\Lambda,I} : BGL(\Lambda) \to B(GL(\Lambda)/GL(\Lambda,I))$$

$$\underline{B}\,\underline{G}^{E}_{\Lambda,I} : BE(\Lambda) \to B(E(\Lambda)/E(\Lambda,I))$$

$$\underline{B}\,\underline{G}^{St}_{\Lambda,I} : BSt(\Lambda) \to B\underline{G}^{St}_{\Lambda,I} .$$

Démonstration. Posons $L_1 = \mathrm{Coker}(K_1(\Lambda) \to K_1(\Lambda/I))$. On a une suite exacte $1 \to GL(\Lambda)/GL(\Lambda,I) \to GL(\Lambda/I) \to L_1 \to 1$. On en déduit la fibration $K(L_1,0) \to B(GL(\Lambda)/GL(\Lambda,I)) \to BGL(\Lambda/I)$. Cette fibration est simple, donc $K(L_1,0) \to B(GL(\Lambda)/GL(\Lambda,I))^+ \to BGL(\Lambda/I)^+$ est encore une fibration. Considérons alors le diagramme de fibrations suivant:

$$\begin{array}{ccccc}
* & \longrightarrow & * & \longrightarrow & K(L_1,0) \\
\downarrow & & \downarrow & & \downarrow \\
U_1 & \longrightarrow & BGL(\Lambda)^+ & \longrightarrow & B(GL(\Lambda)/GL(\Lambda,I))^+ \\
\downarrow & & \downarrow & & \downarrow \\
V_1 & \longrightarrow & BGL(\Lambda)^+ & \longrightarrow & BGL(\Lambda/I)^+
\end{array} .$$

On calcule $\pi_1 U_1 = \pi_1 V_1 = K_1(\Lambda,I)$. Tous les espaces intervenant dans ces fibrations sont simples, donc $\pi_1(U_1) = H_1(U_1) = H_1((B\underline{G}^{GL}_{\Lambda,I})^+) = H_1(B\underline{G}^{GL}_{\Lambda,I})$. On a ainsi montré $K_1(\Lambda,I) \simeq H_1(B\underline{G}^{GL}_{\Lambda,I})$.

Posons $L_2 = \mathrm{Coker}\ (K_2(\Lambda) \to K_2(\Lambda/I))$. On a une suite exacte $1 \to L_2 \to E(\Lambda)/E(\Lambda,I) \to E(\Lambda/I) \to 1$. On en déduit un diagramme de fibrations

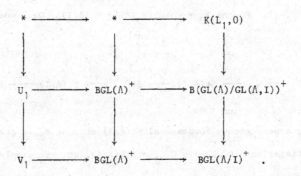

$$\begin{array}{ccccc}
K(L_2,0) & \longrightarrow & * & \longrightarrow & K(L_2,1) \\
\downarrow & & \downarrow & & \downarrow \\
U_2 & \longrightarrow & BE(\Lambda)^+ & \longrightarrow & B(E(\Lambda)/E(\Lambda,I))^+ \\
\downarrow & & \downarrow & & \downarrow \\
V_2 & \longrightarrow & BE(\Lambda)^+ & \longrightarrow & BE(\Lambda/I)^+
\end{array}$$

On calcule $\pi_1 V_2 = L_2$, $\pi_2 V_2 = K_2(\Lambda,I)$, d'où $\pi_1 U_2 = 0$, $\pi_2 U_2 = K_2(\Lambda,I)$. Par le théorème d'Hurewicz, on a $H_2(U_2) = \pi_2 U_2 = K_2(\Lambda,I)$. La comparaison de la suite exacte d'homologie relative de l'application $(\underline{B}\ \underline{G}^{E}_{\Lambda,I})^+$ et de la suite exacte issue de la suite spectrale d'homologie montre que $H_2((\underline{B}\ \underline{G}^{E}_{\Lambda,I})^+) = H_2(U_2)$. On en déduit $K_2(\Lambda,I) \simeq H_2(\underline{B}\ \underline{G}^{E}_{\Lambda,I})$. Posons $L_3 = \mathrm{Coker}(K_3(\Lambda) \to K_3(\Lambda/I))$. On a une suite exacte (cf. [L_3])

$$1 \longrightarrow L_3 \longrightarrow St(\Lambda,I) \xrightarrow{\mu} St(\Lambda) \longrightarrow St(\Lambda/I) \longrightarrow 1 .$$

L'espace $B\ \underline{G}^{St}_{\Lambda,I}$ a pour groupe fondamental $St(\Lambda/I)$ et pour π_2 le groupe $\mathrm{Ker}\,\mu = L_3$. On a donc une fibration $K(L_3,2) \to B\ \underline{G}^{St}_{\Lambda,I} \to BSt(\Lambda/I)$, qui est simple. On en déduit un diagramme de fibrations

$$\begin{array}{ccccc}
K(L_3,1) & \longrightarrow & * & \longrightarrow & K(L_3,2) \\
\downarrow & & \downarrow & & \downarrow \\
U_3 & \longrightarrow & BSt(\Lambda)^+ & \longrightarrow & (B\ \underline{G}^{St}_{\Lambda,I})^+ \\
\downarrow & & \downarrow & & \downarrow \\
V_3 & \longrightarrow & BSt(\Lambda)^+ & \longrightarrow & BSt(\Lambda/I)^+
\end{array}$$

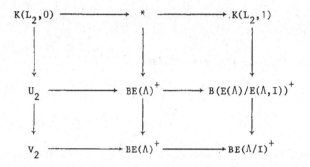

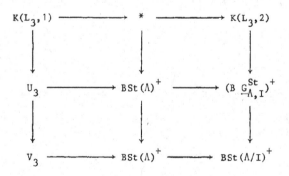

On calcule $\pi_1 V_3 = 0$, $\pi_2 V_3 = L_3$ et $\pi_3 V_3 = K_3(\Lambda,I)$, d'où $\pi_1 U_3 = \pi_2 U_3 = 0$ et $\pi_3 U_3 = K_3(\Lambda,I)$. Par le théorème d'Hurewicz, on a $H_3(U_3) \simeq \pi_3 U_3 = K_3(\Lambda,I)$. La comparaison de la suite exacte d'homologie relative de l'application $(\underline{B}\,\underline{G}^{St}_{\Lambda,I})^+ : BSt(\Lambda)^+ \to (B\,\underline{G}^{St}_{\Lambda,I})^+$ et de la suite spectrale d'homologie montre que $H_3((\underline{B}\,\underline{G}^{St}_{\Lambda,I})^+) \simeq H_3(U_3)$. On en conclut $H_3(\underline{B}\,\underline{G}^{St}_{\Lambda,I}) \simeq K_3(\Lambda,I)$. □

<u>Remarque 1</u>. Les groupes $K_1(\Lambda,I)$ et $K_2(\Lambda,I)$ s'interprètent donc comme des groupes d'homologie relative d'homomorphismes de groupes. Par contre ce n'est plus le cas pour $K_3(\Lambda,I)$ car l'espace $B\,\underline{G}^{St}_{\Lambda,I}$ a pour π_1 le groupe $St(\Lambda/I)$ et pour π_2 le groupe L_3.

<u>Remarque 2.</u> Il est clair d'après la démonstration précédente que

$$H_1(\underline{B}\,\underline{G}^{St}_{\Lambda,I}) = H_2(\underline{B}\,\underline{G}^{St}_{\Lambda,I}) = 0 \text{ et que } H_1(\underline{B}\,\underline{G}^{E}_{\Lambda,I}) = 0.$$

4.b. <u>Les groupes</u> $K_i(\Lambda;I,J)$.

Soient Λ un anneau, I et J deux idéaux bilatères quelconques de Λ. Avec les notations du paragraphe 2, on a trois bicatégories en groupes

$$\underline{\underline{G}}^{GL}_{\Lambda;I,J} = (GL(T);GL(D),GL(D'))$$

$$\underline{\underline{G}}^{E}_{\Lambda;I,J} = (E(T);E(D),E(D'))$$

$$\underline{\underline{G}}^{St}_{\Lambda;I,J} = (St(T)/C(I,J);St(D)/C(I),St(D')/C(J)).$$

Avec les notations introduites au paragraphe 2 concernant l'homologie d'un carré d'espaces on a le

THEOREME 5. Pour tous idéaux bilatères I et J de Λ, on a les isomorphismes

$$K_1(\Lambda;I,J) \simeq H_1(\underline{\underline{B}}\ \underline{\underline{G}}^{GL}_{\Lambda;I,J},\mathbb{Z})$$

$$K_2(\Lambda;I,J) \simeq H_2(\underline{\underline{B}}\ \underline{\underline{G}}^{E}_{\Lambda;I,J},\mathbb{Z})$$

$$K_3(\Lambda;I,J) \simeq H_3(\underline{\underline{B}}\ \underline{\underline{G}}^{St}_{\Lambda;I,J},\mathbb{Z}).$$

Démonstration. Le troisième isomorphisme a déjà été démontré dans le lemme 2.5. Le principe de la démonstration est exactement le même pour les deux autres, le foncteur St étant remplacé par le foncteur E (resp. GL). Les détails sont laissés à la vigilance du lecteur. □

Remarque 1. Les espaces du carré $\underline{\underline{B}}\ \underline{\underline{G}}^{GL}_{\Lambda;I,J}$ sont tous des classifiants de groupes discrets (i.e. des espaces de type (-,1)). Trois des espaces du carré $\underline{\underline{B}}\ \underline{\underline{G}}^{E}_{\Lambda;I,J}$ sont de type (-,1), le quatrième n'a que du π_1 et du π_2. Parmi les espaces du carré $\underline{\underline{B}}\ \underline{\underline{G}}^{St}_{\Lambda;I,J}$ l'un n'a que du π_1, deux n'ont que du π_1 et du π_2, le quatrième n'a que du π_1, π_2 et du π_3.

Remarque 2. On montre aisément que $H_1(\underline{\underline{B}}\ \underline{\underline{G}}^{E}_{\Lambda;I,J},\mathbb{Z}) = 0$ (comparer avec le lemme 2.5).

5. Applications au calcul de groupes K_2 et K_3.

On trouvera des applications de la partie droite de la suite exacte du corollaire (théorème 2) dans [S_2].

5.a. K_2 d'un anneau avec diviseurs de zéro.

Pour tout anneau A on pose $T(A) = A/(ab-ba)$. Le groupe abélien T(A) est isomorphe au groupe additif sous-jacent à l'anneau A lorsque A est commutatif.

PROPOSITION 5.1. [D-K]. Soit A un anneau noethérien régulier. On a $K_2(A[x,y]/(xy)) = K_2(A) \oplus T(A)$. En particulier si A est commutatif, on a $K_2(A[x,y]/(xy)) = K_2(A) \oplus A$, par exemple $K_2(\mathbb{Z}[x,y]/(xy)) = \mathbb{Z}/2\,\mathbb{Z} \oplus \mathbb{Z}$.

Démonstration. On applique le corollaire du théorème 2 à l'anneau $\Lambda = A[x,y]/(xy)$ et aux idéaux $I = \Lambda x$ et $J = \Lambda y$. Il est clair que $I \otimes_{\Lambda^e} J = T(A)$. La surjection $(\Lambda, I) \to (\Lambda/J, I+J/J) = (A[x],(x))$ est scindée, par conséquent on a une suite exacte

$$0 \to T(A) \to K_2(\Lambda, I) \to K_2(A[x],(x)) \to 0.$$

Puisque l'anneau A est noethérien régulier, d'après $[Q_2]$ on a $K_2(A[x],(x)) = 0$. D'autre part, il est clair que $K_2(\Lambda) = K_2(\Lambda, I) \oplus K_2(\Lambda/I)$. Pour la même raison que précédemment $K_2(\Lambda/I) = K_2(A[y]) = K_2(A)$. En définitive, on a $K_2(A[x,y]/(xy)) = K_2(A) \oplus T(A)$. □

Remarque. Dans [Sw] Swan a montré la nontrivialité de $K_2(\mathbb{Z}[x,y]/(xy))/K_2(\mathbb{Z})$. La proposition précédente a été montrée par Dennis et Krusemeyer [D-K] suivant une méthode différente. On peut aussi déduire certains autres résultats de leur article (théorème 3.1, proposition 4.2, théorème 4.7, 4.9) du corollaire du théorème 2.

5.b. K_3 d'une algèbre de groupe.

Nous allons appliquer le corollaire du théorème 2 au "carré de Rim". Soit C_p un groupe cyclique d'ordre p premier, de générateur t. Dans l'algèbre de groupe $\mathbb{Z}[C_p]$, on considère les idéaux $I = (1-t)\,\mathbb{Z}[C_p]$ et $J = (1+t+\ldots+t^{p-1})\,\mathbb{Z}[C_p]$. On a alors le carré cartésien

$$\begin{array}{ccccc} & \mathbb{Z}[C_p] & \xrightarrow{\ t\ \longmapsto\ \zeta\ } & \mathbb{Z}[\zeta] \\ t & \Big\downarrow f & & \Big\downarrow g \\ \downarrow & & & \\ 1 & \mathbb{Z} & \longrightarrow & \mathbb{F}_p \end{array}$$

où ζ est une racine p-ième de l'unité et $\mathbb{F}_p$ le corps à p éléments. L'application f est scindée et on pose $K_i(\mathbb{Z}[C_p]) = K_i(\mathbb{Z}) \oplus \tilde{K}_i(\mathbb{Z}[C_p])$. On constate aisément que le groupe $I \otimes_{\mathbb{Z}[C_p]} J$ est cyclique d'ordre p, de générateur $(1-t) \otimes (1+t+\ldots+t^{p-1})$. Son image dans $K_2(\mathbb{Z}[C_p],I)$ est le symbole $<1-t,1+t+\ldots+t^{p-1}> = <1-t,1>^p = 1$ (cf. [S_2]). Par conséquent, la suite exacte du corollaire s'écrit

(1) $\quad \tilde{K}_3(\mathbb{Z}[C_p]) \to K_3(\mathbb{Z}[\zeta],\ (1-\zeta)) \to \mathbb{Z}/p\mathbb{Z} \to 0.$

Remarquons que $K_3(\mathbb{Z}[\zeta],\ (1-\zeta))$ s'inscrit dans la suite exacte

(2) $\quad 0 \to K_3(\mathbb{Z}[\zeta],\ (1-\zeta)) \to K_3(\mathbb{Z}[\zeta]) \to \mathbb{Z}/(p^2-1)\mathbb{Z}$

car $K_4(\mathbb{F}_p) = 0$ et $K_3(\mathbb{F}_p) = \mathbb{Z}/(p^2-1)\mathbb{Z}$, (cf. [$Q_1$]).

PROPOSITION 4.2 <u>Le groupe $\tilde{K}_3(\mathbb{Z}[C_2])$ est de la forme $\mathbb{Z}/8\,\mathbb{Z} \oplus (?)$ où $(?)$ est un groupe inconnu. De plus, l'homomorphisme $\tilde{K}_3(\mathbb{Z}[C_2]) \to K_3(\mathbb{Z}) \simeq \mathbb{Z}/48\,\mathbb{Z}$ induit par $t \longmapsto -1$ est injectif sur $\mathbb{Z}/8\,\mathbb{Z}$ et trivial sur $(?)$.</u>

<u>Démonstration</u>. En utilisant le résultat de Lee-Szczarba (cf. [L-S]), $K_3(\mathbb{Z}) = \mathbb{Z}/48\,\mathbb{Z}$ et la suite exacte (2), on voit que $K_3(\mathbb{Z},(2)) = (K_3\mathbb{Z})_{(2)} = \mathbb{Z}/16\,\mathbb{Z}$. Ainsi la suite exacte (1) s'écrit

$$\tilde{K}_3(\mathbb{Z}[C_2]) \to \mathbb{Z}/16\mathbb{Z} \to \mathbb{Z}/2\,\mathbb{Z} \to 0.$$

Par conséquent $\tilde{K}_3(\mathbb{Z}[C_2])$ s'envoie surjectivement sur $\mathbb{Z}/8\,\mathbb{Z}$.

Considérons le produit en couronne $C_2 \int \Sigma_\infty$ où Σ_∞ est le groupe symétrique infini. Ce groupe s'identifie canoniquement à un sous-groupe de $GL(\mathbb{Z}[C_2])$ (cf. $[L_2]$). Cette inclusion induit une application $B(C_2 \int \Sigma_\infty)^+ \to BGL(\mathbb{Z}[C_2])^+$, et donc un homomorphisme $\pi_3^s(BC_2) \to \tilde{K}_3(\mathbb{Z}[C_2])$. Considérons le diagramme

$$\begin{array}{ccccc}
\pi_3^s(BC_2) & \longrightarrow & \tilde{K}_3(\mathbb{Z}[C_2]) & \twoheadrightarrow & \mathbb{Z}/8\mathbb{Z} \\
{\scriptstyle tr}\downarrow & (a) & \downarrow {\scriptstyle t \mapsto -1} & (b) & \downarrow \cap \\
\pi_3^s & \longrightarrow & K_3(\mathbb{Z}) & \xrightarrow{\simeq} & \mathbb{Z}/48\mathbb{Z}
\end{array}$$

dans lequel tr est le transfert en homotopie stable. On vient de voir que le carré (b) est commutatif. Le lemme du paragraphe 3 de $[L_2]$ implique que le carré (a) est aussi commutatif. D'après le théorème de Kahn-Priddy [K-P], tr est surjectif sur la 2-torsion. Or d'après la suite spectrale d'Atiyah-Hirzebruch, $\pi_3^s(BC_2)$ est un 2-groupe d'ordre $\leqslant 8$. Par conséquent, $\pi_3^s(BC_2) = \mathbb{Z}/8\mathbb{Z}$ et tr est un isomorphisme sur la 2-torsion.

On utilise alors le fait que $\pi_3^s \to K_3(\mathbb{Z})$ est injectif (cf. $[Q_3]$) pour conclure que le composé $\mathbb{Z}/8\mathbb{Z} = \pi_3^s(BC_2) \to \tilde{K}_3(\mathbb{Z}[C_2]) \to \mathbb{Z}/8\mathbb{Z}$ est un isomorphisme. Ainsi on a $\tilde{K}_3(\mathbb{Z}[C_2]) \simeq \mathbb{Z}/8\mathbb{Z} \oplus (?)$, l'image du groupe inconnu (?) étant triviale dans $K_3(\mathbb{Z},(2))$. A fortiori, l'image de (?) est triviale dans $K_3(\mathbb{Z})$. □

On se propose de calculer le groupe $NK_2(\mathbb{Z}[C_p])$. Rappelons que pour tout anneau Λ on a $NK_2(\Lambda) = K_2(\Lambda[x])/K_2(\Lambda)$, où $\Lambda[x]$ est l'anneau de polynômes à une variable sur Λ.

PROPOSITION 4,3 <u>Pour tout nombre premier</u> p <u>on a un isomorphisme de groupes abéliens</u> $NK_2(\mathbb{Z}[C_p]) \simeq x.(\mathbb{Z}/p)[x]$.

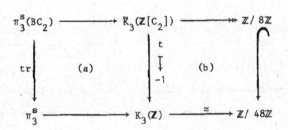

<u>Démonstration</u>. On applique le corollaire du théorème 2 au carré cartesien

$$\begin{array}{ccc} \mathbb{Z}[C_p][x] & \longrightarrow & \mathbb{Z}[\zeta][x] \\ \downarrow & & \downarrow \\ \mathbb{Z}[x] & \longrightarrow & \mathbb{F}_p[x] \end{array}$$

et on trouve $K_2(\mathbb{Z}[C_p][x];I[x],J[x]) \simeq (\mathbb{Z}/p)[x]$.

D'autre part les trois anneaux $\mathbb{Z}$, $\mathbb{F}_p$, $\mathbb{Z}[\zeta]$ sont noethériens réguliers, donc d'après Quillen [Q_2] leur K-théorie est inchangée quand on ajoute une indéterminée. Ainsi $K_2(\mathbb{Z}[C_p][x];I[x],J[x])/K_2(\mathbb{Z}[C_p];I,J) = K_2(\mathbb{Z}[C_p][x])/K_2(\mathbb{Z}[C_p])$

Finalement on trouve:

$$K_2(\mathbb{Z}[C_p][x])/K_2(\mathbb{Z}[C_p]) \simeq (\mathbb{Z}/p)[x]/(\mathbb{Z}/p) = x.(\mathbb{Z}/p)[x]. \quad \square$$

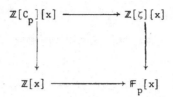

BIBLIOGRAPHIE

[D-K] R. K. DENNIS and M. KRUSEMEYER: $K_2(A[x,y]/(xy))$, a problem of Swan and related computations, J. Pure Applied Algebra 15 (1979), 125-148.

[D-S] R. K. DENNIS and M. R. STEIN: K_2 of radical ideals and semi-local rings revisited in: Algebraic K-Theory II, Lecture Notes in Math. 342 (Springer-Verlag, Berlin, 1973), 281-303.

[G] S. M. GERSTEN: K_3 of a ring is H_3 of the Steinberg group, Proc. Amer. Math. Soc. 37 (1973), 366-368.

[K-P] D. KAHN and S. PRIDDY: Applications of the transfer to stable homotopy, Bull. A.M.S. 78 (1972), 981-987.

[Kv] M. KERVAIRE: Multiplicateurs de Schur et K-théorie dans "Essays in Topology and related topics", Mémoires dédiés à G. de Rahm, pp. 212-225, Springer Verlag, 1970.

[Kn] F. KEUNE: The relativization of K_2, J. Algebra 54 (1978), 159-177.

[L-S] R. LEE and R. H. SZCZARBA: The group $K_3(\mathbb{Z})$ is cyclic of order 48, Ann. Math. 104 (1976), 31-60.

[L_1] J. -L. LODAY: K-théorie algébrique et représentations de groupes, Ann. Sci. Ecole Norm. Sup. 9 (1976), 309-377.

[L_2] J. -L. LODAY: Les matrices monomiales et le groupe de Whitehead Wh_2, dans "Alg. K-theory" (Evanston, 1976), Springer Lecture Notes 551 (1977), 155-163.

[L_3] J. -L. LODAY: Cohomologie et groupe de Steinberg relatifs, J. Algebra 54 (1978), 178-202.

[L_4] J. -L. LODAY: Spaces with finitely many nontrivial homotopy groups (to appear).

[M-S] H. MAAZEN and J. STIENSTRA: A presentation of K_2 of split radical pairs, J. Pure Applied Algebra 10 (1977), 271-294.

[M] J. MILNOR: "Introduction to Algebraic K-theory", Annals of Math. Studies, n° 72, Princeton Univ. Press, Princeton, N.J., 1971.

[Q_1] D. QUILLEN: Cohomology of groups, in "Actes Congres Intern. Math." 1970, t.2, p. 47-51.

[Q_2] D. QUILLEN: Higher algebraic K-theory I, in Alg. K-theory I, Springer Lecture Notes 341 (1972), 85-147.

[Q_3] D. QUILLEN: Letter from Quillen to Milnor on $\mathrm{Im}(\pi_i 0 \xrightarrow{J} \pi_i^s \longrightarrow K_i(\mathbb{Z}))$ in Alg. K-theory (Evanston 1976), Springer Lecture Notes 551 (1976), 182-188.

[S_1] M. STEIN: Relativizing functors on rings and algebraic K-theory, J. Alg. 19 (1971), 140-152.

[S_2] M. STEIN: Excision and K_2 of group rings (preprint).

[Sw] R. G. SWAN: Excision in Algebraic K-theory, J. Pure Applied Algebra 1 (1971), 221-252.

[Wh] J. H. C. WHITEHEAD: Combinatorial homotopy II, Bull. Amer. Math. Soc. 55 (1949), 453-496.

Institut de Recherche Mathématique Avancée (C.N.R.S.)
7 rue R. Descartes
67084 STRASBOURG, FRANCE.

STABILITY FOR K_2 OF DEDEKIND RINGS OF ARITHMETIC TYPE

Wilberd van der Kallen

§1 Introduction

Dunwoody [5] has shown that when R is a euclidean ring the map $K_2(2,R) \to K_2(n,R)$ is surjective for $n \geq 3$. On the other hand Dennis and Stein [4] have given examples where R is a ring of integers in a quadratic imaginary number field and $K_2(2,R) \to K_2(3,R)$ is not surjective. But from the study of K_1 the quadratic imaginary case is known to have particularly bad stability behaviour. (cf. [2], [10], [11]). To be specific, recall that $SL_2(R) = E_2(R)$ when R is a Dedekind ring of arithmetic type with infinitely many units, i.e. when R is the ring of S-integers in a global field, where S is a finite set of places, containing all archimedean places, and $|S| \geq 2$. (See [13]) (Terminology as in Bass-Milnor-Serre [1]). In contrast, there are only five quadratic imaginary number fields whose ring of integers R satisfies $SL_2(R) \neq E_2(R)$. (The five cases are those with R euclidean.) The main result of this paper is that at the K_2 level the situation is similar:

<u>THEOREM 1</u>. Let R be a Dedekind ring of arithmetic type with infinitely many units. Then $K_2(2,R) \to K_2(R)$ is surjective and $K_2(n,R) \to K_2(R)$ is an isomorphism for $n \geq 3$.

From this theorem and its proof one sees that for the rings in question the group $K_2(R)$ is closely related with $E_2(R)$ and therefore also with such topics as the theory of division chains. (cf. [3], [9]). We hope that a further exploration of this connection will give some useful information on $K_2(R)$. Theorem 1 is proved by expanding our earlier proof of the injective stability theorem for K_2 of finite dimensional noetherian rings. ([6]).(For surjectivity see also §5). This time general position arguments are not enough. At a crucial point we need to know, for certain $q,s \in R$, how the "relative elementary subgroup" $E(R,qR)$ of $SL_2(R)$, introduced by Vaserstein in connection with the congruence subgroup problem for SL_2, intersects the congruence subgroup $\ker(SL_2(R) \to SL_2(R/sqR))$. We get the answer from the explicit description (by power norm residue symbols) of the failure of the congruence subgroup property for $SL_2(R)$, as obtained by Vaserstein [13]. (For corrections to the proof see Liehl [8]). So we need some very specific and deep arithmetic information on the ring R in order to get such a sharp bound for the range of stability. We do not need such information in the proof of theorem 2 below (see 2.4). Theorem 2 is a quite general stability theorem for K_2. It is better than the main results in [6] and it is also slightly stronger than the version proved by Suslin and Tulenbayev. (Compare Corollary 2.6). It is no surprise that we recover the Suslin-Tulenbayev Theorem, as we borrow from its proof. But our constructions are described in a different language.

§2 A General Stability Theorem for K_1 and K_2

2.1 Let R be an associative ring with identity. For $n \geq 2$, $q \in R$, we define $U_n(q) = \{a \in R^n$: the column $(1+a_1q, a_2q, \ldots, a_nq)$ is unimodular$\}$.

REMARK. When not stated otherwise, unimodularity will refer to columns, not rows. We should use notations like $(b_1,\ldots,b_n)^T$ for a column, but we simply write $(b_1,\ldots,b_n)$, as in [6]. To get a clear picture the reader has to draw the columns as honest columns anyway.

2.2 We define elementary operations on $U_n(q)$ as follows. (Compare also [1] Ch I §2 and [7] §2).
For $2 \leq i \leq n$, $p \in R$, $a \in U_n(q)$, put
$e_i(p)(a) = (a_1,\ldots,a_{i-1},a_i+p(1+qa_1),a_{i+1},\ldots,a_n)$ and
$e^i(p)(a) = (a_1+pa_i,a_2,\ldots,a_n)$.
These elementary operations $e_i(p)$, $e^i(p)$ generate a group of permutations of $U_n(q)$. The orbit of $(0,\ldots,0)$ under the action of this group is denoted $EU_n(q)$. So an element b of $EU_n(q)$ can be reduced to zero by a finite number of elementary operations. The minimum number that is needed is called the complexity of $(b;q)$, or of b (with respect to q). For instance, $(0,\ldots,0)$ has complexity zero and $(0,\ldots,0,1)$ has complexity one. Several of the constructions below will depend on a choice of the reduction to zero of an element of $EU_n(q)$. We will establish useful properties of these constructions by induction on complexity.

2.3 We say that R satisfies SR_n^2 when the following holds. For any pair of unimodular columns $(a_1,\ldots,a_n)$, $(b_1,\ldots,b_n)$

there are $t_i \in R$ such that both $(a_1+t_1a_n,\ldots,a_{n-1}+t_{n-1}a_n)$ and $(b_1+t_1b_n,\ldots,b_{n-1}+t_{n-1}b_n)$ are unimodular.

2.4 THEOREM 2.

Let $n \geq 2$ and let R satisfy SR^2_{n+1}. Assume that $EU_n(q)$ equals $U_n(q)$ for all $q \in R$. Then

(i) $K_1(n-1,R) \to K_1(R)$ is surjective and

$K_1(m,R) \to K_1(R)$ is an isomorphism for $m \geq n$.

(ii) $K_2(n,R) \to K_2(R)$ is surjective and

$K_2(m,R) \to K_2(R)$ is an isomorphism for $m \geq n+1$.

2.5 REMARKS. The surjectivity in part (i) is well known. The injectivity of $K_1(m,R) \to K_1(R)$ is also known for $m \geq n+1$. For $m = n$ it can be proved in a traditional fashion, but we will only give an outrageously complicated proof here. Namely, we will get the result as an immediate consequence of our proof of injectivity in part (ii). Surjectivity in part (ii) will also come as a corollary of our proof of injectivity, so all efforts are directed at proving this injectivity for K_2. When one only seeks surjectivity for K_2, there is an easier way, not expounded here. Note that in the situation of Theorem 1 we do describe the easier way. (See section 5).

2.6 We will see in the next section that SR_n (cf. [6] 2.1) implies the hypotheses of Theorem 2. Thus we get

COROLLARY (Suslin, Tulenbayev [12]).

Let R satisfy SR_n, $n \geq 2$. Then $K_2(n,R) \to K_2(R)$ is surjective and $K_2(m,R) \to K_2(R)$ is an isomorphism for $m \geq n+1$.

2.7 Note that in the same fashion part (i) of Theorem 2 implies the standard stability theorem for K_1. From the following application one sees that Theorem 2 may yield a better range than the theorem of Suslin and Tulenbayev. First recall that an integral domain is called totally imaginary if its elements are integral over $\mathbb{Z}$ and its field of fractions is a totally imaginary number field. (cf. [1] and [15] §16).

<u>COROLLARY</u>. Let R be a 1-dimensional commutative ring, finitely generated as a $\mathbb{Z}$-algebra. Assume that for each minimal prime ideal P the domain R/P has infinitely many units and is not totally imaginary. Then $K_2(2,R) \to K_2(3,R)$ is surjective and $K_2(m,R) \to K_2(R)$ is an isomorphism for $m \geq 3$.

<u>PROOF</u>. We will see in the next section that R satisfies SR_3^2. (Prop. 3.8). Remains to show $EU_2(q) = U_2(q)$. This equality is equivalent with the equality $E(R,qR) = G(R,qR)$. (notation of [13]). It is instructive to check this. If R is a Dedekind ring of arithmetic type, the conditions of the corollary make that Vaserstein's theorem applies, so that in fact $E(I,J) = G(I,J)$ for any pair of ideals I, J in R. Therefore, in the general case, one may argue in the fashion of [15] §16, with the $E(R,I)$ of [15] replaced by the $E(R,I)$ of [13]. (cf. [8]).

<u>REMARK</u>. This Corollary implies part of Theorem 1. This part is easier than the remainder of Theorem 1.

§3 More about stable Range Conditions

3.1 In this section we collect some technicalities. For unexplained notation and terminology, see [6].

3.2 When $(a_1,\dots,a_n) \in U_n(q)$ there are $b_1,\dots,b_n$ with $b_1(1+a_1q)+b_2a_2q+\dots+b_na_nq = 1$. Put $c = \sum_1^n b_ia_i$. Then $(1-qc)(1+qa_1) - 1 + \sum_2^n qb_ia_i$ has value zero, so the column $(1+qa_1,a_2,\dots,a_n)$ is unimodular too.

3.3 LEMMA. If R satisfies SR_n then $EU_n(q)$ equals $U_n(q)$ for all $q \in R$.

PROOF. Fix an orbit in $U_n(q)$. We seek a convenient choice for a representative $(a_1,\dots,a_n)$ of this orbit. For any choice the column $(1+qa_1,a_2,\dots,a_n)$ is unimodular, hence we can modify the choice so that $(a_2,\dots,a_n)$ is unimodular, next so that $a_1 = 0$, finally so that $a = 0$. Therefore it must have been the orbit $EU_n(q)$ of zero.

3.4 LEMMA. SR_n implies SR^2_{n+1} $(n \geq 2)$.

PROOF. Given unimodular columns a,b in R^{n+1} we have to find $v \in R^n$ so that the columns $(a_1+v_1a_{n+1},\dots,a_n+v_na_{n+1})$, $(b_1+v_1b_{n+1},\dots,b_n+v_nb_{n+1})$ are unimodular. For any $g \in E([n] \times [n+1])$ we may replace the pair a,b by ga,gb. Using SR_n we may thus reduce to the case $a_n = 1$. By [14] Theorem 1 there is a suitable v with $v_n = 0$, in that case.

3.5 LEMMA. Let $M \in GL_m(R)$, $m \geq 2$. Let $(a_1,\dots,a_m)$ be the first column of M and $(b_1,\dots,b_m)$ the last column of M^{-1}. Then $(a_1,\dots,a_{m-1})$ is unimodular if and only if $(b_2,\dots,b_n)$ is unimodular.

PROOF. If $(a_1,\dots,a_{m-1})$ is unimodular, reduce to the case that $a_m = 0$ by multiplying M from the left with a lower triangular matrix. If $(b_2,\dots,b_m)$ is unimodular, reduce to the case that $b_1 = 0$.

3.6 LEMMA. Let R satisfy SR^2_{n+1}, $n \geqslant 2$. Then R satisfies $SR^3_{n+2}(n+2,n+2)$.

PROOF. Let $M_1,M_2,M_3 \in GL_{n+2}(R)$. By the previous lemma we will be done if we show that there is $g \in E([n+2] \times \{1\})$ such that in each of the three matrices gM_i^{-1} the part of the last column consisting of the bottom n+1 entries is unimodular. Let a,b,c denote the last column of $M_1^{-1},M_2^{-1},M_3^{-1}$ respectively. We look for $v_i \in R$ such that the column $(a_2+v_2a_1,\dots,a_{n+2}+v_{n+2}a_1)$ and its two analogues are unimodular. For any $g \in E(\{1\}^* \times [n+2])$ we may replace the triple a,b,c by ga,gb,gc. Using SR_{n+1} we may therefore reduce to the case $a_2 = 1$. Then we want to solve our problem with some v satisfying $v_2 = 0$. We may add multiples of b_2 to $b_3,\dots,b_{n+2}$ and we may also add multiples of c_2 to $c_3,\dots,c_{n+2}$. So we may assume $(b_3,\dots,b_{n+2},b_1)$ and $(c_3,\dots,c_{n+2},c_1)$ are unimodular. Apply SR^2_{n+1}.

3.7 LEMMA. Let $n \geqslant 2$. Let R satisfy SR^2_{n+1} and let $GL_n(R)$ act transitively on unimodular columns of length n. Then R satisfies $SR^2_{n+1}(n+2,n+1)$.

PROOF. Let A,B be n by n+1 matrices, each obtained by deleting the bottom row of some element of $GL_{n+1}(R)$. Let $v,w \in R^n$. We want to find an $x \in R^{n+1}$ such that $v+Ax$, $w+Bx$ are unimodular. Clearly we may replace the system

A,B,v,w by AU,BU,v,w for $U \in GL_{n+1}(R)$ and also by T_1A,T_2B,T_1v,T_2w for $T_i \in GL_n(R)$. From SR^2_{n+1} and Lemma 3.5 it follows that there is $U \in E([n+1] \times \{1\})$ such that the first columns of AU and BU are unimodular. Therefore we may assume A,B have first columns of the form $(1,0,\ldots,0)$ and further that A has first row of the form $(1,0,\ldots,0)$. But then we can choose x so that the first coordinate of $v+Ax$ as well as the second coordinate of $w+Bx$ is equal to one.

3.8 PROPOSITION. Let R be finitely generated as a module over a central subring T whose maximal spectrum is noetherian of dimension $d, d < \infty$. Then R satisfies SR^2_{n+1} for $n \geq \max(2,d+1)$.

PROOF. Recall that R satisfies SR_{n+1} so that $GL_{n+1}(R)$ acts transitively on unimodular columns of length n+1. Using Lemma 3.5 once more we see that SR^2_{n+1} is now equivalent with $SR^2_{n+1}(n+1,n+1)$. By ([6], Theorem 8, pg. 134) it suffices to consider the case that T is a field. But then R satisfies SR_n and Lemma 3.4 applies.

EXERCISE. Give a more direct proof of the Proposition.

§4 Proof of the Theorems

4.1 We start modifying the proof of Theorem 4 of [6] in order to get a proof of Theorems 1 and 2 above. We introduce two sets of hypotheses.

Situation α: $n \geq 2$. The associative ring R satisfies SR^2_{n+1} and $EU_n(q)$ equals $U_n(q)$ for all $q \in R$.

This corresponds with the hypotheses of Theorem 2.

Situation β: $n=2$. The ring R is a Dedekind ring of arithmetic type with infinitely many units.

This corresponds with Theorem 1.

4.2 The proofs for situation α will mostly be simplified versions of those for situation β, but complicated versions of the arguments in [6]. While in [6] the pattern of the proof looks reasonable, the modifications presented here require more perseverance from the reader. From now on we assume that α or β applies. As we are going to use almost all of ([6], sections 2,3,4), we will save some space and refer the reader repeatedly to [6], telling him what to read and when. Of course we now replace the standing assumption $\widetilde{SR}_n$ of [6] by the assumption that α or β holds.

4.3 Read in [6]: All of section 1, 2.1 and 2.2, 3.4 through 3.19. The handwritten $\mathcal{L}$(resp $\mathcal{R}$) of [6] will be denoted by L(resp. R) in this paper.

LEMMA. $E_n(R)$ acts transitively on unimodular clumms of length n.

PROOF. In situation α this follows from the equality of $E\,U_n(1)$ and $U_n(1)$. In situation β it follows from $SL_2(R) = E_2(R)$, which is proved in [13], cf. [8].

4.4 Read [6] 3.20. There is a converse to [6] 3.20:

LEMMA. Let $\langle X,Y\rangle \in C$, $v \in R^{n+1}$. Let $L(x_{n+2}(v))\langle X,Y\rangle$ be defined at the bottom. Choose $T \in St(n+1)$ such that $\langle X,Y\rangle = \langle T\, x_{n+2,1}(*), *\rangle$. Then there is $w \in R^{n+1}$, $r \in R$, with $(w_2,..,w_{n+1})$ unimodular and $x_{n+2}(v)T = Tx_{n+2}(wr)$.

PROOF. Straightforward.

4.5 DEFINITION. When in the situation of lemma 4.4 the column v itself is also unimodular, we say that $L(x_{n+2}(v))\langle X,Y\rangle$ is defined *firmly* at the bottom. Note that then $(w_2r,..,w_{n+1}r)$ is unimodular so that we may replace w by wr and r by 1. Therefore this is the situation of [6] 3.20.

4.6 In the proof of [6] 3.22 it is used that $K_1(n,R) \to K_1(n+1,R)$ is injective. In situation α this property is not yet established. (What is known is that $K_1(m,R) \to K_1(R)$ is injective for $m \geq n+1$. The case m=n is the subject of part (i) of Theorem 2.) Let us therefore give another proof of [6] 3.22. It clearly suffices to show:

LEMMA. Let $v \in R^{n+1}$, $\langle X,Y\rangle \in C$. Let M be the n+1 by 2 matrix whose first column is v and whose second column is obtained from the first column of mat $\langle X,Y\rangle$ by deleting the last entry. Then $L(x_{n+2}(v))\langle X,Y\rangle$ is defined firmly at the bottom if and only if M is completable, by adding columns, to an element of $E(n+1,R)$.

REMARK. Because of lemma 4.3 completability of M is equivalent with unimodularity, i.e. with the existence of a 2 by n+1 matrix N such that NM = id.

PROOF OF LEMMA. Reduce to the case $X = x_{n+2,1}(*)$ and apply lemma 4.3.

4.7 Read [6] 3.23, 3.24, 3.25 and 3.36

LEMMA. Let $q \in R$, $v \in EU_n(q)$. There are $X \in St(\{n+1\}^* \times [n])$, $Y \in St(\{n+1\}^* \times \{1,n+1\}^*)$ such that $x_{1,n+2}(v_1)\ldots x_{n,n+2}(v_n)x_{n+2,1}(q) = X\,Y$ in $St(\{n+1\}^* \times \{n+1\}^*)$.

PROOF. By induction on the complexity of $(v;q)$. In this proof let $x(v)$ stand for $x_{1,n+2}(v_1)..x_{n,n+2}(v_n)$ in $St(\{n+1\}^* \times \{n+1\}^*)$.
If $w = e_i(b)v$ has lower complexity than v with respect to q, note that $x(v)\ x_{n+2,1}(q) = x_{i1}(-bq)\ x(w)\ x_{i1}(bq)\ x_{i,n+2}(-b)\ x_{n+2,1}(q) = x_{i1}(-bq)\ x(w)\ x_{n+2,1}(q)\ x_{i,n+2}(-b)$.
If $w = e^i(b)v$ has lower complexity, note that $x(v)\ x_{n+2,1}(q) = x_{1i}(-b)\ x(w)\ x_{n+2,1}(q)\ x_{n+2,i}(-bq)\ x_{1i}(b)$.

REMARK. This lemma and its proof explain the relevance of $EU_n(q)$ for computing in the chunk. The proof will be needed repeatedly.

4.8 NOTATION. If $s \in R$, $v \in R^n$, then $x_{n+2}(v,s)$ stands for $x_{n+2}(v_1,..,v_n,s)$. (cf. [6] 3.12).

DEFINITION Let $q, s \in R$, $v \in EU_n(q)$, $B \in St(\{n+2\} \times \{1\}^*)$, $T \in St(n+1)$, $U \in$ Up, $w \in R^{n+1}$ such that $x_{n+2}(w)T = Tx_{n+2}(v,s)$.
Then we say that $L(x_{n+2}(w))\ (Tx_{n+2,1}(q)\ B,U)$ is defined and let its value be $L(T)\ L(x_{n+1,n+2}(s))\ \langle X,Y\ B\ U\rangle$, where X,Y are chosen as in lemma 4.7. (Note that $L(x_{n+1,n+2}(s))\ \langle X,Y\ B\ U\rangle$ is defined at the bottom.) To see that $\langle X,Y\ B\ U\rangle$ does not depend on the particular choice of X,Y we may assume $T=1$, $B=1$, $U=1$, $s=0$ and then apply the squeezing principle with $i=n+1$.

REMARKS. Note that this $L(x_{n+2}(w))$ differs from those defined in [6] 3.13, 3.18 in that its argument is in Low × Up rather than C.

Of course the idea is to show eventually that $L(x_{n+2}(w))$ $(Tx_{n+2,1}(q)B,U$ depends only on w and $\langle Tx_{n+2,1}(q)\ B,U\rangle$.
Note further that the definition of $L(x_{n+2}(w))(*,*)$ is algorithmic in nature: Given a reduction to zero of v in $EU_n(q)$, the proof of lemma 4.7 tells us how to construct X and Y. In later proofs we will often need to consider the steps in the algorithm. This makes these proofs more tedious than proofs involving only maps defined at the bottom.

.9 DEFINITION. An element s of R is an irrelevant factor when the following holds. For each $q \in R$, $a \in R^n$ with $as \in EU_n(q)$ we have $a \in EU_n(sq)$. It is an easy exercise to show that units are irrelevant factors. As $EU_n(0) = R^n$, zero is also an irrelevant factor. In situation α it is clear that any element of R is an irrelevant factor. In situation β we have the following corollary to the main results of [1], [13] cf. [8]. Intuitively it says that elements in "general position" are irrelevant factors.

LEMMA. (Situation β). There is a non-zero ideal I of R such that, if $s \in R$ is such that s maps to a unit in the semi-local ring R/I, then s is an irrelevant factor.

PROOF. If $(a,b) \in R^2$, $q \in R$ then $(a,b) \in EU_2(q)$ if and only if, in the notations of [8], [13], the column $\begin{pmatrix} 1+aq \\ bq \end{pmatrix}$ occurs as the first column of some element of E(R,qR). If q=0, $EU_2(q) = R^2$. If $q \neq 0$, we learn from [8], [13] that $(a,b) \in EU_2(q)$ if and only if the power norm residue symbol $\left(\frac{1+aq}{bq}\right)_{r(q)}$ vanishes, where r(q) is an integer that depends in a certain way on valuations $v_{p_1}(q),..,v_{p_m}(q)$, where $p_1,...,p_m$ are fixed prime ideals. Now choose for I a non-zero ideal that is divisible by $p_1,...,p_m$. If s is a unit mod I then $v_{p_i}(sq) =$

$v_{p_i}(q)$ so that for $(as,bs) \in EU_2(q)$, $q \neq 0$, we have $\left(\frac{1+asq}{bsq}\right)_{r(q)} = 1$ hence $\left(\frac{1+asq}{bsq}\right)_{r(sq)} = 1$, hence $(a,b) \in EU_2(sq)$.

REMARK. In the sequel we will not use the fact that the condition on s can be satisfied independent of q. It would suffice to know that for any q there are sufficiently many s that are "irrelevant with respect to q".

10 LEMMA Let $w \in R^{n+1}$, $v \in R^n$, q, $s \in R$, $T \in St(n+1)$ such that $x_{n+2}(w)T = Tx_{n+2}(v,s)$. Let further $A \in St([n] \times \{n+1\})$, $B \in St(\{n+2\} \times \{1\}^*)$, $U \in Up$ and assume that s is an irrelevant factor. If $L(x_{n+2}(w))\ (Tx_{n+2,1}(q)\ B,\ A\ U)$ and $L(x_{n+2}(w))\ (Tx_{n+2,1}(q)\ B\ A,\ U)$ are both defined (see 4.8), then their values are the same.

PROOF. We may move T out of the way (towards the left) and then consider the case T=1. Note that we change w when removing T. In particular, we get $x_{n+2}(w) = x_{n+2}(v,s)$. The reader is expected to take care of such details in the sequel. Say $A = x_{1,n+1}(a_1)\dots x_{n,n+1}(a_n)$. We have v, v-as $\in EU_n(q)$. First we want to reduce to the case v=0, arguing by induction on the complexity of (v;q). When we replace B,U by 1, A^{-1} B A U respectively, the answers don't change. Therefore assume B=1. If $z = e^i(t)\ v$ has lower complexity than v, we use the proof of lemma 4.7 to express $L(x_{n+2}(v,s))\ (x_{n+2,1}(q),\ A\ U)$ in terms of $L(x_{n+2}(z,s))\ (x_{n+2,1}(q),\ A^1U^1)$ with suitable A^1,U^1. We find that $L(x_{n+2}(v,s))\ (x_{n+2,1}(q),\ A\ U) = L(x_{1i}(-t))\ L(x_{n+2}(z,s))\ (x_{n+2,1}(q),\ A^1U^1)$ where $A^1 = x_{1,n+1}(ta_i)\ A$, $U^1 = x_{n+2,i}(-qt)\ x_{n+2,n+1}(-q\ t\ a_i)\ x_{1i}(t)\ U$. By induction hypothesis we may rewrite the expression as $L(x_{1i}(-t))\ L(x_{n+2}(z,s))\ (x_{n+2,1}(q)\ A^1,U^1)$ and a straightforward computation shows that this equals $L(x_{n+2}(v,s))$

$(x_{n+2,1}(q)\ A,U)$. Similarly, if $z = e_i(t)v$ has lower complexity than v, we use the proof of lemma 4.7 to rewrite $L(x_{n+2}(v,s))$ $(x_{n+2,1}(q), A\ U)$ as $L(x_{i1}(-tq))\ L(x_{n+2}(z,s))\ (x_{n+2,1}(q),\ A\ x_{i,n+2}(-t)\ U)$ which equals (by induction hypothesis) $L(x_{i1}(-tq))\ L(x_{n+2}(z,s))$ $(x_{n+2,1}(q)\ A\ x_{i,n+1}(t\ q\ a_1),\ x_{i,n+1}(-t\ q\ a_1)\ x_{i,n+2}(-t)\ U)$. Note that one has to be careful about what the induction hypothesis tells exactly. One really has to move $A\ x_{i,n+1}(t\ q\ a_1)$ over the comma, not just A. With some patience one now finishes the check for this case too. We may further assume $v=0$. Then we have $-as \in EU_n(q)$ and because s is an irrelevant factor we even have $-a \in EU_n(sq)$. When $a=0$ there is nothing to prove. Therefore let us now argue by induction on the complexity of $(-a;sq)$. We have to show that $L(A)\ L(x_{n+2}(-as,s))$ $(x_{n+2,1}(q),\ x_{n+2,n+1}(q\ a_1)\ U)$ equals $L(x_{n+2,n+1}(s))\ \langle x_{n+2,1}(q),\ A\ U\rangle$. If $z = e^i(t)(-a)$ has lower complexity than $-a$, we argue as before, first rewriting $L(x_{n+2}(-as,s))\ (x_{n+2,1}(q),*)$ in terms of z. If $z = e_i(t)(-a)$ has lower complexity than $-a$, the computations are similar but a little longer. (The i-th co-ordinate of $e_i(t)(-a)$ is $-a_i+t-tsqa_1$).

.11 <u>LEMMA</u>. (Additivity, technical form).

Let $q,r,s \in R$, $v,w \in EU_n(q)$, $U \in \underline{Up}$, $B \in St(\{n+2\} \times \{1\}^*)$, $T \in St(n+1)$, $a,b \in R^{n+1}$ such that $x_{n+2}(a)\ T = T\ x_{n+2}(v,r)$ and $x_{n+2}(b)\ T = T\ x_{n+2}(w,s)$. Say $L(x_{n+2}(a))\ (T\ x_{n+2,1}(q)\ B,U) = \langle P,Q\rangle$ and assume that $L(x_{n+2}(b-a))\langle P,Q\rangle$ is defined at the bottom. In situation β assume further that $v_1=0$ and that $s-r-rqw_1$ is an irrelevant factor. Then $L(x_{n+2}(b))\ (T\ x_{n+2,1}(q)\ B,U) = L(x_{n+2}(b-a))\ L(x_{n+2}(a))$ $(T\ x_{n+2,1}(q)\ B,U)$.

<u>PROOF</u>. As we may move T over to the left and B over to U, we further assume T=1, B=1. We start with reducing to the case v=0. In situation α this is done in the same fashion as in the previous proof, using now also some properties of maps defined at the bottom (cf. [6] lemma 3.23). In situation β we have $v_1=0$ so that the reduction to v=0 can be done with steps which do not affect $s-r-rqw_1$. Thus we may reduce to the case v=0 in any situation. Put $y=s-r-rqw_1$.
What we have to show boils down to the equality of $L(x_{n+2}(w,s))(x_{n+2,1}(q), U)$ and $L(x_{n+1,1}(rq))\ L(x_{n+2}(w,y))\ \langle x_{n+2,1}(q),\ x_{n+1,n+2}(r)\ U\rangle$.
Of course we may assume U=1. As $L(x_{n+2}(w,y))\langle x_{n+2,1}(q), *\rangle$ is defined at the bottom, there are $d_i \in R$ such that $w_1=d_2w_2+..+d_nw_n+d_{n+1}y$. It is not difficult to see from this that $L(x_{n+2}(w,y))\ (x_{n+2,1}(q)\ x_{1,n+1}(d_{n+1}),\ x_{1,n+1}(-d_{n+1})\ x_{n+1,n+2}(r))$ is defined. In fact one can evaluate explicitly in this case and one sees, using [6] lemmas 3.23, 3.24, 3.25, that the result is the same as $L(x_{n+2}(w,y))\ \langle x_{n+2,1}(q)\ x_{1,n+1}(d_{n+1}),\ x_{1,n+1}(-d_{n+1})\ x_{n+1,n+2}(r)\rangle$. But we also know, by the previous lemma, that the result equals $L(x_{n+2}(w,y))\ (x_{n+2,1}(q), x_{n+1,n+2}(r))$. Thus it remains to show that $L(x_{n+2}(w,s))\ (x_{n+2,1}(q), 1) = L(x_{n+1,1}(rq))\ L(x_{n+2}(w,y))\ (x_{n+2,1}(q), x_{n+1,n+2}(r))$.
It is easy to see that if the first member in this equation is $\langle Xx_{n+2,1}(q)\ x_{n+1,1}(sq),\ x_{n+1,n+2}(s)\ Y\rangle$ with $X \in St(n)$, $Y \in St(\{n+1\}^* \times \{1,n+1\}^*)$, then the second member is of the form $\langle x_{n+1,1}(*)\ X\ x_{n+2,1}(q)\ x_{n+1,1}(*),\ x_{n+1,n+2}(*)\ Y\ x_{n+1,n+2}(*)\rangle$. Using the semi-direct product structure of $St([n+1] \times [n])$ and $St([n+2] \times \{1,n+1\}^*)$, cf. [6] 3.5, we see that the problem reduces to proving an identity of the form $\langle 1,1\rangle = \langle x_{n+1,1}(*),\ x_{n+1,2}(*)\dots x_{n+1,n+2}(*)\rangle$.
But such identities hold precisely when they hold after applying <u>mat</u>. Now note that at the matrix level the problem has been trivial from the beginning.

4.12 We have to generalize lemma 4.11 to a result like [6] Proposition 3.33. Let us first assume that we are in situation α.

LEMMA (Situation α). Let $v,w \in R^{n+1}$, $X \in$ Low, $Y \in$ Up.
Let $L(x_{n+2}(w))\langle X,Y\rangle$ be defined at the bottom with value $\langle P,Q\rangle$ and let $L(x_{n+2}(v))\langle P,Q\rangle$ be defined at the bottom.
If further $L(x_{n+2}(v+w))\langle X,Y\rangle$ is defined, then its value is $L(x_{n+2}(v))\langle P,Q\rangle$.

PROOF. Say $X = T\,x_{n+2,1}(q)\,B$ with $T \in St(n+1)$, $B \in St(\{n+2\} \times \{1^*\})$.
As usual we move T over to the left and reduce to the case T=1.
By SR^2_{n+1} there is $A \in St([n] \times \{n+1\})$ such that both $L(x_{n+2}(v+w))\langle X A, A^{-1}Y\rangle$ and $L(x_{n+2}(w))\langle X A, A^{-1}Y\rangle$ are defined. Now $L(x_{n+2}(v+w))\langle X,Y\rangle$ equals $L(x_{n+2}(v+w))\langle X A, A^{-1}Y\rangle$ by lemma 4.10. Also $L(x_{n+2}(w))\langle X A, A^{-1}Y\rangle$ equals $L(x_{n+2}(w))\langle X A, A^{-1}Y\rangle$ as one sees by applying lemma 4.11 with v=0, r=0, a=0. Remains to see that $L(x_{n+2}(v))\,L(x_{n+2}(w))\langle X A^{-1}, A\,Y\rangle = L(x_{n+2}(v+w))\langle X A^{-1}, A\,Y\rangle$. But this is the situation of lemma 4.11

4.13 DEFINITION. (Situation α). Let $v \in R^{n+1}$, $X \in$ Low, $Y \in$ Up such that $\mathrm{mat}(x_{n+2}(v)\,X\,Y) \in \mathrm{mat}(C)$. By SR_{n+1} there is $A \in St([n] \times \{n+1\})$ such that $L(x_{n+2}(v))\langle X A, A^{-1}Y\rangle$ is defined. We put $L(x_{n+2}(v))\langle X,Y\rangle = L(x_{n+2}(v))\langle X A, A^{-1}Y\rangle$. To see that this depends only on $\langle X,Y\rangle$ and v, we note that there is $z \in R^{n+1}$ such that both steps in $L(x_{n+2}(-z))\,L(x_{n+2}(v+z))\langle X,Y\rangle$ are defined at the bottom, because of $SR^2_{n+1}(n+2,n+1)$ as in [6] 3.32. By lemma 4.12 we have $L(x_{n+2}(-z))\,L(x_{n+2}(v+z))\langle X,Y\rangle = L(x_{n+2}(v))\langle X A, A^{-1}Y\rangle$ and the left hand side is independent of the choices made in the right hand side. (Similarly the right hand side is independent of the choices made in the left hand side, so both sides are independent of choices). It is clear that our present definition of $L(x_{n+2}(v))\langle X,Y\rangle$ is compatible with the earlier definitions in 4.8, [6] 3.18 and that it is equivalent with [6] 3.31.

.14 PROPOSITION. (Additivity in situation α, cf. [6] 3.33).
$L(x_{n+2}(v))\ L(x_{n+2}(w))\ \langle X,Y\rangle = L(x_{n+2}(v+w))\ \langle X,Y\rangle$ whenever the left hand side is defined.

PROOF. First assume the $L(x_{n+2}(v))$ step is defined at the bottom. Use SR^2_{n+1} to choose a representative (P,Q) of $\langle X,Y\rangle$ so that both $L(x_{n+2}(w))$ (P,Q) and $L(x_{n+2}(v+w))$ (P,Q) are defined. Then lemma 4.11 applies. In the general case we want to get back to this special case by perturbation, as in the proof of [6] 3.30. So we seek $z \in R^{n+1}$ such that in $L(x_{n+2}(z))L(x_{n+2}(v))\ L(x_{n+2}(w))\ \langle X,Y\rangle$ and $L(x_{n+2}(v+z))\ L(x_{n+2}(w))$ $\langle X,Y\rangle$ the steps $L(x_{n+2}(z))$, $L(x_{n+2}(v+z))$ are defined at the bottom. This is an SR^2_{n+1} $(n+2,n+1)$ type problem, cf. [6] 3.30, so z exists.

.15 We now have recovered the results of [6] section 3 in the context of situation α. Therefore we turn to situation β and try to catch up. Note that $n=2$, $n+2=4$. We start with a variation on [6] 3.30. (We will return to situation α in 4.19).

LEMMA. Let $L(x_4(v))\ \langle X,Y\rangle$ be defined at the bottom, with value $\langle P,Q\rangle$. Let $L(x_4(w-v))\ \langle P,Q\rangle$ and $L(x_4(w))\ \langle X,Y\rangle$ be defined firmly at the bottom (cf. 4.5). Then $L(x_4(w))\ \langle X,Y\rangle = L(x_4(w-v))\ L(x_4(v))\ \langle X,Y\rangle$.

PROOF. We may assume $X = x_{41}(q)$, $v_1=0$. First assume $1+qw_1 \neq 0$. We want to modify the situation so that lemma 4.11 applies. To adapt v,w we will choose a suitable $M \in St(\{2,3\} \times \{2,3\})$, multiply the desired equality from the left by $L(M)$ and then move M over to Y. As $L(x_4(w))\ \langle X,Y\rangle$ is defined firmly at the bottom, the column (w_2,w_3) is unimodular. Similarly the column $(w_2-v_2-v_2q\ w_1, w_3-v_3-v_3\ q\ w_1)$ is unimodular. The effect of pushing M through is to transform these two unimodular columns by an element of $E_2(R)$. It is not difficult to see, using the Chinese Remainder Theorem, that M can be chosen such that the new w_3 has an invertible image in R mod $(1+q\ w_1)$ and the new w_3, $w_3-v_3-v_3\ q\ w_1$ are irrelevant factors. (cf. 4.9). Thereafter we can modify further, using an M of the form $x_{23}(*)$, so that w_2-1 becomes a multiple of $1+q\ w_1$. It follows from lemma 4.11 with T=1, a=0, that $L(x_4(w))\ \langle X,Y\rangle = L(x_4(w))\ (X,Y)$. Also it is clear (cf. [6] lemma 3.25) that $L(x_4(v))\ \langle X,Y\rangle = L(x_4(v))\ (X,Y)$. We may thus finish by lemma 4.11. Remains the case that $1+q\ w_1 = 0$. We may assume $v_2=0$. Now if $w_2=0$ the result follows from the squeezing principle with i=2. If $w_2 \neq 0$ we can get back to the case $1+q\ w_1 \neq 0$ by pushing through $M=x_{12}(1)$ in the same fashion as above.

4.16 LEMMA. Let both steps in $L(x_4(u-v))\ L(x_4(v))\ \langle X,Y\rangle$ and both steps in $L(x_4(u-w))\ L(x_4(w))\ \langle X,Y\rangle$ be defined firmly at the bottom. Then the end results agree.

PROOF. First we note that if $L(x_4(y))\ \langle P,Q\rangle$ is defined (firmly) at the bottom, then the same holds for the other step in $L(x_4(-y))\ L(x_4(y))\ \langle P,Q\rangle$. (And the result is $\langle P,Q\rangle$, of course). As in the proof of [6] 3.30 we look for $z \in R^3$ such that $L(x_4(z))\ L(x_4(u-v))\ L(x_4(v))\ \langle X,Y\rangle = L(x_4(z+u-v))\ L(x_4(v))\ \langle X,Y\rangle = L(x_4(z+u-v))\ L(x_4(-z-u+v))\ L(x_4(z+u))\ \langle X,Y\rangle = L(x_4(z+u))\ \langle X,Y\rangle = \cdots = L(x_4(z+u-w))\ L(x_4(w))\ \langle X,Y\rangle = L(x_4(z))\ L(x_4(u-w))\ L(x_4(w))\ \langle X,Y\rangle$. By the previous lemma, 4.6, [6] 3.22, and the remark above, this means that we want the steps $L(x_4(z))$, $L(x_4(z+u-v))$, $L(x_4(z+u-w))$ to be defined firmly at the bottom and the step $L(x_4(z+u))$ defined at the bottom. This amounts to four conditions on z and that is a lot, so we have to analyze closely what they look like. As usual we may assume $X = x_{41}(*)$. As in [6] 3.30 there are three 2 by 3 matrices, say A_1,A_2,A_3, each completable to invertible 3 by 3 matrices, and three vectors a_1,a_2,a_3 in R^2 such that the first three conditions are equivalent to unimodularity of the $a_i+A_i\,z$. To satisfy the fourth condition we need to take z of the form $-u+y\,r$ with $y \in R^3$, (y_2,y_3) unimodular, $r \in R$. So there are four vectors that have to be unimodular: $a_i-A_iu+A_iyr$ $(i=1,2,3)$ and (y_2,y_3). From the considerations in [6] section 2 (cf. [6] 2.11) we see that we can satisfy these requirements provided there is $r \neq 0$ such that for each maximal ideal $\underline{m}$ the conditions $a_i-A_iu+A_iyr \not\equiv 0 \bmod \underline{m}R^2$, $(y_2,y_3) \not\equiv 0 \bmod \underline{m}\,R^2$ can be met simultaneously. ("Local solvability implies global solvability"). For a given $\underline{m}$ the existence of a local solution for y clearly only depends on the vanishing or non-vanishing of r mod $\underline{m}$. By a count as in [6] 2.9 we see that the $\underline{m}$ where there is no solution with non-vanishing r mod $\underline{m}$ are of a very particular type: For such $\underline{m}$ we have $R/\underline{m} \cong \mathbb{F}_2$ and $a_i-A_iu \not\equiv 0 \bmod \underline{m}\,R^2$. So at these $\underline{m}$ there is a solution with vanishing r mod $\underline{m}$. If no such $\underline{m}$ exists, take $r=1$. In the contrary case observe that none of the a_i-A_iu vanishes identically so that there are only finitely many $\underline{m}$ where one needs to keep r mod $\underline{m}$ from vanishing . Therefore r can be chosen suitably.

17 <u>LEMMA</u>. Let $v,w \in R^3$, $\langle X,Y\rangle \in C$ such that both $\underline{mat}(x_4(w)XY)$ and $\underline{mat}(x_4(v+w)XY)$ are in $\underline{mat}(C)$. Then there are $t,u \in R^3$ such that all steps in $L(x_4(t))L(x_4(v-t)L(x_4(u))L(x_4(w-u))\langle X,Y\rangle$ and in $L(x_4(t))L(x_4(v+w-t))\langle X,Y\rangle$ are defined firmly at the bottom. Moreover, the end results agree.

<u>PROOF</u>. By [6] 2.11 condition $SR_3^3(4,3)$ is satisfied. Therefore existence of t,u follows as in [6] 3.32. Now substitute w-u for v, w for u, v+w-t for w in Lemma 4.16.

18 <u>DEFINITION</u>. (Compare [6] 3.31). Let $v \in R^3$, $\langle X,Y\rangle \in C$ such that $\underline{mat}(x_4(v)XY) \in \underline{mat}(C)$. There is $z \in R^3$ such that

both steps in $L(x_4(-z))L(x_4(v+z))\langle X,Y\rangle$ are defined firmly at the bottom. (cf. 4.17). We define $L(x_4(v))\langle X,Y\rangle$ to be equal to the result. It follows from Lemma 4.16 that this does not depend on the choice of z. When $L(x_4(v))\langle X,Y\rangle$ is defined at the bottom, the present definition is consistent with the one in [6] 3.18 by Lemma 4.15. Further "additivity" (cf. [6] 3.33) holds by Lemma 4.17. It easily follows that the present definition is also consistent with [6] 3.13, 3.15, 3.31. Now read [6] 3.34. By induction on complexity one easily shows that $L(x_4(v))\langle X,Y\rangle$ equals $L(x_4(v))(X,Y)$ when the latter is defined.

.19 We have recovered the results of [6] section 3 both for situation α and for situation β. Therefore let us look at [6] section 4. Read [6] 4.1 through 4.7. Note that the computation in [6] 4.6 no longer looks horrendous when compared with the present paper. Our next task is to prove an analogue of [6] Proposition 4.9. First we consider situation α.

LEMMA. (Situation α). Let $v \in R^{n+1}$, $t \in R$, $A \in St(\{n+2\} \times [n+2])$, $B \in St(\{1\} \times [n+2])$ such that $R(x_{21}(t))L(x_{n+2}(v))\langle A,B\rangle$ is defined. Then it equals $L(x_{n+2}(v))R(x_{21}(t))\langle A,B\rangle$.

PROOF. Note that $\underline{mat}(ABx_{21}(t))$ and $\underline{mat}(x_{n+2}(v)ABx_{21}(t))$ are in $\underline{mat}(C)$, so that $L(x_{n+2}(v))R(x_{21}(t))\langle A,B\rangle$ is indeed defined. Write it as LRp and the other version as RLp. (So $p = \langle A,B\rangle$ etc.) It suffices to show that there are $T \in St([n] \times \{n+1\})$, $U \in St(\{1\}^* \times \{n+1\})$ such that $R(U)L(T)LRp = R(U)L(T)RLp$. Because of SR^2_{n+1} we can choose

T,U in such a way that, by pushing $R(U)L(T)$ over to p, we are left with a version of the original problem in which the following holds. Both L_1L_2Rp and RL_1L_2p are defined, where $L_1 = L(x_{n+1,n+2}(v_{n+1}))$, $L_2 = L(x_{n+2}(v_1,\ldots,v_n,0))$. The squeezing principle, with i = n+1 implies that $L_2Rp = RL_2p$, because 4.8 gives a method to evaluate both sides "away from line i+1". Remains to show that $L_1RL_2p = RL_1L_2p$. Using what we know about the shape of L_2p and applying inv, we see that it boils down to the case that, in the original context, v_2 vanishes. Write $A = x_{n+2,1}(q)x_{n+2,2}(a_2)\ldots x_{n+2,n+1}(a_{n+1})$. We now wish to argue by induction on the complexity of $(v_1,v_3,\ldots,v_{n+1})$ with respect to q. The case v = 0 is obvious. When $(v_1+rv_i,v_3,\ldots,v_{n+1})$ has lower complexity, simply multiply by $L(x_{1i}(r))$ and push it through. Remains the case that $(v_1,v_3,\ldots,v_i+r(1+qv_1),\ldots,v_{n+1})$ has lower complexity, $3 \leq i \leq n+1$. Then RLp =

$RL(x_{i2}(-ra_2)x_{i1}(-rq))R(x_{i,n+2}(-r)x_{n+2,3}(a_3)\ldots x_{n+2,n+1}(a_{n+1}))$

$L(x_{n+2}(v_1,0,v_3,\ldots,v_i+r(1+qv_1),\ldots,v_{n+1}))$

$\langle x_{n+2,1}(q)x_{n+2,2}(a_2),x_{12}(*)\ldots x_{1,n+2}(*)\rangle$. Now R can be pushed through all the maps in the right hand side. It follows that $L^{-1}RLp = L(x_{i2}(-ra_2)x_{i1}(-rq))$

$R(x_{i,n+2}(-r)x_{n+2,3}(a_3)\ldots x_{n+2,n+1}(a_{n+1}))L(x_{i,n+2}(r))R\langle x_{n+2,1}(q)$
$x_{n+2,2}(a_2),x_{12}(*)\ldots x_{1,n+2}(*)\rangle$. In this last expression we may move R to the left because $L(x_{i,n+2}(r))$ and $R = R(x_{21}(t))$ slide past each other here. (cf. [6] 4.5, 4.6). It easily follows that $R^{-1}L^{-1}RLp$ equals p.

4.20 PROPOSITION. (Situation α). (cf. [6] 4.9). Let $v = (v_1,\ldots,v_{n+1})$, $w = (w_2,\ldots,w_{n+2})$. Then $L(x_{n+2}(v)) \circ R(x_1(w)) \approx R(x_1(w)) \circ L(x_{n+2}(v))$.

PROOF. Let both composites be defined at $p = \langle X,Y \rangle$. We have to show that the values agree. When $L(x_{n+2}(v))$ p is defined at the bottom we apply inv and get essentially into the situation of Lemma 4.19. In general SR^2_{n+1} (n+2,n+1) implies that there is $z \in R^{n+1}$ such that $L(x_{n+2}(z))$ is defined at the bottom at $R(x_1(w))L(x_{n+2}(v))p$ and $L(x_{n+2}(v+z))$ is defined at the bottom at p. Writing $L_1 = L(x_{n+2}(v))$, $L_2 = L(x_{n+2}(z))$, $R = R(x_1(w))$ we get $L_1Rp = L_2^{-1}L_2L_1Rp = L_2^{-1}RL_2L_1p = RR^{-1}L_2^{-1}RL_2L_1p = RL_2^{-1}R^{-1}RL_2L_1p = RL_1p$, where in the fourth equality we use that L_2^{-1} is defined at the bottom at RL_2L_1p.

.21 We have to prove the analogue of this proposition for situation β too. This is more complicated.

NOTATION. (situation β). When $X \in St(4)$ let us write $\underline{mat}_{ij}(X)$ for the entry of mat(X) at the intersection of the i-th row and the j-th column. So $\underline{mat}_{41}(X)$ denotes the entry in the lower left hand corner.

LEMMA. (situation β). Let $A = x_{41}(q)x_{42}(r)$, $B = x_{13}(*)x_{14}(*)$, $v \in R^3$, $t \in R$, $Z = \underline{inv}(x_4(v)A)$, such that the pairs (q,r) and $(\underline{mat}_{11}(Z),\underline{mat}_{21}(Z))$ are unimodular. Then $L(x_4(v))R(x_{21}(t))\langle A,B \rangle = R(x_{21}(t))L(x_4(v))\langle A,B \rangle$. (Both sides are defined.)

PROOF. To see that $R(x_{21}(t))L(x_4(v))\langle A,B \rangle$ is defined, one inspects the first column of mat $\underline{inv}(x_4(v)ABx_{21}(T))$. Consider the unimodular pair $(1+qv_1+rv_2,v_3)$. When it equals (0,1) the result follows by applying inv to [6] 4.7. Now let V be the set of unimodular pairs such that the lemma holds

whenever $(1+qv_1+rv_2,v_3) \in V$. By lemma 4.3 it suffices to show that V is invariant under $E_2(R)$. Multiplying our problem from the left by $L(x_{13}(a)x_{23}(b))$, $R(x_{23}(-b)x_{43}(*))$ and pushing these two maps over to $\langle A,B\rangle$ we can replace (v_1,v_2) by (v_1+av_3,v_2+bv_3), hence $(1+qv_1+rv_2,v_3)$ by $(1+qv_1+rv_2+(qa+rb)v_3,v_3)$, for any $a,b \in R$. As (q,r) is unimodular this means that V is closed under the operation $(f,g) \mapsto (f+cg,g)$ for any $c \in R$. Further, for $a \in R$ one has $L(x_{34}(a))R(x_{12}(t))\langle A,B\rangle$ = $R(x_{12}(t))L(x_{34}(a))\langle A,B\rangle$ by [6] 4.6. This means that our problem is equivalent to showing that
$R(x_{12}(t))L(x_4(v_1,v_2,v_3-a))\langle P,Q\rangle$ equals
$L(x_4(v_1,v_2,v_3-a))R(x_{12}(t))\langle P,Q\rangle$ with $\langle P,Q\rangle = L(x_{34}(a))\langle A,B\rangle$.
But this last problem boils down to one of the original type with v_3 replaced by $v_3-a-aqv_1-arv_2$. So V is also closed under the operation $(f,g) \to (f,g+af)$.

.22 We say that $R(x_1(w))\langle X,Y\rangle$ is defined (firmly) at the bottom when $L(\underline{\text{inv}}(x_1(w)))$ is defined (firmly) at the bottom at $\underline{\text{inv}}\langle X,Y\rangle$.

<u>LEMMA</u>. (Situation β). Let v,w,X,Y be as usual and assume that $L(x_4(v))\langle X,Y\rangle$ is defined, while $R(x_1(w))$ is defined firmly at the bottom at both $\langle X,Y\rangle$ and $L(x_4(v))\langle X,Y\rangle$. Assume also that $(\underline{\text{mat}}_{41}(XY),\underline{\text{mat}}_{41}(XYx_1(w)))$ is a unimodular pair. Then $R(x_1(w))L(x_4(v))\langle X,Y\rangle = L(x_4(v))R(x_1(w))\langle X,Y\rangle$.

<u>PROOF</u>. We may assume $x_1(w) = x_{21}(t)$, $X = x_{41}(q)x_{42}(r)$, $q,r,t \in R$, $Y = x_{13}(*)x_{14}(*)$. Note that (q,r) is unimodular because $\underline{\text{mat}}_{41}(XY) = q$, $\underline{\text{mat}}_{41}(XYx_1(w)) = q+rt$. Applying <u>inv</u> to Lemma 4.6 we see that all conditions of Lemma 4.21 are satisfied.

23 PROPOSITION. (Situation β). (cf. 4.20). Let v,w,X,Y be as usual and assume that both $L(x_4(v))R(x_1(w))\langle X,Y\rangle$ and $R(x_1(w))L(x_4(v))\langle X,Y\rangle$ are defined. Then their values agree.

PROOF.

CASE 1. $X \in St(\{4\} \times [4])$, $Y \in St(\{1\} \times [4])$, $w = (t,0,0)$, both $\underline{mat}_{41}(XY)$ and $\underline{mat}_{41}(XYx_{21}(t))$ are non-zero and v_3 is in the intersection of those maximal ideals $\underline{m}$ for which $R/m \cong \mathbb{F}_2$. We claim there is z such that $R(x_1(z))L(x_4(v))R(x_{21}(t))\langle X,Y\rangle$ equals $L(x_4(v))R(x_1(z))R(x_{21}(t))\langle X,Y\rangle$ and $R(x_1(z+w))L(x_4(v))\langle X,Y\rangle$ equals $L(x_4(v))R(x_1(z+w))\langle X,Y\rangle$. By Lemma 4.22 this creates an $SR_3^4(4,3)$ type problem, except that one also needs to make $\underline{mat}_{41}(XYx_1(z+w))$ prime to the product of $\underline{mat}_{41}(XY)$ and $\underline{mat}_{41}(XYx_{21}(t))$. As before the problem of finding z is solvable (globally) if it is solvable locally (cf. proof of 4.16 and proof of Theorem 3 in [6] section 2.) Where the residue field has at least three elements there is an easy count as in [6] 2.9. At a place with a residue field with 2 elements one can check that there is a solution of the form $(z_2,z_3,z_4) = (z_2,1,z_4)$. (Apply $\underline{inv}$ to Lemma 4.6).

CASE 2. X,Y,w as in case 1 and both $\underline{mat}_{41}(XY)$ and $\underline{mat}_{41}(XYx_{21}(t))$ are non-zero. We wish to get back to case 1. We may assume $X = x_{41}(q)x_{42}(r)$, $Y = x_{13}(*)x_{14}(*)$. From the proof of Lemma 4.21 we see that we may replace $(v_1,v_2,v_3;q,r,t)$ by $(v_1+av_3,v_2+bv_3,v_3;q,r,t)$ for any $a,b \in R$. Therefore we may assume qv_1+rv_2 is contained in each maximal ideal $\underline{m}$ with $R/\underline{m} \cong \mathbb{F}_2$, $v_3 \notin \underline{m}$. But from the same proof we see that we may replace $(v_1,v_2,v_3;q,r,t)$ by $(v_1,v_2,v_3-v_3(1+qv_1+rv_2);q,r,t)$. Therefore we may reduce to case 1 indeed.

<u>CASE 3</u>. Both $\underline{\text{mat}}_{41}(XY)$ and $\underline{\text{mat}}_{41}(XYx_1(w))$ are non-zero. The proof is similar to the proof of case 1: First one notes that the special case in which $R(x_1(w))\langle X,Y\rangle$ is defined at the bottom is essentially the same as case 2. Then one multiplies by a suitable $R(x_1(z))$ and applies this special case twice. (To prove the existence of a suitable z pick a maximal ideal $\underline{m}$ for which $R/\underline{m}$ has at least a hundred elements, say, and replace the requirement $\underline{\text{mat}}_{41}(XYx_1(w)x_1(z)) \neq 0$ by the stronger requirement $\underline{\text{mat}}_{41}(XYx_1(w)x_1(z)) \not\equiv 0 \bmod \underline{m}$. Then argue as in case 1.)

<u>CASE 4</u>. $L(x_4(v))$ is defined at the bottom at both $\langle X,Y\rangle$ and $R(x_1(w))\langle X,Y\rangle$. We may assume that X,Y are as in case 1 and that $v = (0,0,t), t \in R$. By [6] Lemma 4.7 we have $R(x_1(1,a,b)) L(x_4(v))\langle X,Y\rangle = L(x_4(v))R(x_1(1,a,b))\langle X,Y\rangle$ for all $a,b \in R$. Therefore it is easy to reduce to the case that $\underline{\text{mat}}_{41}(XY)$ is non-zero, without changing $\underline{\text{mat}}_{41}(XYx_1(w))$. Similarly one may also make $\underline{\text{mat}}_{41}(XYx_1(w))$ non-zero so that case 3 applies.

<u>CASE 5</u>. The general case. We choose z such that $L(x_1(z))$ is defined firmly at the bottom at both $R(x_1(w))L(x_4(v))\langle X,Y\rangle$ and $L(x_4(v))\langle X,Y\rangle$, while $L(x_4(z+v))$ is defined firmly at the bottom at both $\langle X,Y\rangle$ and $R(x_1(w))\langle X,Y\rangle$. To see that this can be done one argues as usual, noting in this case that when k is a field one can not fill all of k^3 with four lines of which at least two pass through the origin. (cf. [6] 2.9). The rest is easy. (cf. [6] proof of 4.9).

.24 Now that we have proved the analogue of [6] 4.9 in both situation α and situation β, nothing prevents us from using the remainder of [6] section 4. Therefore $K_2(n+1,R) \to K_2(n+2,R)$ is injective. More generally we see that $K_2(m,R) \to K_2(R)$ is injective for $m \geqslant n+1$. (Recall that enlarging the size only makes things better, cf. 3.3., 3.4). The remainder of Theorems 1 and 2 easily follows from the following.

LEMMA. Let $x \in St(n+1)$ such that the first row and the first column of mat(x) are trivial. Then there is $g \in St(\{1,n+2\}^* \times \{1,n+2\}^*)$ which equals x in St(n+1).

PROOF. Let τ be the isomorphism $St(n+1) \to St(\{1\}^* \times \{1\}^*)$ obtained by substitution of n+2 for the index 1. In St(n+2) we have $\tau(x) = w_{n+2,1}(1)xw_{n+2,1}(1)^{-1} = x$.(cf. [6] proof of 3.10). For $U \in St(\{1\}^* \times \{1\}^*)$ it is easy to see that $\rho(U)\langle 1,1\rangle = \langle 1,U\rangle$. ($\rho$ as in [6] 4.16). So on the one hand $\rho(x)\langle 1,1\rangle = \langle x,1\rangle$ and on the other hand $\rho(x)\langle 1,1\rangle = \rho(\tau(x))\langle 1,1\rangle = \langle 1,\tau(x)\rangle$. This means that there is $g \in$ Med with $(1,\tau(x)) = (xg^{-1},g)$. Now recall the semi-direct product structure of Med and apply mat to see that g comes from $St(\{1,n+2\}^* \times \{1,n+2\}^*)$.

§5 A simpler proof of surjectivity in Theorem 1.

5.1 In the previous section surjectivity came as a byproduct of our proof of injectivity. For those readers who are mainly interested in surjectivity of the map $K_2(2,R) \to K_2(R)$, we now give a proof that is considerably simpler. (R as in Theorem 1).
The idea is to sharpen a proof of ordinary surjective stability for K_2 (we use something like the "transpose" of the proof in [12]). The sharpening is achieved by means of Vasertein's result that $EU_2(q) = U_2(q)$ for suitable q. (cf. 2.7).

5.2 Recall that it suffices to show that $K_2(2,R) \to K_2(3,R)$ is surjective. Therefore we try to write elements of St(3,R) in normal form. As normal form we take the one suggested by 3.5 and [6] 3.36: We say that X can be written in normal form if there are $L_1 \in$ image ($\underline{Low} \to St(3)$), $U \in$ image ($\underline{Up} \to St(3)$), $L_2 \in$ image ($St([3] \times \{1\}) \to St(3)$) such that $X = L_1 U L_2$. (Notations of [6] 3.4, 3.5, 3.6 with n = 1).
Note that if $X \in St(3)$ can be written in normal form and $L \in$ image ($\underline{Low} \to St(3)$), the element LX can also be written in normal form. Similarly, if X can be written in normal form and $L \in$ image ($St(\{1\}^* \times [3]) \to St(3)$), then XL can be written in normal form.
Let P(a,b,q,r,s) denote the property: $x_{13}(a)\, x_{23}(b)\, x_{31}(q)\, x_{32}(r)\, x_{13}(s)$ can be written in normal form. Let Q(a,b,q,r,s,t) denote the property: $x_{13}(a)\, x_{23}(b)\, x_{31}(q)\, x_{32}(r)\, x_{12}(t)\, x_{13}(s)$ can be written in normal form. Note that Q(a,b,q,r,s,0) = P(a,b,q,r,s).

5.3 PROPOSITION. Let $Y = x_{13}(a)\, x_{23}(b)\, x_{31}(q)\, x_{32}(r)\, x_{12}(t)\, x_{13}(s)$ with $a,b,q,r,s,t, \in R$. Then Y can be written in normal form.

PROOF. We have to prove Q(a,b,q,r,s,t).

STEP 1 $P(a-tb,b,q,r+qt,s) \Leftrightarrow Q(a,b,q,r,s,t)$.

This one sees by multiplying Y from the left by $L = x_{12}(-t) \in$ image ($\underline{Low} \to St(3)$) and making some obvious computations. (It helps to write down the corresponding matrices).

STEP 2. $Q(a,b,q,*,*,*) \Leftrightarrow Q(a+pb,b,q,*,*,*)$, for any p.

To prove step 2, multiply Y from the left by $x_{12}(p)$. Etcetera.

STEP 3. $Q(a,b,q,*,*,*) \Leftrightarrow Q(a,b+p(1+qa),q,*,*,*)$, for any p.

To see this, multiply Y from the left by $x_{21}(pq)$, form the right by $x_{32}(-r)\, x_{23}(p)$.

STEP 4. $Q(a,b,q,r,s,t)$ holds if $(a,b) \in EU_2(q)$.

For, by the previous two steps we may assume $(a,b) = (0,0)$. Then it is obvious. (Compare also 4.7).

NOTATION. In the totally imaginary case (cf. 2.7) let m denote the order of the group of roots of unity in R. Otherwise let m be any non-zero non-unit in R.

STEP 5. $P(a,b,q,*,*)$ holds if q is prime to m and $(1+qa,b)$ is unimodular.

Clearly $(a,b) \in U_2(q)$. As q is prime to m we may apply Vaserstein's theorem which tells us that $U_2(q) = EU_2(q)$. (Cf. 2.7, [13], [1]).

STEP 6. $P(a,b,q,r,s) \Leftrightarrow P(a,b-pa,q+rp,r,s)$, for any p.

To see this, multiply Y from the left by $x_{21}(-p)$, from the right by $x_{21}(p)\, x_{23}(ps)$. (Of course $t = 0$ now).

STEP 7. $P(a,b,q,r,s,) \Leftrightarrow P(a-spb,b,q,r+(1+qs)p,s)$, for any p.

Here we multiplied Y from the left by $x_{12}(-sp)$, from the right by $x_{32}(p)$.

STEP 8. $P(a,b,q,r,s) \Leftrightarrow P(a+sr(b+p+pqa),b+p+pqa,q,-qsr,s-srp)$

By step 1 it suffices to prove $Q(a,b,q,r,s,0) \Leftrightarrow$ $Q(a,b+p+pqa,q,0,s-srp,-sr)$. This is done as in step 3.

STEP 9. $P(a,b,q,r,s)$ holds if q,b are prime to m and $1+qa+rb$ is non-zero.

To see this, note that $(1+qa+rb,b,ma)$ is unimodular, with the first entry non-zero, so that there in $y \in R$ such that $(1+qa+rb,b-yma)$ is unimodular. Then $(1+qa+rb-r(b-yma), b-yma) = (1+qa+ryma, b-yma)$ is also unimodular and $P(a,b-yma, q+rym,r,s)$ holds by step 5. Now apply step 6.

STEP 10. $P(a,b,q,r,s)$ holds if q,b are prime to m.

As b is not zero we can use step 7 to reduce to the situation of step 9.

STEP 11. $P(a,b,q,r,s)$ always holds.

Because of the previous step, we wish to get b,q prime to m. This is a local problem, therefore not difficult. We apply step 7 to get r prime to q, then step 6 to get q prime to m. Via step 8 we can make s trivial modulo the primes $\underline{m}$ that divide m but not r. Repeating step 8 we arrive at the situation that some power of r is divisible by m. We still have q prime to m. Computing modulo primes that divide m we easily see that we can get b prime to m by means of steps 6 and 8, while keeping q prime to m. Because of step 1 the proposition follows.

5.4 PROPOSITION. Any element of $St(3)$ can be written in normal form.

PROOF. Let V be the set of elements that can be written in normal form. We want to show that V is invariant under left multiplication by Steinberg generators. The difficult case is left multiplication by $x_{1i}(a)$, $i = 1$ or 2. Given $X \in V$ we need to prove $x_{1i}(a)\,X \in V$. Multiplying from the left by elements from St(2), from the right by elements from $St(\{1\}^{*} \times [3])$, and using the semi-direct product structures of <u>Up</u>, <u>Low</u> (cf. [6] 3.5), one easily reduces to the previous proposition.

5.5 Let $\tau \in K_2(3,R)$. We want to show that τ comes from $K_2(2,R)$. Write τ in normal form $L_1\,U\,L_2$. Replacing τ by $L_2\ \tau\ L_2^{-1}$ we reduce to the case $\tau = L_1U$. As <u>mat</u>(L_1) = <u>mat</u>(U^{-1}) there are $a,b \in R$ with <u>mat</u>$(L_1) = e_{12}(a)\ e_{32}(b)$. Pushing a factor $x_{12}(a)\ x_{32}(b)$ over from L_1 to U, we get to the situation that L_1, U themselves are in $K_2(3,R)$. Because of the semi-direct product structure of <u>Low</u> the element L_1 must come for $St(2,R)$ (use <u>mat</u>), hence from $K_2(2,R)$. Similarly U comes from $St(\{1\}^{*} \times \{1\}^{*})$, hence lies in a conjugate of the (central) image of $K_2(2,R)$, hence comes from $K_2(2,R)$.

REFERENCES

[1] H. Bass, J. Milnor and J.-P. Serre, Solution of the congruence subgroup problem for SL_n $(n \geqslant 3)$ and Sp_{2n} $(n \geqslant 2)$, Publ. I.H.E.S. No. 33 (1967), 421-499.

[2] P.M. Cohn, On the structure of the GL_2 of a ring, Publ. I.H.E.S No. 30 (1966), 5-53.

[3] G. Cooke and P.J. Weinberger, On the construction of division chains in algebraic number rings, with application to SL_2, Comm. Algebra 3 (1975), 481-524.

[4] R.K. Dennis and M.R. Stein, K_2 of discrete valuation rings, Advances in Math. 18 (1975), 182-238.

[5] M.J. Dunwoody, K_2 of a euclidean ring, J. Pure Appl. Algebra 7 (1976), 53-58.

[6] W. van der Kallen, Injective stability for K_2, pp. 77-154 of Lecture Notes in Math. 551, Springer Verlag 1976.

[7] W. van der Kallen, Generators and relations in algebraic K-theory, Proceedings International Conference of Mathematicians at Helsinki 1978, 305-310.

[8] B. Liehl, Die Gruppe SL_2 über Ordnungen von arithmetischem Typ, thesis, München 1979.

[9] C.S. Queen, Some arithmetic properties of subrings of function fields over finite fields, Archiv der Math. 26 (1975), 51-56.

[10] J.-P. Serre, Le problème des groupes de congruence pour SL_2, Annals of Math. (2) 92 (1970), 489-527.

[11] A.A. Suslin, On a theorem of Cohn, in: Rings and Modules, Zap. Naučn. Sem. Leningrad. Otdel. Mat. Inst. Steklov. (LOMI) 64 (1976), 127-130. (Russian).

[12] A.A. Suslin and M.S. Tulenbayev, A Theorem on stabilization for Milnor's K_2 functor, in: Rings and Modules, Zap. Naučn. Sem. Leningrad. Otdel. Mat. Inst. Steklov (LOMI) 64 (1976), 131-152. (Russian).

[13] L.N. Vaserstein, On the group SL_2 over Dedekind rings of arithmetic type, Mat. Sb. 89 (131) (1972), 313-322 = Math. USSR - Sb. 18 (1972), 321-332.

[14] L.N Vaserstein, The stable range of rings and the dimensionality of topological spaces, Funcional. Anal. i Priložen. 5 (1971), 17-27 =
Functional Analysis and its Applications 5 (1971), 102-110.

[15] L.N. Vaserstein and A.A. Suslin, Serre's Problem on Projective modules over polynomial rings, and Algebraic K-theory, Izv. Akad. Nauk SSSR Ser. Mat.Tom 40 (1976) No. 5 = Math. USSR Izvestia Vol. 10 (1976) No. 5, 937-1001.

K-théorie relative d'un idéal bilatère de carré nul: étude homologique en basse dimension.

Christian KASSEL.

Soit A un anneau unitaire (pas nécessairement commutatif) et P un A-bimodule. La formule $(a,p).(b,q) = (ab, aq + pb)$ ($a,b \in A$ et $p,q \in P$) permet de munir le groupe abélien $A\oplus P$ d'une structure d'anneau qu'on notera encore $A\oplus P$. Le facteur $P \simeq \{0\}\times P$ est un idéal bilatère de $A\oplus P$, de carré nul. Dans le cas particulier où on prend pour bimodule l'anneau A lui-même, $A\oplus A$ n'est autre que l'anneau $A[\epsilon]$ des nombres duaux de A ($\epsilon^2 = 0$).

Supposons donné un foncteur F de la catégorie des anneaux unitaires dans celle des groupes. La projection canonique de $A\oplus P$ sur A induit un homomorphisme surjectif et scindé de $F(A\oplus P)$ sur $F(A)$ dont le noyau sera désigné par $F(A\oplus P, P)$. Lorsque $F(A)$ est le groupe $E(A)$ des matrices élémentaires ou le groupe $K_1(A) = GL(A)/E(A)$, on dispose des résultats suivants ([10], Cor. 2.6) exprimés ici avec les notations données plus loin:

$$E(A\oplus P, P) \simeq M'(P) \qquad \text{et} \qquad K_1(A\oplus P, P) \simeq H_0(A,P) .$$

En ce qui concerne le foncteur K_2, Dennis et Krusemeyer (cf. [1], §6), suivant la voie tracée par W. van der Kallen dans son article sur le K_2 des nombres duaux [3], utilisent des résultats de Maazen et Stienstra [7] pour donner une présentation de $K_2(A\oplus P, P)$ dans le cas où A est commutatif et où les structures de A-modules à droite et à gauche coïncident sur P (ce qui, avec nos notations, s'exprime par $[A,A] = [A,P] = 0$). La situation considérée ici est la plus générale possible: aucune restriction n'est imposée à l'anneau ou au bimodule. En utilisant des techniques d'homologie des groupes ainsi que des calculs sur les matrices (ceux-ci sont présentés en appendice), on obtient sur $K_2(A\oplus P,P)$ et sur le groupe de Steinberg relatif $St(A\oplus P,P)$ des résultats qui les déterminent complètement dans plusieurs cas non traités jusqu'ici. Ces méthodes diffèrent totalement de celles des auteurs cités précédemment et évitent en particulier tout calcul dans le groupe de Steinberg.

Notations.

A un anneau, P et Q deux A-bimodules.

$[A,P]$: sous-groupe de P engendré par $ap - pa$, $a \in A$, $p \in P$.

$P \otimes_A Q$ $= P \otimes_{\mathbb{Z}} Q/(ap \otimes q - p \otimes qa,\ pa \otimes q - p \otimes aq)$; c'est un produit tensoriel de bimodules (voir [8], p.288).

$\widetilde{\Lambda}_A P$ $= P \otimes_A P/(p \otimes q + q \otimes p)$; cette notation est inspirée de K. Dennis.

$\Lambda_A^2 P$ $= P \otimes_A P/(p \otimes p)$; il existe une surjection de $\widetilde{\Lambda}_A P$ sur $\Lambda_A^2 P$: c'est un isomorphisme quand tout élément de P est divisible par 2. Lorsque $P = A$, $\widetilde{\Lambda}_A A \simeq (A/[A,A]) \otimes \mathbb{Z}/2$.

$H_i(A,P)$: groupes d'homologie de Hochschild (définis p.288-289 dans [8]). On a $H_0(A,P) \simeq P/[A,P]$ et, si $[A,A] = [A,P] = 0$, $H_1(A,P) \simeq \Omega^1_{A/\mathbb{Z}} \otimes_A P$.

$GL(A)$ $= \varinjlim_n GL_n(A)$, groupe des matrices inversibles.

$E(A)$: le sous-groupe (parfait) des commutateurs de $GL(A)$, engendré par les matrices élémentaires $e_{ij}(a)$ ($a \in A$, $i \neq j$).

$St(A)$: le groupe de Steinberg de A engendré par les générateurs $x_{ij}(a)$ et les relations usuelles.

$M(P)$ $= \varinjlim_n M_n(P)$, groupe abélien des matrices carrées d'ordre fini à coefficients dans P .

$M'(P)$: le noyau de la trace $Tr: M(P) \to P \to P/[A,P] \simeq H_0(A,P)$. $E(A)$ opère par conjugaison sur $M(P)$ et sur $M'(P)$. En tant que $E(A)$-module, $M'(P)$ est engendré par les matrices $E_{ij}(p)$, $p \in P$, $i \neq j$. Le coefficient de $E_{ij}(p)$ situé sur la ligne i et dans la colonne j vaut p ; tous les autres sont nuls.

$St(A,P)$: le "groupe de Steinberg additif". C'est le $St(A)$-module engendré par les générateurs $y_{ij}(p)$ $(p \in P,\ i \neq j \geq 1)$ et les relations

$$y_{ij}(p) + y_{ij}(q) = y_{ij}(p+q) \qquad p,\ q \in P.$$

$$x_{ij}(a).y_{kl}(p) - y_{kl}(p) = \begin{cases} 0 & \text{si } i \neq l \text{ et } j \neq k\,. \\ y_{il}(ap) & \text{si } i \neq l \text{ et } j = k\,. \\ y_{kj}(-pa) & \text{si } i = l \text{ et } j \neq k\,. \end{cases}$$

En associant $E_{ij}(p)$ à $y_{ij}(p)$, on définit une surjection de $St(A)$-modules de $St(A,P)$ sur $M'(P)$ (cf. [4], §2).

$H_i(G)$: homologie du groupe G à coefficients entiers.

1 . Le K_2 relatif.

1.1. Le résultat principal de ce paragraphe inscrit le groupe $K_2(A\oplus P,P)$ dans une suite exacte où apparaissent également deux groupes de K-théorie stable, notés $K_i^s(A,P)$. Lorsque le bimodule P est l'anneau A, ils coïncident avec les groupes $K_i^s(A)$ définis par F. Waldhausen [11]. Le théorème qui suit donne une idée des relations qui lient les groupes de K-théorie relative à la K-théorie stable en basse dimension. Signalons d'ailleurs que de telles relations existent en toute dimension. Pour les besoins de cet article il suffit de savoir que $K_0^s(A,P)$ et $K_1^s(A,P)$ sont respectivement isomorphes aux groupes d'homologie de Hochschild $H_0(A,P)$ et $H_1(A,P)$ ([4], Cor. 1.4) et qu'on dispose des isomorphismes ([5], Prop. 2.3)

$$K_2^s(A,P) \simeq H_2(St(A),M'(P)) \simeq H_2(St(A),St(A,P))\ .$$

1.2. THEOREME. - <u>Soit A un anneau unitaire quelconque et P un A-bimodule, alors il existe une suite exacte de la forme</u>

$$K_2^s(A,P) \xrightarrow{\psi} \tilde{\Lambda}_A P \to K_2(A\oplus P,P) \xrightarrow{\varphi} H_1(A,P) \to 0\ .$$

Si on prend pour bimodule l'anneau lui-même, on a le

1.3. COROLLAIRE. - Pour tout anneau A , la suite

$$K_2^s(A) \to H_o(A,A)/2H_o(A,A) \to K_2(A[\epsilon],(\epsilon)) \to H_1(A,A) \to 0$$

est exacte.

Ces résultats qui seront démontrés au §3, étendent d'un terme vers la gauche des suites analogues obtenues dans le cas commutatif (cf. 6.7 dans [1] et le lemme 6.1 de [9]). Ils permettent de retrouver les deux corollaires suivants.

1.4. COROLLAIRE. - Si tout élément de $H_o(A,A)$ est divisible par 2, on a l'isomorphisme

$$K_2(A[\epsilon],(\epsilon)) \simeq H_1(A,A) .$$

Remarquons que l'hypothèse du corollaire précédent est satisfaite si 2 est inversible dans l'anneau A ; la réciproque est fausse.

1.5. COROLLAIRE. - Pour $A = \mathbb{Z}$, $K_2(\mathbb{Z} \oplus P,P) \simeq \tilde{\Lambda}_{\mathbb{Z}} P \simeq \Lambda^2_{\mathbb{Z}} P \oplus P/2P$.

Il s'agit là du corollaire 6.5 de [1]. On l'obtient ici comme conséquence du théorème 1.2 et de la nullité de $H_1(\mathbb{Z},P)$ et de $K_2^s(\mathbb{Z},P)$ pour tout groupe abélien P ([5], Théorème 2.2 et Cor. 2.3). Le deuxième isomorphisme du corollaire 1.5 est dû à K. Dennis. Il n'est pas canonique.

Dans plusieurs cas, la suite exacte du corollaire 1.3 permet de déterminer entièrement $K_2(A[\epsilon],(\epsilon))$ pour un anneau non commutatif. A titre d'exemple, considérons l'anneau $A = \mathbb{Z}[\mathfrak{S}_3]$ du groupe symétrique $\mathfrak{S}_3$ d'ordre 3.

1.6. PROPOSITION. - $K_2(\mathbb{Z}[\mathfrak{S}_3][\epsilon],(\epsilon)) \simeq H_1(\mathbb{Z}[\mathfrak{S}_3],\mathbb{Z}[\mathfrak{S}_3]) \oplus \mathbb{Z}/2 \oplus \mathbb{Z}/2$.

En effet la surjection de $\mathfrak{S}_3$ sur $\mathbb{Z}/2$ induit des applications entre les deux suites exactes

$$\begin{array}{ccccccc}
H_o(\mathbb{Z}[\mathfrak{S}_3],\mathbb{Z}[\mathfrak{S}_3]) \otimes \mathbb{Z}/2 & \to & K_2(\mathbb{Z}[\mathfrak{S}_3][\epsilon],(\epsilon)) & \to & H_1(\mathbb{Z}[\mathfrak{S}_3],\mathbb{Z}[\mathfrak{S}_3]) & \to & 0 \\
\downarrow & & \downarrow & & \downarrow & & \\
H_o(\mathbb{Z}[\mathbb{Z}/2],\mathbb{Z}[\mathbb{Z}/2]) \otimes \mathbb{Z}/2 & \xrightarrow{i} & K_2(\mathbb{Z}[\mathbb{Z}/2][\epsilon],(\epsilon)) & \to & \Omega^1_{\mathbb{Z}[\mathbb{Z}/2]/\mathbb{Z}} & \to & 0
\end{array}$$

On voit facilement que l'application verticale de gauche est bijective. Un calcul explicite fondé sur la présentation de van der Kallen [3] montre que i est une injection scindée. D'où la proposition.

1.7. Considérons le cas où $A = \mathbb{Z}[G]$ pour un groupe G quelconque. A tout G-module à gauche M, on associe un $\mathbb{Z}[G]$-bimodule $\underline{M}$ de la manière suivante: le groupe abélien sous-jacent à $\underline{M}$ est isomorphe à M ; la multiplication à gauche par un élément de G est donnée par l'action de G ; la multiplication à droite est l'application identique. On sait ([8], p.291) que dans ce cas l'homologie de Hochschild de $\mathbb{Z}[G]$ s'identifie à l'homologie de G, à savoir: $H_i(\mathbb{Z}[G],\underline{M}) \simeq H_i(G,M)$. Dans cette situation, la suite exacte du théorème 1.2 s'écrit

$$K_2^s(\mathbb{Z}[G],\underline{M}) \to \tilde{\Lambda}_{\mathbb{Z}} H_o(G,M) \to K_2(\mathbb{Z}[G] \oplus \underline{M}, \underline{M}) \to H_1(G,M) \to 0 \quad .$$

Donnons quelques exemples de calculs de K_2 rendus possibles par cette méthode. Soit C un groupe abélien; faisons opérer le groupe symétrique $\mathfrak{S}_n$ sur le produit cartésien C^n par permutation des facteurs. Soit C_o^n le sous-groupe invariant de C^n formé par les éléments $(c_1,\ldots,c_n)$ vérifiant $c_1+\ldots+c_n = 0$. Avec ces notations on a

(1.8) $\quad K_2(\mathbb{Z}[\mathfrak{S}_2] \oplus \underline{C}^2,\underline{C}^2) \simeq C \oplus \Lambda^2 C$ si C est divisible par 2.

(1.9) $\quad K_2(\mathbb{Z}[\mathfrak{S}_n] \oplus \underline{C}^n,\underline{C}^n) \simeq 0$ si C est cyclique d'ordre impair et $n \geq 3$.

(1.10) $\quad K_2(\mathbb{Z}[\mathfrak{S}_n] \oplus \underline{C}_o^n,\underline{C}_o^n) \simeq 0$ si $n \geq 5$.

Lorsque $GL(A)$ opère par conjugaison sur le groupe $M'(P)$ des matrices de "trace nulle", on a (voir [4], Cor. 1.2)

(1.11) $\quad K_2(\mathbb{Z}[GL(A)] \oplus \underline{M'(P)}, \underline{M'(P)}) \simeq H_1(A,P)$.

1.12. Remarque. -Il existe une surjection de $K_2^s(A,P)$ sur $H_2(A,P)$. Si l'application ψ du théorème 1.2 se factorisait à travers $H_2(A,P)$, alors il serait possible d'écrire la suite exacte du théorème 1.2 en termes de l'homologie de Hochschild et ainsi de se débarrasser d'un groupe difficilement calculable de K-théorie stable. L'exemple suivant montre qu'en général c'est impossible. En effet prenons $\Lambda = P = \mathbb{F}_2$ (le corps à 2 éléments), alors $H_2(\mathbb{F}_2,\mathbb{F}_2)$ est nul. Cependant ψ est un isomorphisme de $K_2^S(\mathbb{F}_2)$ sur $\tilde{\Lambda}\mathbb{F}_2 \simeq \mathbb{Z}/2$.

2 . Le groupe de Steinberg relatif.

L'objet de ce paragraphe est d'étudier le groupe $St(A\oplus P,P)$ et en particulier d'examiner sous quelles hypothèses il est commutatif. Dans ce but nous déterminons son sous-groupe des commutateurs ainsi que son "abélianisé" $H_1(St(A\oplus P,P))$. Enonçons le

2.1. THEOREME. - a) Le groupe abélien $H_1(St(A\oplus P,P))$ est isomorphe au "groupe de Steinberg additif" $St(A,P)$ en tant que $St(A)$-module.

b) Le sous-groupe des commutateurs de $St(A\oplus P,P)$ est isomorphe au noyau $Ker(\varphi)$ de l'application φ de $K_2(A\oplus P,P)$ sur $H_1(A,P)$ donnée dans le théorème 1.2.

Comme $Ker(\varphi)$ est un sous-groupe de $K_2(A\oplus P,P)$ (sur lequel $St(A)$ opère trivialement) ainsi qu'un quotient de $\tilde{\Lambda}_A P$, on a aussitôt le

2.2. COROLLAIRE. - Le sous-groupe des commutateurs de $St(A\oplus P,P)$ est un groupe abélien sur lequel $St(A)$ opère trivialement. C'est un 2-groupe élémentaire quand $A = P$.

L'énoncé qui suit répond à la question posée au début du paragraphe:

2.3. COROLLAIRE. - $St(A\oplus P,P)$ est abélien si et seulement si l'application $K_2(A\oplus P,P) \xrightarrow{\varphi} H_1(A,P)$ est un isomorphisme.

En utilisant le §1 de ce travail et le §6 de [1], on peut dans de nombreux cas calculer le noyau de φ ou du moins déterminer s'il est nul ou non.

Procédons maintenant à la démonstration des théorèmes 1.2 et 2.1.

3 . Démonstrations.

3.1. Pour prouver les deux théorèmes, nous considérons le diagramme suivant dans lequel toutes les lignes et colonnes sont exactes:

$$\begin{array}{ccccccccc}
 & & 0 & & 0 & & 0 & & \\
 & & \downarrow & & \downarrow & & \downarrow & & \\
0 & \to & K_2(A\oplus P,P) & \to & K_2(A\oplus P) & \to & K_2(A) & \to & 0 \\
 & & \downarrow & & \downarrow & & \downarrow & & \\
1 & \to & St(A\oplus P,P) & \to & St(A\oplus P) & \to & St(A) & \to & 1 \\
 & & \downarrow & & \downarrow & & \downarrow & & \\
1 & \to & E(A\oplus P,P) & \to & E(A\oplus P) & \to & E(A) & \to & 1 \\
 & & \downarrow & & \downarrow & & \downarrow & & \\
 & & 1 & & 1 & & 1 & &
\end{array}$$

Ainsi qu'on l'a noté dans l'introduction, le $E(A)$-module $E(A\oplus P,P)$ est isomorphe au groupe $M'(P)$ des matrices de "trace nulle" ($E(A)$ opère par conjugaison sur $M'(P)$). La suite spectrale de Hochschild-Serre de l'extension scindée $St(A\oplus P) \to St(A)$ ainsi que la nullité des groupes H_1 et H_2 des groupes de Steinberg entraînent le

3.2. LEMME. - <u>Les groupes</u> $H_i(St(A),H_1(St(A\oplus P,P)))$ <u>sont nuls pour</u> $i = 0$ <u>et</u> 1 . <u>De plus il existe une surjection de</u> $H_2(St(A),H_1(St(A\oplus P,P)))$ <u>sur</u> $H_o(St(A),H_2(St(A\oplus P,P)))$.

Examinons maintenant la colonne de gauche du diagramme: la surjection de $St(A\oplus P,P)$ sur le groupe abélien $M'(P)$ induit une surjection de son "abélianisé" $H_1(St(A\oplus P,P))$ sur $M'(P)$ dont le noyau sera noté L . Soit φ la surjection induite de $K_2(A\oplus P,P)$ sur L . Comme $St(A)$ opère trivialement sur $K_2(A\oplus P,P)$, L est un $St(A)$-module trivial et on a le

3.3. LEMME. - L <u>est isomorphe à</u> $H_1(A,P)$.

En effet, appliquons le foncteur $H_o(St(A), ?)$ à la suite exacte de modules

$$(3.4) \qquad 0 \to L \to H_1(St(A\oplus P,P)) \to M'(P) \to 0 .$$

On déduit alors du lemme 3.2 et du théorème 1.1 de [4] que

$$L \simeq H_o(St(A),L) \simeq H_1(St(A),M'(P)) \simeq H_1(A,P) .$$

3.5. Pour identifier $H_1(St(A\oplus P,P))$ à $St(A,P)$, nous nous servons de la terminologie et des résultats de [6] (voir aussi [2], §1). La suite exacte (3.4) donne naissance à une extension relative de (St(A) , $M'(P) \rtimes St(A)$) :

$$0 \to L \to H_1(St(A\oplus P,P)) \to M'(P) \rtimes St(A) \to St(A) \to 1 .$$

Celle-ci est centrale car St(A) opère trivialement sur L . Le fait qu'elle soit universelle s'exprime exactement par la nullité des groupes d'homologie $H_i(St(A),H_1(St(A\oplus P,P)))$ pour $i = 0$ et 1 ([6], Théorème 2).

Par ailleurs, il résulte de travaux non publiés de K. Dennis (cités dans [2] et [4], Théorème 2.4) qu'on a une autre extension relative centrale de (St(A) , $M'(P) \rtimes St(A)$)

$$0 \to H_1(A,P) \to St(A,P) \to M'(P) \rtimes St(A) \to St(A) \to 1 .$$

Elle est également universelle car

$$H_o(St(A),St(A,P)) \simeq H_1(St(A),St(A,P)) \simeq 0$$

(cf. [4], Prop. 2.6), ce qui entraîne l'isomorphisme des deux extensions relatives.

3.6. Le sous-groupe des commutateurs de $St(A\oplus P,P)$ est manifestement isomorphe au noyau de l'application φ de $K_2(A\oplus P,P)$ sur L . Ce sont ces mêmes objets L et φ qu'on retrouve en étudiant la suite spectrale d'homologie

$$E^2_{p,q} = H_p(M'(P),H_q(K_2(A\oplus P,P))) \Rightarrow H_{p+q}(St(A\oplus P,P))$$

associée à la colonne de gauche du diagramme (3.1). La suite spectrale fournit les deux suites exactes

$$0 \to L \to H_1(St(A\oplus P,P)) \to M'(P) \to 0$$

$$H_2(St(A\oplus P,P)) \to H_2(M'(P)) \to K_2(A\oplus P,P) \xrightarrow{\varphi} L \to 0 .$$

Nous appliquons le foncteur $H_o(St(A), ?)$ à la deuxième suite et, sachant que $St(A)$ opère trivialement sur les deux termes de droite, nous obtenons la suite exacte

$$H_o(St(A),H_2(St(A\oplus P,P))) \to H_o(St(A),H_2(M'(P))) \to K_2(A\oplus P,P) \overset{\varphi}{\to} L \to 0 .$$

Utilisons la surjection du lemme 3.2 pour remplacer le groupe de gauche par $H_2(St(A),H_1(St(A\oplus P,P)))$. Or, d'après 1.1,

$$H_2(St(A),H_1(St(A\oplus P,P))) \simeq H_2(St(A),St(A,P)) \simeq K_2^s(A,P) .$$

Pour achever la démonstration du théorème 1.2 , il ne reste plus qu'à déterminer $H_o(St(A),H_2(M'(P)))$. Avant de procéder à ce calcul, signalons que nous aurions également obtenu la suite exacte du théorème 1.2 en examinant la suite spectrale d'homologie associée à l'extension scindée

$$0 \to M'(P) \to E(A\oplus P) \to E(A) \to 1 .$$

3.7. PROPOSITION. - a) <u>Soit</u> P <u>et</u> Q <u>deux</u> A-<u>bimodules</u>, <u>alors</u> $E(A)$ <u>opère diagonalement et par conjugaison sur le produit tensoriel</u> $M'(P) \otimes_{\mathbb{Z}} M'(Q)$. <u>Les coinvariants de ce module sont donnés par</u>

$$H_o(E(A),M'(P) \otimes_{\mathbb{Z}} M'(Q)) \simeq P \otimes_A Q .$$

b) <u>L'action précédente induit une action de</u> $E(A)$ <u>sur</u> $H_2(M'(P)) \simeq \Lambda^2_{\mathbb{Z}} M'(P)$. <u>Il en résulte l'isomorphisme</u>

$$H_o(E(A),\Lambda^2_{\mathbb{Z}} M'(P)) \simeq \widetilde{\Lambda}_A P .$$

3.8. <u>Démonstration de la proposition précédente.</u>

Soit $p = (p_{ij})_{ij}$ une matrice de $M(P)$ et $q = (q_{ij})_{ij}$ une matrice de $M(Q)$. Notons $p \nabla q$ la matrice de $M(P \otimes_{\mathbb{Z}} Q)$ donnée par

$$(p \nabla q)_{ij} = \Sigma_k p_{ik} \otimes q_{kj} .$$

Nous définissons une application bilinéaire I de $M(P) \times M(Q)$ sur $P \otimes_{\mathbb{Z}} Q$ par la formule

$$I(p,q) = Tr(p \nabla q)$$

$$= \sum_{i<j} (p_{ij} \otimes q_{ji} + p_{ji} \otimes q_{ij}) + \sum_{i<j} (p_{ii} \otimes q_{jj} + p_{jj} \otimes q_{ii})$$

$$+ 2 \sum_i p_{ii} \otimes q_{ii} - Tr(p) \otimes Tr(q) \quad .$$

Supposons que $E(A)$ opère sur $M'(P) \otimes_{\mathbb{Z}} M'(Q)$ de la manière décrite dans l'énoncé de la proposition 3.7. Un calcul élémentaire montre que I est $E(A)$-équivariant si et seulement si I est à valeurs dans le quotient de P par le sous-groupe engendré par les éléments $ap \otimes q - p \otimes qa$ et $pa \otimes q - p \otimes aq$ $(a \in A,\ p \in P$ et $q \in Q)$, autrement dit à valeurs dans $P \otimes_A Q$.

D'après les formules données en appendice, tout élément du groupe $H_o(E(A), M'(P) \otimes_{\mathbb{Z}} M'(Q))$ est une combinaison linéaire à coefficients entiers d'éléments de la forme $E_{12}(p) \otimes E_{21}(q)$ $(p \in P\ ,\ q \in Q)$. Dans ce groupe on a aussi les relations

$$E_{12}(ap) \otimes E_{21}(q) = E_{12}(p) \otimes E_{21}(qa)$$

$$E_{12}(pa) \otimes E_{21}(q) = E_{12}(p) \otimes E_{21}(aq) \quad .$$

Dans ces conditions il est clair que $p \otimes q \mapsto E_{12}(p) \otimes E_{21}(q)$ définit une application de $P \otimes_A Q$ dans $H_o(E(A), M'(P) \otimes_{\mathbb{Z}} M'(Q))$, réciproque de I . Ce qui démontre la première partie de la proposition 3.7.

On prouve l'autre partie de manière analogue dès qu'on a remarqué que I se factorise à travers $\Lambda^2_{\mathbb{Z}} M'(P)$ si et seulement si on ajoute à $P \otimes_A P$ les relations $p \otimes q + q \otimes p = 0$ $(p,\ q \in P)$. En effet ces relations sont nécessaires car

$$I(E_{12}(p) + E_{21}(q), E_{12}(p) + E_{21}(q)) = p \otimes q + q \otimes p \quad .$$

Elles sont suffisantes car $I(p,p)$ est somme d'éléments de la forme $m \otimes n + n \otimes m$. On notera le fait suivant: dans $P \otimes_A P$, $Tr(p) \otimes Tr(q)$ est somme d'éléments du type $m \otimes n + n \otimes m$ parce que $Tr(p)$ est lui-même somme d'éléments particuliers, à savoir de "commutateurs" $am - ma$ ($a \in A$, $m \in P$).

Appendice.

A.1. Toute matrice $p = (p_{ij})_{ij}$ de $M'(P)$ s'écrit de manière unique

$$p = \sum_{i\neq j} E_{ij}(p_{ij}) + \sum_{i\geq 2} F_i(p_{ii}) + E_{11}(Tr(p)) \quad ,$$

où on a posé: $F_i(m) = E_{ii}(m) - E_{11}(m)$. Par définition de $M'(P)$, $Tr(p)$ est une somme de "commutateurs" $am - ma$.

Soit P et Q deux A-bimodules. $E(A)$ opère diagonalement et par conjugaison sur $M'(P) \otimes_{\mathbb{Z}} M'(Q)$. Posons

$$g * (p \otimes q) = (gpg^{-1}) \otimes (gqg^{-1}) \qquad \text{avec } g\in E(A),\ p\in M'(P) \text{ et } q\in M'(Q).$$

Notons $\equiv$ la relation d'équivalence sur $M'(P) \otimes_{\mathbb{Z}} M'(Q)$ engendrée par $g * (p \otimes q) - (p \otimes q) \equiv 0$. L'ensemble des classes d'équivalence est le groupe $H_o(E(A), M'(P) \otimes_{\mathbb{Z}} M'(Q))$ déterminé en 3.7.

A.2. FORMULAIRE. - a) $E_{ij}(p) \otimes E_{kl}(q) \equiv 0$ <u>pour</u> $k\neq j$ <u>ou</u> $l\neq i$ ($i\neq j$, $k\neq l$)

b) $E_{ij}(p) \otimes E_{ji}(q) \equiv E_{12}(p) \otimes E_{21}(q)$ ($i\neq j$)

c) $F_k(p) \otimes E_{ij}(q) \equiv E_{ij}(p) \otimes F_k(q) \equiv 0$ ($k\geq 2$, $i\neq j$)

d) $F_i(p) \otimes F_i(q) \equiv 2\, E_{12}(p) \otimes E_{21}(q)$ ($i\geq 2$)

e) $F_i(p) \otimes F_j(q) \equiv E_{12}(p) \otimes E_{21}(q)$ ($i\neq j$)

f) $E_{11}(p) \otimes E_{ij}(q) \equiv E_{ij}(p) \otimes E_{11}(q) \equiv 0$ ($i\neq j$)

g) $E_{11}(p) \otimes F_i(q) \equiv - E_{12}(p) \otimes E_{21}(q)$ ($i\geq 2$)

h) $E_{11}(p) \otimes E_{11}(q) \equiv E_{12}(p) \otimes E_{21}(q)$

i) $E_{12}(ap) \otimes E_{21}(q) \equiv E_{12}(p) \otimes E_{21}(qa)$

j) $E_{12}(pa) \otimes E_{21}(q) \equiv E_{12}(p) \otimes E_{21}(aq)$.

Pour démontrer ces formules, on calcule successivement les expressions suivantes (avec les indices appropriés):

$e_{1n} * [E_{nj}(p) \otimes E_{kl}(q)]$, $w_{2j} * [E_{ij}(p) \otimes E_{ji}(q)]$, $e_{ij} * [F_1(p) \otimes E_{jk}(q)]$,

$e_{ij} * [F_i(p) \otimes E_{jk}(q)]$, $e_{1j} * [F_i(p) \otimes E_{jk}(q)]$, $e_{1i}(a) * [F_i(p) \otimes E_{i1}(q)]$,

$e_{ij}(a) * [F_j(p) \otimes E_{ji}(q)]$, $e_{ij} * [E_{11}(p) \otimes E_{jk}(q)]$, $e_{1j} * [E_{11}(p) \otimes E_{jk}(q)]$,

$e_{1j}(a) * [E_{11}(p) \otimes E_{j1}(q)]$,

avec les conventions: $e_{ij} = e_{ij}(1)$ et $w_{ij} = e_{ij}e_{ji}^{-1}e_{ij}$ $(i \neq j)$.

Références.

[1] R. K. DENNIS et M. I. KRUSEMEYER, $K_2(A[X,Y]/XY)$, a problem of Swan and related computations, J. Pure Appl. Alg. 15 (1979), 125-148.

[2] K. IGUSA, A proof of a theorem by R. K. Dennis, Brandeis, préprint.

[3] W. van der KALLEN, Le K_2 des nombres duaux, C. R. Ac. Sc. Paris 273 (1971), 1204-1207.

[4] Chr. KASSEL, Un calcul d'homologie du groupe linéaire général, C. R. Ac. Sc. Paris 288 (1979), 481-483.

[5] Chr. KASSEL, Homologie du groupe linéaire général et K-théorie stable, C. R. Ac. Sc. Paris 290 (1980), 1041-1044.

[6] J.-L. LODAY, Cohomologie et groupe de Steinberg relatifs, J. of Algebra 54 (1978), 178-202.

[7] H. MAAZEN et J. STIENSTRA, A presentation of K_2 of split radical pairs, J. Pure Appl. Alg. 10 (1977), 271-294.

[8] S. MAC LANE, Homology, Springer Verlag (1963).

[9] J. STIENSTRA, Deformations of the second Chow group, thèse, Utrecht 1978.

[10] R. G. SWAN, Excision in algebraic K-theory, J. Pure Appl. Alg. 1 (1971), 221-252.

[11] F. WALDHAUSEN, Algebraic K-theory of topological spaces I, A.M.S. Proc. Symp. Pure Math. 32 (1978), 35-60.

Institut de Recherche Mathématique Avancée

Université Louis Pasteur

7 rue René Descartes

67084 Strasbourg Cédex (France).

-oOo-

On the boundary map $K_3(\Lambda/I) \to K_2(\Lambda,I)$

Jean-Louis Loday*

Let Λ be a commutative ring and I a two-sided ideal. We compute the image of some elements of $K_3(\Lambda/I)$ under the boundary map $\partial: K_3(\Lambda/I) \to K_2(\Lambda,I)$ in terms of Dennis and Stein's symbols [D-S]. This enables us to establish the non-triviality of some elements of certain K_3 groups (see [L] for notations).

I thank M. Stein for valuable conversations and Northwestern University for its hospitality.

1. Let a,b and c be three elements of Λ such that $1 - abc$ is invertible and ab, bc, ca lie in the ideal I. Their images $\bar{a}$, $\bar{b}$, $\bar{c}$ in Λ/I are such that $\bar{a}\,\bar{b} = \bar{b}\,\bar{c} = \bar{c}\,\bar{a} = 0$. Therefore, the elements $x_{12}^{\bar{a}}$, $x_{23}^{\bar{b}}$ and $x_{31}^{\bar{c}}$ of $St(\Lambda/I)$ commute pairwise and define a group homomorphism $\rho: \mathbb{Z}^3 \to St(\Lambda/I)$. The element $\langle\bar{a},\bar{b},\bar{c}\rangle \in K_3(\Lambda/I)$ is defined to be the image of the standard generator $1 \in \mathbb{Z} = H_3(\mathbb{Z}^3)$ in $H_3(St(\Lambda/I)) = K_3(\Lambda/I)$ under ρ_*, i.e.

$$\langle\bar{a},\bar{b},\bar{c}\rangle \overset{\text{def}}{=} \rho_*(1).$$

Let $K_2(\Lambda,I)$ be the relative K_2-group [L]. The Dennis-Stein symbols $\langle a,bc\rangle$, $\langle b,ca\rangle$ and $\langle c,ab\rangle$ (see below and [K]) are well-defined elements of $K_2(\Lambda,I)$ because $1 - abc$ is invertible and because ab,bc and ca lie in I.

THEOREM 1. <u>With the previous hypotheses and notations, the image of</u> $\langle\bar{a},\bar{b},\bar{c}\rangle \in K_3(\Lambda/I)$ <u>under</u> ∂ <u>is</u>

$$\partial\langle\bar{a},\bar{b},\bar{c}\rangle = \langle a,bc\rangle\langle b,ca\rangle\langle c,ab\rangle \in K_2(\Lambda,I).$$

The proof is given in the next section.

<u>Examples and applications</u>. Let $\mathbb{Z}[\varepsilon]$ be the ring of dual numbers over $\mathbb{Z}$, i.e. $\varepsilon^2 = 0$. Then $\langle\varepsilon,\varepsilon,\varepsilon\rangle$ is an element of $K_3(\mathbb{Z}[\varepsilon])$.

PROPOSITION 1. <u>The element</u> $\langle\varepsilon,\varepsilon,\varepsilon\rangle$ <u>generates an infinite cyclic subgroup of</u> $K_3(\mathbb{Z}[\varepsilon])$.

*Partially supported by an N.S.F. grant.

Proof. Take $\Lambda = \mathbb{Z}[t]/t^3$ and $I = (t^2)$, so that $\Lambda/I = \mathbb{Z}[\varepsilon]$. The relative group $K_2(\Lambda,I)$ is isomorphic to $D(\Lambda,I) \simeq \mathbb{Z}$, with generator $\langle t,t^2\rangle$ (see [M-S] and [St]). By theorem 1 $\partial\langle\varepsilon,\varepsilon,\varepsilon\rangle = 3\langle t,t^2\rangle$, therefore $\langle\varepsilon,\varepsilon,\varepsilon\rangle$ is infinite cyclic. □

<u>Remark.</u> C. Soulé has proved [S] that $\mathrm{rk}(K_3(\mathbb{Z}[\varepsilon]) \otimes \mathbb{Q}) = 1$. The element $\langle\varepsilon,\varepsilon,\varepsilon\rangle$ is a generator of the unique infinite cyclic summand of $K_3(\mathbb{Z}[\varepsilon])$. In fact by proposition 1 it is either a generator or 3 times a generator. But $\langle t,t^2\rangle \in K_2(\Lambda,I)$ has a non trivial image in $K_2(\mathbb{Z}[t]/t^3)$; therefore $\langle\varepsilon,\varepsilon,\varepsilon\rangle$ is a generator.

Let $\mathbb{F}_p$ be a finite field with p elements, p prime. Then $\langle\varepsilon,\varepsilon,\varepsilon\rangle \in K_3(\mathbb{F}_p[\varepsilon])$ is of order p, because $p\langle\varepsilon,\varepsilon,\varepsilon\rangle = \langle p\varepsilon,\varepsilon,\varepsilon\rangle = \langle 0,\varepsilon,\varepsilon\rangle = 0$.

PROPOSITION 2. <u>If</u> $p \neq 3$ <u>the element</u> $\langle\varepsilon,\varepsilon,\varepsilon\rangle \in K_3(\mathbb{F}_p[\varepsilon])$ <u>is a generator of a</u> $\mathbb{Z}/p$-<u>summand</u>.

The proof is the same as for $\mathbb{Z}$. The restriction $p \neq 3$ comes from the fact that $3\langle t,t^2\rangle = 0$ in $K_2(\mathbb{F}_3[t],(t^3))$. □

<u>Remark.</u> Evens and Friedlander [E-F] proved that for $p \neq 2,3$, $K_3(\mathbb{F}_p[\varepsilon]) \simeq K_3(\mathbb{F}_p) \oplus \mathbb{Z}/p \oplus \mathbb{Z}/p$. One can detect a generator ζ for the other summand $\mathbb{Z}/p$ by looking at the birelative group $K_2(\mathbb{Z}[\varepsilon];\ (\varepsilon),(p))$. By [G-W-L], and because $(\varepsilon) \cap (p) = (p\varepsilon)$ is a radical ideal we have $K_2(\mathbb{Z}[\varepsilon],\ (\varepsilon),\ (p)) = D(\mathbb{Z}[\varepsilon],\ (\varepsilon),\ (p)) \simeq \mathbb{Z}/2 \oplus \mathbb{Z}/p$ with generators $\langle\varepsilon,p\varepsilon\rangle$ and $\langle\varepsilon,p\rangle$. The exact sequence

$$K_3(\mathbb{Z}[\varepsilon],(\varepsilon)) \to K_3(\mathbb{F}_p[\varepsilon],(\varepsilon)) \xrightarrow{\partial'} K_2(\mathbb{Z}[\varepsilon];(\varepsilon),(p)) \to K_2(\mathbb{Z}[\varepsilon],(\varepsilon))$$

shows that a generator ζ of the other $\mathbb{Z}/p$ summand maps under ∂' to $\langle\varepsilon,p\rangle$. In particular $K_3(\mathbb{F}_2[\varepsilon],(\varepsilon))$ is of order at least 4.

2. <u>Proof of theorem 1.</u>

We first recall some facts from [L] about crossed modules. There exists an exact sequence

$$(1) \qquad 1 \to L \to St(\Lambda,I) \to St(\Lambda) \to St(\Lambda/I) \to 1$$

where the middle map is a crossed module and $L = \mathrm{Im}(K_3(\Lambda) \to K_2(\Lambda,I))$. We still denote by ∂ the surjection $K_3(\Lambda) \to L$. The invariant of the crossed module (1) is an element of $H^3(St(\Lambda/I), St(\Lambda); L) = Hom(H_3(St(\Lambda/I), St(\Lambda)), L) = Hom(L,L)$. This invariant is id_L.

Consider the commutative diagram

$$\begin{array}{ccc} F_3 & \longrightarrow & \mathbb{Z}^3 \\ \downarrow \tilde{\rho} & & \downarrow \rho \\ St(\Lambda) & \longrightarrow & St(\Lambda/I) \end{array}$$

where F_3 is the free group on three letters A, B and C, and $\tilde{\rho}(A) = x_{12}^a$, $\tilde{\rho}(B) = x_{23}^b$, $\tilde{\rho}(C) = x_{31}^c$. Pulling back the extension (1) we get the crossed module

(2) $$1 \to L \to X \to F_3 \to \mathbb{Z}^3 \to 1 .$$

Its invariant is $\partial \circ \rho_* \in Hom(H_3(\mathbb{Z}^3),L) = H^3(\mathbb{Z}^3,F_3; L)$.

Let

(3) $$1 \to H_3(\mathbb{Z}^3) \to M \to F_3 \to \mathbb{Z}^3 \to 1$$

be the crossed module whose invariant is $id_{H_3(\mathbb{Z}^3)}$. Using $\partial \circ \rho_*: H_3(\mathbb{Z}^3) \to L$ we construct from (3) a crossed module with invariant $\partial \circ \rho_*$. By the classification theorem ([L] theorem 1) this crossed module is isomorphic to (2). Hence we have proved

LEMMA 1. <u>There exists a morphism of crossed modules</u>

$$\begin{array}{ccccccccc} 1 \to & H_3(\mathbb{Z}^3) & \to & M & \to & F_3 & \to & \mathbb{Z}^3 & \to 1 \\ & \downarrow \partial\circ\rho_* & & \downarrow & & \downarrow \tilde{\rho} & & \downarrow \rho & \\ 1 \to & L & \to & St(\Lambda,I) & \to & St(\Lambda) & \to & St(\Lambda/I) & \to 1 . \ \square \end{array}$$

We now describe the image of the standard generator of $H_3(\mathbb{Z}^3)$ in M. Let u (resp. v, resp. w) be any element of M such that its image in F_3 is $[B^{-1},C]$ (resp. $[C^{-1},A]$, resp. $[A^{-1},B]$).

LEMMA 2. <u>The image of</u> $1 \in \mathbb{Z} = H_3(\mathbb{Z}^3)$ <u>in</u> M <u>is</u>

$$U = B\cdot((A\cdot u)u^{-1})\ C\cdot((B\cdot v)v^{-1})\ A\cdot((C\cdot w)w^{-1}),$$

<u>where the dot indicates the action of</u> F_3 <u>on</u> M.

Proof. The image of U in F_3 is trivial by P. Hall's identity. The element U does not depend on the choice of u (resp. v, resp. w) because F_3 acts trivially on $H_3(\mathbb{Z}^3)$ and both u and u^{-1} (resp. v and v^{-1}, resp. w and w^{-1}) occur in U.

In the standard bar resolution the cocycle representing the standard generator of $H_3(\mathbb{Z}^3)$ is

$$(\bar{A},\bar{B},\bar{C}) + (\bar{B},\bar{C},\bar{A}) + (\bar{C},\bar{A},\bar{B}) - (\bar{B},\bar{A},\bar{C}) - (\bar{A},\bar{C},\bar{B}) - (\bar{C},\bar{B},\bar{A})$$

where $^-$ denotes the image in $\mathbb{Z}^3$.

Let R_3' denote the commutator sub-group of $R_3 = \mathrm{Ker}(F_3 \to \mathbb{Z}^3)$. Then the crossed module

$$1 \to H_3(\mathbb{Z}^3) \to M/R_3' \to F_3/R_3' \to \mathbb{Z}^3 \to 1$$

is derived from (3). It has the advantage that M/R_3' is abelian. Using the explicit formula for the 3-cocycle invariant of a crossed module ([L] prop. 1) we compute the the image of the standard generator:

$$0 + B\cdot v' + (-v') - w' - A\cdot(-u') - (u'+C\cdot(-w')) =$$
$$(A-1)\cdot u' + (B-1)\cdot v' + (C-1)\cdot w',$$

where u' (resp. v', resp. w') is a lifting to M/R_3' of [C,B] (resp. [A,C], resp. [B,A]).

We still must compute the image of U in M/R_3'; it is

$$B\cdot((A-1)\cdot u) + C\cdot((B-1)\cdot v) + A\cdot((C-1)\cdot w) =$$
$$(A-1)\cdot u' + (B-1)\cdot v' + (C-1)\cdot w'$$

because $B\cdot u = u'$ (resp. $C\cdot v = v'$, resp. $A\cdot w = w'$). □

We now recall the definition of the relative symbol <,>. Let D be the pullback $\Lambda \times_{\Lambda/I} \Lambda$ and let p_1 (resp. p_2) be the first (resp. second) projection $D \to \Lambda$. It induces a group homomorphism p_{1*}(resp. p_{2*}): $St(D) \to St(\Lambda)$ and $St(\Lambda,I) = \mathrm{Ker}\, p_{1*}/[\mathrm{Ker}\, p_{1*}, \mathrm{Ker}\, p_{2*}]$ (cf. [L]). The map $St(\Lambda,I) \to St(\Lambda)$ is induced by p_{2*}. For any elements r and s of the ring R the (absolute) Dennis-Stein symbol (with a slight sign modification [D-S]) is

$$\langle r,s\rangle = x_{-\alpha}^{s(1-rs)^{-1}}\, x_{\alpha}^{r}\, x_{-\alpha}^{-s}\, x_{\alpha}^{-r(1-rs)^{-1}}\, h_{\alpha}(1-rs)^{-1} \in K_2(R) \subset St(R)$$

where $\alpha = (ij)$, $-\alpha = (ji)$.

For $a' \in I$ and $b' \in \Lambda$, $<(0,a'),(b',b')>$ defines an element of $St(\Lambda,I)$ which in fact lies in $K_2(\Lambda,I)$ and which we denote by $<a',b'>$ if no confusion can arise.

Last step of the proof of theorem 1. We will prove the formula

$$\partial<-\bar{a},-\bar{b},-\bar{c}> = <bc,a><ca,b><ab,c>.$$

This is equivalent to theorem 1 because

$$<-\bar{a},-\bar{b},-\bar{c}> = -<\bar{a},\bar{b},\bar{c}> \text{ and } <x,y> = <y,x>^{-1}$$

(if $K_2(\Lambda,I)$ is written multiplicatively).

By lemma 1 it suffices to compute the image of U in $St(\Lambda,I)$. We use the formula of lemma 2, replacing A(resp. B, resp. C) by $x_{12}^{-(a,a)}$(resp. $x_{23}^{-(b,b)}$, resp. $x_{31}^{-(c,c)}$) and u(resp. v, resp. w) by $x_{21}^{-(0,bc)}$(resp. $x_{32}^{-(0,ca)}$, resp. $x_{13}^{-(0,ab)}$). We check that $x_{21}^{-(0,bc)}$ is a lifting of $[x_{23}^{b},x_{31}^{-c}] = x_{21}^{-bc}$. We rewrite the formula defining the Dennis and Stein's symbol:

$$[x_{-\alpha}^{s},x_{\alpha}^{r}] = <r,s>x_{-\alpha}^{-s\frac{rs}{1-rs}}h_{\alpha}(1-rs)x_{\alpha}^{r\frac{rs}{1-rs}} .$$

Then we compute: $(A\cdot u)u^{-1} = [x_{-\alpha}^{-(a,a)},\ x_{\alpha}^{-(0,bc)}] =$

$$\underbrace{<(0,bc),(a,a)>}_{<bc,a>}\ \underbrace{x_{-\alpha}^{-(0,a\frac{abc}{1-abc})}\ h_{\alpha}(1,1-abc)\ x_{\alpha}^{(0,bc\frac{abc}{1-abc})}}_{E_{\alpha}} ,$$

$\alpha = (21)$ and similarly we obtain $(B\cdot v)v^{-1} = <ca,b>E_{\beta}$, $(C\cdot w)w^{-1} = <ab,c>E_{\gamma}$.

Using the fact that $<-,->$ is central we obtain

$$U = <bc,a><ca,b><ab,c>(x_{\beta}^{-(b,b)}\cdot E_{\alpha})(x_{\gamma}^{-(c,c)}\cdot E_{\beta})(x_{\alpha}^{-(a,a)}\cdot E_{\gamma})$$

The proof that the product of the E's is trivial follows from Dennis and Stein's computation [D-S]. □

3. In this section we compute ∂ of some similar elements in K_3 and derive a formula analogous to $<a,bc><b,ca><c,ab> = 1$ in K_2.

Let $\alpha_1,\alpha_2,\ldots\alpha_n$ be elements of $GL(\Lambda)$ which commute pairwise. To such data is associated an element $\{\alpha_1,\ldots,\alpha_n\}$ of $K_n(\Lambda)$ as follows. The representation

$\alpha: \mathbb{Z}^n \to GL(\Lambda)$ which associates α_i to the $i^{\underline{th}}$ generator induces a map $B\alpha^+: T^n = B(\mathbb{Z}^n) \to BGL(\Lambda)^+$. The space $BGL(\Lambda)^+$ is an infinite loop space, therefore there exists a space Y and a homotopy equivalence $\Omega Y \sim BGL(\Lambda)^+$. By adjunction $B\alpha^+: T^n \to \Omega Y$ gives rise to $f: S^1 \wedge T^n \to Y$. Up to homotopy the suspension of the torus splits into $S^{n+1} \vee$ (wedges of spheres of lower dimensions). Restricting f to S^{n+1} gives $\bar{f}$ and the homotopy class $[\bar{f}] \in \pi_{n+1}(Y) = \pi_n(BGL(\Lambda)^+) = K_n(\Lambda)$ is, by definition, the element $\{\alpha_1,\ldots,\alpha_n\} \in K_n(\Lambda)$.

It is easily seen from this definition that $\{\alpha_1,\alpha_2,\ldots,\alpha_n\} + \{\alpha_1',\alpha_2,\ldots,\alpha_n\} = \{\alpha_1\alpha_1',\alpha_2,\ldots,\alpha_n\}$ and $\{\alpha_{\sigma(1)},\ldots,\alpha_{\sigma(n)}\} = (-1)^{sgn(\sigma)}\{\alpha_1,\ldots,\alpha_n\}$, where σ is a permutation on $1,\ldots,n$.

For n = 1 $\{\alpha_1\}$ is the class of α_1 in $K_1(\Lambda) = GL(\Lambda)^{ab}$.

For n = 2, if $\alpha_1 = \begin{pmatrix} 1 & a \\ 0 & 1 \end{pmatrix}$ and $\alpha_2 = \begin{pmatrix} 1 & 0 \\ b & 1 \end{pmatrix}$ with $ab = ba = 0$, the element $\{\alpha_1,\alpha_2\}$ is $\langle a,b \rangle$. If

$$\alpha_1 = u \quad \text{and} \quad \alpha_2 = v \quad , \text{ with } u,v \in \Lambda^*$$

and Λ abelian, then $\{\alpha_1,\alpha_2\}$ is the Steinberg symbol of u and v, usually denoted $\{u,v\}$.

For n = 3, if

$$\alpha_1 = \begin{pmatrix} 1 & \bar{a} & \\ & 1 & \\ & & 1 \end{pmatrix} \quad \alpha_2 = \begin{pmatrix} 1 & & \\ & 1 & \bar{b} \\ & & 1 \end{pmatrix} \quad \alpha_3 = \begin{pmatrix} 1 & & \\ & 1 & \\ \bar{c} & & 1 \end{pmatrix}$$

with $\bar{a}\,\bar{b} = \bar{b}\,\bar{c} = \bar{c}\,\bar{a} = 0$, then $\{\alpha_1,\alpha_2,\alpha_3\}$ is the element $\langle \bar{a},\bar{b},\bar{c} \rangle$ introduced in the first section.

THEOREM 2. <u>Let</u> α_1,α_2 <u>and</u> $\alpha_3 \in GL(\Lambda)$ <u>and define</u> $\beta_1 = [\alpha_2,\alpha_3]$, $\beta_2 = [\alpha_3,\alpha_1]$ <u>and</u> $\beta_3 = [\alpha_1,\alpha_2]$. <u>If</u> $[\alpha_i,\beta_i] = 1$ <u>for</u> $i = 1, 2$ <u>and</u> 3 <u>the following formula holds in</u> $K_2(\Lambda)$:

$$\{\alpha_1,\beta_1\} + \{\alpha_2,\beta_2\} + \{\alpha_3,\beta_3\} = 0.$$

This theorem is a consequence of its relative version which is the following.

Let I be a 2-sided ideal of Λ. Let α_1, α_2 and α_3 be elements of $E(\Lambda)$ (elementary group). Then we assume that β_i is an element of $E(\Lambda,I) \subset E(\Lambda)$ for $i = 1, 2$ and 3. Denote by $\tilde{\beta}_i \in St(\Lambda,I)$ any lifting to the relative Steinberg group (cf. [L]) of β_i, $i = 1,2,3$. Then the element $\gamma_i = (\alpha_i \cdot \tilde{\beta}_i)\tilde{\beta}_i^{-1}$ is in $K_2(\Lambda,I) \subset St(\Lambda,I)$.

THEOREM 3. <u>With the above hypotheses and notations the following formula holds in</u> $K_2(\Lambda,I)$:

$$\partial\{\bar{\alpha}_1,\bar{\alpha}_2,\bar{\alpha}_3\} = \gamma_1 + \gamma_2 + \gamma_3 \quad (\bar{\alpha}_i = \text{image of } \alpha_i \text{ in } GL(\Lambda/I)).$$

The proof is as in theorem 1. □

References

[D-S] K. Dennis and M. Stein, K_2 of discrete valuation rings, Advances in Math. 18 (1975) 182-238.

[E-F] L. Evens and E. Friedlander, On $K_*(\mathbb{Z}/p^2\mathbb{Z})$ and related homology groups.

[GW-L] D. Guin-Waléry and J-L. Loday, Obstruction à l'excision en K-théorie algébrique, these Proceedings.

[K] F. Keune, The relativization of K_2, J. Algebra 54 (1978), 159-177.

[L] J-L. Loday, Cohomologie et groupe de Steinberg relatifs, J. Algebra 54 (1978), 178-202.

[M-S] H. Maazen and J. Stienstra, A presentation for K_2 of split radical pairs, J. Pure Applied Algebra 10 (1977) 271-294.

[St] J. Stienstra, On K_3 and K_2 of truncated polynomial rings, these Proceedings.

[S] C. Soulé, Rational K-theory of the dual numbers of a ring of algebraic integers, these Proceedings.

I.R.M.A., Strasbourg, France and Northwestern University, Evanston, USA.

ODD INDEX SUBGROUPS OF UNITS IN CYCLOTOMIC FIELDS AND APPLICATIONS

R. James Milgram*

In this note we obtain results on odd index subgroups of the groups of units in certain cyclotomic number fields, and apply them to several questions in algebraic K-theory and topology. For example, we obtain new restrictions on the possible fundamental groups of closed 3-manifolds.

In terms of exposition, though complete proofs are given in most cases, these depend very strongly on results and techniques of previous papers [H-M, M1, M2, Milnor]. Hence the reader should be warned that he may not be able to understand the proofs without adequate preparation.

Let ζ_n be a primitive n^{th} root of unity and $Q(\zeta_n)$ the cyclotomic field generated by ζ_n. Then $Gal(Q(\zeta_n)/Q) = (Z/n)^{\cdot}$, and we denote by K_n (respectively FK_n) the subfield of $Q(\zeta_n)$ with $Gal(K_n/Q)$ the 2-Sylow subgroup of $(Z/n)^{\cdot}$ (respectively the maximal real subfield of K_n).

THEOREM A: Let p,q be odd primes, then the cyclotomic units have odd index in the set of all units in $\mathcal{O}_{K_p}$, or $\mathcal{O}_{K_{pq}}$ if $p,q \equiv 1(4)$, and the Dirichlet symbol $\left(\frac{p}{q}\right) = -1$.

(Here $\mathcal{O}_K$ is the ring of algebraic integers in K.)

We apply theorem A to obtain information about the algebraic K-groups $K_o(ZG)$ for G a 2-hyperelementaty finite group. In particular theorem A implies that for infinite families of such groups we can prove certain elements in $K_o(ZG)$ are non-zero while for other similar families the analogous elements can be shown to vanish.

*Research partially supported by NSF MCS79-06085

As an example of how such questions arise we have

THEOREM B: Let $Q(8a,b,c)$ *be one of the periodic groups (of period 4) studied in* [Milnor], [M1], [M2], *then the vanishing or non-vanishing of the Swan obstruction for* $Q(8a,b,c)$ *depends entirely on the structure of any odd index subgroup of units in* $\coprod_n \mathcal{O}_{K_n}$ *where* n *runs over the set* $(abc,4a,4b,4c)$.

The Swan obstruction is a specific element of $D(Q(8a,b,c))/T \subset K_0(Q(8a,b,c))/T$ which vanishes if and only if $Q(8a,b,c)$ acts freely on a finite complex homotopic to a sphere. $D(G)$ is determined via a Meyer-Vietoris sequence from the global units in $K_1(M_G)$, where M_G is a maximal order, and in general, one would expect a complete knowledge of these units to be required for deciding if the Swan obsturction is 0 or not. But, in fact, Theorem B asserts that only the units in the maximal 2-abelian subfields of the center of $\mathbb{Q}G$ are needed. This is an example of our basic idea, explained in Sections C,D, of using groups of automorphisms of K_* to split off smaller summands which can (a) be treated independently - in particular, shown to be trivial or non-trivial - without having to worry about the rest of K_*, and (b) often be shown to contain the entire answer to a question which originally appeared to require knowledge of all of K_*.

Using theorems A and B we obtain

COROLLARY C: If p,q *are odd primes* $p \equiv 3(4)$, *then the Swan invariant* $\sigma_4(Q(8p,q,1)) = 0$ *if and only if*

a. $q \equiv 1(8)$,

b. $q \equiv 5(8)$ *and* $p^{\omega} \equiv \pm 1(q)$ *with* ω *odd, or*

c. $p \equiv 7(8)$, $q \equiv 5(8)$

(Note that Corollary C gives us an infinite family of distinct prime pairs for which $\sigma_4(Q(8p,q,1)) \neq 0$, as well.)

COROLLARY D: If $p \equiv q \equiv 1(4)$ *are odd primes then* $\sigma_4(Q(8p,q,1)) = 0$ *if* $\left(\frac{p}{q}\right) = -1$.

Also, the techniques of [M2] give considerable information about the surgery obstructions in the cases above. Indeed, we are able to use theorem A to construct certain non-zero elements in the surgery groups $L_3^h(A(8p,q,1))$ and we obtain

THEOREM E: For $p \equiv 7(8)$ *the surgery obstruction does not vanish when* $q \equiv 5(8)$ *satisfies the condition of Corollary C(b), or C(c).*

In particular, these results give new restrictions on the possible finite fundamental groups of compact 3 manifolds. Recall [Milnor], [Lee], that these groups can only be the known ones or the groups $(Q(8a,b,c),d)$ of [Milnor]. We say (p,q) divides (a,b,c) if p divides one of a,b,c and q divides another distinct element in this set. Then we have

COROLLARY F: If p,q *are odd primes with* $p \equiv 7(8)$, $q \not\equiv 1(8)$, *or* $p \equiv 3(8)$, $q \equiv 3(8)$ *or* $p \equiv 3(8)$, $q \equiv 5(8)$, $p^{2s} \equiv -1(q)$ *for some* s, *and* (p,q) *divides* (a,b,c) *then* $(Q(8a,b,c),d)$ *is not the fundamental group of any closed compact 3-manifold.*

Moreover, under the above restrictions $(Q(8a,b,c),d)$ *does not act freely on any sphere* S^{8k+3}, $k \geq 0$.

(Of course $(Q(8a,b,c),d)$ does act freely on S^{8k-1}, $k \geq 1$.)

REMARK G: In [M2] I prove that both the Swan obstruction and the surgery obstruction vanish for the group $G = Q(56,113,1)$, so G acts freely on S^{4n+3} for $n \geq 1$, and acts freely on a homology sphere M^3 as well.

§A. *The proof of theorem A*

Denote by UK_n the units in $\mathcal{O}_{K_n}$. If $|(Z/n)^{\bullet}| = \varphi(n) = 2^r s$ with s odd $s > 1$ then as an Abelian group

1. $$UK_n = Z/2 \times (Z)^{2^{r-1}-1}$$

Denote by U_n the units in $\mathcal{O}_{\mathbb{Q}(\zeta_n)}$, if n is an odd prime then the cyclotomic units in U_n are the subgroup generated by all the quotients

$u_k = \dfrac{\zeta_p^k - \zeta_p^{-k}}{\zeta_p - \zeta_p^{-1}}$, $p|n$, p prime together with the ζ_n^{ℓ}, and if n is composite, the elements $\zeta_m^i - \zeta_m^{-i}$ as well. Here m is composite, $m|n$ and i is a unit mod m [B]. We denote this subgroup of U_n by CU_n.

DEFINITION 2: Let $K \subset Q(\zeta_n)$ be a subfield then the cyclotomic units CK are $UK \cap CU_n$.

LEMMA 3: Let p be an odd prime then CK_p has odd index in UK_p.

PROOF: Up to odd index $UK_p \subset UFK_p$, so lemma 3 is equivalent to the statement that CK_p surjects onto $UFK_p/\text{Squares} = (Z/2)^{2^{r-1}}$. But UFK_p is a module over $Z(\text{Gal}(FK_p/Q)) = Z(Z/2^{r-1})$

by

4. $$(\Sigma a_i g_i)u = \Pi g_i(u^{a_i})$$

so $UFK_p/\text{Squares}$ is a module over $F_2(Z/2^{r-1})$. We have, however

5. $$F_2(Z/2^{r-1}) \cong F_2(x)/x^{2^{r-1}} = 0$$

($x = g + 1$, where g generates $Z/2^{r-1}$). Thus x is nilpotent on

FK_p/Squares and

$$x^{2^{r-1}-1} = 1 + g + g^2 + \cdots + g^{2^{r-1}-1} \tag{6}$$

so $x^{2^{r-1}-1}(u) = \mathrm{Norm}^{Q}_{FK_p}(u) = \pm 1$. By a well known result about the independence of the various powers $x^j(u)$ if $x^n(u) \neq 1$ we see that lemma 3 follows if there is a $u \in U\mathcal{O}_{K_p}$ with $N^{Q}_{FK_p}(u) = -1$. Now, let k be a primitive generator mod p, so $k^{\frac{p-1}{2}} \equiv -1(p)$. Then from the commutative diagram

$$\begin{array}{ccc} \mathcal{O}^{\bullet}_{FK_p} & \xrightarrow{\;N\;} & Z^{\bullet} \\ \Big\downarrow{\mathcal{R}} & & \Big\downarrow{\mathcal{R}'} \\ F_p^{\bullet} & \xrightarrow{\;\tilde{N}\;} & F_p^{\bullet} \end{array} \tag{7}$$

where $\mathcal{R}$ is reduction mod $(2 - \zeta_p - \zeta_p^{-1})$, $\mathcal{R}'$ is reduction mod (p), N is the norm and $\tilde{N}$ is $y \to y^{2^{r-1}}$ we see that

$$N^{FK_p}_{FQ(\zeta_n)}(u_k)$$

has norm -1 in Z, so lemma 3 follows.

We can now complete the proof of theorem A.

Let $\lambda = \zeta_{pq} - \zeta_{pq}^{-1} = \zeta_p^{-1}\zeta_q(\zeta_p^2 - \zeta_q^{-2})$. Let $N_1 : Q(\zeta_{pq}) \to Q(\zeta_p)$ be the norm then

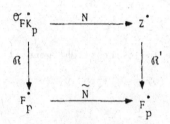

8. $$u_q = \frac{\zeta_p^{-q} - \zeta_p^{-q}}{\zeta_p - \zeta_p^{-1}} = \zeta_p^{1-q}(1 + \zeta_p^2 + \cdots + \zeta_p^{2(q-1)})$$

$$= \zeta_p^{1-q} \prod_{j=1}^{q-1} (\zeta_p^2 - \zeta_q^j)$$

$$= N_1(\lambda) .$$

Now note that $G = \mathrm{Gal}(K_{pq}/Q) = Z/2^r \times Z/2^t$ so

9. $$F_2(G) = F_2(x,y)/x^{2^r} = y^{2^t} = 0) = W .$$

Theorem A now follows from 8 and the following slight generalization of the fact that if x is nilpotent and $x^n(v) \neq 0$ then $v, x(v), \ldots, x^n(v)$ are linearly independent.

LEMMA 10: Let M *be a module over* W *and suppose there is a* $z \in M$ *so that the elements* $y^{2^t-1}(z), y^{2^t-1}x(z), \ldots, y^{2^t-1}x^{2^r-1}(z)$ *are linearly independent then* z *generates a free* W *submodule of* M.

REMARK 11: In [K] Kummer proves that if 2 does not divide the relative class number $h_p^- = |CL(O_{Q(\zeta_p)})|/|CL(O_{Q(\zeta_p+\zeta_p^{-1})})|$ where p is a prime (h_p^- is given by a certain product of Bernoulli numbers [E, p.225]), then 2 does not divide $h_p^+ = |CL(O_{Q(\zeta_p+\zeta_p^{-1})})|$. On the other hand $h_p^+ = |UQ(\zeta_p + \zeta_p^{-1})/CUQ(\zeta_p + \zeta_p^{-1})|$

In fact, in these cases, Kummer shows that

12. $$CQ(\zeta_p+\zeta_p^{-1})/(\text{Squares}) \xrightarrow{\text{sign}} (Z/2)^{(p-1/2)}$$

s an isomorphism. The first cases where $2|h_p^-$ are $p = 29$, 113, and 163. n all 3 cases, the sign map (12) above has a kernel. Specifically,

$$F_2(Z/14) = F_2(Z/2) \oplus F_8(Z/2) \oplus F_8(Z/2)$$

$$F_2(Z/56) = F_2(Z/8) \oplus F_8(Z/8) \oplus F_8(Z/8)$$

$$F_2(Z/81) = F_2 \oplus F_4 \oplus F_{64} \oplus F_{64^3} \oplus F_{64^9}$$

and the kernel in the first two cases are each *one copy* of $\cdot F_8$, while in the third case it is the copy of F_4. In the first cases Kummer shows the F_8 maps non-trivially into $\mathcal{O}_{Q(\zeta_p + \zeta_p^{-1})}/4$, squares, while in the third case he constructs a unit which is not a cyclotomic unit but whose square is. In fact one can now prove

COROLLARY 13: $h^+_{163} = 4\cdot \text{odd}$

(At the primes over 653 in the unique degree 3 extension of Q in $Q(\zeta_{163})$, the image of the norm of the fundamental unit $(-u_2)$ mod 4th powers is $(-1,-1,1)$. Hence since $652 = 4.163$, Kummer's unit cannot itself be a square, and indeed

$$UQ(\zeta_{163} + \zeta_{163}^{-1})/\text{Squares},\ CQ(\zeta_{163} + \zeta_{163}^{-1}) = F_4$$

as a module over $F_2(Z/81)$.)

REMARK 14: The complexities in remark 11 seem to indicate that studying the problem of odd index subgroups of units for fields more complicated than the K_n is going to be extremely complicated.

§B. *The fields* $Q(\zeta_{4p})$

Assume as before that p is an odd prime. Then, if $p - 1 = 2^r s$ with s odd, we have that $Gal(K_{4p}/Q) = Z/2 \times Z/2^r$, and the generators of the cyclotomic units are $\zeta_p^{\gamma} - i\zeta_p^{-\gamma}, i\zeta_p$, and $\zeta_p^{a} - \zeta_p^{-a}/\zeta_p - \zeta_p^{-1}$. The norm of $\zeta_p^{\gamma} - i\zeta_p^{-\gamma}$ under the map from $Q(\zeta_{4p})$ to $Q(\zeta_p)$ is

$$\zeta_p^{2\gamma} + \zeta_p^{-2\gamma} \quad .$$

Let ℓ be the 2 order of 2 at p, i.e. $2^{s2^{\ell}} \equiv 1$ for some odd s but $2^{s2^{\ell-1}} \not\equiv$ (mod p), for any odd s then we have

LEMMA 1: If $\ell > 1$, *then the image under*

$$N: K_{4p}^{\bullet} \to K_p^{\bullet}/(K_p^{\bullet})^2$$

of the cyclotomic units mod squares is isomorphic to the ideal generated by $x^{2^{r-\ell}-1}$ *in* $F_2(Z/2^r)$.

PROOF: $2 = a^{2^{r-\ell}}$ so

$$\frac{\zeta_p^2 - \zeta_p^{-2}}{\zeta_p - \zeta_p^{-1}} = \left(\frac{\zeta_p^{a^{2^{r-\ell}}} - \zeta_p^{-a^{2^{r-\ell}}}}{\zeta_p^{a^{2^{r-\ell}-1}} - \zeta_p^{-a^{2^{r-\ell}-1}}}\right)\left(\frac{\zeta_p^{a^{2^{r-\ell}-1}} - \zeta_p^{-a^{2^{r-\ell}-1}}}{\zeta_p^{a^{2^{r-\ell}-2}} - \zeta_p^{-a^{2^{r-\ell}-2}}}\right) \cdots \frac{\zeta_p^{a} - \zeta_p^{-a}}{\zeta_p - \zeta_p^{-1}}$$

$$= u_a^{(1 + \psi + \psi^2 + \psi^3 + \cdots + \psi^{2^{r-\ell}-1})}$$

but mod 2, $1 + \psi + \cdots + \psi^{2^{r-\ell}-1} = (1+\psi)(1+\psi^2)\cdots(1+\psi^{2^{r-\ell-1}})$

$$= x \cdot x^2 \cdots x^{2^{r-\ell-1}}$$

and the result follows.

COROLLARY 2: Under the assumption of lemma 1, let FK_{4p} *be the real subfield of* K_{4p}*, then the image of the sign homomorphism*

$$UFK_{4p} \to (Z/2)^{2^r}$$

has dimension at least $2^{r-1} + 2^{r-2} - 1$*.*

PROOF: $u = \zeta_p - i\zeta_p^{-1}/\zeta_p^a - i\zeta_p^{-a} \in F\mathbb{Q}(\zeta_{4p})$ and has the form $(\zeta_p - i\zeta_p^{-1})^{(1-\psi)}$. Thus $N(u) = x^{2^{r-\ell}}(u_a)$ where N is the map in lemma 1. Clearly $N(x^V u) = x^V N(u)$, and thus, image N has dimension at least $2^{r-\ell}(2^{\ell-1}-1) - 1$. On the other hand $\operatorname{im} N^2$ is 1, so N, too, is nilpotent and (2) follows easily.

COROLLARY 3: Under the assumption of lemma 1 let $V \subset UFK_{4p}$ *be the subgroup of totally positive units, then*

$$V/(UFK_{4p})^2 = V_1$$

is annihilated by the action of $x^{2^{r-2}+1}$*.*

(Clear.)

§C. *The groups* $K_0(ZG)$ *and* $D(G)$ *for* G *a finite group*

$K_0(ZG)$ is the projective class group of the group ring ZG. Let M_G be a maximal Z order containing ZG in QG. Then $K_0(M_G)$ is calculated using standard techniques from [S-E] in terms of class groups and proper class groups of its center. Moreover [F] the natural map

$$K_0(ZG) \to K_0(M_G)$$

is onto. Its kernel is the group $D(G)$, which may be studied using the Meyer-Vietoris sequence of the pull back diagram

1. $$\begin{array}{ccc} ZG & \longrightarrow & M_G \\ \downarrow & & \downarrow \\ \coprod\limits_{p\,|\,|G|} \hat{Z}_p G & \xrightarrow{\ j\ } & \coprod\limits_{p\,|\,|G|} \hat{Z}_p \otimes M_G \end{array}$$

(See e.g. [M1] for details.) Hence we have an exact sequence

2. $$K_1(M_G) \oplus \coprod_{p\,|\,|G|} K_1(\hat{Z}_p G) \to \coprod_{p\,|\,|G|} K_1(\hat{Z}_p \otimes M_G) \to D(G) \to 0$$

Motivated by 2 we define the local defect groups of G at p as

3. $$LD_p(G) = \mathrm{im}(\varphi_p)/\mathrm{im}\,\varphi_p \cdot j(K_1(\hat{Z}_p G))$$

where $\varphi_p : K_1(\hat{Z}_p \otimes M_G) \to K_1(\hat{Q}_p G))$ is the natural map. In [M1] these groups are calculated for all the G which we consider in this paper. Indeed, fairly general techniques are developed there for studying these groups whenever G is p-hyperelementary. In particular $LD_p(G)$ is finite for each p. Also in [M1] I show

THEOREM 4:

$$D(G) = \coprod_{p \mid |G|} LD_p(G)/im(\varphi K_1(M_G))$$

where $\varphi: K_1(M_G) \to K_1(QG)$ *is the natural map.*

(Note that image (φ) can be identified with the units or positive units in the groups $U(K_\ell)$ where K_ℓ runs over the centers of the simple components of QG.)

Example 5: Let G be the generalized quaternion group $Q(8p,1,1)$ where p is prime $(G = \{x,y \mid (xy)^2 = x^2 = y^{2p}\})$ then

$$LD_2(G) = (Z/2)^2 \oplus \{[Z(\zeta_p + \zeta_p^{-1})/(4)]^\bullet \oplus [Z(\zeta_p + \zeta_p^{-1})/(2)]^\bullet\}$$

$$LD_p(G) = F_p^\bullet \oplus F_p^\bullet \oplus F_p^\bullet$$

Specifically $\hat{Z}_2 G = \hat{Z}_2(Q(8,1,1)) \oplus W$

where

$$W = \hat{Z}_2 \otimes (Z(\zeta_p) \times_T Q(8,1,1))$$

and the local defect splits into 2 pieces in a similar way. The generators of the $(Z/2)^2$ are given by the values -1 at the trivial representations, ones at the remaining representation and 3 at the trivial representation, 1's at the remaining representations. Also

$$\hat{Z}_p G = (\hat{Z}_p(Z/p) \times_T Z/2)^+ \oplus (\hat{Z}_p(Z/p) \times_T Z/2)^-$$
$$\oplus\, \hat{Z}_p(Z/p) \times_T (Z/2)^2/(x^2 = y^2 = (xy)^2 = -1)$$

and each summand above yields one copy of $F_p^{\bullet}$ to $LD_p(G)$.

D. *The action of* $Out(G)$ *on* $K_*(ZG)$ *and* $D(G)$

An action of $Aut(G)$ on Mod_G (the category of isomorphism classes of ZG modules) is defined by

$$\theta(M,u\colon G \times M \to M) = (M,u')$$

where $u'(g,m) = \theta(g)m$. Similarly $Aut(G)$ acts on $GL_n(RG)$ for R any ring. These actions induce actions on $K_o(RG)$, and on $B^+_{GL(RG)}$, hence on the higher $K_*(RG)$ groups as well. Moreover, these actions commute with natural functors such as change of R or the maps in the Meyer-Vietoris sequence C.2.

LEMMA 1: Let $g \in Aut(G)$ *be an inner automorphism, then the action of* g *on* $K_*(RG)$ *is the identity.*

PROOF: For $B^+_{GL(RG)}$ this is well known. For K_o, define $\ell g\colon (M,u) \to (M,u')$ by $\ell g(m) = u(g,m) = g\cdot m$ (where g is given by $g(h) = ghg^{-1}$ for all $h \in G$).

COROLLARY 2: The actions above factor through $Out(G)$ *and are natural with respect to maps and change of rings.*

Restricting attention now to $K_o(ZG)$ note that we have a natural splitting $K_o(ZG) = \tilde{K}_o(ZG) \oplus Z$ where $\tilde{K}_o(ZG)$ is torsion, and this splitting is preserved by the action of $Out(G)$. Hence, writing

3. $$\tilde{K}_o(ZG) = \coprod_{p\text{-prime}} K_o(ZG)_p$$

where M_p denotes the p-primary part of the torsion module M, we have that the action of $Out(G)$ on $K_o(ZG)_p$ induces an action of

$$Z_p(Out(G)) \quad \text{on} \quad K_o(ZG)_p$$

and in particular an action of $Z_2(Out\ G)$ on $K_o(ZG)_2$. Note also that these actions take $D(G)$ to itself.

COROLLARY 4: Let $H \subset Out(G)$ *and suppose* $Z_p(H) = A \oplus B$, *a direct sum of algebras, then* $\tilde{K}_o(ZG)$ *and* $D(G)$ *split as direct sums*

$$\tilde{K}_o(ZG)_A \oplus \tilde{K}_o(ZG)_B, \ D(G)_A \oplus D(G)_B .$$

Moreover, on tensoring $Z_p \otimes K_*(ZG)$ *splits in the same way and these splittings are natural with respect to maps.*

EXAMPLE 5: For example C.5, the generalized quaternion group, the isomorphisms

$$\theta_k: (y \to y^{1+4k}, \ x \to x)$$

$$1 \leq k \leq p - 1$$

are all outer and give a cyclic subgroup $Z/p-1 \subset Out(G)$. Then $p - 1 = 2^r s$ with s odd and

6. $$Z_2(Z/p - 1) = \coprod_{t|s} Z_2(\zeta_t)(Z/2^r)$$

gives a splitting of $Z_2(Z/(p - 1))$. In particular the base block $Z_2(Z/2^r)$ in (6) splits off a particularly important piece of $\tilde{K}_o(ZG)$.

From corollary (4), the base block also splits the 2 torsion of the local defect groups. In fact, since

$$LD_p(G)$$

is invariant under $Out(G)$, the base summand contains it in its entirety. Similarly for the first $Z/2$ in $LD_2(G)$. But $Z/p-1$ acts via the projection $Z/(p-1) \to Z/(\frac{p-1}{2})$ as Galois isomorphisms in $[Z(\zeta_p + \zeta_p^{-1})/(4)]^{\bullet}_{(2)}$. Hence the resulting 2-torsion associated to this base summand is identified with the units in $[(\mathcal{O}_{K_p})/(4)]^{\bullet}$ of the form $(1 + 2x), x \in (\mathcal{O}_{FK_p}/(2))$. Moreover, as a module, this group $(Z/2)^{2^{r-1}}$ is free over $F_2(Z/2^{r-1})$. (This is a standard result in classfield theory, compare [B].)

COROLLARY 7: Let p *satisfy the assumptions of lemma B(1), then the base summand of* $D(Q(8p,1,1)_{(2)}$ *is*

$$Z/2 \oplus Z/2 \oplus [F_p^{\bullet}/(\text{image of positive units in } FK_{4p})]_2 .$$

REMARK 8: Here the first $Z/2$ comes from the $(Z/2)^2$ summand in $LD_2(G)$. The second $Z/2$ comes from the $F_p^{\bullet}$ associated to the block $(\hat{Z}_p(Z/p) \times_T Z/2)^+$.

OTULINE PROOF: The 2 torsion in the base summands for $LD_2(G)$ and $LD_p(G)$ are given by the tables

$$(Z/2)^2 \oplus (Z/2)^{2^{r-1}}$$

$$Z/2^r \oplus Z/2^r \oplus Z/2^r$$

We use -1 at the trivial representation to identify the -1 in $(Z/2)^2$ with the element of order 2 in the first $Z/2^r$, and the powers of the fundamental units $\lambda^i, \tau^{-i}, \lambda$ for $M_2(Q(\lambda_p))^+$, τ for $M_2(Q(\lambda_p))^-$ to identify the first two $(Z/2^r)$'s and then the λ^{2j} to kill the squares. Similarly, the square of the fundamental unit is positive and acts to cancel the squares in the third $Z/2^r$.

Now we use the generating unit in $M_2(Q(\lambda_p)^+$ to identify the *Galois generator* in $(Z/2)^{2^{r-1}}$ with the generator of the first $Z/2^r$. This then gives us the following table

$$Z/2 \ , \ 0$$

$$Z/2 \ , \ 0 \ , \ [\dot{F}_p/\text{image of positive units}]_2$$

and Corollary B(3) shows that the $Z/2$ in the first place in the second row cannot be annihilated. (For details of the structure of the map

$K_1(Z(G)) \to \coprod_p LD_p(G)$ see [M1].)

REMARK 8: Note that $(\dot{F}_p/[\text{image of positive units in } FK_{4p}])_2$ is either $Z/2$ or 0. In particular it is zero if $p \equiv 5(8)$ since then 2 is a non-square and

$$(\zeta_p - i\zeta_p^{-1})(\zeta_p^{-1} + i\zeta_p)$$

$$= 2 + i(\zeta_p^2 - \zeta_p^{-2})$$

is a positive unit with image 2 in $\dot{F}_p$.

E. *The proof of theorem B*

In [M1] a procedure is given for explicitly constructing an element α

n $\coprod_{p \mid |G|} LD_p(G)$ when $G = Q(8a,b,c)$ with $\mathrm{im}\,\alpha$ in $\tilde{K}_0(ZG)$ representing

he Swan obstruction $\sigma_4(G)$. In this section we assume familiarity with those

esults.

Recall that G is the 2-hyperelementary group

1. $$Z/abc \to G \xrightarrow{\ \longleftarrow\ } Q(8,1,1)$$

with Sylow 2-subgroup the ordinary order 8 quaternion group and specified distinct actions on Z/a, Z/b, Z/c. Then $\mathrm{Out}(G)$ contains $(Z/abc)^{\bullet} = V$.

DEFINITION 2: The base summand $B(G)$ of $\tilde{K}_0(ZG)$ is the summand of the 2-primary part of $\tilde{K}_0(ZG)$ corresponding to the summand

$$Z_2(V_2) \subset Z_2(V)$$

where V_2 is the 2-Sylow subgroup of V.

THEOREM 2: The Swan obstruction $\sigma_4(G) \in B(G)$.

PROOF: From [M1] the class of $\sigma_4(G)$ in $LD(G)$ is given by 2 at the usual quaternion representation of $Q(8,1,1)$ and $2 - (\zeta_{p^i} + \zeta_{p^i}^{-1})$ at the representations $M_2(A(\zeta_{p^i} + \zeta_{p^i}^{-1}))$ where $p^i \mid$ one of a,b,c. Now 2 is invariant under V, hence is associated to $B(G)$. Also, writing $Z_2(V) = Z_2(V_2) \oplus W$ the splitting idempotent is

$$e = \frac{1}{\omega}\Big(\big(\sum_{\substack{\text{order } g \text{ odd} \\ g \in V}} g\big) + 1\Big)$$

where $|V| = \omega 2^{\ell}$ with ω odd, and $1-e$ satisfies $\varepsilon(1-e) = 0$ where $\varepsilon\colon Z_2(V) \to Z_2$ is the augmentation. Hence

$$[2 - (\zeta_{p^i} + \zeta^{-1}_{p^i})]^{1-e}$$

is a global cyclotomic unit and represents the trivial class in $D(G)$. In particular, a representative in $LD(G)$ for $\sigma_4(G)$ is given by the classes 2 as before and

$$(2 - (\zeta_{p^i} + \zeta^{-1}_{p^i}))^{c}$$

at the remaining representations.

COROLLARY 3: In the case $G = Q(8p,q,1)$ a representative for $\sigma_4(G)$ is given by the array

1		*1*		*1*
1				*1*
	1		*1*	
1				*-1*
1				*1*
$\frac{1}{4}$				$\frac{1}{4}$

in the notation of [M1, §6.A]

PROOF: In [M1, theorem 6.A.8] a description of $\sigma_4(G)$ is given. In the bottom row it is $(\zeta_p + \zeta_p^{-1})^2$, $(\zeta_q + \zeta_q^{-1})^2$ and in the 4th row it is $(\zeta_p - \zeta_p^{-1})^2$, $(\zeta_q - \zeta_q^{-1})^2$. To change the bottom row we use the unit $(\zeta_p + \zeta_p^{-1})^{-2}$ $(\zeta_q + \zeta_q^{-1})^{-2}$, and to change the unit in the 4th row we use $(\zeta_{pq} - \zeta_{pq}^{-1})^{-1} \cdot (\zeta_p^{-1}\zeta_q - \zeta_p\zeta_q^{-1})^{-1}$. (This amounts to making the procedure in the proof of theorem 2 explicit.)

F. *The proofs of* C *and* D

The Swan obstruction in E.3 for $Q(8p,q,1)$ takes its value in $[\mathcal{O}_{(FK_p \cdot FK_q)/(p \cdot q)}]^2$. In the first summand we can factor out all units while in the second summand we may only factor out positive units.

LEMMA 1: Let $p \equiv 3(4)$, *then* $FK_p \cdot FK_q = FK_q$ *and*

a) $\mathcal{O}_{FK_q/pq} \cong F_q \times (F_{p^{2^\ell}})^{2^\omega}$ *with* $\ell + w = r - 1$, $(q - 1 = 2^r s)$

b) *If* $p^\omega \not\equiv \pm 1(q)$ *with* ω *odd* $(\mathcal{O}_{FK_q/pq})_{(2)}$ */mod units* $= Z/2$

with generator represented by the non-square at F_q.

PROOF: Let α be a non-square at q then u_α and its Galois conjugates generate an odd index subgroup of $U(FK_q)$. Moreover, $N(u_\alpha) = -1$.

Write $\text{im}(u_\alpha) = (a_1, \ldots, a_{2^\omega}) \in (F^{\bullet}_{p^{2^\ell}})^{2^\omega}$ then $\text{im}(N(u_\alpha)) = -1 = (a_1 \ldots a_{2^\omega})^{(p^{2^\ell}-1/p-1)}$, and it follows that $a_1 \ldots a_{2^\omega}$ is a non square in $(F_{p^{2^\ell}})^{\bullet}$ (this is where we require $\ell > 1$ which is equivalent to the condition on p in (b)). In particular, by an argument similar to that in the proof of theorem A $(a_1, \ldots, a_{2^\omega})$ in $F^{\bullet}_{p^{2^\ell}}/\text{squares} = (Z/2)^{2^\omega}$ is spanning under the action of the Galois group. But this implies that given any $\theta \in ((F^{\bullet}_{p^{2^\ell}})^{2^\omega})_2$ there is some unit in FK_q with image θ at (p).

On the other hand an easy calculation shows that

$$\text{Ker}: U(FK_q) \to ((F^{\bullet}_{p^{2^{\ell}}})^{2^{\omega}})_2$$

is spanned over the Galois algebra by

$$(\psi^{2^{\omega}}(u_\alpha)/u_\alpha^p).$$

But the image of this class in $F^{\bullet}_q$ is $\alpha^{1-p} = \alpha^{2\cdot\text{odd}}$, and lemma 1 follows.

COROLLARY 2: For $Q(8p,q,1) = G$ *with* $p \equiv 3(4), \sigma_4(G) = 0$ *if* $q \equiv 1(8)$

PROOF: The obstruction is $(-1,1,4,4)$ and $(-1,1)$ is in the image of global units by lemma 1. Similarly, since $p \equiv 3(4), 4 \equiv 1$ in $(F^{\bullet}_p)_2$ at p and since $q \equiv 1(8)$, 2 is a square at q so $4 \equiv a^4$ at q and hence, again from lemma 1, $(4,1)$ is the image of the square of a unit in FK_q.

COROLLARY 3: For G *as above with* $p \equiv 3(8), \sigma_4(G) \not\equiv 0$ if $q \equiv 5(8)$ and $p^{\omega} \not\equiv \pm 1(q)$ *for some odd* ω, *but* $\sigma_4(G) = 0$ *if* $p^{\omega} \equiv \pm 1(q)$ *with* ω *odd, or* $p \equiv 7(8)$.

PROOF: From lemma 1, $(-1,1)$ is the image w of a unit. $(4,4) \sim (4,1)$ in $(F^{\bullet}_q \times (F^{\bullet}_p)^2)_2$ but since 2 is a non-square at q, $(4,1)$ is not the image of any positive unit in FK_q. Now consider the part of the local defect group

4.

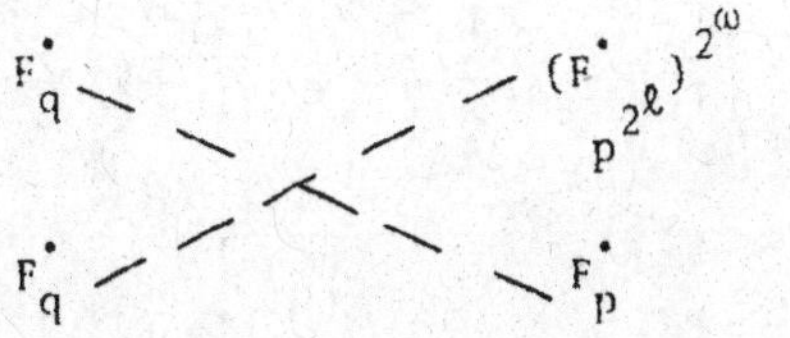

$\ell \geqslant 1$

The units of K_{4p} are generated up to odd index by ± 1, $N_1(\zeta_p - i\zeta_p^{-1})$, and the norm of this second to FK_{4p} is positive and has image 2 at $F_p^{\bullet}$, and $N(\zeta_p + \zeta_p^{-1})^2$ at $F_q^{\bullet}$. Hence, it cannot affect the obstruction class. On the other hand the units of K_{4q} are generated by $\zeta_q - i\zeta_q^{-1}$ up to odd index under the action of the Galois group since 2 is a non-square at q. The only positive unit which is not a square is thus $N_1(2 - i(\zeta_q^{-2} - \zeta_q^{2}))$ which has image 2 at $F_q^{\bullet}$, and -1 in $((F_p^{\bullet})^{2^{\omega}})_2$, so there is no means by which this obstruction vanishes.

Now suppose $p \equiv 7(8)$ so 2 is a non-square mod p. Hence $2^v \equiv 1(p)$ for some odd v, and $N_1(\zeta_p - i\zeta_p^{-1})^v$ has image $N(\zeta_p + \zeta_p^{-1})^{2v}$ at $F_q^{\bullet}$. Hence the obstruction vanishes.

Suppose now that $p^{\omega} \equiv \pm 1(q)$ with ω odd then (4) becomes

$$\begin{array}{ll} F_q^{\bullet} & (F_p^{\bullet})^{2^r} \\ F_q^{\bullet} & F_p^{\bullet} \end{array} \qquad 5.$$

and u_{α}^2 has image (4,1) whence this obstruction vanishes and 3 follows.

PROPOSITION 6: If $p \equiv 3(4)$, $q \equiv 3(4)$ then $\sigma_4(G) \neq 0$ for $G = Q(8p,q,1)$.

PROOF: In this case the obstruction has the form

$$\begin{array}{ccccc} 1 & & 1 & & 1 \\ 1 & & & & 1 \\ & 1 & & 1 & \\ -1 & & & & 1 \\ 1 & & & & 1 \\ 1 & & & & 1 \end{array} \qquad 7.$$

and the only global units available are ± 1's. It is an easy check that in the **four top** rows each -1 has image an even number of -1's and so the obstruction is non-trivial.

The proof of Corollary C is now complete.

To prove Corollary D consider $FK_p \cdot FK_q$ where $(\frac{p}{q}) = -1$ and $p \equiv q \equiv 1(4)$. Then

8. $$(\theta_{FK_p \cdot FK_q})/pq = F_{p^{2^{t-1}}} \times F_{q^{2^{r-1}}}$$

with $p-1 = 2^r s_1$, $q-1 = 2^t s_2$. Moreover, $(F_{p^{2^{t-1}}})^{\bullet}_2 \cong Z/2^{r+t-1} \cong (F_{q^{2^{r-1}}})^{\bullet}_2$.

Now, in $F_{p^{2^t}}$, $N(\zeta_q - \zeta_q^{-1})$ is a non-square, as the degree 2 element of the Galois group takes it to its negative, and the same is true of $N(\zeta_p - \zeta_p^{-1})$ in $F_{q^{2^r}}$.

Thus, $N((\zeta_{pq} - \zeta_{pq}^{-1})(\zeta_p\zeta_q^{-1} - \zeta_p^{-1}\zeta_q)) = u$

in $F_{p^{2^{t-1}}}$ becomes $-(N(\zeta_q - \zeta_q^{-1})^2)$ which is a generator of the 2 subgroup and similarly at q. Finally, an appropriate quotient $\psi_q(u)/u$ represents 1 at p but u^{q-1} at q, and similarly $\psi_p(u)/u$ represents u^{p-1} at p and 1 at q.

The effect of u is to identify $(F_{p^{2^{t-1}}})^{\bullet}_2$, with $(F_{q^{2^{r-1}}})^{\bullet}_2$, and the effect of $\psi_q u/u$ and $\psi_p(u)/u$ is to factor out $im(F_p)^{\bullet}$, and $im(F_q^{\bullet})$ in this common $Z/2^{r+t-1}$. But by Corollary E.3 this contains the Swan obstruction and Corollary D follows.

§G. *The Swan subgroup of* $\widetilde{K}_o(ZG)$

Let G be a finite group, $\Sigma_G \in ZG$ is the element $\sum_{g\in G} g$. Then we have the pull back diagram

1. $$\begin{array}{ccc} ZG & \longrightarrow & ZG/(\Sigma_G) \\ \Big\downarrow \varepsilon & & \Big\downarrow \pi_2 \\ Z & \xrightarrow[\pi_1]{} & Z/|G| \end{array}$$

and we can construct projective modules P_α using the procedure of [Milnor 2] via the pull back diagram

2. $$\begin{array}{ccc} P_\alpha & \longrightarrow & ZG/(\Sigma_G) \\ \Big\downarrow & & \Big\downarrow \pi_2 \\ Z \xrightarrow[\pi_1]{} Z/|G| & \xrightarrow{\cdot\alpha} & Z/|G| \end{array}$$

where α is any element of $(Z/|G|)^{\bullet}$. The classes of these P_α in $\widetilde{K}_o(ZG)$ define the Swan subgroup $T \subset \widetilde{K}_o(ZG)$. T is contained in D(G), and, in fact, in our local defect diagrams ; T is precisely the image of the top row in LD(G). The main importance of the Swan subgroup is that it precisely measures the possible values of the Swan obstruction. That is, the set of all values of $\sigma_n(G)$ is a certain co-set of T in $\widetilde{K}_o(ZG)$.

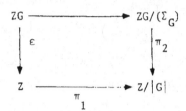

THEOREM 3: Let $G = Q(8p,1,1)$ with $p \equiv 3(4)$ or $p \equiv 1(4)$ and $2^{\omega} \not\equiv \pm 1(p)$ for any odd ω. Let ℓ be an odd number which is a primitive generator mod $p-1$, then there is a unique homotopy type which contains a finite complex $Y^{4\ell-1}$ with fundamental group G and universal cover $X^{4\ell-1} \simeq S^{4\ell-1}$.

PROOF: The homotopy types of $Y^{4\ell-1}$ are determined by a single K-invariant α in $H^{4\ell}(B_G,Z) = Z/8p$. It must be a unit, and by considering homotopy equivalences of B_G the same homotopy type is obtained from $\pm\alpha\beta^{2\ell}$ where β is any other unit. But factoring out this indeterminacy gives $Z/2 \times Z/2$ with generators 3 at 8, and any non-square at p. But by D.7 and D.8 these classes inject into $T \subset \tilde{K}_0(ZG)$ and the result follows. (Compare [W].)

REMARK 4: Since $Q(8p,1,1)$ acts freely on S^{4k+3}, $k \geq 0$, it follows that the unique homotopy type in theorem 3 is actually the homotopy type of a manifold.

COROLLARY 5: Let p,ℓ satisfy the hypotheses of theorem 2, $q \equiv 3(4)$, and suppose $\sigma_4(Q(8p,q,1)) = 0$, then if $X^{4\ell-1}$ is a free finite $Q(8p,q,1)$ complex homotopic to $S^{4\ell-1}$ we have that $X^{4\ell-1}/Q(8p,1,1)$ is the homotopy type of a manifold.

§H. *The proof of theorem E*

We consider a surgery problem over $X^{4\ell-1}$ with ℓ odd of the type considered in [M2]. Corollary G.5 implies we may set it up, and all the results except the one which analyzes the image of the surgery obstruction in $L_3^p(Q(8p,q,1))$ in [M2] continue to hold.

However, the results of §3 of [M2] still are valid and, in particular, we can assume the restriction of the surgery obstruction to $L_3^p(Q(8,1,1))$ is trivial. Thus, a necessary and sufficient condition for the entire obstruction to be non-trivial for all surgery problems over $X^{4\ell-1}$ is that this obstruction be non-trivial.

But if the obstruction in $L_3^p(Q(8p,q,1))$ is trivial, from [M2] we have a single remaining obstruction coming from the class of 2 at $F_q^{\bullet}$ in the bottom row [M1, §6.A] in the local defect group.

THEOREM 1: The class (2,1) in $\mathcal{O}_{FK_q/qp}$ *represents a non-trivial element* γ *in*

$$H^1(Z/2,\tilde{K}_o(G))$$

and the image of γ *in* $L_3^h(G)$ *is not zero where* $G = Q(8p,q,1)$ *with* $p \equiv 7(8)$, $q \equiv 5(8)$.

PROOF: By F.1.b if $p^{\omega} \not\equiv \pm 1(q)$ with ω odd (or by inspection in the remaining case) (2,1) is non-zero mod units in $(\Theta_{FK_q/qp})_{(2)}$ and (4,1) is in the image of positive global units.

LEMMA 2: Let $\tilde{*}$: $\tilde{K}_o(ZG) \to \tilde{K}_o(ZG)$ *be the involution sending* $P \to P^* = Hom_{ZG}(P,ZG)$.

(Here P^* is given a ZG module structure by $\alpha f(x) = f(x)\bar{\alpha}$, $\overline{\Sigma\alpha_i g_i} = \Sigma\alpha_i g_i^{-1}$.) Then $\tilde{*}$ is the identity for the group $Q(8p,q,1)$.

PROOF: From the exact sequence [F] of the pull back diagram

$$\begin{array}{ccc} ZG & \longrightarrow & QG \\ \downarrow & & \downarrow \\ \coprod\limits_{\text{all } p} \hat{Z}_p G & \longrightarrow & \coprod\limits_{p} \hat{Q}_p G \end{array}$$

we have an exact sequence

3. $$K_1(QG) \to \coprod_p LD_p(G) \to \tilde{K}_o \to 0 .$$

Clearly, since the center of QG is fixed under $*$, for $G = Q(8p,q,1)$ and since

$$\begin{aligned} \hat{Q}_p G &= \hat{Q}_p \otimes QG = \coprod \hat{Q}_p \otimes M_{n_i}(D_i) \\ &= \coprod M_{n_i}(\hat{Q}_p \otimes D_i) \end{aligned}$$

we see that $K_1(\hat{Q}_p G)$ is fixed under star as well, whence $LD_p(G)$ is fixed under $*$ and lemma 2 follows.

In particular $H^1(Z/2,\tilde{K}_o(G))$ is precisely the set of elements in $\tilde{K}_o(G)$

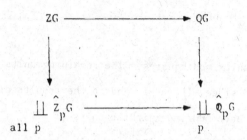

of order exactly 2. Thus (2,1) does represent a non-trivial class in $H^1(Z/2,\tilde{K}_0(G))$. Moreover, the boundaries of Z generators at the maximal quaternion representation cannot hit this element since [H-M] they hit either the elements in

$$CL_+(FK_p \cdot FK_q)/CL(FK_p \cdot FK_q)$$

or the images of non positive units mod squares. The remaining units at FK_{4q} may similarly be excluded since $(\frac{2}{q}) = -1$ and so the results of §B show that the cyclotomic units have odd index in the units of FK_{4q}. Finally, the effect of the elements in image $(L_1^{tor}(ZG))$ in $L_0^P(Z(G))$ is described [Wr] in terms of $\pm$ 1's at primes over which the representations in question are not quaternion. But unless $p^\omega \not\equiv \pm 1(q)$ these elements are already trivial on factoring out by positive units. Hence it remains to consider the case $p^\omega \equiv \pm 1(q)$.

If $p^\omega \equiv +1(q)$ then, in FK_{4q} over p we get 2^{r-1} copies of F_{p^2} and to identify units with the units in $(F_p)^{2^{r-1}}$ at the local defect group we use the norm which takes $\pm$ 1 to 1.

Similarly, if $p^\omega \equiv -1(q)$, once more FK_{4q} over p gives 2^{r-1} copies of F_{p^2}, and the norm works as before.

Thus, only the elements in $L_1^{tor}(ZG)_{(p)}$ which are involved with the faithful representation can affect matters.

But here it is easily checked that the representations at the primes over p are quaternion and theorem 1 follows.

Finally, theorem E follows from theorem 1. Note first that if the image of the surgery obstruction in $L_3^p(G)$ is not zero, then the homotopy type cannot be that of a manifold (since the restriction to $L_3^p(Q_8) = L_3^h(Q_8) = (Z/2)^2$ [H-M] is trivial). But if this image is trivial then [M2, §3.B] shows that the class is represented by (2,1) at the faithful representation.

Now, as we vary over all the possible homotopy types obtained by just changing K-invariants in the kernel of the Swan homomorphism) the above arguments continue to hold and E follows.

Stanford University
Stanford,California

Bibliography

[B] H. Bass, "Generators and relations for cyclotomic units", Nagoya Math. J. 27 (1966), 401-407

[E] H. Edwards, Fermat's Last Theorem, Graduate Texts in Mathematics #50, Springer-Verlag, 1977

[F] A. Fröhlich, "Locally free modules over arithmetic orders", J. Reine Angew. Math. 274/75 (1975), 112-138

[H-M] I. Hambleton, R. J. Milgram, "The surgery obstruction groups for finite 2-groups", Mimeo. Stanford (1979)

[K] E. E. Kummer, Monatsher K. Akad. Wiss. Berlin 1870, 855-880, (see also collected works, Vol 1.)

[Lee] R. Lee, "Semi-characteristic classes", Topology 12 (1973), 183-199

[MI] R. J. Milgram, "The Swan finiteness obstruction for periodic groups", Mimeo. Stanford (1979)

[M2] ____________, "Exotic examples of free group actions on spheres", Mimeo. Stanford (1979)

[Milnor] J. Milnor, "Groups which act on S^n without fixed points", Amer. J. Math. 9 (1957), 623-630

[Milnor 2] Introduction to Algebraic K-theory, Annals of Math Studies #72, Princeton U. Press (1971)

[S-E] R. Swan, K-theory of Finite Groups and Orders, Notes by E. G. Evans, Lecture notes in mathematics #149, Springer-Verlag (1970)

[W] C. T. C. Wall, "Free actions of finite groups on spheres", Proc. Symp. Pure Math. Vol 32 (1974) A. M. S., 115-124

[Wr] C. Wright, Thesis, Stanford (1980)

SK_1 FOR FINITE GROUP RINGS: III

Robert Oliver

If π is a finite group and R a Dedekind domain with quotient field K, define

$$SK_1(R\pi) = \mathrm{Ker}[K_1(R\pi) \to K_1(K\pi)]$$

$$K_1'(R\pi) = \mathrm{Im}[K_1(R\pi) \to K_1(K\pi)] \cong K_1(R\pi)/SK_1(R\pi).$$

These are especially of interest when R is either an n-ring - the ring of integers in some finite extension of $\mathbb{Q}$ - or a p-ring - the ring of integers in some finite extension of $\hat{\mathbb{Q}}_p$. In particular, $SK_1(\mathbb{Z}\pi)$ has been shown by Wall to be precisely the torsion subgroup of $Wh(\pi)$ ([21], Theorem 6.1).

When R is an n-ring, $SK_1(R\pi)$ can be studied by means of localization sequences of Bak [1] or Quillen [7] (see also Section 1 in [12]). In particular, there is a surjection

$$SK_1(R\pi) \twoheadrightarrow \sum_p SK_1(\hat{R}_p\pi)$$

whose kernel, denoted by $Cl_1(R\pi)$, can be studied using K_2. For abelian π, computations of $Cl_1(R\pi) = SK_1(R\pi)$ are summarized in [19]. For non-abelian π, $Cl_1(R\pi)$ was studied in [12], where it was shown, for example, to be generated by induction from elementary subgroups of π.

$SK_1(\hat{R}_p\pi)$, on the other hand, is completely calculated for π a p-group in [13]. More specifically, for any p-ring A and p-group π,

$$SK_1(A\pi) \cong H_2(\pi)/H_2^{ab}(\pi),$$

where $H_2^{ab}(\pi) \subseteq H_2(\pi)$ is the subgroup generated by the $H_2(\rho)$ for all abelian $\rho \subseteq \pi$.

In this paper, the results in [12] and [13] are combined with the induction theory of Dress [6] and some new calculations to give further information on $SK_1(R\pi)$ for R an n- or p-ring. The main results here are:

1. For any n- or p-ring R and any finite group π, $SK_1(R\pi)$ is generated by induction from elementary subgroups of π (Theorems 1 and 3).

2. Formulas are given for $SK_1(A\pi)$ and $\mathrm{tors}(K_1'(A\pi))_{(p)}$ when A is a p-ring and π any finite group (Theorem 2).

3. $SK_1(\mathbb{Z}\pi)_{(2)}$ is calculated for any finite π whose 2-Sylow subgroup is dihedral, quaternionic, or semidihedral (Theorem 6). In particular, this calculates $SK_1(\mathbb{Z}\pi)$ when π has periodic cohomology.

4. $SK_1(\mathbb{Z}\pi) = 0$ if $\pi \cong SL(2,p)$, $PSL(2,p)$ (p prime), $SL(2,2^k)$, Σ_n (the symmetric group), or $Q(4n)$ (the quaternionic group of order $4n$) (Theorem 4). $SK_1(A[A_n]) = 0$ for any p-ring A and alternating group A_n (Proposition 9).

In addition, an example is given to show that $SK_1(\mathbb{Z}\pi)$ is <u>not</u>, in general, detected by restriction to elementary subgroups of π.

As for notation, "trf" and "ind" are used to denote transfer and induction maps, respectively. $G \rtimes H$ always means a semidirect product with G normal. $D(2n)$ denotes the dihedral group of order $2n$, $Q(4n)$ the quaternionic group of order $4n$; and Σ_n and A_n the symmetric and alternating groups, respectively.

SECTION 1.

We start by reviewing the induction theory of Dress. The following description is a summary of Section 1 in [6].

(A) A Mackey functor is a bifunctor $\mathcal{M} = (\mathcal{M}^*, \mathcal{M}_*)$ from the category of finite groups with monomorphisms to the category of abelian groups, such that $\mathcal{M}^*$ is contravariant, $\mathcal{M}_*$ is covariant,

$$\mathcal{M}_*(\pi) = \mathcal{M}^*(\pi) = \mathcal{M}(\pi)$$

for all π, and the following conditions are fulfilled:

(i) $\mathcal{M}_*$ and $\mathcal{M}^*$ send inner automorphisms to the identity.

(ii) For any isomorphism $\alpha: \pi \to \pi'$, $\mathcal{M}_*(\alpha) \circ \mathcal{M}^*(\alpha) = \mathrm{id}$.

(iii) $\mathcal{M}^*$ and $\mathcal{M}_*$ have the Mackey subgroup property: for any groups $\rho, \rho' \subseteq \pi$, the composite

$$\mathcal{M}(\rho') \xrightarrow{\mathcal{M}_*} \mathcal{M}(\pi) \xrightarrow{\mathcal{M}^*} \mathcal{M}(\rho)$$

is equal to the sum, over all double cosets $\rho g \rho' \subseteq \pi$, of the composites

$$\mathcal{M}(\rho') \xrightarrow{\mathcal{M}^*} \mathcal{M}(g^{-1}\rho g \cap \rho') \xrightarrow{\text{conj.}} \mathcal{M}(\rho \cap g\rho' g^{-1}) \xrightarrow{\mathcal{M}_*} \mathcal{M}(\rho).$$

(B) A Green functor $\mathcal{G}$ is a Mackey functor together with a ring structure on $\mathcal{G}(\pi)$ for all π, satisfying Frobenius reciprocity. Namely, for any inclusion $\alpha: \rho \hookrightarrow \pi$,

$$\alpha^*(x \cdot y) = \alpha^*(x) \cdot \alpha^*(y) \qquad \text{for } x, y \in \mathcal{G}(\pi)$$

$$x \cdot \alpha_*(y) = \alpha_*(\alpha^*(x) \cdot y) \qquad \text{for } x \in \mathcal{G}(\pi),\ y \in \mathcal{G}(\rho)$$

$$\alpha_*(x) \cdot y = \alpha_*(x \cdot \alpha^*(y)) \qquad \text{for } x \in \mathcal{G}(\rho),\ y \in \mathcal{G}(\pi)$$

(where $\alpha^* = \mathcal{G}^*(\alpha)$, $\alpha_* = \mathcal{G}_*(\alpha)$).

(C) A <u>Green module</u> $\mathcal{M}$ over $\mathcal{G}$ is a Mackey functor together with a $\mathcal{G}(\pi)$-module structure on $\mathcal{M}(\pi)$ for all π, such that the same relations hold as in (B) (but with $y \in \mathcal{M}(\pi)$ or $\mathcal{M}(\rho)$ instead).

(D) A <u>defect group</u> for a Green functor $\mathcal{G}$ is a group π such that the map

$$\mathcal{G}_*: \sum_{\rho \subsetneqq \pi} \mathcal{G}(\rho) \to \mathcal{G}(\pi)$$

is not onto.

(E) (Propositions 1.1' and 1.2 in [6]). Let $\mathcal{M}$ be a Green module over a Green functor $\mathcal{G}$, let π be a finite group, and D the set of subgroups of π which are defect groups for $\mathcal{G}$. Then

$$\mathcal{M}(\pi) \cong \varinjlim_{\rho \in D} \mathcal{M}(\rho) \cong \varprojlim_{\rho \in D} \mathcal{M}(\rho)$$

where the limits are taken with respect to inclusion and conjugation among subgroups in D.

The following proposition lists the Green modules which will be used here:

<u>Proposition 1</u>. Let p be a fixed prime. Then

(i) For any Dedekind domain R with quotient field K of characteristic zero,

$$SK_1(R\pi)_{(p)}, \quad K_1(R\pi)_{(p)}, \quad \text{and} \quad K_1'(R\pi)_{(p)}$$

are modules over the Green functor $G_0(R\pi)_{(p)}$, whose defect groups are p-K-elementary groups.

(ii) For any n-ring R, $Cl_1(R\pi)_{(p)}$ is a module over the Green functor $G_0(R\pi)_{(p)}$.

(iii) $H_n(\pi)_{(p)}$ $(n \geq 1)$ and $H_2(\pi)/H_2^{ab}(\pi)$ are modules over the Green functor

$$\hat{H}^0(\pi)_{(p)} \cong \mathbb{Z}/|\pi_p|;$$

whose defect groups are the p-groups.

<u>Proof</u>. (i) To check that $K_1(R\pi)$ has the Mackey subgroup property for any $\rho, \rho' \subseteq \pi$, note that the composite

$$K_1(R\rho') \xrightarrow{\text{ind}} K_1(R\pi) \xrightarrow{\text{trf}} K_1(R\rho)$$

is induced by tensoring with the bimodule ${}_{R\rho}R\pi_{R\rho'}$. This splits as a sum of bimodules, one summand corresponding to each double coset $\rho g \rho' \subseteq \pi$; and tensoring with $R(\rho g \rho')$ induces the composite

$$K_1(R\rho') \xrightarrow{\text{trf}} K_1(R[g^{-1}\rho g \cap \rho']) \xrightarrow{\text{conj}} K_1(R[\rho \cap g\rho' g^{-1}]) \xrightarrow{\text{ind}} K_1(R\rho).$$

(The details are the same as those in the corresponding proof for $K_0(K\pi)$; see, e.g. Section 44 of [5].)

That $K_1(R\pi)$ and $SK_1(R\pi)$ are $G_0(R\pi)$-modules satisfying Frobenius reciprocity is shown in [9]. $G_0(K\pi)_{(p)}$ is generated by induction from p-K-elementary subgroups by Theorem 28 (p.112) in [17]; and the corresponding result for $G_0(R\pi)_{(p)}$ then follows from Proposition XI.3.3 in [2].

(ii) $Cl_1(R\pi)$ is clearly a $G_0(R\pi)$-submodule of $SK_1(R\pi)$.

(iii) For $H_n(\pi)_{(p)}$, this is shown in Section XII.9 of [4]. That $H_2^{ab}(\pi)$ is preserved under the transfer follows easily from the Mackey subgroup property. □

For any field K of characteristic zero and $n > 1$, let $K\zeta_n$ denote the extension by the nth roots of unity and regard $Gal(K\zeta_n/K)$ as a subgroup of $(\mathbb{Z}_n)^*$ or $Aut(\mathbb{Z}_n)$. By definition, a p-hyper-elementary group $\pi = \mathbb{Z}_n \rtimes \tilde{\pi}$ ($p \nmid n$, $\tilde{\pi}$ a p-group) is p-K-elementary if and only if $\tilde{\pi}$ conjugates $\mathbb{Z}_n$ via automorphisms in $Gal(K\zeta_n/K)$. This is equivalent to the condition that the $\tilde{\pi}$-action on $K[\mathbb{Z}_n]$ leave all field summands invariant; in other words

$$K[\pi] = \sum_i K_i[\bar{\pi}]^t$$

($K_i[\bar{\pi}]^t$ the appropriate twisted group ring) when $K[\mathbb{Z}_n] = \sum_i K_i$ is the decomposition.

The next proposition greatly simplifies the limits involved in applying Dress induction to calculate $SK_1(R\pi)$ and $Cl_1(R\pi)$ (for R an n-ring); and $SK_1(A\pi)$ and $K_1'(A\pi)$ (for A a p-ring). For any group π and field K, two elements $g, h \in \pi$ will be called K-conjugate if h is conjugate to g^a for some $a \in Gal(K\zeta_n/K)$ $(n = |g|)$. For any cyclic $\sigma = \langle g \rangle \subseteq \pi$, set

$$N_\pi^K(\sigma) = N_\pi^K(g) = \{x \in \pi \mid xgx^{-1} = g^a, \text{ some } a \in Gal(K\zeta_n/K),\ n = |g|\}.$$

We define the category of "rings with bimodule morphisms" to be the category whose objects are rings (with identity), and where $Mor(R,S)$ consists of all isomorphism classes of bimodules $_SM_R$ such that M is finitely generated and projective as an S-module. Composition of morphisms is defined by tensor product. The usual category of rings with homomorphisms is mapped to this category by sending any $f: R \to S$ to the appropriate bimodule $_SS_R$ (i.e., where $s_1(s_2)r = s_1s_2 \cdot f(r)$). Note that the K_n can be regarded as functors on this category.

Proposition 2. Let p be a fixed prime, and R a Dedekind domain with quotient field K of characteristic zero. Let X be a covariant functor from the category of R-orders in finite dimensional semisimple K-algebras with bimodule morphisms to the category of $\mathbb{Z}_{(p)}$-modules. Let

$$ind = X({}_{R\pi}R\pi_{R\rho}): X(R\rho) \to X(R\pi)$$

and

$$trf = X({}_{R\rho}R\pi_{R\pi}): X(R\pi) \to X(R\rho)$$

for any pair $\rho \subseteq \pi$ of finite groups, and assume that this makes $X(R\pi)$

into a Green module all of whose defect groups are p-K-elementary.

Assume furthermore:

(A) For any p-K-elementary group $\pi = \mathbb{Z}_n \rtimes \rho$ ($p \nmid n$ and ρ a p-group), the natural projections induce an isomorphism

$$X(R\pi) \xrightarrow{\cong} \sum_i X(R_i[\rho]^t);$$

where $K[\mathbb{Z}_n] = \Sigma K_i$ (a sum of fields), and $R_i \subseteq K_i$ is the integral closure of R.

(B) For any $m|n$ prime to p and having the same prime divisors, write

$$R^n = R \otimes_{\mathbb{Z}} \mathbb{Z}\zeta_n \quad \text{and} \quad R^m = R \otimes_{\mathbb{Z}} \mathbb{Z}\zeta_m.$$

Then for any p-group ρ and homomorphism $t: \rho \to \mathrm{Gal}(K\zeta_n/K)$, with $R^n[\rho]^t$ and $R^m[\rho]^t$ the induced twisted group rings, the composite

$$X(R^m[\rho]^t) \xrightarrow{\text{ind}} X(R^n[\rho]^t) \xrightarrow{\text{trf}} X(R^m[\rho]^t)$$

(induced by the obvious bimodules) is multiplication by n/m.

Then, for any finite π, letting $g_1,\ldots,g_k$ be K-conjugacy class representatives for elements of order prime to p, and setting $n_i = |g_i|$,

$$X(R\pi) \cong \sum_{i=1}^{k} \varinjlim\{X(R\zeta_{n_i}[\rho]^t): \rho \subseteq N_\pi^K(g_i),\ \rho \text{ a p-group}\}.$$

Here, limits are taken with respect to inclusion, and conjugation by elements of $N_\pi^K(g_i)$. $R\zeta_n$ denotes the integral closure of R in $K\zeta_n$.

<u>Proof</u>. Note that by (A), the above formula for $X(R\pi)$ holds when π is p-K-elementary. Since for general π,

$$X(R\pi) \cong \varinjlim\{X(P\rho): \rho \subseteq \pi,\ \rho \text{ p-K-elementary}\};$$

the main problem is to rearrange the decomposition of the $X(R\rho)$ given by (A), so as to be consistent with induction maps.

For any cyclic $\sigma \subseteq \pi$ of order prime to p, and any p-subgroup $\rho \subseteq N_{\pi}^{K}(\sigma)$, define

$$X_{\sigma}(\rho) = \mathrm{Coker}[\Sigma\mathrm{ind}: \sum_{\tau \subsetneq \sigma} X(R[\tau \rtimes \rho]) \to X(R[\sigma \rtimes \rho])]$$

(regarding $\tau \rtimes \rho$, $\sigma \rtimes \rho$ as subgroups of π). In steps 1 to 3 below, we obtain decompositions

$$\Psi(\sigma,\rho): X(R[\sigma \rtimes \rho]) \xrightarrow{\cong} \sum_{\tau \subseteq \sigma} X_{\tau}(\rho),$$

natural in ρ, and such that for $\sigma' \subseteq \sigma$ the square

$$\begin{array}{ccc} X(R[\sigma' \rtimes \rho]) & \xrightarrow{\mathrm{ind}} & X(R[\sigma \rtimes \rho]) \\ \cong\downarrow \Psi(\sigma',\rho) & & \cong\downarrow \Psi(\sigma,\rho) \\ \sum_{\tau \subseteq \sigma'} X_{\tau}(\rho) & \xhookrightarrow{\mathrm{incl}} & \sum_{\tau \subseteq \sigma} X_{\tau}(\rho) \end{array}$$

commutes. Then, in Step 4, $X_{\sigma}(\rho)$ will be shown isomorphic to a sum of copies of $X(R\zeta_n[\rho]^t)$ $(n = |\sigma|)$; and the final formula will be obtained in Step 5.

<u>Step 1</u>. We first write down the abstract algebraic conditions necessary for constructing a splitting of the type described above. Fix an integer n, let $\{E_m\}_{m|n}$ be a collection of abelian groups, with homomorphisms

$$i_k^m: E_k \to E_m \quad \text{and} \quad t_k^m: E_m \to E_k$$

for each pair $k|m|n$. Assume that

(1) $i_k^m \circ i_d^k = i_d^m$ and $t_d^k \circ t_k^m = t_d^m$ for all $d|k|m|n$

(2) $t_k^{[k,m]} \circ i_m^{[k,m]} = i_{(k,m)}^k \circ t_{(k,m)}^m$ for all k,m dividing n (where $[k,m]$ and (k,m) denote the l.c.m. and g.c.d., respectively)

(3) $t_k^m \circ i_k^m$ is an isomorphism for all $k|m|n$.

Then for all $m|n$,

$$E_m = \sum_{k|m} i_k^m(E_k');$$

where

$$E_k' = \mathrm{Ker}[\Sigma t_d^k\colon E_k \to \Sigma\{E_d\colon d|k,\ d \neq k\}].$$

In particular

$$E_m' \cong \mathrm{Coker}[\Sigma i_k^m\colon \Sigma\{E_k\colon k|m,\ k \neq m\} \to E_m].$$

The proof of this result is straightforward, by induction on the number of distinct primes dividing n.

Step 2. From here through Step 4, we fix a cyclic subgroup $\sigma \subseteq \pi$ of order prime to p, set $n = |\sigma|$, and fix a p-subgroup $\rho \subseteq N_\pi^K(\sigma)$. For any $m|n$, let $\mathbb{Z}_m$ denote the order m subgroup of σ. Decompose

$$K[\mathbb{Z}_n] \cong \sum_{m|n} \sum_{i=1}^{r_m} K_i^m,$$

a sum of fields; where the action of $\mathbb{Z}_n$ on K_i^m (as a $\mathbb{Z}_n$-representation) has order m. (In particular, $K_i^m \cong K\zeta_m$). Let $R_i^m \subseteq K_i^m$ denote the integral closure of R.

For any $m|n$, we again write $R^m = R \otimes_{\mathbb{Z}} \mathbb{Z}\zeta_m$. Define

$$Y_m = \mathrm{Im}[X(R[\mathbb{Z}_m \rtimes \rho]) \to X(R^m[\rho]^t)],$$

where the map is induced by some fixed projection of $\mathbb{Z}[\mathbb{Z}_m]$ onto $\mathbb{Z}\zeta_m$: By (A), the composite of projections

$$X(R[\mathbb{Z}_m \rtimes \rho]) \to \sum_{k|m} X(R^k[\rho]^t) \to \sum_{k|m} \sum_{i=1}^{r_k} X(R_i^k[\rho]^t)$$

is an isomorphism; and so the first map induces an isomorphism

$$\Gamma_m\colon X(R[\mathbb{Z}_m \rtimes \rho]) \xrightarrow{\cong} \sum_{k|m} Y_k.$$

(Note that by (A), the projection of $X(R[\mathbb{Z}_m \rtimes \rho])$ to $X(R[\mathbb{Z}_k \rtimes \rho])$ is a surjection for all $k|m$).

For $k|m|n$, consider the following squares:

$$\begin{array}{ccccc}
X(R[\mathbb{Z}_k \rtimes \rho] & \xrightarrow{\text{ind}} & X(R[\mathbb{Z}_m \rtimes \rho] & \xrightarrow{\text{trf}} & X(R[\mathbb{Z}_k \rtimes \rho]\ ) \\
\downarrow & & \downarrow & & \downarrow \\
X(R^k[\rho]^t) & \xrightarrow{\text{ind}} & X(R^m[\rho]^t) & \xrightarrow{\text{trf}} & X(R^k[\rho]^t)
\end{array}$$

where the vertical maps are induced by projections of rings. The left-hand square is induced by a commutative square of ring homomorphisms, and hence commutes for any $k|m$. When k and m have the same prime divisors, multiplication in $R^m[\rho]^t$ induces a bimodule isomorphism

$$R^k[\rho]^t \otimes_{R[\mathbb{Z}_k \rtimes \rho]} R[\mathbb{Z}_m \rtimes \rho] \xrightarrow{\cong} R^m[\rho]^t$$

(just compare bases); and so the right-hand square commutes in this case.

For convenience, for $m|n$, we write $\bar{m}$ for the product of the distinct primes dividing m. By the above squares, the induction and transfer maps induce homomorphisms

$$i_k^m: Y_k \to Y_m \ (\text{for } k|m|n) \quad \text{and} \quad t_k^m: Y_m \to Y_k \ (\text{for } \bar{m}|k|m|n).$$

We now check that condition (1) to (3) in Step 1 hold for the t_k^m and i_k^m (when defined):

(1) These follow immediately from the naturality of the induction and transfer maps among the $X(R^m[\rho]^t)$.

(2) Fix $k,m|n$, write $D = [k,m]$ and $d = (k,m)$, and assume that k and D have the same prime divisors. Checking that

$$t_k^D \circ i_m^D = i_d^k \circ t_d^m$$

amounts to showing that multiplication in $R^D[\rho]^t$ induces an isomorphism of bimodules

$$R^k[\rho]^t \otimes_{R^d[\rho]^t} R^m[\rho]^t \xrightarrow{\cong} R^D[\rho]^t.$$

This is easily seen to be onto (look at roots of unity); and is shown to be one-to-one by counting dimensions.

(3) If $k|m|n$ and k,m have the same prime divisors, then $t_k^m \circ i_k^m$ is multiplication by m/k by (B). Since X takes values among $\mathbb{Z}_{(p)}$-modules (and $p\nmid n$), this is an isomorphism.

Step 1 can now be applied, with $E_m = Y_m$ if $\bar{n}|m|n$ and $E_m = 0$ if $\bar{n}\nmid m|n$. Upon setting

$$Y'_m = \mathrm{Ker}[\Sigma t_k^m : Y_m \to \Sigma\{Y_k : \bar{m}|k|m,\ k \neq m\}]$$

for $m|n$ (and noting that $\bar{m} = \bar{n}$ if $\bar{n}|m$) we get

$$Y_n = \sum_{\bar{n}|m|n} i_m^n(Y'_m).$$

<u>Step 3</u>. For any $k|m|n$, let

$$I_k^m : X(R[\mathbb{Z}_k \rtimes \rho]) \to X((R[\mathbb{Z}_m \rtimes \rho])$$

denote the induction map. We must study how it splits up under the identifications Γ_k and Γ_m. The main problem here is developing a reasonable notation.

For any $m > 0$, let $\delta(m)$ denote the set of (positive) divisors of m. For $k|m$, we define maps

$$\alpha_k^m : \delta(m) \to \delta(k) \quad \text{and} \quad \beta_k^m : \delta(k) \to \delta(m)$$

as follows. If m is a prime power, set

$$\alpha_k^m(d) = \begin{cases} d\cdot\frac{k}{m} & \text{if } m|dk \\ 1 & \text{otherwise} \end{cases} \quad \text{and} \quad \beta_k^m(d) = \begin{cases} d\cdot\frac{m}{k} & \text{if } d > 1 \\ 1 & \text{if } d = 1. \end{cases}$$

This is then extended to arbitrary m by requiring that if $m = m_1m_2$ with $(m_1,m_2) = 1$, and $k_i = (k,m_i)$, $d_i = (d,m_i)$; then

$$\alpha_k^m(d) = \alpha_{k_1}^{m_1}(d_1)\cdot\alpha_{k_2}^{m_2}(d_2), \text{and } \beta_k^m(d) = \beta_{k_1}^{m_1}(d_1)\cdot\beta_{k_2}^{m_2}(d_2).$$

Note in particular that $\alpha_k^m \circ \beta_k^m = \mathrm{id}$, and that for any $d|k$, d and $\beta_k^m(d)$ have the same prime divisors.

The commutativity of the following square, for any $k|m|n$, now follows from the definitions and the functoriality of X:

$$\begin{array}{ccc} X(R[\mathbb{Z}_k \rtimes \rho]) & \longrightarrow & X(R[\mathbb{Z}_m \rtimes \rho]) \\ \cong \downarrow \Gamma_m & & \cong \downarrow \Gamma_m \\ \sum_{d|k} Y_d & \xrightarrow[\sum_{d|m} i^d_{\alpha(d)}]{} & \sum_{d|m} Y_d \end{array} \qquad (\alpha = \alpha_k^m).$$

We define maps T_k^m in the opposite direction as the composites

$$T_k^m : X(R[\mathbb{Z}_m \rtimes \rho]) \xrightarrow[\cong]{\Gamma_m} \sum_{d|m} Y_d \xrightarrow{\sum_{d|k} t_d^{\beta(d)}} \sum_{d|k} Y_k \xrightarrow[\cong]{\Gamma_k^{-1}} X(R[\mathbb{Z}_k \rtimes \rho])$$

(where $\beta = \beta_k^m$).

We check that the conditions of Step 1 hold for these maps:

(1) That $I_k^m \circ I_d^k = I_d^m$ and $T_d^k \circ T_k^m = T_d^m$ (for $d|k|m|n$) follows from the relations

$$\alpha_d^k \circ \alpha_k^m = \alpha_d^m \quad \text{and} \quad \beta_k^m \circ \beta_d^k = \beta_d^m$$

(checked by considering one prime at a time).

(2) That $T_k^{[k,m]} \circ I_m^{[k,m]} = I_{(k,m)}^k \circ T_{(k,m)}^m$ follows from the corresponding relations for the t_k^m and i_k^m (Step 2); upon noting that for any $d|k$,

$$\alpha_m^{[k,m]} \circ \beta_k^{[k,m]}(d) = \beta_{(k,m)}^m \circ \alpha_{(k,m)}^k(d) = d'$$

and

$$\alpha_{(k,m)}^k(d) = (d,d') \quad \text{and} \quad \beta_k^{[k,m]}(d) = [d,d'].$$

It suffices to prove these when m and k are powers of the same prime, in which case either the α's or the β's are identity maps.

(3) For any $k|m|n$,

$$T^m_k \circ I^m_k = \Gamma^{-1}_k \circ \Big(\sum_{d|k} t^{\beta(d)}_d \circ i^{\beta(d)}_d \Big) \circ \Gamma_k$$

(where $\beta = \beta^m_k$). Since for all $d|k$, d and $\beta^m_k(d)$ have the same prime divisors, $t^{\beta(d)}_d \circ i^{\beta(d)}_d$ is an isomorphism (Step 2). Thus

$$\sum_{d|k} t^{\beta(d)}_d \circ i^{\beta(d)}_d : \sum_{d|k} Y_d \to \sum_{d|k} Y_d$$

is an isomorphism, and so is $T^m_k \circ I^m_k$.

By Step 1, we now get canonical isomorphisms

$$\Psi(\mathbb{Z}_m, \rho): X(R[\mathbb{Z}_m \rtimes \rho]) = \sum_{k|m} I^m_k(X'_k) \xrightarrow{\cong} \sum_{k|m} X_{\mathbb{Z}_k}(\rho)$$

for all $m|n$. Here,

$$X'_k = \operatorname{Ker}[\Sigma T^k_d: X(R[\mathbb{Z}_k \rtimes \rho]) \to \Sigma\{X(R[\mathbb{Z}_d \rtimes \rho]): d|k,\ d \neq k\}]$$

$$\cong \operatorname{Coker}[\Sigma I^k_d: \Sigma\{X(R[\mathbb{Z}_d \rtimes \rho]): d|k,\ d \neq k\} \to X(R[\mathbb{Z}_k \rtimes \rho])] = X_{\mathbb{Z}_k}(\rho).$$

By construction, $\Psi(\mathbb{Z}_m, \rho)$ is natural with respect to inclusions of p-K-elementary subgroups of π, as well as conjugation by elements of $N(\mathbb{Z}_m)$.

<u>Step 4</u>. We must now identify $X_{\mathbb{Z}_n}(\rho)$. From Step 3,

$$X_{\mathbb{Z}_n}(\rho) \cong \operatorname{Ker}[\Sigma T^n_m: X(R[\mathbb{Z}_n \rtimes \rho]) \to \Sigma\, X(R[\mathbb{Z}_m \rtimes \rho]): m|n,\ m \neq n\}]$$

$$\cong \operatorname{Ker}\Big[\sum_m \sum_k t_k^{\beta^n_m(k)} : \sum_{d|n} Y_d \to \sum_{\substack{m|n \\ m \neq n}} \sum_{k|m} Y_k\Big].$$

For convenience, for any $d|n$, set

$$S(d) = \{k|d: d = \beta^n_m(k) \text{ for some } m|n,\ m \neq n\}.$$

We then get

$$X_{\mathbb{Z}_n}(\rho) \cong \sum_{d|n} \operatorname{Ker}[\Sigma t^d_k: Y_d \to \sum_{k \in S(d)} Y_k].$$

As before, $\bar{n}$ denotes the product of the distinct primes dividing n. If $\bar{n}\nmid d|n$, let $m|n$ be the largest divisor with the same prime factors as d; then $\beta^n_m(d) = d$ and so $d \in S(d)$. If $\bar{n}|d|n$, we easily check that

$$S(d) = \{k|d: \bar{n}|k \neq d\}.$$

Applying the results of Step 2 now gives a natural isomorphism

$$X_{\mathbb{Z}_n}(\rho) \cong \sum_{\bar{n}|d|n} \mathrm{Ker}[\Sigma t^d_k: Y_d \to \Sigma\{Y_k: \bar{n}|k|d,\ k\neq d\}] = \sum_{\bar{n}|d|n} Y'_d \cong Y_n.$$

We again consider the decomposition of $K[\mathbb{Z}_n]$ (see Step 2):

$$K[\mathbb{Z}_n] \cong \sum_{m|n} \sum_{i=1}^{r_m} K^m_i,$$

where $\mathbb{Z}_n$ acts on any K^m_i with order m (and so $K^m_i \cong K\zeta_m$). From Step 2 we have isomorphisms

$$X(R[\mathbb{Z}_n \rtimes \rho]) \xrightarrow{\cong} \sum_{m|n} Y_m \xrightarrow{\cong} \sum_{m|n} \sum_{i=1}^{r_m} X(R^m_i[\rho]^t)$$

where $R^m_i \subseteq K^m_i$ is the integral closure of R. Hence

$$X_{\mathbb{Z}_n}(\rho) \cong Y_n \cong \sum_{i=1}^{r_n} X(R^n_i[\rho]^t) \cong \sum^{r_n} X(R\zeta_n[\rho]^t).$$

($R\zeta_n$ was defined to be the algebraic closure of R in $K\zeta_n$).

Note that

$$\dim_K(\sum_{i=1}^{r_n} K^n_i) = \dim_{\mathbb{Q}}(\mathbb{Q}\zeta_n) = \varphi(n)$$

(Euler φ-function). It follows that $r_n = \varphi(n)/[K\zeta_n:K]$, the number of K-conjugacy classes (in $\mathbb{Z}_n$) of generators of $\mathbb{Z}_n$.

<u>Step 5</u>. We have now constructed, for each cyclic $\sigma \subseteq \pi$ of order prime to p and each p-subgroup $\rho \subseteq N^K_\pi(\sigma)$, a decomposition

$$\Psi(\sigma,\rho): X(R[\sigma \rtimes \rho]) \xrightarrow{\cong} \sum_{\tau\subseteq\sigma} X_\tau(\rho);$$

natural in ρ and such that for $\sigma' \subseteq \sigma$ the composite

$$\Psi(\sigma,\rho)\circ \mathrm{ind}\circ \Psi(\sigma',\rho)^{-1}: \sum_{\tau\subseteq\sigma'} X_\tau(\rho) \to \sum_{\tau\subseteq\sigma} X_\tau(\rho)$$

is just the inclusion of factors. Letting $\sigma_1,\ldots,\sigma_s$ be conjugacy class representatives for cyclic subgroups of order prime to p, we now get

$$X(R\pi) \cong \varinjlim\{X(R\bar{\pi}): \bar{\pi}\subseteq\pi,\ \bar{\pi}\ \text{p-K-elementary}\}$$

$$\cong \sum_{i=1}^{s} \varinjlim\{X_{\sigma_i}(\rho): \rho\subseteq N_\pi^K(\sigma_i),\ \rho\ \text{a p-group}\}. \quad (1)$$

Here the limits in (1) are taken with respect to inclusion, and conjugation by elements of $N(\sigma_i)$. Note that it doesn't matter whether they are taken over all $\rho\subseteq N_\pi^K(\sigma_i)$ or over $\rho\subseteq N_\pi^K(\sigma_i)/\sigma_i$.

It remains to study the limits in (1). We saw in Step 4 that for any cyclic $\sigma\subseteq\pi$ with $p\nmid n = |\sigma|$, and any p-subgroup $\rho\subseteq N_\pi^K(\sigma)$, there is a natural isomorphism

$$X_\sigma(\rho) \cong \sum^{r} X(R\zeta_n[\rho]^t)$$

where r is the number of K-conjugacy classes (in σ) of generators of σ. Conjugation by any $g\in N(\sigma)$ permutes these factors, permuting them trivially exactly when $g\in N_\pi^K(\sigma)$. So

$$\varinjlim\{X_\sigma(\rho): \rho\subseteq N_\pi^K(\sigma),\ \rho\ \text{a p-group}\}$$

$$\cong \sum^{r'} \varinjlim\{X(R\zeta_n[\rho]^t): \rho\subseteq N_\pi^K(\sigma),\ \rho\ \text{a p-group}\},$$

where the new limit is taken with respect to inclusion and conjugation in $N_\pi^K(\sigma)$, and r' is the number of K-conjugacy classes (in π) of generators of σ. Putting these results together now gives the desired formula for $X(R\pi)$. □

The limits in the formula in Proposition 2 have the advantage that the indexing sets have maximum elements. More specifically, for $g \in \pi$ with $p \nmid n = |g|$, let $\bar{\rho} \subseteq N_\pi^K(g)$ be a p-Sylow subgroup. Then the map

$$X(R\zeta_n[\bar{\rho}]^t) \to \varinjlim\{X(R\zeta_n[\rho]^t) : \rho \subseteq N_\pi^K(g),\ \rho \text{ a p-group}\}$$

is onto, and its kernel can be computed using the same formulas as for group homology (e.g., Section XII.9 of [4]).

SECTION 2.

For the first part of this section, we fix an unramified extension $B \supseteq A$ of p-rings, a pair $\rho \triangleleft \pi$ of p-groups, and an isomorphism

$$t: \pi/\rho \xrightarrow{\cong} \mathrm{Gal}(B/A).$$

Let $B[\pi]^t$ denote the corresponding twisted group ring. We must first find a formula for $SK_1(B[\pi]^t)$; afterwards Proposition 2 can be applied to compute $SK_1(A\pi)$ for arbitrary finite π.

First consider $K_1(B[\pi]^t)$.

Lemma 3. (i) The induction map

$$\mathrm{ind}: K_1(B\rho) \to K_1(B[\pi]^t)$$

is onto.

(ii) For any $\sigma \subseteq \rho$, σ normal in π, the induction map

$$\mathrm{Ker}[K_1(B\rho) \to K_1(B[\rho/\sigma])] \to \mathrm{Ker}[K_1(B[\pi]^t) \to K_1(B[\pi/\sigma]^t)]$$

is onto.

Proof. (ii) Since surjections of p-adic orders induce surjections of K_1, it suffices to prove this when $|\sigma| = p$. Let $z \in \sigma$ be a generator; we need only show the surjectivity of

$$i_*: K_1(B\rho, (1-z)) \to K_1(B[\pi]^t, (1-z)).$$

First consider the maps

$$i_k: K_1(B\rho/(1-z)^{k+1}, (1-z)^k) \to K_1(B[\pi]^t/(1-z)^{k+1}, (1-z)^k)$$

for $k \geq 1$. $\mathrm{Coker}(i_k)$ is generated by elements

$$[1+(1-z)^k\lambda g] \qquad (\lambda \in B,\ g \in \pi-\rho).$$

Fix some $\mu \in B$ which reduces to a generator $\bar{\mu} \in \bar{B}^*$ ($\bar{B}$ the residue field); then $g\bar{\mu} \neq \bar{\mu}$ and so $(\mu^{-1}g(\mu)-1)$ is a unit in B. Setting

$$\lambda' = (\mu^{-1}g(\mu)-1)^{-1}\lambda,$$

we get

$$[1+(1-z)^k\lambda g] = [1+(1-z)^k\lambda'(\mu^{-1}g(\mu)-1)g]$$

$$= \mu^{-1}[1+(1-z)^k\lambda' g]\mu[1+(1-z)^k\lambda' g]^{-1} = 1 \text{ in } K_1(B[\pi]^t/(1-z)^{k+1},(1-z)^k).$$

Hence i_k is onto for all k, and comparing exact sequences for ideals gives that

$$K_1(B\rho/(1-z)^k, (1-z)) \to K_1(B[\pi]^t/(1-z)^k,(1-z))$$

is onto for all k. So by Theorem 10 in [20] (see also Lemma 1 in [13]),

$$K_1(B\rho/p^k,(1-z)) \to K_1(B[\pi]^t/p^k,(1-z))$$

is onto for all k. So by Theorem 10 in [20],

$$K_1(B\rho,(1-z)) \to K_1(B[\pi]^t,(1-z))$$

is the inverse limit of a sequence of surjections of finite groups, and hence onto.

(i) This follows from (ii), once we check that $K_1(B)$ surjects onto $K_1(B[\pi/\rho]^t)$. But $B[\pi/\rho]^t$ is a matrix algebra over A, the composite

$$K_1(B) \xrightarrow{\text{ind}} K_1(B[\pi/\rho]^t) \cong K_1(A)$$

is the transfer (i.e., norm) map, and this is onto since B/A is unramified ([22], Proposition VIII.1.3). □

We now need to know that $K_1'(B\rho)$ is cohomologically trivial under the induced π/ρ-action. For this, some of the machinery from [13] is needed. For any p-group σ, define

$$I(B\sigma) = \mathrm{Ker}[B\sigma \to B]; \quad \overline{I(B\sigma)} = I(B\sigma)/\langle x - gxg^{-1} \mid g \in \sigma,\ x \in I(B\sigma)\rangle.$$

It was shown in Proposition 2 and 3 in [13] that the p-adic logarithm induces a monomorphism

$$\log\colon Wh'(B\sigma) \to \overline{I(B\sigma)} \otimes \mathbb{Q}.$$

Lemma 4. Let $\sigma \subseteq \rho$ be any subgroup generated by central commutators of order p. Then log restricts to an isomorphism

$$\log_0\colon \mathrm{Ker}[K_1'(B\rho) \to K_1'(B[\rho/\sigma])] \xrightarrow{\cong} \mathrm{Ker}[\overline{I(B\rho)} \to \overline{I(B[\rho/\sigma])}].$$

Proof. By Lemma 14 and Proposition 15 in [13], the map

$$SK_1(B\rho) \to SK_1(B[\rho/\sigma])$$

is onto. Letting $\{z_1,\ldots,z_k\}$ be commutators generating σ, it follows that

$$\mathrm{Ker}[K_1'(B\rho) \to K_1'(B[\rho/\sigma])]$$

is generated by the groups $1+(1-z_i)B\rho$ for $1 \le i \le k$. Since

$$\{(1-z_i)B\rho\colon 1 \le i \le k\} \quad \text{generates} \quad \mathrm{Ker}[\overline{I(B\rho)} \to \overline{I(B[\rho/\sigma])}];$$

and the logarithm induces surjections

$$1+(1-z_i)B\rho \twoheadrightarrow \mathrm{Ker}[\overline{I(B\rho)} \to \overline{I(B[\rho/z_i])}]$$

(Lemma 11 in [13]); $\log_0$ is onto (and into). Note that $\rho^{ab} = (\rho/\sigma)^{ab}$, so

$$\mathrm{Ker}[K_1'(B\rho) \to K_1'(B[\rho/\sigma])] = \mathrm{Ker}[Wh'(B\rho) \to Wh'(B[\rho/\sigma])];$$

and $\log_0$ is one-to-one by Proposition 3 in [13]. □

This applies to show:

Lemma 5. $K_1'(B\rho)$, with the π/ρ-action induced by

$$\pi/\rho \to \mathrm{Aut}(B) \times \mathrm{Out}(\rho),$$

is cohomologically trivial.

Proof. Set $\bar{\pi} = \pi/\rho$, and fix $\lambda \in B$ with $\mathrm{Tr}_{B/A}(\lambda) = 1$ ([22], Proposition VIII.1.3). For any $B[\bar{\pi}]^t$-module M, multiplication by λ has the property

$$\sum_{g\in\bar{\pi}} g\lambda g^{-1} = \sum_{g\in\bar{\pi}} g(\lambda) = \mathrm{id}: M \to M;$$

so $\hat{H}^*(\bar{\pi};M) = 0$ for any such M ([4], Prop.XII.2.4). In particular, if ρ is abelian, this applies to any power of the Jacobson radical $J \subseteq B\rho$; so

$$\hat{H}^*(\bar{\pi};(1+J^k)/(1+J^{k+1})) \cong \hat{H}^*(\bar{\pi};J^k/J^{k+1}) = 0$$

for all $k \geq 1$, $(B\rho)^*/(1+J^k)$ is cohomologically trivial for all k, and $K_1'(B\rho) = (B\rho)^*$ is cohomologically trivial.

If ρ is nonabelian, let $\sigma \subseteq \rho$ be the subgroup generated by all central commutators of order p. Then $\sigma \lhd \pi$, and $\sigma \neq 1$ by Lemma 12 in [13]. Assume inductively that $K_1'(B[\rho/\sigma])$ is cohomologically trivial. Then

$$\mathrm{Ker}[K_1'(B\rho) \to K_1'(B[\rho/\sigma])] \cong \mathrm{Ker}[\overline{I(B\rho)} \to \overline{I(B[\rho/\sigma])})$$

by Lemma 4, $\overline{I(B\rho)}$ and $\overline{I(B[\rho/\sigma])}$ are $B[\bar{\pi}]^t$-modules, and so

$$\hat{H}^*(\bar{\pi};K_1'(B\rho)) \cong \hat{H}^*(\bar{\pi};\mathrm{Ker}[K_1'(B\rho) \twoheadrightarrow K_1'(B[\rho/\sigma])]) = 0. \quad \square$$

One more lemma is needed before completing the calculation of $SK_1(B[\pi]^t)$.

Lemma 6. There exists an extension $1 \to \sigma \to \tilde{\pi} \overset{\alpha}{\to} \pi \to 1$ of p-groups, such that $SK_1(B\tilde{\rho}) = 0$ $(\tilde{\rho}=\alpha^{-1}(\rho))$.

Proof. By Lemma 17 in [13], there is a central extension

$$1 \to \sigma \to \bar{\rho} \overset{\bar{\alpha}}{\to} \rho \to 1$$

of p-groups such that the homomorphism

$$\delta^{\bar{\alpha}}: H_2(\rho) \to \sigma$$

(in the notation of Section 3 of [13]) is one-to-one. Set $q = |\pi/\rho|$, and consider the map of wreath products

$$\bar{\alpha} \wr \Sigma_q : \bar{\rho} \wr \Sigma_q \to \rho \wr \Sigma_q .$$

Regarding π as a subgroup of $\rho \wr \Sigma_q$, set

$$\tilde{\pi} = (\bar{\alpha} \wr \Sigma_q)^{-1}(\pi) \qquad \text{and} \qquad \tilde{\rho} = (\bar{\alpha} \wr \Sigma_q)^{-1}(\rho).$$

Then the extension $1 \to \sigma^q \to \tilde{\rho} \overset{\tilde{\alpha}}{\to} \rho \to 1$ is still central, and $\delta^{\tilde{\alpha}}$ still one-to-one. By Lemma 22(i) in [13],

$$SK_1(B\tilde{\rho}) \cong H_2(\tilde{\rho})/H_2^{ab}(\tilde{\rho}) = 0. \quad \square$$

We can now finally show:

<u>Proposition 7</u>. Let $B \supseteq A$ be any unramified extension of p-rings, $\rho \triangleleft \pi$ p-groups,

$$t: \pi/\rho \xrightarrow{\cong} \mathrm{Gal}(B/A)$$

an isomorphism, and $B[\pi]^t$ the corresponding twisted group ring. Then the induction map

$$\mathrm{ind}: SK_1(B\rho) \to SK_1(B[\pi]^t)$$

is onto, and induces an isomorphism $SK_1(B[\pi]^t) \cong H_o(\pi/\rho; SK_1(B\rho))$.

<u>Proof</u>. Using Lemma 6, choose an extension $1 \to \sigma \to \tilde{\pi} \overset{\alpha}{\to} \pi \to 1$, with $\tilde{\rho} = \alpha^{-1}(\rho)$, such that $SK_1(B\tilde{\rho}) = 0$. The composites

$$K_1(B\tilde{\rho}) \xrightarrow{\mathrm{ind}} K_1(B[\tilde{\pi}]^t) \xrightarrow{\mathrm{trf}} K_1(B\tilde{\rho})$$

$$K_1'(B\rho) \xrightarrow{\mathrm{ind}} K_1'(B[\pi]^t) \xrightarrow{\mathrm{trf}} K_1'(B\rho)$$

are induced by tensoring with $B[\tilde{\pi}]^t$ or $B[\pi]^t$ as bimodules, and are hence the norm maps for the π/ρ-actions. So $\mathrm{Ker(ind)} \subseteq \mathrm{Ker(Norm)}$ in both cases. Since $K_1(B\tilde{\rho})$ $(\cong K_1'(B\tilde{\rho}))$ and $K_1'(B\rho)$ are cohomologically trivial (Lemma 5) and the induction maps are onto (Lemma 3), this implies that

$$H_0(\pi/\rho;K_1(B\tilde{\rho})) \cong K_1(B[\tilde{\pi}]^t) \quad \text{and} \quad H_0(\pi/\rho;K_1'(B\rho)) \cong K_1'(B[\pi]^t).$$

Now consider the following diagrams with exact rows, where ν_1, ν_2, ν_3, and ν_3' are induced by inclusion:

$$\begin{array}{ccccccc}
\mathrm{Ker}[K_1(B\tilde{\rho}) \to K_1(B\rho)] & \to & H_0(\pi/\rho;K_1(B\tilde{\rho})) & \to & H_0(\pi/\rho;K_1(B\rho)) & \to & 0 \\
\downarrow \nu_1 & & \downarrow \nu_2 & & \downarrow \nu_3 & & \\
\mathrm{Ker}[K_1(B[\tilde{\pi}]^t) \to K_1(B[\pi]^t)] & \to & K_1(B[\tilde{\pi}]^t) & \to & K_1(B[\pi]^t) & \to & 0
\end{array}$$

$$\begin{array}{ccccccccc}
0 & \to & H_0(\pi/\rho;SK_1(B\rho)) & \to & H_0(\pi/\rho;K_1(B\rho)) & \to & H_0(\pi/\rho;K_1'(B\rho)) & \to & 0 \\
 & & \downarrow & & \downarrow \nu_3 & & \downarrow \nu_3' & & \\
0 & \to & SK_1(B[\pi]^t) & \to & K_1(B[\pi]^t) & \to & K_1'(B[\pi]^t) & \to & 0.
\end{array}$$

We have just seen that ν_2 and ν_3' are isomorphisms, and ν_1 is onto by Lemma 3. So ν_3 is also an isomorphism, and hence

$$H_0(\pi/\rho;SK_1(B\rho)) \cong SK_1(B[\pi]^t). \ \square$$

$SK_1(A\pi)$ can now be studied for arbitrary finite π. Before deriving an actual formula, we consider some general consequences of Proposition 7.

<u>Theorem 1</u>. Let A be a p-ring and π a finite group. Then

(i) $SK_1(A\pi)$ and $K_1(A\pi)_{(p)}$ are generated by induction from p-elementary subgroups of π.

(ii) If $B \supseteq A$ is a totally ramified extension of p-rings, then the inclusion $A\pi \subseteq B\pi$ induces an isomorphism

$$\mathrm{ind}\colon SK_1(A\pi) \to SK_1(B\pi).$$

(iii) If $B \supseteq A$ is unramified, then the transfer map

$$\text{trf}: SK_1(B\pi) \to SK_1(A\pi)$$

is onto.

<u>Proof</u>. $SK_1(A\pi)$ is a p-group by Theorem 2.5 in [21]. Hence, we may assume throughout that $\pi = \mathbb{Z}_n \rtimes \bar{\pi}$ is p-A-elementary ($p \nmid n$ and $\bar{\pi}$ a p-group). Decompose $A\pi$ as a sum

$$A\pi = \sum_i A_i[\bar{\pi}]^t$$

where $A[\mathbb{Z}_n] = \Sigma A_i$ and the A_i are p-rings. For each i, let ρ_i denote the kernel of the twisting in $A_i[\bar{\pi}]^t$.

(i) For each i, let $\mathbb{Z}_{m_i} \subseteq \mathbb{Z}_n$ be the largest subgroup centralizing ρ_i (note that $(m_i, n/m_i) = 1$). Then $A_i[\rho_i]$ is a summand of $A[\mathbb{Z}_{m_i} \times \rho_i])$, and by Proposition 7,

$$SK_1(A_i[\bar{\pi}]^t) \subseteq \text{Im}[SK_1(A[\mathbb{Z}_{m_i} \times \rho_i]) \to SK_1(A\pi)].$$

The proof for $K_1(A\pi)_{(p)}$ is similar (using Lemma 3).

(ii) A_i is unramified over A for all i, so $B_i = B \otimes_A A_i$ is a p-ring and

$$\rho_i = \text{Ker}[t: \bar{\pi} \to \text{Gal}(A_i/A) \cong \text{Gal}(B_i/B)].$$

By Proposition 7, and Proposition 15 in [13], the inclusion $A_i[\bar{\pi}]^t \subseteq B_i[\bar{\pi}]^t$ induces an isomorphism

$$SK_1(A_i[\bar{\pi}]^t) \cong H_0(\bar{\pi}; SK_1(A_i[\rho_i])) \cong H_0(\bar{\pi}; SK_1(B_i[\rho_i])) \cong SK_1(B_i[\bar{\pi}]^t).$$

(iii) By (i), it suffices to do this when $\pi = \mathbb{Z}_n \times \bar{\pi}$ is p-elementary. Then the transfer for $A\pi \subseteq B\pi$ is a sum of transfer maps

$$\text{trf}: B_i[\bar{\pi}] \to A_j[\bar{\pi}]$$

for $B_i \supseteq A_j$ unramified over B and A. Those are all onto by Proposition 21 in [13]. □

<u>Theorem 2</u>. Let π be a finite group, and A a p-ring with quotient field K. Let $g_1,\ldots,g_k$ be K-conjugacy class representatives for elements in π of order prime to p, and set

$$N_i = N_\pi^K(g_i); \qquad Z_i = Z_\pi(g_i).$$

Then

$$\text{(i)}\quad SK_1(A\pi) \cong \sum_{i=1}^{k} H_o(N_i; H_2(Z_i)/H_2^{ab}(Z_i))_{(p)}$$

$$\text{(ii)}\quad \mathrm{tors}(K_1'(A\pi))_{(p)} \cong [\mathrm{tors}(A^*)_{(p)}]^k \oplus \sum_{i=1}^{k} H^o(N_i; Z_i^{ab})_{(p)}.$$

<u>Proof</u>. $SK_1(A\pi)$ and $\mathrm{tors}(K_1'(A\pi))$ are Green modules over $G_o(A\pi)$ by Proposition 1. Condition (A) of Proposition 2 holds trivially for both of these, since $A\pi$ itself splits as a sum of twisted group rings when π is p-K-elementary.

It suffices to prove condition (B) for $X = K_1$: SK_1 and $\mathrm{tors}(K_1')$ both being subquotients. Fix $m|n$ prime to p and having the same prime divisors, let $\bar{\pi}$ be a p-group, and

$$t\colon \bar{\pi} \to \mathrm{Gal}(K\zeta_n/K)_{(p)} \cong \mathrm{Gal}(K\zeta_m/K)_{(p)}$$

a homomorphism with kernel ρ. The diagram

$$\begin{array}{ccccc} K_1(A\zeta_m[\rho]) & \xrightarrow{\mathrm{ind}} & K_1(A\zeta_n[\rho]) & \xrightarrow{\mathrm{trf}} & K_1(A\zeta_m[\rho]) \\ \downarrow i_1 & & \downarrow i_2 & & \downarrow i_1 \\ K_1(A\zeta_m[\bar{\pi}]^t) & \xrightarrow{\mathrm{ind}} & K_1(A\zeta_n[\bar{\pi}]^t) & \xrightarrow{\mathrm{trf}} & K_1(A\zeta_m[\bar{\pi}]^t) \end{array}$$

commutes, and i_1 and i_2 are onto by Lemma 3. The composite in the top row is multiplication by $[K\zeta_n: K\zeta_m]$, and so the same holds in the bottom row. Condition (B) now follows by taking sums.

For any $A\zeta_n[\bar{\pi}]^t$ with $\rho = \mathrm{Ker}(t)$, the composite

$$K_1'(A\zeta_n[\rho]) \xrightarrow{\mathrm{ind}} K_1'(A\zeta_n[\bar{\pi}]^t) \xrightarrow{\mathrm{trf}} K_1'(A\zeta_n[\rho])$$

is the norm map for the $\bar{\pi}/\rho$-action. $K_1'(A\zeta_n[\rho])$ is cohomologically trivial (Lemma 5) and the induction map is onto (Lemma 3); hence the transfer induces an isomorphism

$$K_1'(A\zeta_n[\bar{\pi}]^t) \cong H^0(\bar{\pi};K_1'(A\zeta_n[\rho])).$$

So by Proposition 4.1 in [21],

$$\mathrm{tors}(K_1'(A\zeta_n[\bar{\pi}]^t))_{(p)} \cong H^0(\bar{\pi};\mathrm{tors}(A\zeta_n)^*\times\rho^{ab})_{(p)} \cong \mathrm{tors}(A^*)_{(p)}\times H^0(\bar{\pi};\rho^{ab}).$$

Applying Proposition 2 and 7, it remains only to show that

$$\varinjlim_{\rho\in\mathcal{P}(N_i)} H_0(\rho;H_2(\rho\cap Z_i)/H_2^{ab}(\rho\cap Z_i)) \cong H_0(N_i;H_2(Z_i)/H_2^{ab}(Z_i))_{(p)}$$

and

$$\varinjlim_{\rho\in\mathcal{P}(N_i)} H^0(\rho;(\rho\cap Z_i)^{ab}) \cong H^0(N_i;Z_i^{ab})_{(p)}$$

for each i ($\mathcal{P}(N_i)$ denoting the set of p-subgroups). H_1 and H_2/H_2^{ab} being Green modules (Proposition 1), we are reduced to proving:

<u>Lemma 8</u>. Let $\mathcal{M}$ be a Green module over a Green functor $\mathcal{G}$, let $H \triangleleft G$ be a pair of finite groups, and $D(G)$ the set of subgroups of G which are defect groups for $\mathcal{G}$. Then

$$\varinjlim_{\pi\in D(G)} H_0(\pi;\mathcal{M}(\pi\cap H)) \cong H_0(G;\mathcal{M}(H)),$$

$$\varinjlim_{\pi\in D(G)} H^0(\pi;\mathcal{M}(\pi\cap H)) \cong H^0(G;\mathcal{M}(H)),$$

and similarly for inverse limits.

<u>Proof</u>. As in [6], define bifunctors $\hat{\mathcal{M}}$ and $\hat{\mathcal{G}}$ from finite G-sets to abelian groups: for any finite G-set S,

$$\hat{\mathcal{M}}(S) = \Big[\sum_{s\in S}\mathcal{M}(G_s)\Big]^G \quad \text{and} \quad \hat{\mathcal{G}}(S) = \Big[\sum_{s\in S}\mathcal{G}(G_s)\Big]^G.$$

In particular, $\hat{\mathcal{M}}(G/\pi) \cong \mathcal{M}(\pi)$ and $\hat{\mathcal{G}}(G/\pi) \cong \mathcal{G}(\pi)$ for any $\pi \subseteq G$.

For any G-set S, make $G/H\times S$ into a G-set by setting

$$g(aH,s) = (ag^{-1}H,gs) \qquad (a,g\in G,\ s\in S).$$

This defines a new bifunctor $\hat{\mathcal{M}}_{G/H}(S) = \mathcal{M}(G/H\times S)$; and the transfer map

$$\hat{\mathcal{G}}^*(pr_2):\ \hat{\mathcal{G}}(S) \to \hat{\mathcal{G}}(G/H \times S)$$

makes it into a $\hat{\mathcal{G}}(S)$-module. The Mackey subgroup property for $\hat{\mathcal{M}}_{G/H}$ follows easily from its characterization in [6] in terms of pullback squares; and similarly for Frobenius reciprocity for $\hat{\mathcal{M}}_{G/H}$ as a $\hat{\mathcal{G}}$-module.

The left action of G/H on $G/H \times S$ is G-equivariant, inducing a G/H-action on $\hat{\mathcal{M}}_{G/H}(S)$. Multiplication in $\hat{\mathcal{M}}_{G/H}(S)$ by any $x \in \hat{\mathcal{G}}(S)$ is G/H-equivariant, so the functors

$$H_o(G/H; \hat{\mathcal{M}}_{G/H}) \qquad \text{and} \qquad H^o(G/H; \hat{\mathcal{M}}_{G/H})$$

are also $\hat{\mathcal{G}}$-modules. But for $\pi \subseteq G$,

$$H^o(G/H; \hat{\mathcal{M}}_{G/H}(G/\pi)) \cong H^o(G/H; \hat{\mathcal{M}}(G/H \times G/\pi)) \cong H^o(H\pi/H; \hat{\mathcal{M}}(G/\pi \cap H))$$

$$\cong H^o(\pi/\pi \cap H;\ \mathcal{M}(\pi \cap H)),$$

and similarly for $H_o(G/H;\ \hat{\mathcal{M}}_{G/H}(G/\pi))$. The limits now follow from Propositions 1.1' and 1.2 in [6] (applied to $S = \coprod_{\pi \in D(G)} G/\pi$). □

To finish the section, we now list some corollaries of Theorems 1 and 2.

<u>Proposition 9</u>. Let p be a fixed prime.

(i) A finite group π has the property that $SK_1(A\pi) = 0$ for all p-rings A, if and only if

$$[H_2(Z_\pi(g))/H_2^{ab}(Z_\pi(g))]_{(p)} = 0$$

for any $g \in \pi$ of order prime to p.

(ii) Let π be a finite group whose p-Sylow subgroup has a normal abelian subgroup with cyclic quotient. Then $SK_1(A\pi) = 0$ for any p-ring A.

(iii) For any p-ring A and $n > 1$

$$SK_1(A[\Sigma_n]) \cong SK_1(A[A_n]) = 0.$$

Proof. (i) This follows immediately from Theorem 2. Note that for fixed π and A sufficiently large,

$$SK_1(A\pi) \cong \sum_{i=1}^{k} [H_2(Z_\pi(g_i))/H_2^{ab}(Z_\pi(g_i))]_{(p)};$$

where $g_1,\ldots,g_k$ are conjugacy class representatives for $g \in \pi$ with $p \nmid |g|$.

(ii) This follows immediately from (i), and Proposition 9 in [13].

(iii) Note first that for any $g \in \Sigma_n$, $Z_{\Sigma_n}(g)$ has the form

$$Z_{\Sigma_n}(g) = \mathbb{Z}_{m_1} \wr \Sigma_{n_1} \times \ldots \times \mathbb{Z}_{m_k} \wr \Sigma_{n_k}.$$

where $m_i \mid |g|$ and $\mathbb{Z}_m \wr \Sigma_s$ denotes the wreath product. By (i), it will suffice to show that $H_2(\pi) = H_2^{ab}(\pi)$ whenever π is a product of symmetric groups, or of index 2 in such a product.

$H_2(\Sigma_n)$ and $H_2(A_n)$ are calculated in [16] (Abschnitt 1). It follows from the calculations there that for any $n \geq 4$, the induction maps

$$H_2(\mathbb{Z}_2 \times \mathbb{Z}_2) \to H_2(A_4) \to H_2(A_n)_{(2)} \to H_2(\Sigma_n)$$

are all isomorphisms. Furthermore, $H_2(A_n)$ is a 2-group unless $n = 6$ or 7,

$$H_2(A_6) \cong H_2(A_7) \cong \mathbb{Z}_6,$$

and A_6 and A_7 have abelian 3-Sylow subgroups. So for all n,

$$H_2(A_n)/H_2^{ab}(A_n) \cong H_2(\Sigma_n)/H_2^{ab}(\Sigma_n) \cong 0.$$

If π is a product of symmetric or alternating groups, then $H_2(\pi)/H_2^{ab}(\pi) = 0$ since H_2/H_2^{ab} is multiplicative:

$$H_2(\pi_1 \times \pi_2) \cong H_2(\pi_1) \oplus H_2(\pi_2) \oplus (\pi_1^{ab} \otimes \pi_2^{ab})$$

for any π_1 and π_2, and $\pi_1^{ab} \otimes \pi_2^{ab} \subseteq H_2^{ab}(\pi_1 \times \pi_2)$. If π is a semi-direct product

$$\pi = (A_{n_1} \times \ldots \times A_{n_k}) \rtimes \mathbb{Z}_2^{k-1} ,$$

then (by the spectral sequence) $H_2(\pi)$ is generated by

$$H_2(A_{n_1} \times \ldots \times A_{n_k}), \quad H_2(\mathbb{Z}_2^{k-1}), \quad \text{and} \quad H_1(\mathbb{Z}_2^{k-1};\ A_{n_1}^{ab} \times \ldots \times A_{n_k}^{ab}).$$

The first two are contained in $H_2^{ab}(\pi)$. The third vanishes, since A_n^{ab} has odd order for any n. □

SECTION 3.

We now list some results on $SK_1(R\pi)$ for n-rings R; combining results in the first two sections with those in [12] and [13] and the short exact sequences

$$0 \to Cl_1(R\pi)_{(p)} \to SK_1(R\pi)_{(p)} \to SK_1(\hat{R}_p\pi) \to 0.$$

The following follows directly from Theorem 1 here and Theorem 1 in [12]:

Theorem 3. For any n-ring R and finite group π, $SK_1(R\pi)_{(p)}$ is generated by induction from p-elementary subgroups of π. □

For integral group rings we get, for example:

Theorem 4. (i) $SK_1(\mathbb{Z}\pi) = 0$ if $\pi \cong PSL(2,p)$ or $SL(2,p)$ (p prime), $SL(2,2^n)(n \geq 1)$, or $Q(4n)(n \geq 1)$.

(ii) $Wh(\Sigma_n) = SK_1(\mathbb{Z}[\Sigma_n]) = 0$ for all n.

Proof. (i) The only non-cyclic elementary subgroups of these groups are dihedral, quaternionic, or elementary abelian 2-groups (see Theorem II.8.27 in [8]). But for any n,

$$SK_1(\mathbb{Z}[\mathbb{Z}_2^n]) \cong SK_1(\mathbb{Z}[D(2^n)]) \cong SK_1(\mathbb{Z}[Q(2^n)]) = 0$$

by results in [19] (Section 2) and [13] (corollary to Proposition 9).

(ii) $\mathbb{Q}[\Sigma_n]$ is a sum of matrix algebras over $\mathbb{Q}$ ([5], p.196). Hence $Wh(\Sigma_n)$ is finite ([2], Theorem X.3.5), and $Cl_1(\mathbb{Z}[\Sigma_n]) = 0$ by Proposition 13 in [12]. Applying Theorem 6.1 in [21],

$$Wh(\Sigma_n) \cong SK_1(\mathbb{Z}[\Sigma_n]) \cong \sum_p SK_1(\hat{\mathbb{Z}}_p[\Sigma_n]);$$

and this is zero by Proposition 9. □

Note that Magurn [10] has shown that $SK_1(\mathbb{Z}[D(2n)]) = 0$ for any dihedral group $D(2n)$.

We now apply Proposition 2 to $SK_1(\mathbb{Z}\pi)$:

Theorem 5. Let π be any finite group, fix a prime p, and let $\sigma_1,\ldots,\sigma_k$ be conjugacy class representatives for cyclic subgroups $\sigma \subseteq \pi$ of order prime to p. For each i, set

$$n_i = |\sigma_i|, \quad N_i = N(\sigma_i),$$

and let $\mathcal{P}(N_i)$ be the set of p-subgroups of N_i. Then

$$SK_1(\mathbb{Z}\pi)_{(p)} \cong \sum_{i=1}^{k} \varinjlim_{\rho\in\mathcal{P}(N_i)} SK_1(\mathbb{Z}\zeta_{n_i}[\rho]^t)_{(p)}$$

and

$$Cl_1(\mathbb{Z}\pi)_{(p)} \cong \sum_{i=1}^{k} \varinjlim_{\rho\in\mathcal{P}(N_i)} Cl_1(\mathbb{Z}\zeta_{n_i}[\rho]^t)_{(p)}.$$

Proof. We must check that conditions (A) and (B) in Proposition 2 hold.

(A) For any $\pi = \mathbb{Z}_n \rtimes \bar{\pi}$ ($p \nmid n$ and $\bar{\pi}$ a p-group), $\hat{\mathbb{Z}}_p\pi$ splits as a sum of twisted group rings. So it suffices to check that (A) holds for Cl_1, namely that

$$\Sigma\, pr_{d*}: Cl_1(\mathbb{Z}[\mathbb{Z}_n \rtimes \bar{\pi}]) \to \sum_{d|n} Cl_1(\mathbb{Z}\zeta_d[\bar{\pi}]^t)$$

is an isomorphism. Surjectivity follows from Lemma 1 in [12] (the map being induced by an inclusion of orders); and injectivity from Mayer-Victoris sequences and Lemma 5 in [12].

(B) Here, it suffices to consider SK_1. Fix $m|n$ prime to p having the same prime divisors, let $\bar{\pi}$ be a p-group, and

$$t: \bar{\pi} \to Gal(\mathbb{Q}\zeta_n/\mathbb{Q})_{(p)} \cong Gal(\mathbb{Q}\zeta_m/\mathbb{Q})_{(p)}$$

a homomorphism with kernel ρ. The vertical maps in the diagram

$$\begin{array}{ccccc} SK_1(\mathbb{Z}\zeta_m[\rho]) & \xrightarrow{\text{ind}} & SK_1(\mathbb{Z}\zeta_n[\rho]) & \xrightarrow{\text{trf}} & SK_1(\mathbb{Z}\zeta_m[\rho]) \\ \downarrow & & \downarrow & & \downarrow \\ SK_1(\mathbb{Z}\zeta_m[\bar{\pi}]^t) & \xrightarrow{\text{ind}} & SK_1(\mathbb{Z}\zeta_n[\bar{\pi}]^t) & \xrightarrow{\text{trf}} & SK_1(\mathbb{Z}\zeta_m[\bar{\pi}]^t) \end{array}$$

are onto, so we need only show that the composite in the top row is multiplication by n/m. But $\mathbb{Z}\zeta_n$ is a free $\mathbb{Z}\zeta_m$-module of rank n/m (just compare bases), and we are done. □

As an example, Theorem 5 will be applied to compute the 2-torsion in $SK_1(\mathbb{Z}\pi)$ for any π with dihedral, quaternionic, or semidihedral 2-Sylow subgroup. This of course first requires the computation of SK_1 for some twisted group rings. The following lemma seems to be of independant interest.

<u>Lemma 10</u>. For $\Lambda \subseteq \mathbb{Q}$ or $\Lambda = \hat{\mathbb{Z}}_p$, let $\mathfrak{A}$ be a hereditary Λ-order in a semisimple $\mathbb{Q}$- or $\hat{\mathbb{Q}}_p$-algebra A, and let $\mathfrak{M} \supseteq \mathfrak{A}$ be a maximal order. Then the inclusion induces an isomorphism $K_2(\mathfrak{A}) \cong K_2(\mathfrak{M})$.

<u>Proof</u>. First note that for any prime $p \notin \Lambda^*$, $\hat{\mathfrak{A}}_p/J$ (J the Jacobson radical) is a sum of matrix rings over finite fields. Hence $K_2(\hat{\mathfrak{A}}_p/J) = 0$, and the Quillen localization sequence [14]:

$$\sum_{p\notin\Lambda^*} K_2(\hat{\mathfrak{A}}_p/J) \to K_2(\mathfrak{A}) \to K_2(A) \to \sum_{p\notin\Lambda^*} K_1(\hat{\mathfrak{A}}_p/J)$$

shows that $SK_2(\mathfrak{A}) = 0$. Similarly, $SK_2(\mathfrak{M}) = 0$, and if $\Lambda = \hat{\mathbb{Z}}_p$ the result now follows from Lemma 3 in [12].

If $\Lambda \subseteq \mathbb{Q}$, comparing the above localization sequences for $\mathfrak{A}$, $\hat{\mathfrak{A}}_p$, $\mathfrak{M}$, and $\hat{\mathfrak{M}}_p$ produces the following diagram with exact rows:

$$\begin{array}{ccccccc} 0 \to & K_2(\mathfrak{A}) & \to & K_2(A) & \to & \sum_{p\notin\Lambda^*} \mathrm{Coker}[K_2(\hat{\mathfrak{A}}_p) \to K_2(\hat{A}_p)] \\ & \downarrow f_1 & & \downarrow \mathrm{id} & & \downarrow f_2 \\ 0 \to & K_2(\mathfrak{M}) & \to & K_2(A) & \to & \sum_{p\notin\Lambda^*} \mathrm{Coker}[K_2(\hat{\mathfrak{M}}_p) \to K_2(\hat{A}_p)]. \end{array}$$

Since f_2 is an ismorphism by the result on p-adic orders, so is f_1. □

In fact, using the description of p-adic hereditary orders in [15] (Theorem 39.14), this shows that $K_{2n}(\mathfrak{A}) \cong K_{2n}(\mathfrak{M})$ for all $n \geq 1$.

<u>Proposition 11</u>. Let R be the ring of integers in an algebraic number field K where 2 is unramified. Let π be a dihedral, quaternionic, or semidihedral 2-group, let

$$t: \pi \to \mathrm{Gal}(K/\mathbb{Q})$$

be a homomorphism with kernel ρ, and let $R[\pi]^t$ be the corresponding twisted group ring. Then

$$\begin{aligned} Cl_1(R[\pi]^t)_{(2)} &\cong \mathbb{Z}_2 && \text{if } \rho \text{ is non-abelian and } K^\pi \text{ totally imaginary} \\ &\cong 0 && \text{otherwise.} \end{aligned}$$

<u>Proof</u>. By Theorem 1 in [12], the induction map

$$Cl_1(R[\rho]) \to Cl_1(R[\pi]^t)_{(2)}$$

is onto. If ρ is non-abelian, then by Theorem 4 in [12]:

$$\begin{aligned} Cl_1(R\rho) &\cong \mathbb{Z}_2 && \text{if } K \text{ is totally imaginary} \\ &\cong 0 && \text{if } K \text{ has a real imbedding.} \end{aligned}$$

Furthermore, $Cl_1(R\rho) = 0$ if ρ is abelian (i.e., cyclic or $\mathbb{Z}_2 \times \mathbb{Z}_2$); so we need only consider the case where ρ is non-abelian, K totally imaginary, and $t \neq 0$. In particular, $\mathrm{Im}(t) \cong \mathbb{Z}_2$ and $|\pi| \geq 16$ in this case. Set $K' = K^\pi$, and let R' be its ring of integers.

(i) Assume first that K' has a real embedding: we must show that $Cl_1(R[\pi]^t) = 0$. Set $|\rho| = 2^s$ and consider the Mayer-Victoris sequences:

$$\begin{array}{ccccc}
K_2(R[\rho]/\Sigma_\rho) \oplus K_2(R) & \longrightarrow & K_2(R/2^s) & \xrightarrow{\partial_1} & K_1(R[\rho]) \\
\downarrow & & \downarrow & & \downarrow \\
K_2(R[\pi]^t/\Sigma_\rho) \oplus K_2(R[\mathbb{Z}_2]^t) & \rightarrow & K_2(R/2^s[\mathbb{Z}_2]^t) & \xrightarrow{\partial_2} & K_1(R[\pi]^t)
\end{array}$$

(Σ_ρ being the sum of the elements in ρ). Since $Cl_1(R\rho) \subseteq Im(\partial_1)$ (see the proof of Proposition 12 in [12]), it will suffice to show that $\partial_2 = 0$; or $K_2(R[\mathbb{Z}_2]^t)$ surjects onto $K_2(R/2^s[\mathbb{Z}_2]^t)$.

Let $\mathfrak{M} \supseteq R[\mathbb{Z}_2]^t$ be a maximal order. Since 2 is unramified in R, $R[\mathbb{Z}_2]^t$ is hereditary and has odd index in $\mathfrak{M}$ ([15], Theorem 40.13 and 40.14). So

$$R/2^s[\mathbb{Z}_2]^t = \mathfrak{M}/2^s \quad \text{and} \quad K_2(R[\mathbb{Z}_2]^t) = K_2(\mathfrak{M})$$

(Lemma 11); and we must show that

$$K_2(\mathfrak{M}) \rightarrow K_2(\mathfrak{M}/2^s)$$

is onto. By Morita equivalence (using Theorem 21.6 in [15]):

$$K_2(\mathfrak{M}) = K_2(R') \quad \text{and} \quad K_2(\mathfrak{M}/2^s) = K_2(R'/2^s).$$

But $SK_1(R', 2^sR') = 0$ [3], and we are done.

(ii) Now assume R' is totally imaginary. Since surjections of orders induce surjections of Cl_1 (Lemma 1 in [12]), it suffices to consider the case $|\pi| = 16$. Fix $z \in \pi$ central of order 2, note that $\pi/z \cong D(8)$, $\rho/z \cong \mathbb{Z}_2^2$, and consider the following pullback squares:

$$\begin{array}{ccc}
R[\pi]^t & \longrightarrow & R[D8]^t \\
\downarrow & & \tilde{\alpha}\downarrow \;\searrow^{\tilde{\alpha}_1}\; R/4[D8]^t \\
R[\pi]^t/\langle 1+z\rangle & \xrightarrow{\tilde{\beta}} & R/2[D8]^t \;\swarrow_{\tilde{\alpha}_2}
\end{array}
\qquad
\begin{array}{ccc}
R[\rho] & \longrightarrow & R[\mathbb{Z}_2^2] \\
\downarrow & & \alpha\downarrow \;\searrow^{\alpha_1}\; R/4[\mathbb{Z}_2^2] \\
R[\rho]/\langle 1+z\rangle & \xrightarrow{\beta} & R/2[\mathbb{Z}_2^2] \;\swarrow_{\alpha_2}
\end{array}$$

Define maps

$$\varepsilon: R/2[\mathbb{Z}_2^2] \rightarrow R/2 \quad \text{and} \quad \tilde{\varepsilon}: R/2[D8]^t \rightarrow R'/2;$$

where ε is the augmentation map and $\tilde{\varepsilon}$ the augmentation map composed with the trace. Then the following diagram commutes:

$$\begin{array}{llllllll}
\Phi: K_2(R/2[\mathbb{Z}_2^2]) & \xrightarrow{\partial} & K_1(R\rho/\langle 1+z\rangle,2) \rightarrow & & & & & \\
 & & & K_1(R\rho/\langle 4,1+z\rangle,2) & \cong R/2[\mathbb{Z}_2^2] & \xrightarrow{\varepsilon} R/2 & \xrightarrow{Tr} & \mathbb{Z}_2 \\
\downarrow & & \downarrow & \downarrow & \downarrow & \downarrow Tr & & \downarrow id \\
\tilde{\Phi}: K_2(R/2[D8]^t) & \xrightarrow{\partial} & K_1(R\pi^t/\langle 1+z\rangle,2) & & & & & \\
 & & \rightarrow & K_1(R\pi^t/\langle 4,1+z\rangle,2) & \cong \dfrac{R/2[D8]^t}{\langle x-gxg^{-1}\rangle} & \xrightarrow{\tilde{\varepsilon}} R'/2 & \xrightarrow{Tr} & \mathbb{Z}_2
\end{array}$$

where Φ and $\tilde{\Phi}$ denote the compositions of the two rows. $Cl_1(R\rho)$ was shown non-zero in ([12], Proposition 17) by proving that Φ is onto and $\Phi \circ \alpha_* = \Phi \circ \beta_* = 0$. Since $\tilde{\Phi} \circ \tilde{\beta}_* = 0$ by construction, we will be done upon showing that $\tilde{\Phi} \circ \tilde{\alpha}_* = 0$.

Now let $\mathfrak{M} \cong R^4 \supseteq R[\mathbb{Z}_2^2]$ and $\tilde{\mathfrak{M}} \supseteq R[D8]^t$ be maximal orders, and consider the following diagram:

$$\begin{array}{lllllll}
\Psi: K_2(R/4[\mathbb{Z}_2^2]) & \rightarrow & K_2(\mathfrak{M}/4) & \xrightarrow{\partial} & SK_1(\mathfrak{M},4\mathfrak{M}) \cong SK_1(R,4R)^4 & \xrightarrow{\sigma} & \mathbb{Z}_2 \\
\downarrow & & \downarrow & & \downarrow j_* & & \downarrow id \\
K_2(R/4[D8]^t) & \rightarrow & K_2(\tilde{\mathfrak{M}}/4) & \xrightarrow{\partial} & SK_1(\tilde{\mathfrak{M}},4\tilde{\mathfrak{M}}) & \xrightarrow{\tilde{\sigma}} & \mathbb{Z}_2
\end{array}$$

where σ denotes the sum $(SK_1(R,4R) \cong \mathbb{Z}_2)$ and Ψ is the composite of the top row. Clearly $\Psi \circ \alpha_{1*} = 0$, and it was shown in [12] (Proposition 17) that $\Psi = \Phi \circ \alpha_{2*}$. To show that $\tilde{\Phi} \circ \tilde{\alpha}_* = 0$, we first construct $\tilde{\sigma}$ such that $\tilde{\sigma} \circ j_* = \sigma$, thus constructing

$$\tilde{\Psi}: K_2(R/4[D8]^t) \rightarrow \mathbb{Z}_2$$

with $\Psi = \tilde{\Psi} \circ i_*$ and $\tilde{\Psi} \circ \tilde{\alpha}_{1*} = 0$. That $\tilde{\Psi} = \tilde{\Phi} \circ \tilde{\alpha}_{2*}$ (and hence $\tilde{\Phi} \circ \tilde{\alpha}_* = 0$) will then follow upon showing that i_* is onto.

Fix generators $a,b \in \mathbb{Z}_2^2$ and $x \in D(8)$ so that $K[D8]^t$ can be regarded as a twisted group ring:

$$K[D8]^t = K[\mathbb{Z}_2^2][x]^t\colon\ x^2=1,\ x\lambda x^{-1} = \bar{\lambda}\ (\lambda \in K),\ xax^{-1}=a,\ xbx^{-1}=ab$$

$(\lambda \rightarrow \bar{\lambda}$ denoting the Galois involution fixing $K')$. The twisting action leaves two of the summands of $K[\mathbb{Z}_2^2] \cong K^4$ invariant and permutes the other two. The inclusion

$$K^4 = K[\mathbb{Z}_2^2] \to K[D8]^t \cong M_2(K') \oplus M_2(K') \oplus M_2(K)$$

thus sends two factors to maximal subfields of $M_2(K')$ and the other two onto the diagonal in $M_2(K)$.

So by Morita equivalence, j_* takes the form

$$j_*\colon SK_1(R,4R)^4 \to SK_1(\tilde{\mathfrak{m}},4\tilde{\mathfrak{m}}) \cong SK_1(R',4R')^2 \oplus SK_1(R,4R);$$

where two factors $SK_1(R,4R)$ are mapped via the transfer and the other two via the identity. Upon comparing [3] with the exact sequences for ideals,

$$trf\colon SK_1(R,4R) \to SK_1(R',4R') \cong \mathbb{Z}_2$$

is seen to be an isomorphism. Letting

$$\tilde{\varepsilon}\colon SK_1(\tilde{\mathfrak{m}},4\tilde{\mathfrak{m}}) \to \mathbb{Z}_2$$

be the sum of the three $\mathbb{Z}_2$-terms, we thus get $\tilde{\varepsilon} \circ j_* = \varepsilon$.

To show that

$$i_*\colon K_2(R/4[\mathbb{Z}_2^2]) \to K_2(R/4[D8]^t)$$

is onto, again regard $R[D8]^t = R[\mathbb{Z}_2^2][x]^t$ as a twisted group ring over $R[\mathbb{Z}_2^2]$. Set $S = R[\mathbb{Z}_2^2]^x$. Since 2 is unramified in R, there is $\lambda \in \hat{R}_2$ with $\lambda+\bar{\lambda} = 1$. Then $\{\lambda,\bar{\lambda}\}$ is an $\hat{R}_2'$-basis for $\hat{R}_2$ and an $\hat{S}_2$-basis for $\hat{R}_2[\mathbb{Z}_2^2]$; defining a homomorphism

$$\hat{R}_2[D8]^t \to \mathrm{End}_S(\hat{R}_2[\mathbb{Z}_2^2]) \cong M_2(\hat{S}_2)$$

which is easily checked to be an isomorphism.

The composite

$$K_1(R/4[\mathbb{Z}_2^2]) \xrightarrow{\ incl_*\ } K_1(R/4[D8]^t) \cong K_1(S/4)$$

is the transfer map, and is onto by Lemma 3. $S/4$ is additively generated by units, so $K_2(S/4)$ is generated by symbols [18]. Using the reciprocity law ([11], Theorem 14.1):

$$\mathrm{trf}\{x, \mathrm{ind}(y)\} = \{\mathrm{trf}(x), y\} \qquad (x \in K_1(R/4[\mathbb{Z}_2^2]),\ y \in K_1(S/4))$$

the composite

$$i_*: K_2(R/4[\mathbb{Z}_2^2]) \xrightarrow{\mathrm{trf}} K_2(S/4) \cong K_2(R/4[D8]^t)$$

is now seen to be onto. □

Combining this with Theorem 5 (and Proposition 9(ii)) now gives:

<u>Theorem 6</u>. Let π be any finite group whose 2-Sylow subgroup is dihedral, quaternionic, or semidihedral. Then

$$SK_1(\mathbb{Z}\pi)_{(2)} \cong \mathbb{Z}_2^k,$$

where k is the number of conjugacy classes of cyclic subgroups $\sigma \subseteq \pi$ such that

(i) $|\sigma|$ is odd

(ii) The 2-Sylow subgroup of $Z_\pi(\sigma)$ is non-abelian.

(iii) There is no $g \in N(\sigma)$ with $gxg^{-1} = x^{-1}$ for all $x \in \sigma$. □

In particular, when π is a periodic group, $SK_1(\mathbb{Z}\pi)$ is a 2-group by Theorem 2 in [12], and hence computed by the above formula.

At the beginning of the section, we saw that $SK_1(\mathbb{Z}\pi)$ is generated by induction from elementary subgroups. Much more useful, of course, would be to know that $SK_1(\mathbb{Z}\pi)$ were detected by restriction to elementary subgroups, but Proposition 12 gives counterexamples to this.

As an example, fix a homomorphism

$$t: D(16) \to \mathrm{Aut}(\mathbb{Z}_{15}) \cong \mathrm{Gal}(\mathbb{Q}\zeta_{15}/\mathbb{Q})$$

such that $Ker(t) = D(8)$ and $Im(t)$ leaves $\mathbb{Z}_3$ fixed. By Proposition 12, the groups

$$SK_1(\mathbb{Z}\zeta_{15}[D(8)]) \xrightarrow{i_*} SK_1(\mathbb{Z}\zeta_{15}[D(16)]^t) \xrightarrow{trf_1} SK_1(\mathbb{Z}\zeta_{15}[D(8)])$$

$$SK_1(\mathbb{Z}\zeta_{15}[D(8)]) \xrightarrow{i_*} SK_1(\mathbb{Z}\zeta_{15}[D(16)]^t) \xrightarrow{trf_2} SK_1(\mathbb{Z}\zeta_3[D(16)])$$

all have order 2, and i_* is an isomorphism. Both composites are zero: the first is a norm map and the second factors through $SK_1(\mathbb{Z}\zeta_3[D(8)])$ (whose inclusion into $D(16)$ is zero). So $trf_1 = trf_2 = 0$, thus

$$trf\colon SK_1(\mathbb{Z}[\mathbb{Z}_{15}\rtimes D(16)]) \to SK_1(\mathbb{Z}[\mathbb{Z}_{15}\times D(8)]) \oplus SK_1(\mathbb{Z}[\mathbb{Z}_3\times D(16)])$$

is not one-to-one, and $SK_1(\mathbb{Z}[\mathbb{Z}_{15}\rtimes D(16)])$ is not detected by restriction to elementary subgroups.

REFERENCES

1. A. Bak, Surgery and K-theory groups of quadratic forms over finite groups and orders (Preprint).

2. H. Bass, Algebraic K-theory, Benjamin (1968).

3. H. Bass, J. Milnor, and J.-P. Serre, Solution of the congruence subgroup problem for SL_n $(n\geq 3)$ and Sp_{2n} $(n\geq 2)$, Inst. Hautes Etudes Sci.Publ.Math.No.33(1967), 59-137.

4. H. Cartan and S. Eilenberg, Homological algebra, Princeton Univ. Press(1956).

5. C. Curtis and I. Reiner, Representation theory of finite groups and associated algebras, Interscience(1962).

6. A. Dress, Introduction and structure theorems for orthogonal representations of finite groups, Annals of Math.102(1975), 291-325.

7. D. Grayson, Higher algebraic K-theory II, Lecture Notes in Math., vol.551, Springer-Verlag(1956), 217-240.

8. B. Huppert, Endliche Gruppen, Springer-Verlag(1967).

9. T.-Y. Lam, Introduction theorems for Grothendieck groups and Whitehead groups of finite groups, Ann.Sci.Ecole Norm. Sup.1(1968), 91-148.

10. B. Magurn, SK_1 of dihedral groups, J.Algebra 51(1978), 399-415.

11. J. Milnor, Introduction to algebraic K-theory, Princeton Univ.Press (1971).

12. R. Oliver, SK_1 for finite group rings: I, Invent.math. 57(1980), 183-204.

13. R. Oliver, SK_1 for finite group rings: II, Math. Scand. (to appear)

14. D. Quillen, Higher algebraic K-theory I, Lecture Notes in Math., vol.341, Springer-Verlag(1973), 85-147.

15. I. Reiner, Maximal orders, Academic Press (1975).

16. J. Schur, Über die Darstellung der symmetrischen und der alternierenden Gruppe durch gebrochene lineare Substitutionen, Journal reine ang.Math.139(1911), 155-250.

17. J.-P. Serre, Représentations linéaires des groupes finis, Hermann (1967).

18. M. Stein, Surjectivity stability in dimension 0 for K_2 and related functors, Trans.Amer.Math.Soc.178(1973), 165-191.

19. M. Stein, Whitehead groups of finite groups, Bull.Amer.Math.Soc. 84(1978), 201-212.

20. C.T.C. Wall, On the classification of hermetian forms, III: Complete semilocal rings, Invent.Math.19(1973), 59-71.

21. C.T.C. Wall, Norms of units in group rings, Proc. London Math.Soc. 29(1974),593-632.

22. A. Weil, Basic Number Theory, Springer-Verlag (1967).

On a conjecture concerning $K_*(\mathbb{Z}/p^2)$

by

Stewart Priddy

Let $F(a,b)$ be the homotopy theoretic fiber of $\Psi^a - \Psi^b$: $BU \longrightarrow BU$. Then $\pi_{2i-1}F(a,b) = \mathbb{Z}/(a^i-b^i)$ and $\pi_{2i}F(a,b) = 0$. In his paper on the algebraic K-theory of finite fields, Quillen [Q] established a homotopy equivalence $BGL(\mathbb{F}_q)^+ \simeq F(q,1)$ and thus computed $K_*(\mathbb{F}_q) = \pi_* BGL(\mathbb{F}_q)^+$. Since then, there has been speculation as to whether $K_*(\mathbb{Z}/p^k)$ could be computed in a similar manner. For example, Karoubi has asked if

(1) $$BGL(\mathbb{Z}/p^k)^+ \simeq F(p^k, p^{k-1}) \qquad p > 2$$

This is consistent with the calculations $K_1(\mathbb{Z}/p^k) = \mathbb{Z}/(p^k - p^{k-1})$ and $K_2(\mathbb{Z}/p^k) = 0$ for $p > 2$ (Milnor [M]).

Recently, Evens and Friedlander [E] have computed $K_3(\mathbb{Z}/p^2) = \mathbb{Z}/p^4-p^2$ and $K_4(\mathbb{Z}/p^2) = 0$ for $p > 3$. This is also consistent with (1). In fact, based on these calculations, Friedlander and I were lead to conjecture (1) independently of Karoubi.

The purpose of this note is to show (1) is false in general. More precisely, we prove that $BGL(\mathbb{Z}/p^2)^+$ is not homotopy equivalent to $F(p^2, p)$ for $p > 3$ by proving the

<u>Proposition</u> Any map f: $BGL(\mathbb{Z}/p^2) \longrightarrow BU$ has second Chern class $c_2(f) = 0 \bmod p$ $(p > 3)$.

Using the obvious fibration $U \longrightarrow F(p^2, p) \longrightarrow BU$, one easily computes $H^*(F(p^2, p); \mathbb{Z}/p) = E[x_1,x_3,x_5,\ldots] \otimes P[c_1,c_2,c_3,\ldots]$. Thus the falsity of (1) follows immediately.

Proof of the Proposition proceeds by reducing the question to the group $SL_2(\mathbb{Z}/p^2)$. One then uses the character tables of Praetorius [P] or Rohrbach [R] to show $c_2(\rho) = 0 \bmod p$ for any irreducible representation ρ. The vanishing of $c_2(f)$ then follows easily from Atiyah's isomorphism $R(G)^\wedge \longrightarrow K^\circ(BG)$.

It is a pleasure to thank L. Evens, E. Friedlander, and C. Giffen for many useful conversations on this subject and to thank B. Srinivasan for telling us about [P], [R].

<u>Reduction to</u> $SL_2(\mathbb{Z}/p^2)$: Since, $K_3(\mathbb{Z}/p^2) \approx H_3(St(\mathbb{Z}/p^2);\ \mathbb{Z})$ and $K_2(\mathbb{Z}/p^2) = 0$, the central extension

$$0 \longrightarrow K_2(\mathbb{Z}/p^2) \longrightarrow St(\mathbb{Z}/p^2) \longrightarrow SL(\mathbb{Z}/p^2) \longrightarrow 1$$

shows that $K_3(\mathbb{Z}/p^2) \approx H_3(SL(\mathbb{Z}/p^2);\ \mathbb{Z})$. Let $\mathbb{Z}_{(p)}$ denote the integers localized at p. The basic homological calculation of Evens and Friedlander [E; Theorem 3.4] is $H_k(SL(\mathbb{Z}/p^2);\ \mathbb{Z}_{(p)}) = \mathbb{Z}/p^2,\ 0$ for $k = 3,4$. Thus by the universal coefficient theorem we have the isomorphism

$$H^4(SL(\mathbb{Z}/p^2);\ \mathbb{Z}_{(p)}) \approx \mathbb{Z}/p^2 \tag{2}$$

Now let A be the subgroup of order p in $SL_2(\mathbb{Z}/p^2)$ generated by $a = \begin{pmatrix} 1+p & 0 \\ 0 & 1-p \end{pmatrix}$.

<u>Lemma</u> The restriction homomorphism

(3) $H^4(SL_2(\mathbb{Z}/p^2);\ \mathbb{Z}_{(p)}) \longrightarrow H^4(A;\ \mathbb{Z}_{(p)}) \approx \mathbb{Z}/p$ is surjective.

<u>Proof</u>: Let $V = V_2$ be the set of 2 x 2 matrices over $\mathbb{Z}/p$ of trace zero. Then the coefficient sequence $0 \longrightarrow \mathbb{Z}/p \xrightarrow{i} \mathbb{Q}/\mathbb{Z}_{(p)} \xrightarrow{xp} \mathbb{Q}/\mathbb{Z}_{(p)} \longrightarrow 0$ induces a homomorphism between restrictions

$$\begin{array}{ccc} H^3(SL_2(\mathbb{Z}/p^2);\ \mathbb{Z}/p) & \xrightarrow{i_*} & H^3(SL_2(\mathbb{Z}/p^2);\ \mathbb{Q}/\mathbb{Z}_{(p)}) \\ \downarrow & & \downarrow \\ H^3(V;\ \mathbb{Z}/p) & \xrightarrow{i_*} & H^3(V;\ \mathbb{Q}/\mathbb{Z}_{(p)}) \\ \downarrow & & \downarrow \\ H^3(A;\ \mathbb{Z}/p) & \xrightarrow[\approx]{i_*} & H^3(A;\ \mathbb{Q}/\mathbb{Z}_{(p)}) \end{array}$$

In the notation of [E; §12], there is an element $4\chi_1 + \chi_2 \in H^3(V;\ \mathbb{Z}/p)$ which maps under i_* to an element in the image of the restriction $H^3(SL_2(\mathbb{Z}/p^2);\ \mathbb{Q}/\mathbb{Z}_{(p)}) \longrightarrow H^3(V;\ \mathbb{Q}/\mathbb{Z}_{(p)})$. (In [E] this element actually restricts from a supergroup $\overline{SL_2}(\mathbb{Z}/p^2)$ of $SL_2(\mathbb{Z}/p^2)$). Moreover $4\chi_1 + \chi_2$ restricts to $2h^1 \cdot \delta h^1$ in $H^3(A;\ \mathbb{Z}/p)$ where h^1 generates $H^1(A;\ \mathbb{Z}/p)$

and δ is the mod p Bockstein so that $2h^1 \cdot \delta h^1$ generates $H^3(A; \mathbb{Z}/p)$. Thus the composite restriction $H^3(SL_2(\mathbb{Z}/p^2); \mathbb{Q}/\mathbb{Z}_{(p)}) \longrightarrow H^3(A; \mathbb{Q}/\mathbb{Z}_{(p)})$ is surjective. The Lemma follows using the natural isomorphism $H^3(-; \mathbb{Q}/\mathbb{Z}_{(p)}) \longrightarrow H^4(-; \mathbb{Z}_{(p)})$ which holds for any finite group.

Irreducible representations of $SL_2(\mathbb{Z}/p^2)$: We will now show that if ρ is any representation of $SL_2(\mathbb{Z}/p^2)$ then $c_2(\rho|_A) = 0 \bmod p$. This is the key step in the proof of the Proposition.

We have reproduced below part of the character table for $SL_2(\mathbb{Z}/p^2)$, $p > 2$ (see [R; p. 93], [P; p. 389]).

	$\chi_{1,\gamma}$	χ_2	χ_3	χ_4
a	$p(\omega^{\gamma}+\omega^{-\gamma})$	$\frac{p-1}{2}$	$\frac{p-1}{2}$	0
deg	$p(p+1)$	$\frac{p^2-1}{2}$	$\frac{p^2-1}{2}$	$p(p-1)$

Here ω is a primitive $p^{\underline{th}}$ root of unity, a generates A, and $\gamma = 1,2,\ldots,(p-1)/2$. Let $s\colon A \longrightarrow U(1)$ be the standard irreducible representation defined by $s(a) = \omega$ and let $reg\colon A \longrightarrow U(p)$ be the regular representation. In terms of their characters $\chi_s(a) = \omega$ and $\chi_{reg}(a) = 1+\omega+\cdots+\omega^{p-1} = 0$. Thus upon restriction to A we have

$$\chi_{1,\gamma} = p(\chi_s^{\gamma} + \chi_s^{-\gamma}) + (p-1)\chi_{reg}$$

$$\chi_2 = (p-1)/2\ \chi_{reg} + (p-1)/2$$

$$\chi_3 = \chi_2$$

$$\chi_4 = (p-1)\chi_{reg}.$$

Let $c = 1 + c_1 + c_2 + \cdots$ be the total Chern class reduced mod p. Then $c(\chi_s) = 1 + y$ where $y \in H^2(A; \mathbb{Z}/p) = \mathbb{Z}/p) = \mathbb{Z}/p$ is a generator. Similarly

$c(\chi_{reg}) = (1+y)(1+2y)\cdots(1+(p-1)y) = 1-y^{p-1}$. Thus (upon restriction to A) we have

$$c(\chi_{1,\gamma}) = (c(\chi_s^{\gamma}) \cdot c(\chi_s^{-\gamma}))^p \, (c(\chi_{reg}))^{p-1}$$

$$= ((1+\gamma y)(1-\gamma y))^p \, (1-y^{p-1})^{p-1}$$

$$= (1-\gamma^2 y^{2p})(1-y^{p-1})^{p-1}$$

$$c(\chi_2) = c(\chi_3) = c(\chi_{reg})^{\frac{p-1}{2}} = (1-y^{p-1})^{\frac{p-1}{2}}$$

$$c(\chi_4) = c(\chi_{reg})^{p-1} = (1-y^{p-1})^{p-1}$$

Thus in all cases $c_2 = 0 \bmod p$ for $p > 3$.

The general case: Let $f\colon BSL(\mathbb{Z}/p^2) \longrightarrow BU$; we will show $c_2(f) = 0 \bmod p$. If X is a CW-complex then $K^\circ(X)$ is filtered as usual by $K_i^\circ(X) = \ker(K^\circ(X) \longrightarrow K^\circ(X^{i-1}))$ and $K_{2i-1}^\circ(X) = K_{2i}^\circ(X)$. By Atiyah's theorem [A] if G is a finite group then $\hat{\alpha}\colon R(G)^\wedge \xrightarrow{\approx} K^\circ(BG)$ where α assigns to each representation the homotopy class of its classifying map. Thus the filtration induced on $R(G)$ by α has the property that $R_{2i-1}(G) = R_{2i}(G)$.

Now let $G = SL_2(\mathbb{Z}/p^2)$ and let $f'\colon BG \longrightarrow BU$ be the restriction of f and consider $[f'] \in K_2^\circ(BG)/K_4^\circ(BG) \approx R_2(G)/R_4(G)$. Since

$$c_1\colon R_2(G)/R_4(G) \xrightarrow{\approx} H^2(G;\, \mathbb{Z})$$

[A] and $H^2(G;\, \mathbb{Z}_{(p)}) = 0$ [E; §3] we may assume $[f'] = 0$ (or at least a prime to p multiple of $[f']$ is zero). Thus $[f'] \in K_4^\circ(BG)/K_6^\circ(BG) \approx R_4(G)/R_6(G)$. Let $\rho \in R(G)$ be a virtual representation of G such that $[f']$ corresponds to $[\rho] \in R_4(G)/R_6(G)$ and consider

$$R_4(G)/R_6(G) \xrightarrow{c_2} H^4(G;\, \mathbb{Z}).$$

Working mod p we have $c_2(\rho) = 0$ by §2. Thus $c_2[\rho] = 0$ and so $c_2[f'] = 0$. Since $c_2K_6^\circ(BG) = 0$ we have $c_2(f') = 0$. It now follows from the isomorphism (2) that $c_2(f) = 0 \bmod p$.

References

[A] M. Atiyah, Characters and cohomology of finite groups, Inst. Hautes Etudes Sci. Publ. Math., 9 (1961), 23-64.

[E] L. Evens and E. Friedlander, On $K_*(\mathbb{Z}/p^2)$ and related homology groups, (to appear).

[M] J. Milnor, Introduction to Algebraic K-Theory. Ann. of Math. Studies No. 72, Princeton University Press, Princeton, 1971.

[P] H. Praetorius, Die Charaktere der Modulargruppen der Stufe q^2, Hamberger Abhanglungen (1930), 365-394.

[Q] D. Quillen, On the cohomology and K-theory of the general linear groups over a finite field, Ann. of Math. 96 (1972), 552-586.

[R] H. Rohrbach, Die Charaktere du binären Kongruenzgruppen mod p^2, Berlin Universität, Instituts für angewandte Mathematik, Schriften, Band 1, Heft 2 (1932).

K'-THEORY OF NOETHERIAN SCHEMES

Clayton Sherman*

The purpose of this paper is to establish, for a Noetherian scheme X, some results concerning the groups $K'_*(X)$, the K-theory of coherent sheaves on X. The principal tool in this theory is a spectral sequence defined by Quillen. Results of Gersten and Quillen have shown that a knowledge of the properties of this spectral sequence could be useful in applications of algebraic K-theory to algebraic geometry; of particular importance is a conjecture of Gersten concerning the K-theory of regular local rings.

Section 1 recalls definitions and contains some elementary remarks about the notion of very clean ring introduced in [15]; in particular, we give an example showing that a localization of a very clean ring need not be very clean.

The main result of Section 2 is a technical one describing the behavior of the spectral sequence with respect to finite morphisms. As the first of several applications, we prove that if $F \to F'$ is an extension of finite fields, then the transfer maps $K_n(F'(t)) \to K_n(F(t))$ are surjective for $n \geq 2$. We also prove that the canonical maps $K_n(F(t)) \to K_n(F'(t))$ are injective for $n \geq 0$.

In Section 3 we prove Gersten's Conjecture for certain discrete valuation rings with residue class field $k(t)$, where k is a finite field. We also point out that although the conjecture is far from settled in general, one can derive useful information from the one-dimensional cases of the conjecture which have already been established.

In Section 4 we formulate, for a regular scheme X, conjectures concerning a certain filtration on $K_.(X)$. We establish certain special cases of these conjectures; in particular we prove that, for

*This material is based upon work supported by the National Science Foundation under Grant No. MCS-7903084.

many Dedekind rings R, $(SK_{\cdot}(R))^2 = 0$ in the graded ring $K_{\cdot}(R)$.

Although Gersten's Conjecture per se only concerns regular local rings, in Section 5 we show that its conclusion is valid for many (non-regular) one-dimensional integral domains, and also for some non-regular two-dimensional local domains. This section also contains the computation of $K'_*(X)$ for certain singular curves X.

All schemes considered in this paper will be assumed to be Noetherian and separated. I would like to thank Keith Dennis for some useful conversations concerning the proof of Theorem 3.1.

1. Very Clean Rings.

All rings considered in this paper will be assumed to be commutative and Noetherian. Given such a ring R, let $\mathcal{P}(R)$ denote the category of finitely generated projective R-modules, and $\mathcal{M}(R)$ the category of all finitely generated R-modules. Recall that $K_n(R)$ is by definition $K_n(\mathcal{P}(R))$, and that $K'_n(R)$ is $K_n(\mathcal{M}(R))$; recall further that if R is regular, then the Cartan map $K_n(R) \to K'_n(R)$ is an isomorphism.

$\mathcal{M}(R)$ is filtered by "codimension of support": define $\mathcal{M}^i(R)$ to be the (abelian) subcategory consisting of those modules M for which $M_{\mathfrak{p}} = 0$ for all primes $\mathfrak{p}$ with ht $\mathfrak{p} < i$. $\mathcal{M}^j$ is a Serre subcategory of $\mathcal{M}^i$ for $j > i$, so we may form the quotient abelian categories $\mathcal{M}^i/\mathcal{M}^j$. Quillen's Localization Theorem ([11], Thm. 5) gives long exact sequences

$$\cdots \to K_{n+1}(\mathcal{M}^i/\mathcal{M}^{i+1}) \to K_n(\mathcal{M}^{i+1}) \to K_n(\mathcal{M}^i) \to K_n(\mathcal{M}^i/\mathcal{M}^{i+1}) \to \cdots$$

By splicing these sequences together in a standard way, Quillen obtains a 4th quadrant sequence of cohomological type

$$E_1^{ij}(R) = K_{-i-j}(\mathcal{M}^i/\mathcal{M}^{i+1}) \Rightarrow K'_{-i-j}(R).$$

It is convergent whenever R has finite Krull dimension, and is functorial with respect to flat homomorphisms. It is not hard to show that $K_n(\mathcal{M}^i/\mathcal{M}^{i+1}) = \coprod_{\mathrm{ht}\,\mathfrak{p}=i} K_n k(\mathfrak{p})$ ($k(\mathfrak{p})$ denotes the residue class field of the prime $\mathfrak{p}$), so we may also write

$$E_1^{ij}(R) = \coprod_{\mathrm{ht}\,\mathfrak{p}=i} K_{-i-j} k(\mathfrak{p}).$$

In [2], R was defined to be <u>clean</u> if $K_0(\mathcal{M}^{i+1}) \to K_0(\mathcal{M}^i)$ is zero for all i; in [15], we extended this by defining R to be <u>very clean</u> if $K_n(\mathcal{M}^{i+1}) \to K_n(\mathcal{M}^i)$ is zero for all $i, n \geq 0$. As shown in [11] (Prop. 7.5.6) this property is equivalent to the condition that the sequence

$$0 \to K'_n(R) \to \coprod_{\mathrm{ht}\,\mathfrak{p}=0} K_n k(\mathfrak{p}) \xrightarrow{d_1} \coprod_{\mathrm{ht}\,\mathfrak{q}=1} K_{n-1} k(\mathfrak{q}) \to \cdots$$

be exact for all $n \geq 0$. Recall that <u>Gersten's Conjecture</u> is that regular local rings are very clean. As shown in [2] and [11], a verification of this conjecture could have important consequences for relating the K-theory and intersection theory of a regular scheme.

Aside from its geometric applications, Gersten's Conjecture and, more generally, the concept of very clean ring, are of interest in themselves for understanding the K'-theory of rings, and, consequently, for understanding the K-theory of regular rings. In particular, we have the following basic result.

Proposition 1.1. Suppose that R is a very clean integral domain with field of fractions F. Then the map $K_n'(R) \to K_n(F)$ is injective for all $n \geq 0$; consequently, if R is regular, then the map $K_n(R) \to K_n(F)$ is injective. Suppose further that $R_{\mathfrak{p}}$ is very clean for each height 1 prime $\mathfrak{p}$ of R. Then $K_n'(R) = \bigcap_{ht\mathfrak{p}=1} K_n'(R_{\mathfrak{p}})$ (as subgroups of $K_n(F)$). In particular, if R is also regular, then $K_n(R) = \bigcap_{ht\mathfrak{p}=1} K_n(R_{\mathfrak{p}})$.

Proof. As remarked above, if R is very clean, then there is an exact sequence $0 \to K_n'R \to K_nF \xrightarrow{d_1} \coprod_{ht\mathfrak{p}=1} K_{n-1}k(\mathfrak{p})$, which, in particular, establishes the first statement. Next, suppose that $R_{\mathfrak{p}}$ is very clean for each height 1 prime $\mathfrak{p}$ of R. Then for each such $\mathfrak{p}$, there is also an exact sequence

$$0 \to K_n'R_{\mathfrak{p}} \to K_nF \xrightarrow{(d_1)_{\mathfrak{p}}} K_{n-1}k(\mathfrak{p}).$$

But by Prop. 1.4 of [15], the $\mathfrak{p}$-th component of d_1 may be identified with $(d_1)_{\mathfrak{p}}$. The other statements follow immediately from these remarks. #

Gersten proved that the class of clean rings is closed under polynomial extension and localization with respect to arbitrary multiplicatively closed subsets [2]. (Cf. [18] for other results on clean rings.) In [15] the author proved that the class of very clean rings is also closed under polynomial extension. If one could prove that the class of very clean rings were closed under localization, this would establish Gersten's Conjecture for a large class of regular local rings. Here, however, we give an example showing that this need not be true for arbitrary localizations, though it might still hold for localizations at prime ideals. The key to the example is the following result.

Proposition 1.2. Let R be a Noetherian catenary integral domain of finite Krull dimension. Let f be a non-zero element of R. Suppose that R is very clean. Then R/(f) is very clean if and only if R_f is very clean.

Proof. This is an immediate consequence of Prop. 1.1c of [15], and Thm. 2.1 of [17], which relates the E_2 terms of the spectral sequences for R, R/(f), and R_f via long exact sequences. #

Now let k be a field, char $k \neq 2$, $R = k[X,Y]$, $f = Y^2 - X^3 + X$. Then $A = R/(f)$ is Dedekind, but not a UFD (cf. [5], Ex. I.6.2). Hence Pic $A \neq 0$. But Pic A is the kernel of $K_0(A) \to K_0(F)$, where F is the field of fractions of A. Hence A is not even clean. Since k[X,Y] is very clean ([15], Thm. 2.4), the result above implies that R_f is not.

2. Finite Morphisms.

In this section we describe the behavior of the spectral sequence with respect to finite extensions. We shall actually prove our result in the more general setting of Noetherian schemes and finite surjective morphisms. This complements a result from [15], where the case of a closed immersion was considered.

Given a Noetherian scheme X, in analogy with Section 1, Quillen defines $K_n(X) = K_n(\mathcal{P}(X))$, where $\mathcal{P}(X)$ is the category of vector bundles on X, and $K'_n(X) = K_n(\mathcal{M}(X))$, where $\mathcal{M}(X)$ is the category of coherent sheaves on X. As before, $\mathcal{M}(X)$ is filtered by the categories $\mathcal{M}^p(X)$, where $\mathcal{M}^p(X)$ consists of those sheaves $\mathcal{F}$ with codim (Supp $\mathcal{F}$) $\geq$ p. The spectral sequence E(X) has

$$E_1^{pq}(X) = K_{-p-q}(\mathcal{M}^p/\mathcal{M}^{p+1}) \cong \coprod_{x\in X^{(p)}} K_{-p-q}k(x)$$

where $X^{(p)}$ is the set of points of codimension p.

We need a technical lemma. Suppose that Z is a Noetherian scheme and that $z \in Z^{(p)}$. Let $\overline{\{z\}}$ denote the reduced induced closed subscheme supported on the closure of z. By Prop. 1.3 of [15], the closed immersion $\overline{\{z\}} \to Z$ induces a morphism of spectral sequences $E(\overline{\{z\}}) \to E(Z)$ augmenting the filtration degree by p. That result further shows that we have the following.

Lemma 2.1. The map $K_nk(z) \overset{\cong}{\leftarrow} E_1^{0,-n}(\overline{\{z\}}) \to E_1^{p,n-p}(Z) \overset{\cong}{\to} \coprod_{z\in Z^{(p)}} K_nk(z)$ is the z-th canonical injection. #

Now suppose that f: $Y \to X$ is a finite surjective morphism of Noetherian schemes. Note that this implies that $\dim Y = \dim X$. Note also that, if $y \in Y^{(p)}$, then $f(y) \in X^{(p')}$, where $p' \geq p$ ([7], Thm. 20). Furthermore, k(y) is a finite extension of k(f(y)), so the transfer maps $K_nk(y) \to K_nk(f(y))$ are defined ([11], p. 111).

Now, as Quillen shows in [11], direct image induces a map f: $K'(Y) \to K'(X)$. We extend this result:

Theorem 2.2: f induces a morphism of spectral sequences $E(Y) \to E(X)$

compatible with the map on the abutments f_*. The components of the map on the E_1 level may be described as follows: Given $x \in X^{(p)}$, $y \in Y^{(p)}$, the component of $E_1^{pq}(Y) \to E_1^{pq}(X)$ from $K_n k(y)$ to $K_n k(x)$ is

$$\begin{cases} 0 & \text{if } x \neq f(y) \\ \text{the transfer map } K_n k(y) \to K_n k(x) & \text{if } x = f(y) \\ & \text{(where } n = -p-q) \end{cases}$$

Proof. Under our hypothesis on f, direct image restricts to exact functors $\mathcal{M}^p(Y) \to \mathcal{M}^p(X)$, hence induces exact functors $\mathcal{M}^p/\mathcal{M}^{p+1}(Y) \to \mathcal{M}^p/\mathcal{M}^{p+1}(X)$, which gives rise to the required morphism $E(Y) \to E(X)$.

Now, we have a commutative diagram

$$\begin{array}{ccc} \overline{\{y\}} & \to & \overline{\{f(y)\}} \\ \downarrow & & \downarrow \\ Y & \to & X \end{array}$$

where the horizontal morphisms are finite and surjective, and the vertical morphisms are closed immersions. This gives commutativity (up to natural isomorphism) in the lower square of the following diagram of exact functors:

$$\begin{array}{ccc} \mathcal{P}(k(y)) & \to & \mathcal{P}(k(f(y))) \\ \uparrow & & \uparrow \\ \mathcal{M}^0/\mathcal{M}^1(\overline{\{y\}}) & \to & \mathcal{M}^0/\mathcal{M}^1(\overline{\{f(y)\}}) \\ \downarrow & & \downarrow \\ \mathcal{M}^p/\mathcal{M}^{p+1}(Y) & \to & \mathcal{M}^p/\mathcal{M}^{p+1}(X) \end{array}$$

The arrows of the bottom square are defined by taking direct images. (Here we are using the fact that $\text{codim}(\overline{\{f(y)\}}) = p' \geq p$, in order to define the right-hand morphism.) The top vertical arrows are defined by taking stalks at y and f(y), respectively, while the top horizontal arrow is defined by restriction of scalars. Since f is finite, it follows easily that the top square commutes up to natural isomorphism, too. Applying K_n, we get a commutative diagram

$$\begin{array}{ccc}
K_n k(y) & \to & K_n k(f(y)) \\
\uparrow\cong & & \uparrow\cong \\
E_1^{0,-n}(\overline{\{y\}}) & \to & E_1^{0,-n}(\overline{\{f(y)\}}) \\
\downarrow & & \downarrow \\
E_1^{pq}(Y) & \to & E_1^{pq}(X) \\
\downarrow\cong & & \downarrow\cong \\
\coprod_{y\in Y^{(p)}} K_n k(y) & \to & \coprod_{x\in X^{(p)}} K_n k(x)
\end{array}$$

It follows from this diagram and from the lemma (applied to y) that the composition $K_n k(y) \to \coprod_{y\in Y^{(p)}} K_n k(y) \to \coprod_{x\in X^{(p)}} K_n k(x)$ is equal to the composition $K_n k(y) \to K_n k(f(y)) \stackrel{\cong}{\to} E_1^{0,-n}(\overline{\{f(y)\}}) \to E_1^{pq}(X) \stackrel{\cong}{\to} \coprod_{x\in X^{(p)}} K_n k(x)$. If $f(y) \in X^{(p)}$, then the lemma (applied to f(y)) gives the result. If $f(y) \in X^{(p')}$ with $p' > p$, then $\mathcal{M}^0/\mathcal{M}^1(\overline{\{f(y)\}}) \to \mathcal{M}^p/\mathcal{M}^{p+1}(X)$ factors through $\mathcal{M}^{p'}/\mathcal{M}^{p'+1}(X)$. Since $\mathcal{M}^{p'}/\mathcal{M}^{p'+1}(X) \to \mathcal{M}^p/\mathcal{M}^{p+1}(X)$ is zero, it follows that $E_1^{0,-n}(\overline{\{f(y)\}}) \to E_1^{pq}(X)$ is zero in this case. This completes the proof. #

Now suppose that $f: Y \to X$ is an arbitrary finite morphism of Noetherian schemes. By factoring f through its scheme-theoretic image, and using the result on closed immersions established in [15], we obtain:

<u>Corollary 2.3</u>. f induces a natural morphism of spectral sequences $E(Y) \to E(X)$ augmenting the filtration by $d = \mathrm{codim}(f(Y),X)$. Given $y \in Y^{(p)}$, $x \in X^{(p+d)}$, the component of $E_1^{pq}(Y) \to E_1^{p+d,q-d}(X)$ from $K_n k(y)$ to $K_n k(x)$ may be described as in the preceding result. #

<u>Remark</u>. A related result may be found in [3] (Thm. 7.22).
As a typical application of these results, we have

<u>Proposition 2.4</u>. Let $F \to F'$ be an extension of finite fields, and let $i: F(t) \to F'(t)$ denote the corresponding (finite) extension of rational function fields. Then the transfer maps $i: K_n F'(t) \to K_n F(t)$ are surjective for all $n > 1$.

Proof. By the theorems above there is a morphism of localization sequences

$$\begin{array}{ccccccccc}
\cdots \to K_{n+1}F'(t) & \to & \coprod_{\substack{\underline{m}' \\ \text{maximal}}} K_n k(\underline{m}') & \to & K'_n F'[t] & \to & K_n F'(t) & \to & \cdots \\
\downarrow & & \downarrow & & \downarrow & & \downarrow & & \\
\cdots \to K_{n+1}F(t) & \to & \coprod_{\substack{\underline{m} \\ \text{maximal}}} K_n k(\underline{m}) & \to & K'_n F[t] & \to & K_n F'(t) & \to & \cdots
\end{array}$$

By Theorem 1.4 of [2], F[t] and F'[t] are very clean, so each row breaks up into short exact sequences (in fact, by Theorem 4.5 of [14], split short exact sequences):

$$\begin{array}{ccccccccc}
0 \to & K'_n F'[t] & \to & K_n F'(t) & \to & \coprod_{\underline{m}'} K_{n-1} k(\underline{m}') & \to 0 \\
& \downarrow j & & \downarrow i & & \downarrow \ell & \\
0 \to & K'_n F[t] & \to & K_n F(t) & \to & \coprod_{\underline{m}} K_{n-1} k(\underline{m}) & \to 0
\end{array}$$

There is a diagram, commutative up to natural isomorphism:

$$\begin{array}{ccc}
\mathcal{P}(F') & \to & \mathcal{M}(F'[t]) \\
\downarrow & & \downarrow \\
\mathcal{P}(F) & \to & \mathcal{M}(F[t])
\end{array}$$

where the horizontal arrows are defined by extension of scalars, and the vertical arrows are defined by restriction of scalars. This gives rise to a commutative diagram

$$\begin{array}{ccc}
K_n F' & \to & K'_n F'[t] \\
h \downarrow & & \downarrow j \\
K_n F & \to & K'_n F[t]
\end{array}$$

By Quillen's homotopy theorem ([11], Thm. 8), the horizontal maps are isomorphisms. On the other hand, Quillen's computation of the K-theory of finite fields shows that h is surjective for $n > 0$ ([10], Remark following Thm. 8). It follows that j is surjective.

Now, given a maximal ideal $\underline{m}$ of F[t], pick any $\underline{m}'$ lying over it. Since $k(\underline{m}) \to k(\underline{m}')$ is again an extension of finite fields, it follows from Quillen's result and the theorems above that ℓ is surjective for $n > 1$. Finally, it follows from the snake lemma that i_* is surjective for such n. #

For completeness, we prove

<u>Proposition 2.5</u>. With notation as above, the map $i^*: K_nF(t) \to K_nF'(t)$ is injective for all $n \geq 0$. Similarly, $K_nF(t) \to K_n\overline{F}(t)$ is injective for all $n \geq 0$.

<u>Proof</u>. Since K-groups commute with filtered inductive limits, the second statement obviously follows from the first. To prove the first, observe that $F'[t]$ is flat over $F[t]$. By the functoriality of the localization sequence with respect to flat morphisms, we have a commutative diagram with (split) short exact rows:

$$\begin{array}{ccccccccc} 0 \to & K_n'F'[t] & \to & K_nF'(t) & \to & \coprod_{\underline{m}'} K_{n-1}k(\underline{m}') & \to 0 \\ & \uparrow s & & \uparrow i & & \uparrow t & \\ 0 \to & K_n'F[t] & \to & K_nF(t) & \to & \coprod_{\underline{m}} K_{n-1}k(\underline{m}) & \to 0 \end{array}$$

This time, extension of scalars gives a commutative diagram

$$\begin{array}{ccc} K_nF' & \to & K_n'F'[t] \\ \uparrow u & & \uparrow s \\ K_nF & \to & K_n'F[t] \end{array}$$

As noted above, the horizontal maps are isomorphisms. On the other hand, Quillen has shown that u is injective ([10], Thm. 8). Hence s is injective. To analyze the map t, we need Prop. 1.2 of [15]. That result shows that, for a given $\underline{m}$, the only non-trivial components of $K_{n-1}k(\underline{m}) \to \coprod_{\underline{m}'} K_{n-1}k(\underline{m}')$ correspond to those $\underline{m}'$ dividing $\underline{m}$, and that for such an $\underline{m}'$, the map $K_{n-1}k(\underline{m}) \to K_{n-1}k(\underline{m}')$ is that induced by the finite extension $k(\underline{m}) \to k(\underline{m}')$. (One needs to observe that, by separability, there is no ramification.) It follows again from Quillen's result that t is injective. The snake lemma then shows that i is injective. #

3. Gersten's Conjecture for Some Regular Local Rings.

Let R be a discrete valuation ring (DVR) with maximal ideal $\underline{m}$. Note that Gersten's Conjecture is valid for R if and only if the map $g: K_n(R/\underline{m}) \to K'_n(R)$, induced by restriction of scalars, is zero for all $n \geq 0$. We recall that the conjecture is known to be valid for any equicharacteristic DVR [13]. In the unequal characteristic case, the only cases which have been established so far are for DVR's with finite residue class field ([2], Thm. 1.3), and for the local rings of a polynomial ring in one variable over such a ring ([16], Thm. 2.4).

We can extend these results to prove that the conjecture is valid for any DVR whose residue class field is either an algebraic extension of a finite field or a rational function field in one variable over such a field. The techniques involved are totally different from those used in this paper and deserve consideration in a separate article, so we shall not present the proofs here. We can, however, establish special cases of the second result by using the sort of arguments employed in the proofs of Propositions 2.4 and 2.5:

Proposition 3.1. Let R be a DVR of characteristic 0, with residue class field $\mathbb{F}_{p^r}(t)$ (p a prime); let e denote its index of ramification over $Z_{(p)}$. Then $es(\operatorname{Im} g) = 0$ for some s dividing r. If R is Henselian, then we can take s = 1; in particular, the conjecture is valid in this case if R is unramified. For arbitrary R, Gersten's Conjecture will be valid for R if e and r are powers of p.

Proof. Let $f \in \mathbb{Z}[X]$ be a monic polynomial of degree r whose reduction mod p, $\bar{f}$, is irreducible; hence f is irreducible over $\mathbb{Z}$. Let f_1 be any monic irreducible factor of f in R[X], of degree s, say; we have $s|r$.

Define $A = \mathbb{Z}_{(p)}[X]/(f)$, $R_1 = R[X]/(f_1)$. Now, $\bar{f} \in \mathbb{F}_p[X]$ is irreducible, but splits into distinct linear factors in $\mathbb{F}_{p^r}[X]$. Hence $\overline{f_1} \in \mathbb{F}_{p^r}(t)[X]$ splits into distinct linear factors also.

A standard result (cf. [12], I-6) then shows that A is an unramified DVR with residue class field $\mathbb{F}_p[X]/(\bar{f}) \cong \mathbb{F}_{p^r}$, and that R_1 is a semi-local Dedekind domain with s maximal ideals $\underline{m}_1,\ldots,\underline{m}_s$, say, corresponding to the s linear factors of $\overline{f_1}$; furthermore, $R_1/\underline{m}_i \cong R/m \cong \mathbb{F}_{p^r}(t)$ for each i. The canonical map $A \to R_1$ is an injection, so we may regard A as a subring of R_1. (This is a consequence of the following standard argument: If a pair of polynomials are relatively prime over $\mathbb{Z}_{(p)}[X]$, then they are relatively prime in $\mathbb{Q}[X]$, and thus relatively prime in L[X] (L = quotient field of R), hence relatively prime in R[X].)

Now, R_1 is finite over R, so by the main result of the preceding section, there is a commutative diagram

$$\begin{array}{ccc} \coprod_{i=1}^{s} K_n(\mathbb{F}_{p^r}(t)) & \xrightarrow{g_1} & K'_n(R_1) \\ \downarrow h & & \downarrow \\ K_n(\mathbb{F}_{p^r}(t)) & \xrightarrow{g} & K'_n(R) \end{array}$$

where h just adds up components.

Next, lift $t \in R/\underline{m}$ to an element T of $R \subset R_1$. It is easily seen that T is algebraically independent of A, so that the subring A[T] is a polynomial ring in one variable over A. It is also clear that elements of A[T] - (p) are units in R_1, so that R_1 contains $R_2 = (A[T])_{(p)}$. R_2 is a DVR, so, since R_1 is torsion-free as an R_2-module, R_1 is flat over R_2. By the functoriality of the localization sequence with respect to flat morphisms, there is a commutative diagram

$$\begin{array}{ccc} K_n(\mathbb{F}_{p^r}(t)) & \xrightarrow{g_2} & K'_n(R_2) \\ \downarrow d & & \downarrow \\ \coprod_{i=1}^{s} K_n(\mathbb{F}_{p^r}(t)) & \xrightarrow{g_1} & K'_n(R_1) \end{array}$$

where, by Prop. 1.2 of [15], d is the diagonal map composed with multiplication by e.

Now, $g_2 = 0$ by the main result of [13], so, putting the two diagrams together, we see that g annihilates $es \cdot (K_n(\mathbb{F}_{p^r}(t)))$, as claimed. If R is Henselian, then, since $\bar{f}$ splits into distinct monic factors, it follows that f splits over R, and we can take s = 1.

For any DVR, the maps $K_n(R/\underline{m}) \to K'_nR = K_nR$ are zero for n = 0,1, so in proving the last statement we may assume n > 1. Now, for any n > 0 we have

$$K_n(\mathbb{F}_{p^r}(t)) = K_n(\mathbb{F}_{p^r}) \oplus \coprod_{\substack{h \text{ monic,} \\ \text{irreducible}}} K_{n-1}(\mathbb{F}_{p^r}[t]/(h))$$

([14], Thm. 4.5). But, if n > 1, Quillen's computation of the K-theory of finite fields [10] then shows that $K_n(\mathbb{F}_{p^r}(t))$ is the direct sum of finite cyclic groups whose orders are of the form $p^u - 1$. Hence if e and s are powers of p, then $K_n(\mathbb{F}_{p^r}(t))$ is divisible by es, from which it follows that Im g = 0 in this case. #

<u>Remark</u>. As is clear from the proof, if one knows that R dominates a DVR with residue class field $\mathbb{F}_{p^r}$ (as in the Henselian case), then one can take s = 1; if R is known to be unramified over this DVR, then one can conclude that Gersten's Conjecture is valid for R.

Although Gersten's Conjecture is still far from settled, especially in the unequal characteristic case, the following result shows that we can already extract some useful information about regular local rings of higher dimension from the one-dimensional cases which have already been established.

<u>Proposition 3.2</u>. Let R be a regular local ring of dimension greater than 1, and $\underline{m}$ its maximal ideal. Then the transfer maps $K_n(R/m) \to K_n(R)$ are zero under any of the following hypotheses:

(1) R is equicharacteristic;

(2) $R/\underline{m}$ is algebraic over a finite field or a rational function field in one variable over such a field;

(3) R is of unequal characteristic and unramified.

<u>Proof</u>. Choose a regular system of parameters $\{x_1, \ldots, x_n\}$. Let $\underline{p} = (x_1, \ldots, x_{n-1})$. Then $\underline{p}$ is a prime ideal of height n - 1, and $R/\underline{p}$ is

a DVR with field of fractions $k(\underline{p})$ and residue class field $R/\underline{m}$. Now consider the connecting homomorphism in the localization sequence for $\mathcal{M}^{n-1}$ and $\mathcal{M}^{n}$ ([11], Thm. 5):

$$\begin{array}{ccccc} \cdots \to K_{j+1}(\mathcal{M}^{n-1}/\mathcal{M}^{n}) & \overset{\partial}{\to} & K_j(\mathcal{M}^{n}) & \to & K_j(\mathcal{M}^{n-1}) \to \cdots \\ \downarrow\cong & & \downarrow\cong & & \\ \coprod_{ht\,\underline{p}=n-1} K_{j+1}k(\underline{p}) & & K_j(R/\underline{m}) & & \end{array}$$

By Prop. 1.4 of [15], the $\underline{p}$-th component of this map, $\partial_{\underline{p}}$, may be identified with the connecting homomorphism in the corresponding localization sequence for the DVR $R/\underline{p}$:

$$\cdots \to K_{j+1}k(\underline{p}) \xrightarrow{\partial_{\underline{p}}} K_j(R/\underline{m}) \to K_j(R/\underline{p}) \to \cdots$$

(cf. the proof of Prop. 1.1).

In cases (1) and (2), $R/\underline{p}$ satisfies the same hypothesis that R does, so, as remarked at the beginning of this section, $R/\underline{p}$ is known to be very clean. Thus $\partial_{\underline{p}}$ is surjective, hence so is ∂. It follows from the localization sequence that the composition $K_j(R/\underline{m}) \to K_j(\mathcal{M}^{n-1}) \to K_j'(R) \overset{\cong}{\leftarrow} K_j(R)$ is zero.

We proceed in the same way in case (3), except that, by the hypothesis, we can arrange that $x_1 = p$ (where p is the characteristic of $R/\underline{m}$). Then $R/\underline{p}$ contains the field $\mathbb{F}_p$, and the argument used in case (1) applies. #

We shall give an application of this result in the next section.

4. The Topological Filtration of $K_\cdot(X)$

Let R be a Dedekind ring, with field of fractions F. Put $SK_n(R) = \ker(K_nR \to K_nF)$. From the localization sequence we see that $SK_n(R) = \bigcap_{\underline{m}} \mathrm{Im}(K_n(R/\underline{m}) \to K_nR)$, where $\underline{m}$ ranges over the maximal ideals of R. Thus R is very clean if and only if $SK_n(R) = 0$ for all $n \geq 0$. Now, $SK_0(R) \cong \tilde{K}_0(R)$ may be identified with the ideal classgroup C(R), so R is clean if and only if it is a PID. However, even if R is a PID, an example of Bass shows that one can have $SK_1(R) \neq 0$ [1]. Thus a Dedekind ring need not be very clean even if it is a PID. There is strong evidence, however, that Dedekind rings satisfy a somewhat weaker condition, which generalizes part of Cor. 1.11 of [8]. More generally, let X be a one-dimensional regular scheme, with function field k(X), and put $SK_n(X) = \ker(K_nX \to K_nk(X))$.

Conjecture A. $(SK_\cdot(X))^2 = 0$ in the graded ring $K_\cdot(X)$.

As we shall show, this is a special case of a more general conjecture, which, for K_0, was first considered by Grothendieck. Before formulating this, we point out that, despite the result of Bass mentioned above, for some PID's R, one can show that $SK_\cdot(R) = 0$. One case is that in which R is the polynomial ring in one variable over a field ([2], Thm. 1.4). We present a related case below; in fact, we prove a little more.

Theorem 4.1. Let R be a PID containing a field k, such that the canonical map $k \to R/\underline{m}$ is an isomorphism for all maximal ideals $\underline{m}$. Then the localization sequence for R breaks up into split short exact sequences. In particular, $SK_\cdot(R) = 0$ (i.e., R is very clean).

Proof. The key here is that we can resolve $R/\underline{m}$-vector spaces in an exact functorial way. For let $\underline{m}$ be generated by the element a, and consider the short exact sequence

$$(*) \qquad 0 \to R \xrightarrow{a} R \to R/\underline{m} \to 0$$

where a· denotes multiplication by a. Given an $R/\underline{m}$-vector space V, we may tensor (*) over k with V to obtain an exact sequence of R-modules:

(**) $$0 \to R \otimes_k V \to R \otimes_k V \to (R/\underline{m}) \otimes_k V \to 0.$$

Since $k \to R/\underline{m}$ is an isomorphism, we may identify the module on the right with V. Let r: $\mathcal{P}(R/\underline{m}) \to \mathcal{M}(R)$ be defined by restriction of scalars; let i: $\mathcal{P}(R) \to \mathcal{M}(R)$ be the inclusion; and let f: $\mathcal{P}(R/\underline{m}) \to \mathcal{P}(R)$ be the exact functor $V \mapsto R \otimes_k V$. Then (**) represents an exact sequence of exact functors: $0 \to if \to if \to r \to 0$. The desired conclusion then follows from Thm. 4.7 of [14]. #

Suppose now that X is any scheme. Tensor product defines a bi-exact functor $\mathcal{P}(X) \times \mathcal{P}(X) \to \mathcal{P}(X)$, which, as a consequence of Waldhausen's construction of products in K-theory ([20]; cf. also [4]), gives rise to a graded ring structure on $K_.(X)$. Similarly, if X is Noetherian, the pairing $\mathcal{P}(X) \times \mathcal{M}(X) \to \mathcal{M}(X)$ makes $K'_.(X)$ a graded module over $K_.(X)$.

Given a Noetherian scheme X, the images of the maps $K_n(\mathcal{M}^i(X)) \to K'_n(X)$ define a filtration of $K'_n(X)$. If X is regular, then by composing with the inverse of the Cartan isomorphism $K_n(X) \to K'_n(X)$, we obtain a filtration of $K_n(X)$: $K_n(X) = F^0K_n(X) \supset F^1K_n(X) \supset \cdots$. For $n = 0$, Grothendieck has termed this the <u>topological filtration</u> [22], and we shall adopt this terminology for higher n, as well. Note that if X is very clean, then $F^iK_n(X) = 0$ for $i > 0$, $n \geq 0$.

In the study of intersection theory on an algebraic variety X, one forms the Chow group A(X) - this is the graded abelian group of cycles modulo rational equivalence. If X is smooth and quasi-projective, then, as a consequence of Chow's Moving Lemma, this group has a natural graded ring structure induced by intersection of cycles. One of the goals of the research culminating in the publication of <u>SGA 6</u> was to find a suitable substitute for the Chow ring for an arbitrary regular scheme X. It was hoped that the topological filtration could be shown to be compatible with the multiplication in $K_.(X)$, so that the associated graded group would have a graded ring structure, which might provide the desired substitute for the Chow ring. However, Gothendieck was only able to establish this for schemes for which the Moving Lemma is valid - in other words, schemes for which one already has the Chow ring at hand! Consequently, a second filtration of $K_.(X)$ was introduced, in some ways less satisfactory than the first, and it is the associated graded group of this filtration that is employed in SGA 6. (It is now known that a cohomology theory defined in terms of higher K-theory may provide the appropriate generalization of the Chow group; cf. [3].)

We generalize Grothendieck's question to higher n:

Conjecture B: If X is a regular scheme, then
$F^iK_m(X) \cdot F^jK_n(X) \subset F^{i+j}K_{m+n}(X)$.

As remarked above, Grothendieck established the conjecture whenever X is a smooth quasi-projective algebraic variety and $m = n = 0$. Note that if $\dim X = 1$, then, as a consequence of Quillen's localization sequence, $F^1K_.(X) = SK_.(X)$. Since X is one-dimensional, $F^2K_.(X) = 0$, so Conjecture B in this case reduces to Conjecture A.

In order to establish some cases of Conjecture B, we need to introduce another conjecture which, as will be clear from the theorem below, is related to Gersten's Conjecture. Given a morphism $f: Y \to X$ of regular schemes, inverse image defines an exact functor $f^!: \mathcal{P}(X) \to \mathcal{P}(Y)$, and thus a map $f^*: K_.(X) \to K_.(Y)$.

Conjecture C: $f^*(F^iK_m(X)) \subseteq F^iK_m(Y)$.

In [22], Grothendieck verified Conjecture C for $m = 0$, whenever X and Y are smooth quasi-projective varieties; again, the proof depends on the Moving Lemma. The following is a partial verification of Conjecture C in the case of a closed immersion:

<u>Theorem 4.2.</u> Let X be a regular scheme, Y a regular closed subscheme, $f\colon Y \to X$ the inclusion, y the generic point of Y. Let $m \geq 0$ and $j > 0$ be given integers, and assume that $F^jK_m(\mathcal{O}_{X,y}) = 0$. Then $f^*(F^j K_m(X)) \subset F^1K_m(Y)$.

Proof. Consider the following diagram of K-groups:

$$\begin{array}{ccccccccc}
 & & & & & & & & K_m(\mathcal{M}^1(Y)) \\
 & & & & & & & & \downarrow \\
K_m(\mathcal{M}^j(X)) & \to & K'_m(X) & \xleftarrow{\cong} & K_m(X) & \xrightarrow{f^*} & K_m(Y) & \xrightarrow{\cong} & K'_m(Y) \\
\downarrow & & \downarrow & & \downarrow & & \downarrow & & \downarrow \\
K_m(\mathcal{M}^j(\mathcal{O}_{X,y})) & \to & K'_m(\mathcal{O}_{X,y}) & \xleftarrow{\cong} & K_m(\mathcal{O}_{X,y}) & \to & K_m(k(y)) & = & K_m(k(y))
\end{array}$$

The horizontal maps in the left-hand square are those induced by the inclusions $\mathcal{M}^j \to \mathcal{M}^0$; the horizontal maps in the second and fourth squares are the Cartan maps; the horizontal maps in the third square are induced by taking inverse images. The vertical maps in the four squares are induced by taking stalks at y. (As observed by Quillen [[11], 7.5.2), a flat morphism carries $\mathcal{M}^j$ to $\mathcal{M}^j$, so the left-most map is defined.) The right-most column is part of the localization sequence associated to $\mathcal{M}^1(Y) \to \mathcal{M}^0(Y)$, hence is exact. The diagram is clearly commutative.

By hypothesis, the bottom horizontal arrow in the first square is the zero map. But then commutativity of the diagram shows that $f^*(F^jK_m(X))$ lies in the kernel of the map $K_m(Y) \to K_m(k(y))$; by commutativity of the right-hand square and exactness of the right-most column, this kernel is precisely $F^1K_m(Y)$. #

Before using this theorem to derive certain cases of Conjecture B, we need to discuss an appropriate version of the projection formula. For our purposes, it will suffice to consider the following situation. Let $f\colon Y \to X$ be a finite morphism of regular schemes. Since f is finite, direct image induces an exact functor $f_!\colon \mathcal{P}(Y) \to \mathcal{M}(X)$, hence a map $K_.(Y) \to K'_.(X)$ (cf. [11], 7.2.8). Since X is regular, we may compose this map with the inverse of the Cartan isomorphism to obtain a map $f_*: K_.(Y) \to K_.(X)$. The projection formula is the statement that f_* is a homomorphism of graded $K_.(X)$-modules. In other words, if $a \in K_n(Y)$, $b \in K_m(X)$, then in $K_{n+m}(X)$, we have

$$f_*(a) \cdot b = f_*(a \cdot f^*(b)).$$

This formula is an immediate consequence of the functorial properties of Waldhausen's constructions, and the following diagram of functors, which is commutative up to natural isomorphism:

$$\begin{array}{ccccc}
\mathcal{P}(Y) \times \mathcal{P}(X) & \xrightarrow{f_! \times 1} & \mathcal{M}(X) \times \mathcal{P}(X) & \leftarrow & \mathcal{P}(X) \times \mathcal{P}(X) \\
\downarrow {\scriptstyle 1 \times f^!} & & & & \\
\mathcal{P}(Y) \times \mathcal{P}(Y) & & \downarrow \otimes & & \downarrow \otimes \\
\downarrow \otimes & & & & \\
\mathcal{P}(Y) & \xrightarrow{f_!} & \mathcal{M}(X) & \longleftarrow & \mathcal{P}(X)
\end{array}$$

We first note a special case of Conjecture B.

<u>Lemma 4.3</u>. Let X be a regular scheme. For all $m, n, i \geq 0$, $F^i K_m(X) \cdot K_n(X) \subset F^i K_{m+n}(X)$.

<u>Proof</u>. Tensor product defines a biexact functor $\mathcal{M}^i \times \mathcal{P} \to \mathcal{M}^i$, which fits into the following commutative diagram:

$$\begin{array}{ccccc}
\mathcal{M}^i \times \mathcal{P} & \to & \mathcal{M} \times \mathcal{P} & \leftarrow & \mathcal{P} \times \mathcal{P} \\
\downarrow & & \downarrow & & \downarrow \\
\mathcal{M}^i & \to & \mathcal{M} & \leftarrow & \mathcal{P}
\end{array}$$

The conclusion is a consequence of this diagram and the functorial nature of Waldhausen's constructions. #

<u>Lemma 4.4</u>. Let X be a regular scheme, Y a regular closed subscheme of codimension i, $f_*: K_m(Y) \to K_m(X)$ the map defined above. For all $m, \ell \geq 0$, $f_*(F^\ell K_m(Y)) \subset F^{i+\ell} K_m(X)$.

<u>Proof</u>. Direct image defines an exact functor $\mathcal{M}(Y) \to \mathcal{M}(X)$ - we have already considered the restriction of this functor to $\mathcal{P}(Y)$ in defining the map f_*. Furthermore, this functor restricts to an exact functor $\mathcal{M}^\ell(Y) \to \mathcal{M}^{i+\ell}(X)$. These functors fit into a commutative diagram, from which the desired result follows:

$$\begin{array}{ccc} \mathcal{M}^{\ell}(Y) & \to & \mathcal{M}(Y) \\ \downarrow & & \downarrow \\ \mathcal{M}^{i+\ell}(X) & \to & \mathcal{M}(X) \end{array}$$ #

We introduce some convenient notation. Let X be a regular scheme, Y a regular closed subscheme of codimension i. Let us put $F^i_Y K_n(X) = f_*(K_n(Y))$. Note that it follows from Lemma 4.4 that $F^i_Y K_n(X) \subset F^i K_n(X)$. Our main result on Conjecture B is:

Theorem 4.5. Let $f: Y \to X$ be a closed immersion of regular schemes, with Y of codimension i in X. Let $m, j \geq 0$ be given integers, and assume that there exists an $\ell \geq 0$ such that $f^*(F^j K_m(X)) \subset F^{\ell} K_m(Y)$. Then $F^i_Y K_n(X) \cdot F^j K_m(X) \subset F^{i+\ell} K_{n+m}(X)$ for all $i, n \geq 0$. If, furthermore, $f^*(F^j K_m(X)) = 0$, then $F^i_Y K_n(X) \cdot F^j K_m(X) = 0$.

Proof. Let $a \in K_n(Y)$, $b \in F^j K_m(X)$. We need to prove that $f_*(a) \cdot b \in F^{i+\ell} K_{n+m}(X)$. By the projection formula, $f_*(a) \cdot b = f_*(a \cdot f^*(b))$. By hypothesis, $f^*(b) \in F^{\ell} K_m(Y)$, so it follows from Lemma 4.3 that $a \cdot f^*(b) \in F^{\ell} K_m(Y)$. Lemma 4.4 then gives the required result. The same argument also proves the second statement. #

As a first consequence of this, we see that Conjecture C almost implies Conjecture B:

Corollary 4.6. If Conjecture C is valid, then, with notation as above, $F^i_Y K_n(X) \cdot F^j K_m(X) \subset F^{i+j} K_{n+m}(X)$. #

Next we note that, by Theorem 4.2, Gersten's Conjecture gives partial information about Conjecture C, hence partial information about Conjecture B:

Corollary 4.7. With notation as above, if Gersten's Conjecture is valid for the local rings of X, then $F^i_Y K_n(X) \cdot F^1 K_m(X) \subset F^{i+1} K_{n+m}(X)$. In particular, this is true if X is a nonsingular algebraic variety. #

In certain cases, we do not need the full force of Gersten's Conjecture in order to draw useful conclusions:

Corollary 4.8. Let X be a regular scheme, x a closed point of codimension i; let $\{x\}$ denote the corresponding reduced induced closed subscheme. Assume that $\mathcal{O}_{X,x}$ is equicharacteristic, or that $k(x)$ is either an algebraic extension of a finite field or a rational function field in one variable over such a field. Then for any $j > 0$, $m,n \geq 0$, $F^i_{\{x\}}K_n(X) \cdot F^j K_m(X) = 0$. Consequently, if $\dim X = d$, and if each closed point of codimension d satisfies one of the conditions above, then $F^d K_n(X) \cdot F^j K_m(X) = 0$.

Proof. $F^1(\{x\}) = 0$, so the first statement follows from Prop. 3.2, Theorem 4.2, and the second part of Theorem 4.5. If $\dim X = d$, then $\mathfrak{m}^{d+1}(X) = 0$, so, as noted in Section 2, $K_n(\mathfrak{m}^d(X)) \cong \coprod_{x\in X^{(d)}} K_n k(x)$. It follows that $F^d K_n(X)$ is generated by the subgroups $F^d_{\{x\}}K_n(X)$ as x ranges over $X^{(d)}$; thus the second statement follows from the first. #

Finally, we specialize to the one-dimensional case to get a partial confirmation of Conjecture A:

Corollary 4.9. Let X be a one-dimensional regular scheme. Assume that each closed point x satisfies one of the conditions of Cor. 4.8. Then $(SK_.(X))^2 = 0$. For arbitrary X, we at least have $SK_.(X) \cdot SK_m(X) = 0$ for $m \leq 2$.

Proof. The first part is a special case of Cor. 4.8. For the second part, note that for any discrete valuation ring R, $F^1K_m(R) = 0$ for $m \leq 2$. For $m = 0,1$, this is simply a consequence of the fact that R is local; for $m = 2$, it is a result of Dennis and Stein [21]. The result then follows from Theorems 4.2 and 4.5, as in the proof of Cor. 4.8. #

As the arguments above indicate, in order to establish Conjecture A for a given one-dimensional regular scheme X, it suffices to establish the conclusion of Thm. 4.2 for $Y = \{x\}$, x any closed point of X; this is equivalent to verifying Conjecture C for the map $\{x\} \to X$. Our last result shows that this problem reduces to a local one.

Proposition 4.10. Let X be a one-dimensional regular scheme; x a closed point; f: $\{x\} \to X$ the inclusion. Let p: Spec $k(x) \to$ Spec $\mathcal{O}_{X,x}$ denote the canonical map. Then Conjecture C is valid for f if and only if it is valid for p.

Proof. We consider the commutative diagram employed in the proof of Theorem 4.2, which, in this case, takes the following form:

$$\begin{array}{ccccccc}
\coprod_{y\in X^{(1)}} K_n k(y) & \to & K'_n(X) & \xleftarrow{\cong} & K_n(X) & \xrightarrow{f^*} & K_n(\{x\}) \\
\downarrow & & \downarrow & & \downarrow & & \downarrow\cong \\
K_n k(x) & \to & K'_n(\mathcal{O}_{X,x}) & \xleftarrow{\cong} & K_n(\mathcal{O}_{X,x}) & \xrightarrow{p^*} & K_n(k(x))
\end{array}$$

Note that Conjecture C for f (resp., p) is equivalent to the statement that the composition along the top (resp., bottom) row is zero. Since it is clear that, for $y \neq x$, the y-th component of the top composition is zero, and since the map on the left is projection on the x-th component, the equivalence of the two statements is clear. #

In terms of the notation introduced earlier, Conjecture C for p is the statement that $p^*p_* = 0$. This is weaker than Gersten's Conjecture for $\mathcal{O}_{X,x}$ (which is that $p_* = 0$). On the other hand, at the moment we do not know how to establish this result in any cases where Gersten's Conjecture itself has not been established (cf. Cor. 4.9).

(Added in proof) Gillet ("Comparison of K-theory Spectral Sequences, with Applications," these proceedings), has independently considered the questions of this section. By using different methods, he is able to prove (cf. Cor. 2.4) that our conjectures are implied by Gersten's Conjecture. This is an improvement on our Thm. 4.2 and Cor. 4.7.

5. Gersten's Conjecture for Some Nonregular Rings.

In a remark preceding his proof of Gersten's Conjecture for regular semilocal rings essentially of finite type over a field, Quillen states that it seems unreasonable to suppose that the conjecture is valid for any larger class of local rings than the regular local rings. Our first result shows, however, that in the one-dimensional case, examples of interesting non-regular local domains for which the conjecture is valid, are common. (We should point out that it is easy to construct examples of one-dimensional local domains which are not even clean ([18], Example 1).)

We establish the following notation. R will denote a one-dimensional Noetherian local integral domain, with maximal ideal $\underline{m}$, and field of fractions F. $\bar{R}$ will denote the integral closure of R in F. It is well known that $\bar{R}$ must be Noetherian, semilocal, and one-dimensional ([9], 33.2,33.10); consequently, it is a semilocal PID. We shall denote by $\underline{m}_1,\ldots,\underline{m}_s$ the maximal ideals of $\bar{R}$. If R is complete, another basic result ([9], 32.1,17.8,16.8,30.5) shows that $\bar{R}$ is complete and local, with maximal ideal $\underline{m}'$, say.

Theorem 5.1. R is very clean if it satisfies any of the following hypotheses:

(1) R is essentially of finite type over a field k, and for some i, the canonical map $R/\underline{m} \to R/\underline{m}_i$ is an isomorphism;

(2) R is essentially of finite type over $\mathbb{Z}$ or over a finite field, and $\mathrm{GCD}(\dim_{R/\underline{m}} R/\underline{m}_i) = 1$;

(3) R is complete, $R/\underline{m} \to \bar{R}/\underline{m}'$ is an isomorphism, and either R is equicharacteristic or $R/\underline{m}$ is finite.

Proof. The key here is that in each case, $\bar{R}$ is known to be finite over R by results of Nagata ([9], Theorems 37.5 and 32.1), so we can use the theorem proved in Section 2 to compare the localization sequences of R and $\bar{R}$:

$$\begin{array}{ccccccc}
\cdots \to K_{n+1}(F) & \xrightarrow{\bar{\partial}} & \coprod_{i=1}^{s} K_n(\bar{R}/\underline{m}_i) & \to & K'_n(\bar{R}) & \to & K_n(F) \to \cdots \\
\| & & \downarrow t & & \downarrow & & \| \\
\cdots K_{n+1}(F) & \xrightarrow{\partial} & K_n(R/\underline{m}) & \to & K'_n(R) & \to & K_n(F) \to \cdots
\end{array}$$

In case (1), Quillen has shown that $\bar{R}$ is very clean ([11], Thm. 7.5.11). Thus, $\bar{\partial}$ is surjective. It follows from the hypothesis and from the explicit description of the map t given in Theorem 2.2, that t is surjective, and thus that ∂ is surjective.

In case (2), Gersten has shown that $\bar{R}$ is very clean in [2]. (Gersten only establishes the local case, but as he points out in the Addendum, an older result of Swan shows that the theorem is valid more generally in the semilocal case.) The residue class fields are finite fields in this case, so by Theorem 2.2 and a result of Quillen ([10], Remark following Thm. 8), t is surjective for $n > 0$. The hypothesis that $GCD(\dim_{R/\underline{m}} R/\underline{m}_i) = 1$ is obviously equivalent to the condition that t be surjective for $n = 0$.

In case (3), as pointed out above, $\bar{R}$ is local. If R is equicharacteristic, then $\bar{R}$ is very clean by [13]. If $R/\underline{m}$ is finite, Gersten's result applies. The hypothesis that $R/\underline{m} \to \bar{R}/\underline{m}'$ is an isomorphism then finishes the argument. #

<u>Remark</u>. A number of related results can be proved in this way. For example, in cases (1) and (2), one could obviously allow R to be semi-local. As another example, if $\bar{R}$ is known to be local (as in case (3)), then, under appropriate hypotheses, $\bar{\partial}$ will have a section [14], so, if t is an isomorphism, ∂ will have a section, as well.

<u>Corollary 5.2</u>. Let R be a local ring of any curve defined over an algebraically closed field. Then R is very clean. #

As another example along the same lines, it follows from (1) that, for any field k, the local rings of the plane cuspidal cubic $y^2 = x^3$ and the plane nodal cubic $y^2 = x^2(x+1)$ are very clean; as indicated in the remark above, in the first case one can also show that the localization sequence breaks up into split short exact sequences. It follows from (3) that the Conjecture is valid for the completions of the local rings of the first curve; again, one can obtain a splitting result, as well.

We can use the same technique to derive an analogous global result for certain affine curves. Part of this result was used in Section 3 of [18].

<u>Proposition 5.3</u>. Let k be a field, X an affine curve defined over k. Assume that the normalization of X is isomorphic to the affine line $\mathbb{A}^1_k$, and that for each point $x \in X$, the canonical map of residue class fields $k(x) \to k(\bar{x})$ is an isomorphism for some $\bar{x}$ lying over x. Then X is very clean. If, furthermore, there is only one point of $\mathbb{A}^1_k$ lying over each point of X, then the localization sequence for X breaks up into split short exact sequences, and the canonical map $K_n(k) \to K'_n(X)$ is an isomorphism for all $n \geq 0$.

<u>Proof</u>. $\mathbb{A}^1_k$ is finite over X, so Theorem 2.2 gives a diagram with exact rows and commutative squares:

$$\begin{array}{ccccccccc}
\cdots \to & K_{n+1}(k(t)) & \to & \coprod_{\bar{x} \in \mathbb{A}^1_k} K_n k(x') & \to & K'_n k[t] & & \to & K_n(k(t)) \to \cdots \\
 & \| & & \downarrow f & & \downarrow g \;\; \nwarrow h & K_n k & \nearrow & \| \\
 & & & & & \swarrow i & \downarrow v & \searrow & \\
\cdots \to & K_{n+1}(k(t)) & \to & \coprod_{x \in X} K_n k(x) & \to & K_n' X & & \to & K_n(k(t)) \to \cdots
\end{array}$$

$\mathbb{A}^1_k$ is very clean by Theorem 1.4 of [2]. Under our basic hypothesis, the map f will be surjective, so that the argument used in the proof of the preceding theorem shows that X is very clean.

If the additional hypothesis is satisfied, then f will be an isomorphism, and by the Five Lemma, so will g. Three of the triangles in the right-hand square commute by functoriality of K'_n with respect to flat morphisms. Since h is an isomorphism by Quillen's homotopy theorem ([11], Thm. 8), it suffices to show that the remaining triangle is commutative. Since X is very clean, v is monic, so it suffices to show that $vi = vhg$. But this follows from commutativity of the diagram. #

Remarks. 1) By using the latter half of Theorem 4.5 and a similar argument, we see that if X is an affine curve over an algebraically closed field, and if the coordinate ring of the normalization of X is a PID, then X is very clean. 2) Let R be any one-dimensional Noetherian integral domain whose field of fractions F is a global field. One can show that $\bar{R}$ is finite over R, so we may compare the localization sequences. Soulé has proved that $SK_n(\bar{R}) = 0$ for $n > 0$. (This is essentially Thm. 3 of [19]. However, there the result is only established up to 2-torsion, and, moreover, there is a mistake in the proof. Soulé asserts that the mistake can be remedied, and in such a way that the argument can be completed, as well.) By the same argument used in the proof of part (2) of Theorem 5.1, it follows that $K'_n(R) \to K_n(F)$ is injective for $n > 0$; if $\bar{R}$ is a PID, then sometimes this is valid for $n = 0$, also.

As examples, we see that for any field k, the rings $R_1 = k[X,Y]/(Y^2 - X^3)$ and $R_2 = k[X,Y]/(Y^2 - X^2(X+1))$ are very clean. Furthermore $K_n k \to K'_n(R_1)$ is an isomorphism for all $n \geq 0$.

Finally, we construct an example of a two-dimensional local integral domain which is very clean, but not regular. Let k be an algebraically closed field, and R any one-dimensional, non-normal, local integral domain essentially of finite type over k. Let $\underline{m}$ denote the maximal ideal of R. Put $A = R[t]_M$, where M is any maximal ideal of R[t] of height 2.

Theorem 5.4. A is very clean.

Proof. The proof is a simple modification of the proof of the main result of [16], so we shall only offer a sketch. A reference to that argument shows that it suffices to do the following: (1) Given any height 2 prime M' of R[t], find a height 1 prime $\underline{p} \subset M'$ such that $R[t]_{M'}/\underline{p}R[t]_{M'}$ is very clean, and such that M' is the only height 2 prime containing $\underline{p}$. (2) Show that R is very clean.

We have $M' = (\underline{m}, f)$, where f is a monic polynomial irreducible modulo $\underline{m}$. M' contains some minimal prime $\underline{p}$ of (f); we have ht $\underline{p} = 1$ by Krull's Principal Ideal Theorem. Clearly, M' is the only height 2 prime containing $\underline{p}$. Furthermore, both R and $R[t]_{M'}/\underline{p}R[t]_{M'}$ are one-dimensional local integral domains essentially of finite type over k. By part (a) of Theorem 5.1, each of them is very clean. #

The examples provided by Theorem 5.4 are somewhat special, since they are simply the local rings of the "cylinder" over a one-dimensional example. A ring A obtained in this way will never be normal, because R itself isn't. It would be interesting to know whether there are two-dimensional normal (but not regular) examples.

References

1. H. Bass, Some problems in "classical" algebraic K-theory in "Classical" Algebraic K-Theory and Connections with Arithmetic, Lecture Notes in Mathematics, Vol. 342, Springer-Verlag, New York, 1973.

2. S. Gersten, Some Exact Sequences in the Higher K-theory of Rings, in Higher K-Theories, Lecture Notes in Mathematics, Vol. 341, Springer-Verlag, New York, 1973.

3. H. Gillet, Riemann Roch Theorems for Higher Algebraic K-Theory, preprint.

4. D. Grayson, Products in K-theory and intersecting algebraic cycles, Inv. Math. 47(1978), 71-84.

5. R. Hartshorne, Algebraic Geometry, Springer-Verlag, New York, 1978.

6. J. L. Loday, K-théorie algébrique et représentations de groupes, Ann. Sci. Ecole Norm. Sup. (4), 9(1976), 309-377.

7. H. Matsumura, Commutative Algebra, W. A. Benjamin, New York, 1970.

8. J. Milnor, Introduction to Algebraic K-Theory, Ann. of Math. Study #72, Princeton University Press, Princeton, 1971.

9. M. Nagata, Local Rings, John Wiley, New York, 1962.

10. D. Quillen, On the cohomology and K-theory of the general linear groups over a finite field, Ann. of Math. 96(1972), 552-586.

11. D. Quillen, Higher algebraic K-theory I, in Higher K-Theories, Lecture Notes in Mathematics, Vol. 341, Springer-Verlag, New York, 1973.

12. J. P. Serre, Corps Locaux, Hermann, Paris, 1968.

13. C. Sherman, The K-theory of an equicharacteristic discrete valuation ring injects into the K-theory of its field of quotients, Pac. J. of Math., 74(1978), 497-499.

14. C. Sherman, Some Splitting Results in the K-Theory of Rings, Amer. J. Math., 101(1979), 609-632.

15. C. Sherman, K-Cohomology of Regular Schemes, Comm. in Alg. 7(10) (1979), 999-1027.

16. C. Sherman, Gersten's Conjecture for Arithmetic Surfaces, J. Pure and Appl. Alg. 14(1979), 167-174.

17. C. Sherman, Some Theorems on the K-Theory of Coherent Sheaves, Comm. in Alg. 7(14) (1979), 1489-1508.

18. C. Sherman, Cartan Maps, Clean Rings, and Unique Factorization, to appear in J. of Alg.

19. C Soulé, K-theorie des anneaux d'entiers de corps de nombres et cohomologie étale, Inv. Math. 55(1979), 251-295.

20. F. Waldhausen, Algebraic K-theory of generalized free products, I and II, Ann. of Math. 108(1978), 135-204, 205-256.

21. K. Dennis and M. Stein, K_2 of discrete valuation rings, Adv. in Math. 18 (1975), 182-238.

22. A. Grothendieck, Classes de faisceaux et theoreme de Riemann-Roch, in Theorie des Intersections et Theoreme de Riemann-Roch (SGA 6), Lecture Notes in Mathematics, Vol. 225, Springer-Verlag, New York, 1971.

On higher p-adic regulators

By Christophe Soulé

C.N.R.S., Paris VII(*)

The comparison between K-theory and étale cohomology makes it possible to exhibit torsion classes in the K-groups of rings of integers of number fields [17]. We show here that it can also be used to produce classes of infinite order. We get this way an explicit construction of the classes of Borel by means of units in cyclotomic extensions.

The construction that is used (products and transfer in K-theory with coefficients) is valid in a more general context. We get by this method a free part in the K-theory (with ℓ-adic coefficients) of a local field. In a forthcoming paper we shall show in this way that the K-theory (with ℓ-adic coefficients, ℓ big enough) of an abelian variety with complex multiplication is of rank at least the order of the zeroes of its zeta function at negative integers.

The construction used here allows us to compare the global situation to the local one. If p is an odd prime, F an abelian number field of degree prime to p, and F_p the product of the completions of F above p. We show that the group $K_{2i-1}(F) \otimes \mathbb{Q}_p$ maps injectively into $K_{2i-1}(F_p) \otimes \mathbb{Q}_p$ as soon as some p-adic L-function is nonzero at the point i. This result, which fits with conjectures of Coates [5], is proved in a way very analogous to the approach to transcendental regulators due to Bloch for K_3 of a cyclotomic field [2].

I wish to thank J. Coates, B. Coleman, S. Lichtenbaum and B. Mazur for helpful conversations. I am also grateful to the

(*) Partially supported by an NSF Grant.

mathematics departments of Harvard and Cornell Universities for their hospitality while this work was being done.

1. The global Case:

Let F be a number field, $\mathcal{O}_F$ its ring of integers. A. Borel [3] has computed the rational K-theory of $\mathcal{O}_F$. If r_1(resp. r_2) denotes the number of real (resp. complex) places of F, one has

$$K_m(F) \otimes \mathbb{Q} = \begin{cases} \mathbb{Q}^{r_1+r_2-1} & \text{if } m=1 \\ \mathbb{Q}^{r_2} & \text{if } m\equiv 3 \pmod 4 \\ \mathbb{Q}^{r_1+r_2} & \text{if } m\equiv 1 \pmod 4 \text{ and } m\neq 1 \\ 0 & \text{if } m \text{ is even} > 0 . \end{cases}$$

In [17] some morphisms

$$\bar{c}_{i,k} : K_{2i-k}(A;\mathbb{Z}/\ell^n) \to H^k(\text{Spec } A, \mu_{\ell^n}^{\otimes i})$$

were defined, for any prime number ℓ, any integer $n \geq 1$, and any abelian ring A where ℓ is invertible.
The target of these maps $\bar{c}_{i,k}$ is the étale cohomology of the scheme Spec A with coefficients in the sheaf of roots of unity, tensored i times with itself.

For any ring A, define

$$K_{2i-k}(A;\mathbb{Z}_\ell) = \lim_{\substack{\leftarrow \\ n}} K_{2i-k}(A;\mathbb{Z}/_{\ell}n)$$

and, when A contains $1/\ell$,

$$H^k(\text{Spec } A, \mathbb{Z}_\ell(i)) = \lim_{\substack{\leftarrow \\ n}} H^k(\text{Spec } A, \mu_{\ell^n}^{\otimes i})$$

and, by taking the projective limit of the $\bar{c}_{i,k}$'s,

$$c_{i,k}\colon K_{2i-k}(A;\mathbb{Z}_\ell) \to H^k(\mathrm{Spec}\ A;\mathbb{Z}_\ell(i)) \quad .$$

Remark that, since $K_*(\mathcal{O}_F)$ is of finite type [14],

$$K_m(\mathcal{O}_F;\mathbb{Z}_\ell) = K_m(\mathcal{O}_F)\otimes\mathbb{Z}_\ell \ .$$

Furthermore, if $m > 1$ and if $A = \mathcal{O}_F[1/\ell]$ denotes the ring $\mathcal{O}_F$ localized outside ℓ,

$$K_m(\mathcal{O}_F)\otimes\mathbb{Z}_\ell = K_m(A)\otimes\mathbb{Z}_\ell \ .$$

Theorem 1:

Let ℓ be an odd prime number and $A = \mathcal{O}_F[1/\ell]$. The map

$$c_{i,k}\otimes 1\colon K_{2i-k}(\mathcal{O}_F)\otimes\mathbb{Q}_\ell \to H^k(\mathrm{Spec}\ A,\mathbb{Z}_\ell(i))\otimes\mathbb{Q}_\ell$$

is an isomorphism when $k = 1$ or 2, and $2i > k+1$.

Using the result of Borel that was recalled above, the proof is in two steps:

i) The ℓ-adic cohomology of Spec A has the correct rank.

ii) The morphism $c_{i,k}\otimes 1$ is surjective.

Proof of i):

To simplify notations we write A instead of Spec A.

Define

$$H^k(A,\mathbb{Q}_\ell/\mathbb{Z}_\ell(i)) = \lim_{\substack{\to\\ n}} H^k\ (A,\mu_{\ell^n}^{\otimes i})\ .$$

For any $n \geq 1$, there is an exact sequence

$$\cdots \to H^k(A,\mu_{\ell^n}^{\otimes i}) \to H^k(A,\mathbb{Q}_\ell/\mathbb{Z}_\ell(i)) \xrightarrow{x\ell^n} H^k(A,\mathbb{Q}_\ell/\mathbb{Z}_\ell(i)) \to H^{k+1}\ldots$$

Taking the projective limit one recovers $H^k(A,\mathbb{Z}_\ell(i))$ from $H^k(A,\mathbb{Q}_\ell/\mathbb{Z}_\ell(i))$. It will be proved that:

a) $H^0(A,\mathbb{Q}_\ell/\mathbb{Z}_\ell(i))$ is finite

b) $H^1(A,\mathbb{Q}_\ell/\mathbb{Z}_\ell(i)) = (\mathbb{Q}_\ell/\mathbb{Z}_\ell)^{r_1+r_2}$ + (finite group) if i is odd >1

$(\mathbb{Q}_\ell/\mathbb{Z}_\ell)^{r_2}$ + (finite group) if i is even.

c) $H^2(A,\mathbb{Q}_\ell/\mathbb{Z}_\ell(i)) = 0$ if $i \geq 2$.

It implies that $H^k(A,\mathbb{Z}_\ell(i))\otimes\mathbb{Q}_\ell$ and $K_{2i-k}(A)\otimes\mathbb{Q}_\ell$ have the same rank when $k = 1$ or 2 .

The facts a) and c) have been proved in [17] (Theorem 5 and footnote).

Let us prove b). Call F_∞ the maximal ℓ-cyclotomic extension of F (obtained by adjoining to it all the ℓ^n-th roots of unity, $n \geq 1$, in some algebraic closure). Let $G_\infty = \mathrm{Gal}(F_\infty : F)$ the Galois group of F_∞ over F and $F_0 = F(\mu_\ell)$. The action of G_∞ on the roots of unity gives rise to an imbedding $\varkappa$ of G_∞ into $\mathbb{Z}_\ell^*$ defined by the formulas

$$g \cdot \zeta = \zeta^{\varkappa(g)} \quad \text{when} \quad g \in G_\infty \quad \text{and} \quad \zeta^{\ell^n} = 1 .$$

The group $\Gamma = \mathrm{Gal}(F_\infty : F_0)$ is isomorphic to the additive group $\mathbb{Z}_\ell$ and one denotes by F_n the subfield of F_∞ fixed by Γ^{ℓ^n}. Let A_n (resp. A_∞) be the integral closure of A in F_n (resp. F_∞).

The Hochschild-Serre spectral sequence:

$$E_2^{p,q} = H^p(G_\infty, H^q(A_\infty, \mathbb{Q}_\ell/\mathbb{Z}_\ell(i)) \Longrightarrow H^{p+q}(A, \mathbb{Q}_\ell/\mathbb{Z}_\ell(i))$$

degenerates. Actually, when $i \geq 2$, one gets

$$H^q(A_\infty, \mathbb{Q}_\ell/\mathbb{Z}_\ell(i)) = 0 \text{ if } q \geq 2, \quad \text{since } cd_\ell A_\infty = 1,$$

$$E_2^{p,q} = 0 \qquad \text{if } p \geq 2, \quad \text{since } cd_\ell G_\infty = 1,$$

and $H^1(G_\infty, \mathbb{Q}_\ell/\mathbb{Z}_\ell(i)) = 0$

Therefore

$$E_2^{0,0} = H^0(A, \mathbb{Q}_\ell/\mathbb{Z}_\ell(i)),$$

$$E_2^{0,1} = H^1(A, \mathbb{Q}_\ell/\mathbb{Z}_\ell(i)) = H^1(A_\infty, \mathbb{Q}_\ell/\mathbb{Z}_\ell(i))^{G_\infty},$$

$$E_2^{1,1} = H^2(A, \mathbb{Q}_\ell/\mathbb{Z}_\ell(i)) = H^1(A_\infty, \mathbb{Q}_\ell/\mathbb{Z}_\ell(i))_{G_\infty} = 0 \text{(by c)},$$

and $E_2^{p,q} = 0$ in the remaining cases.

Here M^{G_∞} (resp. M_{G_∞}) denotes the group of invariants (resp. coinvariants) of the G_∞-module M.

Let $\mathcal{M}$ be the G_∞-module $H^1(A_\infty, \mathbb{Q}_\ell/\mathbb{Z}_\ell(1))$. Call $\mathbb{Z}_\ell(1)$ the Tate module. As an additive group it is isomorphic to $\mathbb{Z}_\ell$, but it is acted on by G_∞ via the character κ:

$$g \cdot u = \kappa(g)u, \quad u \in \mathbb{Z}_\ell(1), \; g \in G_\infty,$$

where κ is the cyclotomic character defined above. When M is a G_∞-module, call $M(i)$ the tensor product $M \otimes \mathbb{Z}_\ell(1)^{\otimes i}$ (with diagonal action). One has isomorphisms of G_∞-modules (since A_∞^*

contains μ_{ℓ^n} for any $n \geq 1$)

$$\mathcal{M}(i-1) = H^1(A_\infty, \mathbb{Q}_\ell/\mathbb{Z}_\ell)(i) = H^1(A_\infty, \mathbb{Q}_\ell/\mathbb{Z}_\ell(i)).$$

Therefore

$$\mathcal{M}(i-1)^{G_\infty} = H^1(A, \mathbb{Q}_\ell/\mathbb{Z}_\ell(i))$$

and

$$\mathcal{M}(i-1)_{G_\infty} = H^2(A, \mathbb{Q}_\ell/\mathbb{Z}_\ell(i)) = 0 \quad (i \geq 2) .$$

The module $\mathcal{M}$ was first studied by Iwasawa (its Pontryagin dual is the Galois group of the maximal ℓ-abelian extension of F_∞ unramified outside ℓ). Using the Kummer exact sequences of sheaves

$$0 \to \mu_{\ell^n} \to G_m \xrightarrow{x\ell^n} G_m \to 0$$

one gets an exact sequence of G_∞-modules ([10], Lemma 10)

$$0 \to \mathcal{E} \to \mathcal{M} \to \mathcal{O}\!\mathcal{l} \to 0$$

where $\mathcal{E} = (\varinjlim_n A_n^*) \otimes \mathbb{Q}_\ell/\mathbb{Z}_\ell$

and $\mathcal{O}\!\mathcal{l} = \varinjlim_n \mathrm{Pic}\,(A_n)$,

$\mathrm{Pic}(A_n)$ being the ideal class group of A_n. Since the cohomological dimension of G_∞ is one, there are exact sequences

$$0 \to \mathcal{E}(i-1)^{G_\infty} \to \mathcal{M}(i-1)^{G_\infty} \to \mathcal{O}\!\mathcal{l}(i-1)^{G_\infty} \to$$

$$\mathcal{E}(i-1)_{G_\infty} \to \mathcal{M}(i-1)_{G_\infty} \to \mathcal{O}\!\mathcal{l}(i-1)_{G_\infty} \to 0 .$$

The group $\mathcal{M}(i-1)_{G_\infty}$ being zero, the same is true of $\mathcal{O}\!\mathcal{l}(i-1)_{G_\infty}$. This implies ([6], Lemma 6.2) that $\mathcal{O}\!\mathcal{l}(i-1)^{G_\infty}$ is finite. Therefore to prove b) it will be enough to show that

$$\mathcal{E}(i-1)^{G_\infty} = (\mathbb{Q}_\ell/\mathbb{Z}_\ell)^{r_2} + \text{(finite group) when } i \text{ is even}$$

$$(\mathbb{Q}_\ell/\mathbb{Z}_\ell)^{r_1+r_2} + \text{(finite group) when } i \text{ is odd, } i \neq 1 .$$

From now on <u>assume</u> that F contains the ℓ-th roots of unity. In fact $G_0 = \mathrm{Gal}(F_0:F)$ is of order prime to ℓ, therefore

$$K_*(A;Z_\ell) = K(A_0;Z_\ell)^{G_0} \quad \text{and} \quad H^*(A,Z_\ell(i)) = H^*(A_0,Z_\ell(i))^{G_0}$$

(to see this, remark that the two composites of the transfer with the map induced by $A \to A_0$ are the product by the order of G_0) , so the theorem 1 will be proved in general whenever it is proved for F_0 .

Since ℓ is odd and $F = F_0$, the field F is totally imaginary, $r_1 = 0$ and $r_2 = d/2$, where $d = [F:\mathbb{Q}]$ is the degree of F over $\mathbb{Q}$. The algebra $\mathbb{Z}_\ell[[G_\infty]]$ of the profinite group $G_\infty = \Gamma \simeq \mathbb{Z}_\ell$ is isomorphic to the power series ring $\Lambda = \mathbb{Z}_\ell[[T]]$. The element $1+T$ corresponds to a generator γ of Γ , and acts on the roots of unity by the formulas

$$(1+T)\cdot\zeta = \zeta^c \quad \text{if} \quad \zeta^{\ell^n} = 1$$

where $c = \kappa(\gamma) \in \mathbb{Z}_\ell^*$ is a constant, congruent to one modulo ℓ.

The module $\mathcal{E}$ is thus given a structure of Λ-module. Iwasawa shows ([10], Theorem 15) that its Pontryagin dual

$$\hat{\mathcal{E}} = \mathrm{Hom}(\mathcal{E}, \mathbb{Q}_\ell/\mathbb{Z}_\ell)$$

is the kernel of an exact sequence

$$0 \to \hat{\mathcal{E}} \to \Lambda^{d/2} \oplus M \to \Phi \to 0$$

where Φ is a finite Λ-module, and M is the sum of finitely many modules of type $\Lambda/(\xi_\alpha(T))$, $\alpha \geq 1$, where $\xi_\alpha(T) = ((1+T)^{\ell^\alpha}-1)/((1+T)^{\ell^{\alpha-1}}-1)$. The module $\Lambda(1)$ (i.e., the Tate twist of Λ) is isomorphic to Λ by the change of variable:

$$T \text{ goes to } c^{-1}(1+T)-1.$$

Therefore $M(1-i)_{G_\infty}$ is finite, since $\xi_\alpha(c^{i-1}-1) \neq 0$ when $i \geq 2$. On the other hand

$$\Lambda(1-i)_{G_\infty} \simeq \Lambda_{G_\infty} = \mathbb{Z}_\ell[[T]]/(T) = \mathbb{Z}_\ell \,.$$

So (if $\wedge$ denotes the Pontryagin dual) one has

$$(\mathcal{E}(i-1)^{G_\infty})^\wedge = \hat{\mathcal{E}}(1-i)_{G_\infty} = \Lambda(1-i)^{d/2}_{G_\infty} + \text{(finite group)}$$
$$= \mathbb{Z}_\ell^{d/2} + \text{(finite group)}.$$

This finishes the computation of $H^k(A,\mathbb{Z}_\ell(i)) \otimes \mathbb{Q}_\ell$ (*).

<u>Proof of ii)</u>: The proof of i) given above suggests that the free part of $K_*(A) \otimes \mathbb{Z}_\ell$ could be obtained by means of units in the cyclotomic extensions of F.

Let $(\zeta_n) \in \varprojlim_n \mu_{\ell^n} = \mathbb{Z}_\ell(1)$ be a generator of the Tate module: ζ_n is a primitive ℓ^n-th root of unity, and $\zeta_{n+1}^\ell = \zeta_n$. To the element ζ_n is <u>canonically</u> attached an element

$$a_n \in K_2(A_n;\mathbb{Z}/\ell^n)$$

in the group K_2 of the ring A_n with coefficients $\mathbb{Z}/\ell^n$. The Bockstein morphism

(*) For another proof of i) (assuming $H^2(A,\mathbb{Q}_\ell/\mathbb{Z}_\ell(i)) = 0$, $i \geq 2$) see [15].

$$K_2(A_n,\mathbb{Z}/\ell^n) \to K_1(A_n)_{(\ell^n)} = \mu_{\ell^n}$$

maps α_n to ζ_n (here $X_{(\ell^n)}$ denotes the group of elements of order ℓ^n in the abelian group X). To define α_n consider the isomorphisms

$$\pi_2(BGL_1(A_n);\mathbb{Z}/\ell^n) = \pi_1(BGL_1(A_n))_{(\ell^n)} = \mu_{\ell^n},$$

where $BGL_1(A_n)$ is the classifying space of the group $GL_1(A_n) = A_n^*$. The element α_n is defined by stabilization as the image in

$$\pi_2(BGL(A_n);\mathbb{Z}/\ell^n) = K_2(A_n;\mathbb{Z}/\ell^n)$$

of the element of $\pi_2(BGL_1(A_n);\mathbb{Z}/\ell^n)$ corresponding to $\zeta_n \in \mu_{\ell^n}$ by the isomorphisms above.

On the other hand, let

$$(u_n) \in \lim_{\substack{\leftarrow \\ n}} A_n^* = E$$

be a projective system of units (or rather p-units) for the norm maps $N_{n+1,n}$ from F_{n+1} to F_n: $N_{n+1,n}(u_{n+1}) = u_n$. The element u_n belongs to $K_1(A_n) = A_n^*$, and one can consider the product

$$u_n \cdot \alpha_n^{i-1} \in K_{2i-1}(A_n;\mathbb{Z}/\ell^n)$$

and its image

$$N_n(u_n \cdot \alpha_n^{i-1}) \in K_{2i-1}(A;\mathbb{Z}/\ell^n)$$

by the transfer map N_n from A_n to A.

Lemma 1:

i) The elements $N_n(u_n \cdot \alpha_n^{i-1})$ form, for the different values of n, a projective system with respect to the morphisms

$$K_{2i-1}(A;\mathbb{Z}/\ell^{n+1}) \to K_{2i-1}(A;\mathbb{Z}/\ell^{n}).$$

ii) One can define this way a morphism

$$\varphi : E(i-1)_{G_\infty} \to K_{2i-1}(\mathcal{O}_F;\mathbb{Z}_\ell).$$

Proof of Lemma 1:

i) Since $N_n \circ N_{n+1,n} = N_{n+1}$ it suffices to show that, if r_n denotes the reduction of coefficients from $\mathbb{Z}/\ell^{n+1}$ to $\mathbb{Z}/\ell^n$, one has

$$r_n(N_{n+1,n}(u_{n+1} \cdot \alpha_{n+1}^{i-1})) = u_n \cdot \alpha_n^{i-1} .$$

Let j_n be the morphism induced by the injection $A_n \to A_{n+1}$. Note that $j_n(\alpha_n) = r_n(\alpha_{n+1})$ as can be seen from the following commutative diagram:

$$\begin{array}{ccccc}
\pi_2(BGL_1(A_{n+1});\mathbb{Z}/\ell^{n+1}) & \longrightarrow & (A_{n+1}^*)_{(\ell^{n+1})} & = & \mu_{\ell^{n+1}} \\
\big\downarrow r_n & & \big\downarrow \times\ell & & \\
\pi_2(BGL_1(A_{n+1});\mathbb{Z}/\ell^{n}) & \longrightarrow & (A_{n+1}^*)_{(\ell^{n})} & = & \mu_{\ell^{n}} \\
\big\uparrow j_n & & \big\uparrow \mathrm{id} & & \\
\pi_2(BGL_1(A_{n});\mathbb{Z}/\ell^{n}) & \longrightarrow & (A_{n}^*)_{(\ell^{n})} & = & \mu_{\ell^{n}}
\end{array} .$$

Therefore

$$r_n(N_{n+1,n}(u_{n+1}\cdot\alpha_{n+1}^{i-1})) = N_{n+1,n}(u_{n+1}\cdot r_n(\alpha_{n+1})^{i-1})$$

$$= N_{n+1,n}(u_{n+1}\cdot j_n(\alpha_n^{i-1}))$$

$$= N_{n+1,n}(u_{n+1})\cdot\alpha_n^{i-1} \quad \text{by the adjunction formula [13]}$$

$$= u_n\cdot\alpha_n^{i-1}\ .$$

ii) Given a generator (ζ_n) of $\mathbb{Z}_\ell(1)$ one defines

$$\varphi: E\otimes\mathbb{Z}_\ell(1)^{\otimes i-1} = E(i-1) \to K_{2i-1}(A;\mathbb{Z}_\ell)$$

by the formula

$$\varphi(u_n\otimes\zeta_n^{\otimes(i-1)}) = (N_n(u_n\cdot\alpha_n^{i-1})).$$

The equalities below show that φ commutes with the action of G_∞ (trivial action on the K-theory of A):

$$\varphi((gu_n)\otimes(g\zeta_n)^{i-1}) = \kappa(g)^{i-1}\varphi((gu_n)\otimes\zeta_n^{\otimes(i-1)})$$

$$= \kappa(g)^{i-1}(N_n((gu_n)\cdot\alpha_n^{i-1})) = (N_n((gu_n)\cdot(g\alpha_n)^{i-1})$$

$$= (N_n(u_n\cdot\alpha_n^{i-1})),\ \text{for any}\ \ g\in G_\infty.$$

This ends the proof of Lemma 1.

To finish the proof of Theorem 1 it will be enough to show that the composite morphism $c_{i,1}\circ\varphi$ is rationally an isomorphism (i.e., is an isomorphism after tensoring with $\mathbb{Q}_\ell$). Let

$$\tilde{\varphi}: E(i-1)_{G_\infty} \to H^1(A, \mathbb{Z}_\ell(i))$$

be the map defined by the formula

$$\tilde{\varphi}(u_n \otimes \zeta_n^{\otimes i-1}) = (N_n(u_n \cup a_n^{i-1})),$$

where $a_n \in H^0(A_n, \mu_{\ell^n})$ is the element corresponding to ζ_n, and

$$u_n \in A_n^*/(A_n^*)^{\ell^n} \subset H^1(A, \mu_{\ell^n}).$$

The multiplication formula for Chern classes ([17], Theorem 1), and the fact that $ic_{i,1}$ commutes with transfer ([17], Theorem 2) show that

$$i(c_{i,1} \circ \varphi) = (i!)\tilde{\varphi} .$$

It is then enough to show that $\tilde{\varphi}$ is rationally an isomorphism. For this write additively $E_n = A_n^*$ the group of p-units in F_n, and assume again that F contains the ℓ-th roots of unity $(F = F_0)$. Since the generator γ of $\Gamma = \mathrm{Gal}(F_\infty : F_0)$ corresponds to $1+T$ in $\Lambda = \mathbb{Z}_\ell[[T]]$ (see the proof of i)), the field F_n is fixed by $(1+T)^{\ell^n}$. Therefore the image of E_n/ℓ^n in $\mathcal{E}_{(\ell^n)} = A_\infty^* \otimes \mathbb{Z}/\ell^n$ is contained in the kernel $\mathcal{E}^{\omega_n}_{(\ell^n)}$ of the multiplication by $\omega_n = (1+T)^{\ell^n} - 1$. The following diagrams commute $(n \geq 0)$:

$$\begin{array}{ccc}
E_{n+1}/\ell^{n+1} & \longrightarrow & \mathcal{E}^{\omega_{n+1}}_{(\ell^{n+1})} \\
{\scriptstyle N_{n+1,n}}\big\downarrow & & \big\downarrow{\scriptstyle \ell\omega_{n+1}/\omega_n} \\
E_n/\ell^n & \longrightarrow & \mathcal{E}^{\omega_n}_{(\ell^n)}
\end{array}$$

Let E' be the projective limit of the groups $\mathcal{E}^{\omega_n}_{(\ell^n)}$. One can check that the diagram below commutes:

$$\begin{array}{ccccc}
E(i-1)_{G_\infty} & \xrightarrow{\tilde{\varphi}} & H^1(A,\mathbf{Z}_\ell(i)) & & \\
 & & & \searrow^{\beta} & \\
\downarrow & & & & \varprojlim\limits_n H^1(A,\mathbf{Q}_\ell/\mathbf{Z}_\ell(i))_{(\ell^n)} \\
 & & & \nearrow_{\beta'} & \\
E'(i-1)_{G_\infty} & \xrightarrow{\varphi'} & \varprojlim\limits_n \mathcal{E}(i-1)^{G_\infty}_{(\ell^n)} & &
\end{array}$$

We saw in proving i) that β is rationally an isomorphism. The map β' is induced by the injection of $\mathcal{E}$ into $\mathcal{M}$, and is therefore (see i)) rationally an isomorphism. The map φ' is defined by the formula

$$\varphi'(e_n) = \frac{\omega_n}{\omega_0} \cdot e_n$$

if

$$e_n \in \mathcal{E}^{\omega_n}_{(\ell^n)}$$

(note that $\omega_0 \cdot \varphi'(e_n) = 0$, i.e., $\varphi'(e_n)$ is invariant by G_∞).

The following two lemmas will conclude the proof of Theorem 1:

<u>Lemma 2</u>: <u>The map</u> $E(i-1)_{G_\infty} \otimes \mathbf{Q}_\ell \to E'(i-1)_{G_\infty} \otimes \mathbf{Q}_\ell$ <u>is an isomorphism.</u>

<u>Proof of Lemma 2</u>: From Iwasawa ([10], Lemma 7 and Theorem 12) we get exact sequences

$$0 \to E_n/\ell^n \to \mathcal{E}^{\omega_n}_{(\ell^n)} \to \Omega_n \to 0$$

with $\Omega_n = \mathrm{Ker}(\mathrm{Pic}(A_n) \to \mathrm{Pic}(A_\infty))_{(\ell^n)}$. The Λ-module $\Omega = \varprojlim\limits_n \Omega_n$

is contained in $\varprojlim_n \mathrm{Pic}(A_n)$ (the projective limit being taken with respect to the norms). The module $\varprojlim_n \mathrm{Pic}(A_n)$ is pseudo-isomorphic to the Pontryagin dual of the module $\mathfrak{A}$ considered in i) above (cf. [10], Theorem 11). Since $\mathfrak{A}(i-1)_{G_\infty}$ and $\mathfrak{A}(i-1)^{G_\infty}$ are finite (cf. i)), we see that $\Omega(i-1)^{G_\infty}$ and $\Omega(i-1)_{G_\infty}$ are finite. Q.E.D.

<u>Lemma 3</u>: <u>The map</u> φ' <u>is rationally an isomorphism.</u>

<u>Proof of Lemma 3</u>: Consider the Pontryagin dual $\widehat{\varphi'}$ of the map φ':

$$\widehat{\varphi'}:(\hat{\mathcal{E}}(1-i)/\omega_0) \otimes \mathbb{Q}_\ell/\mathbb{Z}_\ell \to \hat{E}'(1-i)^{\omega_0},$$

with

$$\hat{E}' = \varinjlim_n(\hat{\mathcal{E}}/(\ell^n, \omega_n)).$$

The exact sequence

$$0 \to \hat{\mathcal{E}} \to \Lambda^{d/2} \oplus M \to \Phi \to 0$$

used to prove i) gives at the n-th level

$$\Phi^{\omega_n} \to \hat{\mathcal{E}}/\omega_n \to (\Lambda^{d/2} \oplus M)/\omega_n \to \Phi/\omega_n \to 0.$$

This allows us, to study $\widehat{\varphi'}$, to replace $\hat{\mathcal{E}}$ by $\Lambda^{d/2} \oplus M$. Using the fact that M does not contribute to the free part (see i)) and that $\Lambda(1-i) \simeq \Lambda$, one is left with studying the map

$$(\Lambda/\omega_0) \otimes \mathbb{Q}_\ell/\mathbb{Z}_\ell \xrightarrow{\widehat{\varphi'}} (\varinjlim_n(\Lambda/(\ell^n, \omega_n)))^{\omega_0}$$

defined by

$$\widehat{\varphi'}(\lambda \otimes 1) = (\frac{\omega_n}{\omega_0} \cdot \lambda).$$

It is an isomorphism, as it can be seen at the finite levels. Therefore the kernel and cokernel of the original map $\widehat{\varphi'}$ are finite. Q.E.D.

2. The local case:

Let L be a finite extension of the p-adic field $\mathbb{Q}_p$.

Theorem 2: *For any odd prime ℓ, and any integer $i \geq 1$, the cokernel of the map*

$$c_{i,k}: K_{2i-k}(L;\mathbb{Z}_\ell) \to H^k(L,\mathbb{Z}_\ell(i)) \ , \ k = 1 \text{ or } 2,$$

is killed by multiplication by $i!$ (*).

Proof of Theorem 2:

For $\ell \neq p$ or $k = 2$ this result is proved in [17] (Theorems 4, 6 and 8). When $k = 1$ and $\ell = p$ we shall use the analogue of the proof of Theorem 1 above. Let L_∞ be the maximal p-cyclotomic extension of L, let $\Gamma = \mathrm{Gal}(L_\infty : L)$, and choose an isomorphism $\mathbb{Z}_p[[\Gamma]] \simeq \mathbb{Z}_p[[T]] = \Lambda$. Call L_n the subfield of L_∞ fixed by $\omega_n = (1+T)^{p^n} - 1$. Let $X_n = L_n^*$ be the group of invertible elements in L_n, and $X = \varprojlim_n X_n$ be the projective limit of these groups for the norms. One can define as in the proof of Theorem 1, ii), morphisms

$$\varphi: X(i-1)_\Gamma \longrightarrow K_{2i-1}(L;\mathbb{Z}_p)$$

and

$$\tilde{\varphi}: X(i-1)_\Gamma \longrightarrow H^1(L,\mathbb{Z}_p(i))$$

(*) The recent theory of B. Dwyer and E. Friedlander modifies $c_{i,k}$ in such a way that the factorial $i!$ is not needed in this statement (the same is true for the Theorem 3 below).

such that $i(c_{i,1} \cdot \varphi) = (i!)\tilde{\varphi}$ and $\tilde{\varphi}(u_n \otimes \zeta_n^{\otimes i-1}) = (N_n(u_n \cup a_n^{i-1}))$. The following proposition will then imply Theorem 2:

<u>Proposition 1:</u> <u>If</u> $i \geq 2$ <u>the map</u>

$$\tilde{\varphi} : X(i-1)_\Gamma \to H^1(L, \mathbb{Z}_p(i))$$

<u>is an isomorphism.</u>

<u>Proof of Proposition 1:</u>

One uses the duality theorem for the Galois cohomology of a local field ([16], II. 5.2., Theorem 2). If $\bar{L}$ is a separable closure of L, M a finite $\mathrm{Gal}(\bar{L}:L)$-module, and $M' = \mathrm{Hom}(M, \bar{L}^*)$, the cup-product

$$H^k(L,M) \times H^{2-k}(L,M') \to H^2(L,G_m) = \mathbb{Q}/\mathbb{Z}$$

is a perfect duality of finite groups. By this duality the norm map between two local fields is dual to the map induced by the inclusion.

Therefore one gets duality isomorphisms

$$X_n/_{p^n} = H^1(L_n, \mu_{p^n}) \xrightarrow{D} H^1(L_n, \mathbb{Z}/_{p^n})^\wedge$$

$$X = \lim_{\substack{\leftarrow \\ n}} H^1(L_n, \mu_{p^n}) \xrightarrow{D} (\lim_{\substack{\rightarrow \\ n}} H^1(L_n, \mathbb{Z}/_{p^n}))^\wedge = H^1(L_\infty, \mathbb{Q}_p/\mathbb{Z}_p)^\wedge$$

$$H^1(L, \mathbb{Z}_p(i)) \xrightarrow{D} H^1(L, \mathbb{Q}_p/\mathbb{Z}_p(1-i))^\wedge .$$

Furthermore the following diagram commutes

$$\begin{array}{ccc}
\varprojlim_n H^1(L_n, \mu_{p^n}^{\otimes(i-1)})_\Gamma & \xrightarrow{\tilde{\varphi}} & H^1(L, \mathbb{Z}_p(i)) \\
\downarrow D & & \downarrow D \\
(H^1(L_\infty, \mathbb{Q}_p/\mathbb{Z}_p(1-i))^\Gamma)^\wedge & \xrightarrow{\hat{\psi}} & H^1(L, \mathbb{Q}_p/\mathbb{Z}_p(1-i))^\wedge ,
\end{array}$$

where the vertical maps are isomorphisms and $\hat{\psi}$ is Pontryagin dual to the morphism ψ induced by the inclusion $L \to L_\infty$. By the Hochschild-Serre spectral sequence, one gets that the kernel of this map

$$\psi: H^1(L, \mathbb{Q}_p/\mathbb{Z}_p(1-i)) \to H^1(L_\infty, \mathbb{Q}_p/\mathbb{Z}_p(1-i))^\Gamma$$

is the group of coinvariants $(\mathbb{Q}_p/\mathbb{Z}_p(1-i))_\Gamma$. But if $i \geq 2$ this group is zero. So ψ is an isomorphism. q.e.d.

Remarks:

1) The structure of the Λ-module X is well-known ([10], Theorem 25). This leads, as in the part i) of the proof of Theorem 1, to the following result:

$$\dim_{\mathbb{Q}_p} H^1(L, \mathbb{Q}_p(i)) = d \quad \text{if} \quad i > 0,$$

where d is the degree of L over $\mathbb{Q}_p$.

2) Unlike the case of $\mathcal{O}_F$ considered in Theorem 1, one cannot (a priori) replace $K_m(L;\mathbb{Z}_p)$ by $K_m(L) \otimes \mathbb{Z}_p$ (*).

(*) When F is a number field, the example of K_2 [18] shows that in general $\varprojlim_n (K_mF)/\ell^n = K_m(F;\mathbb{Z}_\ell) \neq (K_mF) \otimes \mathbb{Z}_\ell$. The Theorem 6i) of [17] is thus wrong as stated . To correct it one must replace $K_{2i-2}(F) \otimes \mathbb{Z}_\ell$ by $K_{2i-2}(F;\mathbb{Z}_\ell)$.

3) J. Tate proved in [19] that, when $m = 2$ and L is contained in a cyclotomic p-extension of $\mathbb{Q}_p$, $K_2(L) = K_2^{top}(L) + $ (uniquely divisible group), where $K_2^{top}(L)$ is the (finite) group of roots of unity in L. It is possible that the following decomposition always occurs:

$$K_m(L) = K_m^{top}(L) + \text{(uniquely divisible group)},$$

where $K_m^{top}(L)$ is the sum of a finite group and finitely many copies of $\mathbb{Z}_p$. One would then have

$$K_m^{top}(L) \simeq \varprojlim_q K_m(L; \mathbb{Z}/q) \overset{\text{def.}}{=\!=} K_m(L; \hat{\mathbb{Z}}).$$

4) B. Wagoner defined in [20] topological K-groups:

$$K_m^W(\mathcal{O}_L) = \varprojlim_{\alpha} K_m(\mathcal{O}_L/\pi^{\alpha})$$

where $\mathcal{O}_L$ is the ring of integers in L and π a uniformizing element, and

$$K_m^W(L) = K_m^W(\mathcal{O}_L) \oplus K_{m-1}(\mathcal{O}_L/\pi).$$

Since the groups $K_m(\mathcal{O}_L/\pi^{\alpha})$ are finite [20], one checks that the map

$$K_m(L) \to K_m^W(L)$$

factors through $K_m(L; \hat{\mathbb{Z}})$. It would be interesting to know if the morphisms $c_{i,k}$ factor through $K_{2i-k}^W(L)$. One knows already [20] that

$$K_m^W(L) \otimes \mathbb{Q}_p = \mathbb{Q}_p^d \text{ if } m \text{ is odd, and zero if not.}$$

3. Regulators (*)

3.0. Notations:

In this paragraph F will be an abelian totally real number field of degree d over $\mathbb{Q}$, $\mathcal{O}_F$ will be its ring of integers, p will denote an odd prime number which does not divide d, and $F_p = \prod_{v|p} F_v$ the product of the completions of F at the places dividing p.

As above, let F_∞ be the maximal p-cyclotomic extension of F, let $F_0 = F(\mu_p)$ and let F_n be the n-th step of the $\mathbb{Z}_p$-extension F_∞ of F_0. Call $G_\infty = \mathrm{Gal}(F_\infty:F)$, $G_0 = \mathrm{Gal}(F_0:F)$ and $\Delta = \mathrm{Gal}(F_0:\mathbb{Q})$. Finally, let ω be the character of Δ obtained by its action on the p-th roots of unity: $\delta\cdot\zeta = \zeta^{\omega(\delta)}$ when $\zeta^p = 1$ and $\delta \in \Delta$.

3.1. Theorem 3: To any odd integer $i \geq 3$ is attached a morphism

$$\rho_i : K_{2i-1}(F_p;\mathbb{Z}_p) \to \mathbb{Z}_p^d$$

and a sub-$\mathbb{Z}_p$-module $K_{2i-1}^{\mathrm{cycl}}(\mathcal{O}_F)$ of $K_{2i-1}(\mathcal{O}_F) \otimes \mathbb{Z}_p$ (described explicitly below in 3.1.2) such that

1) The volume of the image of ρ_i in $\mathbb{Z}_p^d$ divides $((i-1)!)^d$.

2) The image of $K_{2i-1}^{\mathrm{cycl}}(\mathcal{O}_F)$ in $\mathbb{Z}_p^d$ via the composition of ρ_i with the natural morphism $j : K_{2i-1}(\mathcal{O}_F) \otimes \mathbb{Z}_p \to K_{2i-1}(F_p;\mathbb{Z}_p)$ has the following volume:

(*) Classically, the regulator of a number field F is the volume of the lattice obtained by imbedding its units $(=K_1(\mathcal{O}_F))$ into $\mathbb{R}^{r_1+r_2-1}$ via the logarithms of the archimedean places. This number comes in the Dirichlet formula for the residue at one of the complex zeta function of F. A p-adic analog of this is due to Leopoldt. See for instance [11], 3.3. and 4.3.

$$\mathrm{vol}(\rho_i \cdot j(K_{2i-1}^{cycl}(\mathcal{O}_F))) = \underline{C} \cdot L_p(F, \omega^{1-i}, i),$$

where $L_p(F, \omega^{1-i}, s)$, $s \in \mathbb{Z}_p$, denotes the Kubota-Leopoldt p-adic L-function attached to F and ω^{1-i}, and C is a non-zero constant (given below, up to a p-adic unit, in 3.1.5).

Proof of Theorem 3:

3.1.1. This result is a translation, by means of the Theorems 1 and 2 above, of the classical result about the quotient of local units by the cyclotomic units ([9],[11]). We shall use the more general version of it given by R. Gillard in [8].

For any irreducible character Φ of Δ over $\mathbb{Q}_p$ we choose an absolutely irreducible component ψ of Φ (with values in an algebraic closure of $\mathbb{Q}_p$). Let $A = A_\Phi$ be the ring of values of ψ. To the character Φ is attached an idempotent element

$$e_\Phi = \frac{1}{d} \sum_{\delta \in \Delta} \Phi(\delta^{-1})\delta$$

in $\mathbb{Z}_p[\Delta]$, and ψ induces an isomorphism

$$(*) \qquad e_\Phi \mathbb{Z}_p[\Delta] \xrightarrow{\sim} A$$

(see [8], 1 or [12], 2).

Let M be a $\mathbb{Z}_p$-module given with a continuous action of $\mathrm{Gal}(F_\infty : \mathbb{Q})$. The module M can then be decomposed according to the action of Δ:

$$M = \bigoplus_\Phi e_\Phi M .$$

Each component $e_\Phi M$ inherits a structure of $A[[T]]$-module. The ring A acts via the isomorphism (*) above, and the action of the group $\Gamma = \mathrm{Gal}(F_\infty : F_0) \simeq \mathbb{Z}_p$ gives rise to an action of the power series:

$$(1 + T) \cdot \zeta = \zeta^c, \text{ when } \zeta^{p^n} = 1 .$$

The constant $c = c_{\psi} = 1 + q_0$, is such that q_0 is the greatest common multiple of p and the conductor of ψ ([8],1).

Examples of such modules M are given by

$$U = \lim_{\substack{\leftarrow \\ w}} \left(\prod_{w|p} U_n^w \right),$$

where U_n^w is the group of units in the completion of F_n at the place w (dividing p) which are congruent to one modulo the maximal ideal, and by

$$C = \lim_{\substack{\leftarrow \\ n}} C_n ,$$

the subgroup of $E \otimes \mathbb{Z}_p$ consisting of projective limits of cyclotomic units ([8],4) (tensored with $\mathbb{Z}_p$).

Following [8], when $\Phi \neq 1$, one can define

- An epimorphism $\Lambda_{\Phi}: e_{\Phi}U \to A[[T]]$ ([8],Lemma 3)
- A generator $\theta(\Phi)$ of the module $e_{\Phi}C$ ([8],4.2), depending on the choice of primitive roots of unity ζ_n, such that the image of $\theta(\Phi)$ via the composite morphism

$$e_{\Phi}C \longrightarrow e_{\Phi}U \xrightarrow{\Lambda_{\Phi}} A[[T]]$$

is a power series $f'(\dot{T}, \psi)$, $\dot{T} = c(1+T)^{-1}-1$, which can be related to the p-adic L-function of the primitive Dirichlet character attached to ψ (see 3.1.5 below).

When $\Phi = 1$, Iwasawa showed ([9],3.2, 2.3 and Prop.9) that there exists an exact sequence of G_{∞}-modules

$$0 \longrightarrow e_1C \longrightarrow e_1U \xrightarrow{\epsilon} \mathbb{Z}_p \longrightarrow 0,$$

and an isomorphism $e_1U \simeq \mathbb{Z}_p[[T]]$ (ϵ being the reduction modulo T). One can then take $f'(\dot{T},1) = T$ to get the same properties as above.

3.1.2. Let $\varphi: E(i-1)_{G_\infty} \to K_{2i-1}(\mathcal{O}_F) \otimes \mathbb{Z}_p$ be the morphism defined in the proof of Theorem 1, ii). Define

$$K_{2i-1}^{cycl}(\mathcal{O}_F) = \varphi(C(i-1)_{G_\infty}).$$

The Φ-component $e_\Phi K_{2i-1}^{cycl}(\mathcal{O}_F)$ of this Δ-module is then generated over A by

$$v_\Phi = \varphi(\theta(\Phi) \otimes (\zeta_n)^{\otimes(i-1)}) .$$

3.1.3. The Proposition 1 shows that there is an isomorphism

$$\tilde{\varphi}: X(i-1)_{G_\infty} \to H^1(F_p, \mathbb{Z}_p(i)),$$

where X is the projective limit of the groups $\prod_{w|p} F_{n,w}^*$. The module U is a direct factor in X. Furthermore Λ_Φ induces maps

$$e_\Phi U(i-1)_{G_\infty} \to A(i-1)_{G_0}$$

which give rise to an epimorphism

$$\pi: X(i-1)_{G_\infty} \to U(i-1)_{G_\infty} \to \bigoplus_\Phi A(i-1)_{G_0} \simeq \mathbb{Z}_p[\Delta/G_0] \simeq \mathbb{Z}_p^d$$

(note that $A_\Phi(i-1)$ and $A_{\Phi\omega^{i-1}}$ are isomorphic). The regulator map ρ_i is defined as the composite

$$\rho_i = \pi \circ \tilde{\varphi}^{-1} \circ c_{i,1} ,$$

where $c_{i,1}$ is the Chern class

$$c_{i,1} : K_{2i-1}(F_p; \mathbb{Z}_p) \to H^1(F_p, \mathbb{Z}_p(i)).$$

Let ρ_i^{Φ} be the Φ-component of ρ_i.

The properties of $c_{i,1}$ imply that the composite morphism

$$\tilde{\varphi}^{-1} \circ c_{i,1} \circ j \circ \varphi : E(i-1)_{G_\infty} \to X(i-1)_{G_\infty}$$

is the product of the usual map by $(i-1)!$.

3.1.4. The volume of the image of ρ_i divides $((i-1)!)^d$ since π is surjective (cf. Theorem 2).

By what has been recalled of [8], one has

$$\rho_i^{\Phi}(j(v_{\Phi})) = ((i-1)!)\,\varepsilon_i(f'(T, \psi)),$$

where ε_i is the projection

$$\varepsilon_i : A[[T]] \to A(i-1)_{G_0}$$

which makes commutative the following diagram (recall that a generator (ζ_n) of the Tate module has been chosen)

$$\begin{array}{ccc} e_{\Phi}U & \xrightarrow{\Lambda_{\Phi}} & A[[T]] \\ \downarrow & & \downarrow \varepsilon_i \\ e_{\Phi}U(i-1)_{G_\infty} & \longrightarrow & A(i-1)_{G_0} \end{array}$$

One has

$$A(i-1)_{G_0} = A_{\Phi\omega^{i-1}}$$

and this group is trivial unless $\Phi\omega^{i-1}$ is trivial on G_0.

Since $(1+T)\cdot\zeta_n = \zeta_n^c$ one gets $\epsilon_i(f'(\dot{T},\psi)) = f'(c^i-1,\psi)$ when $\psi\omega^{i-1}$ is trivial on G_0, and zero if not.

For any element a in A one has

$$\rho_i^{\Phi}(j(a\cdot v_{\Phi})) = a\cdot\rho_i^{\Phi}(j(v_{\Phi})) = ((i-1)!)f'(c^i-1, \psi^a)$$

or zero, where $\psi^a(x) = \psi(ax)$ $(x \in e_{\Phi}\mathbb{Z}_p[\Delta])$. One can choose a basis $\mathcal{B}$ of A over $\mathbb{Z}_p$ such that the elements ψ^a, $a \in \mathcal{B}$, are all the absolutely irreducible components of Φ. Therefore, up to a p-adic unit, one has

$$\mathrm{vol}(\rho_i\circ j(K_{2i-1}^{cycl}(\mathcal{O}_F))) = ((i-1)!)^d \prod_{\chi\in\widehat{\mathrm{Gal}}(F:\mathbb{Q})} f'(c^i-1,\chi\omega^{1-i}),$$

where $\widehat{\mathrm{Gal}}(F:\mathbb{Q})$ is the group of characters of $\mathrm{Gal}(F:\mathbb{Q})$.

3.1.5. Following [8] (1(*) and 5.2) one has, up to a unit, when $\chi\omega^{1-i} \neq 1$,

$$f'(c^i-1,\chi\omega^{1-i}) = \begin{cases} L_p(\chi\omega^{1-i},i) & \text{if } \underline{\chi}\omega^{-1}(p) \neq 1 \\ L_p(\underline{\chi}\omega^{1-i},i)/(c^i-1) & \text{if } \underline{\chi}\omega^{-1}(p) = 1 \end{cases}$$

where $\underline{\chi}$ is the primitive Dirichlet character attached to χ and $L_p(\underline{\chi}\omega^{1-i},s)$, $s \in \mathbb{Z}_p$, the p-adic L-function of Kubota-Leopoldt. When $\underline{\chi}\omega^{1-i} = 1$, the number $L_p(1,i)$ is, up to a unit, equal to $c_1^{i-1}-1$ (with $c_1 = 1+p$). By definition, the p-adic L-function $L_p(F,\omega^{1-i},s)$ is the product over all characters χ of $\mathrm{Gal}(F:\mathbb{Q})$ of the functions $L_p(\underline{\chi}\omega^{1-i},s)$. Its values at negative values are, up to Euler factors which are units ([12],Prop.1.1),

$$L_p(F,\omega^{1-i},-i') = \zeta_F(-i') ,$$

when $i + i' \equiv 0$ (modulo $(F_0 : F)$), where $\zeta_F(s)$, $s \in \mathbb{C}$, is the usual zeta function of F. The formula of the theorem is then proved with the following constant:

$$\underline{C} = ((i-1)!)^d (\prod_{\chi(p)=\omega(p)} (c_\chi^i - 1)^{-1}) .$$

3.2. Corollary : Under the hypothesis of Theorem 3, when $L_p(F,\omega^{1-i},i)$ is nonzero, the morphism $K_{2i-1}(\mathcal{O}_F)$ / torsion $\to K_{2i-1}(F_p)$/torsion is injective.

Proof of the corollary: Assuming $L_p(F,\omega^{1-i},i) \neq 0$ the group $K_{2i-1}^{cycl}(\mathcal{O}_F)$ contains a lattice of rank $d = r_1$ whose image via $\rho_i \circ j$ is of finite index in $\mathbb{Z}_p^d$. Since $K_{2i-1}(\mathcal{O}_F)$/torsion $\simeq \mathbb{Z}^d$ the corollary follows.

3.3. Since all the constants c are congruent to one modulo p, one has

$$f'(c^s-1, \psi) \equiv f'(0, \psi) \text{ (modulo } p),$$

for any $s \in \mathbb{Z}_p$. Let i' be an integer such that $i \equiv i'$ modulo $(F_0:F)$, and $i' < 0$. Since $\omega^{1-i} \equiv \omega^{1-i'}$, one gets that $L_p(F,\omega^{1-i},i)$ is nonzero as soon as $\zeta_F(i')$ is not divisible by p. When F is contained in $\mathbb{Q}(\mu_p)$ this will be the case for instance when p-1 divides i-1, or when p is a regular prime.

3.r. One can think that $L_p(F,\omega^{1-i},i)$ is never trivial (for an odd number $i \geq 3$). This can be seen as some analogue of Leopoldt's conjecture about the nonvanishing of the residue at one of the p-adic zeta function of F, a totally real number field. One knows that this conjecture is equivalent to the fact that the map $K_1(\mathcal{O}_F) \otimes \mathbb{Q}_p \to K_1(F_p) \otimes \mathbb{Q}_p$ is injective. It was proved for F abelian using the linear independence of p-adic logarithms [4] .

B. Coleman [7] expressed the value of $L_p(F,\omega^{1-i},s)$ at $s = i$ in terms of p-adic i-logarithms.

3.5. In [15], P. Schneider studies the group

$$R_i(F) = \ker(H^1(A, \mathbb{Q}_p/\mathbb{Z}_p(i)) \to H^1(F_p, \mathbb{Q}_p/\mathbb{Z}_p(i))) .$$

The theorem 3 also proves that the non-vanishing of $L_p(F,\omega^{1-i},i)$ implies that $R_i(F)$ is finite (and $H^2(A, \mathbb{Q}_p/\mathbb{Z}_p(1-i)) = 0$), $i \geq 2$, ([15], §5 Cor. 4 and §4 Lemma 2 i)).

3.6. The "main conjecture" on cyclotomic fields was recently proved by B. Mazur and A. Wiles for the field $\mathbb{Q}$ of rational numbers. If one knew that $L_p(\mathbb{Q},\omega^{1-i},i)$ was non-zero, this would give an exact formula for its valuation, in terms of etale cohomology (see [5] and [15]) (*) .

3.7. L. Villemot (These de 3eme cycle, Orsay) recently extended the results of Gillard [8] to any totally real abelian field F . As a consequence, the Corollary 3.2. is true without assuming that the degree of F over Q is prime to p .

(*) Another consequence of the "main conjecture" is the one of Lichtenbaum relating $\zeta_F(1-i)$ to the ℓ-adic cohomology of $\mathcal{O}_F[1/\ell]$ for i even ([17], I.1) . As a consequence, for $F = \mathbb{Q}$, the hypothesis that ℓ is properly irregular is not needed anymore in [17], Theorem I.2.2.

References

[1] Bayer, P., Neukirch J.: On values of zeta functions and ℓ-adic Euler characteristics. Inv. Math., 50, 1978, pp. 35-64.

[2] Bloch S.: Higher regulators, algebraic K-theory, and zeta functions of elliptic curves, preprint.

[3] Borel A.: Stable real cohomology of arithmetic groups. Ann. Scient. Ec. Norm. Sup., 4ième série, 7, 1974, pp. 235-272.

[4] Brumer A.: On the units of algebraic number fields, Mathematika, 14, 1967, pp.121-124.

[5] Coates J.: On the values of the p-adic zeta functions at the odd positive integers. Unpublished.

[6] Coates J., Lichtenbaum S: On ℓ-adic zeta functions. Ann. of Maths., 98, 1973, pp.498-550.

[7] Coleman B.: P-adic analogues of the multi-logarithms. Preprint

[8] Gillard R.: Unités cyclotomiques, unités semi-locales et $\mathbb{Z}_\ell$-extensions II, Ann. Inst. Fourier, Grenoble, 29, 1979, pp. 1-15.

[9] Iwasawa K.: On some modules in the theory of cyclotomic fields, J. Math. Soc. Japan 16, 1, 1964, pp.42-82.

[10] Iwasawa K.: On $\mathbb{Z}_\ell$-extensions of algebraic number fields, Ann. of Math. 98, 1973, pp. 246-326.

[11] Lang S.: Cyclotomic fields I, Graduate Texts in Maths., 59, 1978; Berlin-Heidelberg-New York. Springer-Verlag.

[12] S. Lichtenbaum: On the values of zeta and L-functions I, Ann. of Maths., 96, 1972, pp. 338-360.

[13] Loday J.-L.: K-théorie et représentations de groupes. Ann. Scient. Ec. Norm. Sup., 4ième série, 9, 1976, pp. 309-377.

[14] Quillen D.: Finite generation of the groups K_i of rings of algebraic integers. Lec. Notes in Maths. n° 341, 1973, pp. 179-210. Berlin-Heidelberg-New York. Springer-Verlag.

[15] Schneider P.: Über gewisse Galoiscohomologiegruppen. Math. Zeitschrift, 168, 1979, pp. 181-205.

[16] Serre J.-P.: Cohomologie galoisiemne. Lec. Notes in Maths. n° 5, 1964. Berlin-Heidelberg-New York. Springer-Verlag.

[17] Soulé C.: K-théorie des anneaux d'entiers de corps de nombres et cohomologie étale, Inv. Math., 55, 1979, pp. 251-295.

[18] Tate J.: Relations between K_2 and Galois cohomology. Inv. Math., 36, 1976, pp. 257-274.

[19] Tate J.: On the torsion in K_2 of fields, in Algebraic Number Theory symposium, Kyoto, 1977, pp. 243-261. S. Iyanaga Ed.

[20] Wagoner J. B.: Continuous cohomology and p-adic K-theory. Lec. Notes in Maths. 551, 1976, pp. 241-248. Berlin-Heidelberg-New York. Springer Verlag.

Rational K-theory of the dual numbers of a ring of algebraic integers*

by C. Soulé,** C.N.R.S. , Paris VII.

0. Let F be a number field, $d=[F:\mathbb{Q}]$ its degree, $\mathcal{O}$ its ring of integers. Denote by $A=\mathcal{O}[\varepsilon]$, $\varepsilon^2=0$, the ring of dual numbers on $\mathcal{O}$. We shall compute the rational K-theory of A.

__Theorem:__ __For any integer $n\geq 0$, the group $K_n(A)$ is finitely generated. It is the direct sum of $K_n(\mathcal{O})$ with a group R_n satisfying:__

$$R_n \otimes \mathbb{Q} = \begin{cases} \mathbb{Q}^d & \text{if } n \text{ is odd} \\ 0 & \text{if } n \text{ is even.} \end{cases}$$

Recall that the rational K-theory of $\mathcal{O}$ is known by a result of Borel [1].

__Proof:__

1. The injection $\mathcal{O}\to A$ given by the constant term admits a splitting $A\to\mathcal{O}$ (mapping ε to zero). Therefore $K_n(\mathcal{O})$ is a direct summand in $K_n(A)$.

2. Consider the general linear group $GL_m(A)$, $m\geq 1$. The reduction modulo ε yields (split) exact sequences:

---

* I am grateful to the referee for giving many accurate corrections to the first version of this paper.

** Partially supported by NSF.

$$1 \to 1 + \varepsilon M_m(\mathcal{O}) \to GL_m(A) \to GL_m(\mathcal{O}) \to 1 ,$$

where $M_m(\mathcal{O})$ is the additive group of m by m matrices over $\mathcal{O}$. The splitting of this extension gives an action of $GL_m(\mathcal{O})$ on the kernel $1+\varepsilon M_m(\mathcal{O})$. The $GL_m(\mathcal{O})$-module $1+\varepsilon M_m(\mathcal{O})$ obtained this way is isomorphic to the additive group $M_m(\mathcal{O})$, on which $GL_m(\mathcal{O})$ acts by conjugation:

$$(1+\varepsilon m).(1+\varepsilon m') = 1 + \varepsilon(m+m')$$
$$g(1+\varepsilon m)g^{-1} = 1 + \varepsilon(gmg^{-1}) , \quad m,m' \in M_m(\mathcal{O}), \ g \in GL_m(\mathcal{O}).$$

In other words, $GL_m(A)$ is isomorphic to the semi-direct product $GL_m(\mathcal{O}) \ltimes M_m(\mathcal{O})$. It is thus an arithmetic group, and its homology is finitely generated [2]. It was proved in [6] that the homology of $GL_m(A)$ is stable (with respect to m). Therefore $K_n(A)$ is finitely generated.

3. To compute $K_n(A)\otimes\mathbb{Q}$ we shall use the same method as Borel for $K_n(\mathcal{O})\otimes\mathbb{Q}$. First we remark that $K_1(A)=\mathcal{O}^*\oplus\mathcal{O}$, so the theorem is true for n=1.

The graded vector space $\mathbb{Q}\oplus \bigoplus_{n\geq 2} K_n(A)\otimes\mathbb{Q}$ is the primitive part of the Hopf algebra $\bigoplus_{n\geq 0} H_n(SL(A);\mathbb{Q})$ (by the Milnor-Moore theorem applied to the H-space $BSL^+(A)$). It will then be enough to know the rational cohomology groups $H^n(SL_m(A);\mathbb{Q})$ for big values on m.

Let SM_m denote the group of m by m matrices of trace zero. The group $SL_m(A)$ is a semi-direct product $SL_m(\mathcal{O}) \ltimes SM_m(\mathcal{O})$. Let $G_m = Res_{F/\mathbb{Q}}SL_m$ be the semi-simple group over $\mathbb{Q}$ obtained by restriction of scalars from F to $\mathbb{Q}$. Similarly let $H_m = Res_{F/\mathbb{Q}}SM_m$. In particular $(G_m \ltimes H_m)(\mathbb{Q})=SL_m(F[\varepsilon])$, and we have injections $SL_m(A) \to (G_m \ltimes H_m)(\mathbb{R})$.

For the real Lie group $(G_m \ltimes H_m)(\mathbb{R})$, we can consider the continuous cohomology H^n_{cont} with complex coefficients (computed with continuous cochains). The reader is referred to ([3],IX) for the description of this cohomology. If we forget the topology, we get a map

$$H^n_{cont}((G_m \ltimes H_m)(\mathbb{R});\mathbb{C}) \to H^n((G_m \ltimes H_m)(\mathbb{R});\mathbb{C})$$

that we compose with the restriction to $SL_m(A)$.

Lemma 1: The map

$$\Phi : H^n_{cont}((G_m \ltimes H_m)(\mathbb{R});\mathbb{C}) \to H^n(SL_m(A);\mathbb{C})$$

defined above is an isomorphism whenever m is big enough with respect to n.

Proof of Lemma 1:

Since $SL_m(A)$ (resp. $(G_m \ltimes H_m)(\mathbb{R})$) is an extension of $SL_m(\mathcal{O})$ by $SM_m(\mathcal{O})$ (resp. $G_m(\mathbb{R})$ by $H_m(\mathbb{R})$), its cohomology (resp. continuous cohomology) is the limit of a Hochschild-Serre spectral sequence $'E_r^{pq}$ (resp $''E_r^{pq}$) whose second term is

$$'E_2^{pq} = H^p(SL_m(\mathcal{O});H^q(SM_m(\mathcal{O});\mathbb{C}))$$

(resp. $''E_2^{pq}=H^p_{cont}(G_m(\mathbb{R});H^q_{cont}(H_m(\mathbb{R});\mathbb{C}))$, see [3], IX, 5.8). There exists a morphism between these two spectral sequences which converges to Φ. At the level of the E_2 terms it is induced by the morphism

$$H^q_{cont}(H_m(\mathbb{R});\mathbb{C}) \to H^q(SM_m(\mathcal{O});\mathbb{C}).$$

This map is easily seen to be an isomorphism (it concerns abelian groups).

Let V be a finite dimensional complex representation of a semi-simple algebraic group G over $\mathbb{Q}$, and Γ an arithmetic subgroup of $G(\mathbb{Q})$. One knows that the map

$$H^p_{cont}(G(\mathbb{R});V) \to H^p(\Gamma;V)$$

is an isomorphism when the rank of G is big enough with respect to p (see [1] and [4]). Applying this to $V=H^q_{cont}(H_m(\mathbb{R});\mathbb{C})$, $q\geq 0$, we get that the map $''E_2^{pq} \to {}'E_2^{pq}$ is an isomorphism whenever m is big enough with respect to p. Passing to the limit in the spectral sequences, this proves Lemma 1.

4. In this paragraph the integer m will be fixed.

A theorem of Van-Est ([3], IX 5.6.) asserts that the continuous cohomology of a real Lie group can be computed in terms of (relative) Lie algebra cohomology. Here consider the following (real) Lie algebras: $\mathfrak{g}=\mathrm{Lie}_{\mathbb{R}} G_m(\mathbb{R})$, $\mathfrak{h}=\mathrm{Lie}_{\mathbb{R}} H_m(\mathbb{R})$, $\mathfrak{L}=\mathfrak{g}+\mathfrak{h}$ with $\mathfrak{g}$ acting on $\mathfrak{h}$ via the adjoint representation, and

$$\mathfrak{k} = \mathrm{Lie}_{\mathbb{R}}(\text{maximal compact subgroup of } G_m(\mathbb{R})).$$

To be explicit $\mathfrak{g}=\mathfrak{sl}_m(\mathbb{C})^{r_2}\times \mathfrak{sl}_m(\mathbb{R})^{r_1}$ and $\mathfrak{k}=\mathfrak{su}_m^{r_2}\times \mathfrak{so}^{r_1}$, where $\mathfrak{sl}$ (resp. $\mathfrak{su}$, $\mathfrak{so}$) is the Lie algebra of the special linear (resp. unitary, resp. orthogonal) group, and r_1 (resp. r_2) is the number of real (resp. complex) places of F. The Lie algebra $\mathfrak{h}$ is abelian, and isomorphic to $\mathfrak{g}$ as $G_m(\mathbb{R})$-module. By Van-Est theorem (loc. cit.) we have:

$$H^n_{cont}((G_m \ltimes H_m)(\mathbb{R}),\mathbb{C}) = H^n(\mathfrak{L},\mathfrak{k};\mathbb{C}).$$

Furthermore the Hochschild-Serre spectral sequence $''E_r^{pq}$ considered above in 3. for continuous cohomology is isomorphic to one in Lie algebra cohomology ([3], I 6.5.) whose second term is

$$E_2^{pq} = H^p(\mathfrak{g},\mathfrak{k};H^q(\mathfrak{h};\mathbb{C}))$$

and converging to $H^{p+q}(\mathfrak{L},\mathfrak{k};\mathbb{C})$.

Since $H^q(\mathfrak{h};\mathbb{C})$ is finite dimensional and fully reducible under $G(\mathbb{R})$, and since the relative Lie algebra cohomology of $\mathfrak{g}$ modulo $\mathfrak{k}$ in an irreducible non trivial finite dimensional $\mathfrak{g}$-module is trivial ([3], I 4.2.), we get

$$E_2^{pq} = H^p(\mathfrak{g},\mathfrak{k};\mathbb{C}) \otimes H^q(\mathfrak{h};\mathbb{C})^{\mathfrak{g}} .$$

<u>Lemma 2</u>: <u>The spectral sequence above degenerates</u>: $E_2^{pq} = E_\infty^{pq}$.

<u>Proof of Lemma 2</u>:

The cohomology of $\mathfrak{L}$ modulo $\mathfrak{k}$ is the one of the standard complex $([\Lambda^n(\mathfrak{L}/\mathfrak{k})^*]^{\mathfrak{k}},\partial)$, where V^* denotes the dual of a complex vector space V, and $\Lambda^n V$ its n-th exterior product. The coboundary map ∂ is defined by the formula

$$(*) \qquad (\partial\omega)(x) = \sum_i (-1)^i x_i.\omega(x_0\wedge\ldots\wedge\hat{x}_i\wedge\ldots\wedge x_n)$$

$$+ \sum_{i<j} (-1)^{i+j}\omega([x_i,x_j]\wedge x_0\wedge\ldots\wedge\hat{x}_i\wedge\ldots\wedge\hat{x}_j\wedge\ldots\wedge x_n).$$

where ω is in $[\Lambda^n(\mathfrak{L}/\mathfrak{k})^*]^{\mathfrak{k}}$, and $x=x_0\wedge\ldots\wedge x_n$ is in $\Lambda^{n+1}(\mathfrak{L}/\mathfrak{k})$. The first sum is zero here, since the action of $\mathfrak{L}$ on the coefficients is trivial. The decomposition $\mathfrak{L} = \mathfrak{g}+\mathfrak{h}$ leads to

$$\Lambda^n(\mathfrak{L}/\mathfrak{k})^* = \bigoplus_{p+q=n} \Lambda^p(\mathfrak{g}/\mathfrak{k})^* \otimes \Lambda^q(\mathfrak{h})^*$$

and the coboundary map ∂ is a direct sum $\partial = \bigoplus_{r\geq 1} \partial_r$, where ∂_r sends $[\Lambda^p(\mathfrak{g}/\mathfrak{k})^* \otimes \Lambda^q(\mathfrak{g})^*]^{\mathfrak{k}}$ to $[\Lambda^{p+r}(\mathfrak{g}/\mathfrak{k})^* \otimes \Lambda^{q+1-r}(\mathfrak{g})^*]^{\mathfrak{k}}$. This is the origin of the spectral sequence E_r^{pq}.

This spectral sequence admits a multiplicative structure, given by cup-products in cohomology. Therefore, to prove $E_2^{pq} = E_\infty^{pq}$, it is enough to prove that all higher differentials d_r, $r\geq 2$, are zero on E_2^{on}, $n\geq 0$. For this it will be enough to show that the subspace $\mathbb{C}\otimes[\Lambda^n(\mathfrak{h})^*]^{\mathfrak{g}}$ of $[\Lambda^n(\mathfrak{L}/\mathfrak{k})^*]^{\mathfrak{k}}$ is killed by ∂. Actually, $\partial_1 = 0$ will imply that this subspace is a set of representatives for E_2^{on}, and $\partial_r = 0$, $r\geq 2$, will show that the higher derivatives d_r are trivial.

Now look at the formula (*) above, and, for $r\geq 1$, assume that the elements $x_0,\ldots,x_{r-1}$ are in $\mathfrak{g}/\mathfrak{k}$, $x_r,\ldots,x_n$ are in $\mathfrak{h}$, and ω is in $\mathbb{C}\otimes[\Lambda^n(\mathfrak{h})^*]^{\mathfrak{g}}$. For $r=1$, since ω is $\mathfrak{g}$-invariant, we have

$$(\partial\omega)(x) = \sum_j \omega(x_1\wedge\ldots\wedge[x_0,x_j]\wedge\ldots\wedge x_n) = 0 .$$

For $r\geq 2$, it is easy to see that

$$\omega([x_i,x_j]\wedge\ldots\wedge\hat{x}_i\wedge\ldots\wedge\hat{x}_j\wedge\ldots\wedge x_n) = 0 ,$$

since the term in parentheses is not contained in $\Lambda^n(\mathfrak{h})$. q.e.d.

5. Let us denote by $I(H^\cdot)$ (resp. $I^n(H^\cdot)$) the indecomposable quotient of a graded algebra $H^\cdot$ (resp. of H^n). We get from Lemma 2

$$I(H^\cdot_{cont}(G_m \ltimes H_m)(\mathbb{R});\mathbb{C}) = I(H^\cdot(\mathfrak{g},\mathfrak{k};\mathbb{C})) \oplus I((\Lambda^\cdot(\mathfrak{h})^*)^{\mathfrak{g}}).$$

We know from paragraph 3. above that $K_n(A)\otimes\mathbb{C}$ has the same dimension over $\mathbb{C}$ as $I^n(H^\cdot_{cont}(G_m \ltimes H_m)(\mathbb{R});\mathbb{C})$ for $m>>n>1$. Furthermore ([1]) the dimension of $K_n(\mathcal{O})\otimes\mathbb{C}$ is that of $I^n(H^\cdot_{cont}(\mathfrak{g},\mathfrak{k};\mathbb{C}))$ for m big enough. So $\dim_{\mathbb{C}}(R_n\otimes\mathbb{C}) = \dim_{\mathbb{C}}(I^n((\Lambda^\cdot(\mathfrak{h})^*)^{\mathfrak{g}}))$ for m big enough. One knows (see for instance [5], 9.2.) that $(\Lambda^n(\mathfrak{g})^*)^{\mathfrak{g}} \simeq H^n(\mathfrak{g};\mathbb{C})$ $= H^n(\mathfrak{sl}_m^d(\mathbb{C});\mathbb{C})$. For this algebra the indecomposable quotient has dimension d in odd degrees n satisfying $1<n<2m$, and is zero otherwise. This concludes the proof of the theorem.

6. Remarks:

- J. -L. Loday (in these Proceedings) describes a generator of the free summand of $K_3(\mathbb{Z}[\varepsilon]), \varepsilon^2=0$, in terms of (generalized) symbols.
- The referee observed that the same proof as above applies to the groups ${}_{\eta}L(A)$ of Karoubi, $\eta=\pm 1$. Using analogous notation, one gets that $R_n \otimes \mathbb{Q}$ has dimension d when $n \equiv 3 \pmod 4$, and is zero otherwise.
- Maybe one could extend the method given here to the rings $A = \mathcal{O}[t]/(t^{\alpha})$, $\alpha \geqslant 3$.

References

[1] A. Borel: Stable real cohomology of arithmetic groups. Ann. Scient. Ec. Norm. Sup., 4è série, 7,235-272, 1974.

[2] A. Borel, J. -P. Serre: Corners and arithmetic groups. Comm. Math Helv., 48, 244-297, 1974.

[3] A. Borel, N. Wallach: Continuous cohomology; discrete subgroups, and representations of reductive groups. Ann. of Math. Studies. Princeton University Press, no.94.

[4] F. T. Farrell, W. C. Hsiang: On the rational homotopy groups of the diffeomorphisms groups of discs, spheres and aspherical manifolds, Proc. of Symposia in Pure Maths., Vol. 32, 325-337, 1978.

[5] J. -L. Koszul: Homologie et cohomologie des algèbres de Lie: Bull. SMF 78, 65-127, 1950.

[6] W. Van der Kallen: Homology stability for linear groups, Preprint.

On K_2 and K_3 of truncated polynomial rings

by

Jan Stienstra*

Introduction. K-theory of truncated polynomial rings is the study of the groups $K_i(R[t]/(t^n), (t))$, or rather of the projective systems

$$K_i(R[t]/(t^2),(t)) \leftarrow K_i(R[t]/(t^3)),(t)) \leftarrow \ldots \leftarrow K_i(R[t]/(t^n),(t)) \leftarrow \ldots$$

For i = 1 it is the study of (generalized) Witt vectors [2], [15], since

$$K_1(R[t]/(t^n), (t)) \cong (1 + tR[t]/(t^n))^{\times}$$

Witt vectors have been studied thoroughly, because of their role in number theory and the theory of formal groups. K-theory profits from this.

For i = 2 the study started with Van der Kallen's K_2 of the dual numbers [12]:

$$K_2(R[t]/(t^2), (t)) \cong \Omega^1_{R/\mathbb{Z}} \text{ if } \tfrac{1}{2} \in R.$$

Graham extended this to K_2 of truncated polynomials over $\mathbb{Q}$-algebras [7] and Roberts and Geller did the case $R = \mathbb{Z}$ [19]:

$$K_2(R[t]/(t^n), (t)) \cong [\Omega^1_{R/\mathbb{Z}}]^{\oplus(n-1)} \text{ if } \mathbb{Q} \subset R$$

$$K_2(\mathbb{Z}[t]/(t^n), (t)) \cong \bigoplus_{l=2}^{n} \mathbb{Z}/_{l\,\mathbb{Z}}$$

The deepest and most complicated results in this field were obtained in Bloch's study of K_2 of truncated polynomial rings and the symbol part in the higher K's in characteristic $\geqslant 3$ [2]. He found a fantastic application of these results to crystalline cohomology. Later Deligne and Illusie set up another machinery (the De Rham-Witt complex) with the same applications but without K-theory [10], [11].

*Partially supported by an N.S.F. grant.

For $i \geq 3$ it is the usual story in K-theory: one has a definition, one has to devise new techniques and till now one has only very few explicit results. Evens and Friedlander computed

$$K_3(\mathbb{F}_p[t]/(t^2), (t)) \cong \mathbb{Z}/p \oplus \mathbb{Z}/p$$
$$K_4(\mathbb{F}_p[t]/(t^2), (t)) = 0$$

for $p \geq 5$ [6] and Snaith showed (using the lower bounds found in this paper)

$$K_3(\mathbb{F}_q[t]/(t^2), (t)) \cong \mathbb{F}_q \oplus \mathbb{F}_q$$

if $q = 2^r$ [21]. Soulé computed the rank of the group $K_i(\mathcal{O}[t]/(t^2), (t))$ when $\mathcal{O}$ is the ring of integers in a number field F [22]:

$$\dim_{\mathbb{Q}} K_i(\mathcal{O}[t]/(t^2),(t)) \otimes \mathbb{Q} = \begin{cases} [F:\mathbb{Q}] & \text{if } i \text{ odd} \\ 0 & \text{if } i \text{ even} \end{cases}$$

Between K_2 and K_3 are the relative K_2's i.e. the groups $K_2(R[t]/(t^q), (t^n))$. The presentation given by Maazen and myself [16] and by Keune [13] makes relative K_2's accessible for systematic, brute force computations. Results apply on the one hand to $K_2(R[t]/(t^q), (t))$ (take $n = 1$) and on the other hand via the exact sequence

$$K_3(R[t]/(t^q), (t)) \to K_3(R[t]/(t^n), (t)) \to K_2(R[t]/(t^q), (t^n)) \to K_2(R[t]/(t^q), (t)) \to K_2(R[t]/(t^n), (t))$$

they give information about $K_3(R[t]/(t^n), (t))$.
These relative K_2's and their relation with (absolute) K_2 and K_3's are the subject of this paper. In §1 we recall a few facts about the presentation, find a smaller set of generators and derive a relation (1.9), (1.10) which turns out to be extremely important for the subsequent sections. In §2 we study for certain rings the groups

$K_2(R[t]/(t^q), (t))$ and compare them with the more general $K_2(R[t]/(t^q), (t^n))$. Doing this we get another proof of Bloch's results for K_2, in which we can avoid crystalline d log and Ω^2 by using Frobenius operations. The kernel of the map $K_2(R[t]/(t^q), (t^n)) \to$ $\to K_2(R[t]/(t^q), (t))$ gives information about the part of $K_3(R[t]/(t^n), (t))$ which can not be lifted to $K_3(R[t]/(t^q), (t))$. Note that symbols in K_3 can be lifted and that therefore our results are not covered by Bloch's. In §3 we organize the elements of this kernel. If R is a $\mathbb{Q}$-algebra or $R = \mathbb{Z}$ we show for $q \geq 2n$

$$R^{n-1} \xrightarrow{\sim} \ker[K_2(R[t]/(t^q), (t^n)) \to K_2(R[t]/(t^q), (t))]$$

In the $R = \mathbb{Z}$ case the thus found surjection

$$K_3(\mathbb{Z}[t]/(t^2), (t)) \to \mathbb{Z}$$

splits, in a canonical way as Loday can show, and according to Soulé's result mentioned above $K_3(\mathbb{Z}[t]/(t^2), (t))$ is the direct sum of this $\mathbb{Z}$ and a torsion group. If R is an algebra over a perfect field k of characteristic $p > 0$, which can be lifted to a smooth algebra over W(k) (the Witt vectors of k), we construct for q large enough a surjection

$$\begin{array}{l} K_1(R[t]/(t^{2n}), (t))/\{(1-at^n) \mid a \in R\} \\ \quad \Big\downarrow \Delta_n \\ \qquad \ker[K_2(R[t]/(t^q), (t^n)) \to K_2(R[t]/(t^q), (t))] \end{array}$$

Though I expected, on the basis of all kinds of computations in [23], already for a long time a relation between K_1 and that kernel, I could write down the map Δ_n only after learning from an observation by Raynaud about the De Rham-Witt complex. For those readers familiar with De Rham-Witt and its relation to K-theory we mention Raynaud's result: the sequence

$$0 \longrightarrow W\Omega^{i-1} \xrightarrow[\binom{F^n d}{F^n}]{} W\Omega^{i} \oplus W\Omega^{i-1} \xrightarrow[(V^n,-dV^n)]{} W\Omega^{i} \longrightarrow W_n\Omega^{i} \longrightarrow 0$$

is exact.

I conjecture that Δ_n is actually an isomorphism. This is equivalent to

<u>Conjecture</u>. If R is a perfect field then for all n and $q \gg n$

$$K_1(R[t]/(t^{2n}),\ (t))/\{(1-at^n)\,|\,a \in R\} \underset{\Delta_n}{\tilde{\to}} K_2(R[t]/(t^q),\ (t^n))$$

It seems likely that an explicit formula for the inverse of Δ_n involves variations on the dilogarithm function (which has already come into K-theory via work on regulator maps and tame symbols (Bloch, Deligne, Beilinson) and norm residue symbols (Coleman)). Unfortunately, I have not yet found the correct formula, except for $n < \frac{p}{2}$ or $n = p$. The conjecture will be proved for this case in §4. Note that as a consequence of the conjecture we get a surjection

$$K_3(R[t]/(t^n),\ (t)) \twoheadrightarrow K_1(R[t]/(t^{2n}),\ (t))/\{(1-at^n)\,|\,a \in R\}$$

if R is an algebra over a perfect field k of characteristic $p > 0$, which is liftable to a smooth W(k)-algebra. The kernel of this map consists of all elements of $K_3(R[t]/(t^n),\ (t))$ which can be lifted to $K_3(R[t]/(t^q),\ (t))$ for every $q > n$. Incidently, the map is not compatible with the obvious maps on K_3 and K_1 which arise when n varies. Taking a special case in which the conjecture is true, we find for $p \neq 3$ a surjection

$$K_3(\mathbb{F}_p[t]/(t^2),\ (t)) \twoheadrightarrow \mathbb{F}_p \oplus \mathbb{F}_p$$

This gives a lower bound for $K_3(\mathbb{F}_p[t]/(t^2),\ (t))$. As said, Evens and Friedlander for $p \geqslant 5$ and Snaith for $p = 2$ have shown that it is actually an equality. As for the missing case $p = 3$, the conjecture implies a surjection

$$K_3(\mathbb{F}_3[t]/(t^2),\ (t)) \twoheadrightarrow \mathbb{Z}/9$$

Maybe this is even an isomorphism. Who knows?

The research for this paper was done partly for the author's Ph.D. thesis [23] and partly during his instructorship at the University of Chicago. I want to thank Ans van Hoof of the mathematics department in Utrecht for typing the manuscript.

§1

For an ideal I in a ring S (always commutative with 1) one can define relative K-groups $K_i(S, I)$ so that there is a long exact sequence

$$(1.1)\quad \ldots \to K_3(S) \xrightarrow{\pi_3} K_3(S/I) \to K_2(S, I) \to K_2(S) \xrightarrow{\pi_2}$$

$$\xrightarrow{\pi_2} K_2(S/I) \to K_1(S, I) \to K_1(S) \xrightarrow{\pi_1} K_1(S/I) \to \ldots$$

(see [13], [14]). If I is contained in the Jacobson radical of S, one has the following facts, proved in [1], [16], [13]

(1.2) The maps π_1 and π_2 are surjective

(1.3) $K_1(S, I) = (1 + I)^\times$

We reformulate this as follows

(1.3.a) $K_1(S, I)$ is the abelian group given in a presentation with generators $\langle a\rangle$, one for every $a \in I$, and defining relations

$\langle a\rangle + \langle b\rangle = \langle a + b - ab\rangle$

Of course, $\langle a\rangle$ corresponds to $1 - a \in 1 + I$.

(1.4) $K_2(S, I)$ is the abelian group which has a presentation with generators $\langle a, b\rangle$, one for every $(a, b) \in S \times I \cup I \times S$, and defining relations

(D1) $\langle a, b\rangle = - \langle b, a\rangle$	if $a \in I$
(D2) $\langle a, b\rangle + \langle a, c\rangle = \langle a, b + c - abc\rangle$	if $a \in I$ or $b, c \in I$
(D3) $\langle a, bc\rangle = \langle ab, c\rangle + \langle ac, b\rangle$	if $a \in I$.

(1.5) <u>Lemma</u>. If I is an ideal in the Jacobson radical of S, the following relations hold in $K_2(S, I)$

(1) $\langle a, 1\rangle = \langle 1, a\rangle = 0$ for $a \in I$

$\langle a, 0\rangle = \langle 0, a\rangle = 0$ for $a \in S$

(2) $\langle a, b^m\rangle = m\langle ab^{m-1}, b\rangle$

for all $m \in \mathbb{N}$ and $(a, b) \in S \times I \cup I \times S$

(3) $m\langle a, b\rangle = \langle \sum_{i=1}^{m} (-1)^{i-1}\binom{m}{i} a^i b^{i-1}, b\rangle$

for all $m \in \mathbb{N}$ and $(a, b) \in S \times I \cup I \times S$

(4) In particular, if S is an $\mathbb{F}_p$-algebra,

$p^r\langle a, b\rangle = \langle a^{p^r} b^{p^r - 1}, b\rangle$

for all $r \in \mathbb{N}$ and $(a, b) \in S \times I \cup I \times S$

(5) If S is an $\mathbb{F}_p$-algebra and I is nilpotent, every element of $K_2(S, I)$ is killed by a power of p.

Let $\tilde{S}$ be a ring containing S. If $a \in I$ and $b \in S \cap \tilde{S}^*$, then the image of $\langle a, b\rangle$ under the map $K_2(S, I) \to K_2(\tilde{S})$ is the Steinberg symbol $\{1 - ab, b\}$.

Proof. See [16] □

If S is a truncated polynomial ring, say $S = R[t]/(t^q)$, and I is the ideal generated by a power of t, say $I = (t^n)$, one can find generators of $K_2(S, I)$ which look nicer and are better to handle than the monstrous expressions

$$\langle a_0 + a_1 t + \dots + a_{q-1}t^{q-1}, b_n t^n + \dots + b_{q-1}t^{q-1}\rangle,$$

which (1.4) gives. This will be shown in (1.6) and (1.7). Similar things are well known for K_1 (i.e. for Witt vectors). We will mention those in passing, without proof.

(1.6) Lemma. Let I and J be ideals in the Jacobson radical of S, with $J \subset I$. Then the following two sequences (with the obvious maps) are exact

$$K_2(S, J) \to K_2(S, I) \to K_2(S/J, I/J) \to 0$$

$$0 \to K_1(S, J) \to K_1(S, I) \to K_1(S/J, I/J) \to 0$$

<u>Proof</u>. The only thing which may not be a priori obvious is

$$\mathrm{image}[K_2(S, J) \to K_2(S, I)] \supset \ker[K_2(S, I) \to K_2(S/J, I/J)].$$

Let $a \in R$, $b \in I$, c, $d \in J$. Then one has in $K_2(S, I)$

$$\langle a + c, b + d\rangle = \langle a, b + d\rangle + \langle c(1 - ab - ad)^{-1}, b + d\rangle$$

$$= \langle a, b\rangle + \langle a, d(1 - ab)^{-1}\rangle + \langle c(1 - ab - ad)^{-1}, b + d\rangle$$

The last two terms on the right are in the image of $K_2(S, J)$. This shows that one can define a homomorphism from $K_2(S/J, I/J)$ to $\mathrm{coker}[K_2(S, J) \to K_2(S, I)]$ by sending a generator $\langle \bar{a}, \bar{b}\rangle$ to the coset containing $\langle a, b\rangle$, where a and b are any old liftings of $\bar{a}$ and $\bar{b}$ respectively. This map is clearly inverse to the obvious one from $\mathrm{coker}[K_2(S, J) \to K_2(S, I)]$ to $K_2(S/J, I/J)$. □

(1.7) <u>Proposition</u>. For every ring R and all integers $q > n \geqslant 1$ the group $K_2(R[t]/(t^q), (t^n))$ is generated by the elements $\langle at^i, t\rangle$ and $\langle at^i, b\rangle$ with $a, b \in R$ and $n \leqslant i < q$. Similarly, $K_1(R[t]/(t^q), (t^n))$ is generated by the elements $\langle at^i\rangle$ with $a \in R$ and $n \leqslant i < q$.

<u>Proof</u> for K_2. Use induction on $q - n$. If $q = n + 1$ the group $K_2(R[t]/(t^q), (t^n))$ is generated by the expressions $\langle at^n, b_0 + b_1t + b_2t^2 + \ldots + b_nt^n\rangle$. Now make the following computation:

$$\langle at^n, b_0+b_1t+b_2t^2+\ldots+b_nt^n\rangle = \langle at^n, b_0\rangle + \langle at^n, b_1t\rangle + \langle at^n, b_2t^2+\ldots+b_nt^n\rangle$$

by twice (D2). Then (D3) and (1.5) show

$$\langle at^n, b_1t\rangle = \langle ab_1t^n, t\rangle$$

$$\langle at^n, b_2t^2 + \ldots + b_nt^n\rangle = 0$$

Thus one sees that $K_2(R[t]/(t^{n+1}), (t^n))$ is generated by the elements $\langle at^n, b\rangle$ and $\langle at^n, t\rangle$ with $a, b \in R$. The induction step is provided by (1.6) plus the fact that there is one very trivial way to lift

the named generators from $K_2(R[t]/(t^{q-1}), (t^n))$ to $K_2(R[t]/(t^q), (t^n))$.
The K_1-case is left as a simple exercise. □

(1.8) <u>Proposition</u>. Let m be an integer which is invertible in the ring R. Then for all $q > n \geq 1$ the groups $K_2(R[t]/(t^q), (t^n))$ and $K_1(R[t]/(t^q), (t^n))$ are m-divisible i.e. multiplication by m is an automorphism.

<u>Proof</u>. For K_2. Fix q and n. The group $K_2(R[t]/(t^q), (t^n))$ carries a finite filtration with Fil^0 = the whole group, and for $r > 1$

$$\mathrm{Fil}^r = \text{subgroup generated by the } \langle at^s, b\rangle \text{ and } \langle at^{s-1}, t\rangle \text{ with } s \geq n + r \text{ and } a, b \in R$$

It suffices to show that multiplication by m is an automorphism of $\mathrm{Fil}^r/\mathrm{Fil}^{r+1}$ for $r = 0, 1, \ldots$. Well, the substitution $t \mapsto mt$ induces an automorphism on $K_2(R[t]/(t^q), (t^n))$ because m is invertible in R. One can easily see that this automorphism preserves the filtration and hence induces an automorphism of each $\mathrm{Fil}^r/\mathrm{Fil}^{r+1}$. We compare this with the map which we actually want i.e. multiplication by m. Fix r and put $s = n + r$. From the fact that in $K_2(R[t]/(t^{s+1}), (t^n))$ one has $\langle a(mt)^s, b\rangle = m^s\langle at^s, b\rangle$, one deduces with (1.6)

$$\langle a(mt)^s, b\rangle \equiv m^s\langle at^s, b\rangle \text{ in } \mathrm{Fil}^r/\mathrm{Fil}^{r+1}.$$

As for the other generators, in $K_2(R[t]/(t^{s+1}), (t^n))$ one computes with (D3) and (1.5):

$$\begin{aligned}
\langle a(mt)^{s-1}, mt\rangle &= \langle am^s t^{s-1}, t\rangle + \langle am^{s-1}t^s, m\rangle \\
\langle am^s t^{s-1}, t\rangle &= m^s\langle at^{s-1}, t\rangle - \langle -\tbinom{m^s}{2}a^2t^{2s-1}, t\rangle \\
\langle am^{s-1}t^s, m\rangle &= m\langle am^{s-1}t^s, 1\rangle - \langle am^{s-1}t^s, -\tbinom{m}{2}am^{s-1}t^s\rangle \\
&= -s\langle -\tbinom{m}{2}a^2m^{2s-2}t^{2s-1}, t\rangle
\end{aligned}$$

From these computations one concludes

$$\langle a(mt)^{s-1}, mt\rangle \equiv m^s\langle at^{s-1}, t\rangle \text{ in } \mathrm{Fil}^r/\mathrm{Fil}^{r+1}$$

So the automorphism which the substitution $t \mapsto mt$ induces on Fil^r/Fil^{r+1} is multiplication by m^s. Hence multiplication by m induces an automorphism on each Fil^r/Fil^{r+1} and therefore also on $K_2(R[t]/(t^q), (t^n))$.

A similar argument works for K_1, but there is also an alternative. Namely, an element of $K_1(R[t]/(t^q), (t^n))$ is a truncated polynomial of the form $1 + t^n f(t)$ and there is an explicit "Taylor series" for its m-th root:

$$(1 + t^n f(t))^{1/m} = \sum_{i=0}^{"\infty"} \binom{1/m}{i} (t^n f(t))^i \qquad \square$$

Relations (2) and (4) of (1.5) are very powerful tools, but they have a serious limitation: one of the arguments should be in the ideal I. It is not sufficient that one can write down both sides of a relation like $\langle a, b^m\rangle = m\langle ab^{m-1}, b\rangle$. In [16] p. 276 + 279 we pointed already at some of these troubles. The present paper is devoted to a more thorough treatment. For instance (2) and (4) of (1.5) and (D1) say that $\langle a, b^m\rangle + m\langle b, ab^{m-1}\rangle$ and $\langle a^{p^r} b^{p^r-1}, b\rangle + \langle a^{p^r-1} b^{p^r}, a\rangle$ are zero if a or b is in I. This happens more often, however. What conditions are really necessary? The following two lemmas give a hint.

(1.9) <u>Lemma</u>. Let R be a ring and $a \in R$. Let n, q, m, k be non-negative integers satisfying $1 \leq n < q$ and $m + k \geq 2n$. Then the relation

$$\langle t^m, at^k\rangle + m\langle at^{m+k-1}, t\rangle = 0$$

is valid in $K_2(R[t]/(t^q), (t^n))$.

<u>Proof</u>. Since for every integer i with $0 \leq i \leq m - 1$ at least one of the inequalities $k + i \geq n$ or $m - i - 1 \geq n$ holds, one is allowed to write

$$\langle at^{k+i}, t^{m-i}\rangle = \langle at^{k+i+1}, t^{m-i-1}\rangle + \langle at^{k+m-1}, t\rangle$$

Induction does the rest. $\square$

(1.10) Lemma. Let R be an $\mathbb{F}_p$-algebra. Let $a \in R$. Let n, q, r, s, m be integers satisfying $1 \leqslant n < q$, $0 \leqslant s \leqslant r$ and

$$mp^{\frac{1}{2}(r+s)} \geqslant 2n \text{ if } r + s \text{ is even}$$
$$mp^{\frac{1}{2}(r+s-1)} \geqslant n \text{ if } r + s \text{ is odd.}$$

Then the relation

$$\langle a^{p^{r-s}-1}t^{mp^r}, a\rangle + m\langle a^{p^r}t^{mp^{r+s}-1}, t\rangle = 0$$

is valid in $K_2(R[t]/(t^q), (t^n))$.

Proof. First the case with r + s even. Put $k = \frac{1}{2}(r - s)$. So $r - s = 2k$, $r + s = 2r - 2k$ and $mp^{r-k} \geqslant 2n$. One can calculate:

$$\begin{aligned}\langle a^{p^{2k}-1}t^{mp^r}, a\rangle &= p^k\langle a^{p^k-1}t^{mp^{r-k}}, a\rangle \\ &= \langle t^{mp^{r-k}}, a^{p^k}\rangle \\ &= -mp^{r-k}\langle a^{p^k}t^{mp^{r-k}-1}, t\rangle \\ &= -m\langle a^{p^r}t^{mp^{2r-2k}-1}, t\rangle\end{aligned}$$

This proves the claim when r + s is even.

Next the case r + s odd. Put $k = \frac{1}{2}(r - s - 1)$. So $r - s = 2k + 1$, $r + s = 2r - 2k - 1$ and $mp^{r-k-1} \geqslant n$. Now the following computation proves the claim:

$$\begin{aligned}\langle a^{p^{2k+1}-1}t^{mp^r}, a\rangle &= p^{k+1}\langle a^{p^k-1}t^{mp^{r-k-1}}, a\rangle \\ &= p\langle t^{mp^{r-k-1}}, a^{p^k}\rangle \\ &= -\langle a^{p^{k+1}}(t^{mp^{r-k-1}})^{p-1}, t^{mp^{r-k-1}}\rangle \\ &= -mp^{r-k-1}\langle a^{p^{k+1}}t^{mp^{r-k}-1}, t\rangle \\ &= -m\langle a^{p^r}t^{mp^{2r-2k-1}-1}, t\rangle\end{aligned}$$

□

(1.11) Theorem. For every ring R there is an exact sequence

$$R/2R \xrightarrow{\rho} K_2(R[t]/(t^2),\ (t)) \xrightarrow{\sigma} \Omega^1_{R/\mathbb{Z}} \to 0$$

with $\rho(a \bmod 2) = \langle at, t\rangle$ and $\sigma\langle at, b + ct\rangle = adb$ and for all integers n, q with $2 \leqslant n < q < 2n$ there is an isomorphism

$$\tau\colon R^{q-n} \oplus (\Omega^1_{R/\mathbb{Z}})^{\oplus(q-n)} \to K_2(R[t]/(t^q),\ (t^n))$$

with $\tau(a_1,\dots,a_{q-n};\ b_1dc_1,\dots,b_{q-n}dc_{q-n}) =$

$$= \langle a_1t^n, t\rangle + \langle a_2t^{n+1}, t\rangle + \dots + \langle a_{q-n}t^{q-1}, t\rangle +$$

$$+ \langle b_1t^n, c_1\rangle + \langle b_2t^{n+1}, c_2\rangle + \dots + \langle b_{q-n}t^{q-1}, c_{q-n}\rangle$$

Proof. (Sketch). 1. ρ is well-defined because $\langle 2xt, t\rangle = 2\langle xt, t\rangle = \langle x, t^2\rangle = 0$; 2. ρ is a homomorphism because of the relations (D1) and (D2); 3. Put $\sigma\langle at, b + ct\rangle = adb$ and $\sigma\langle b + ct, at\rangle = -adb$ and check that this respects the relations (D1) - (D3); 4. Use $\langle at, b + ct\rangle = \langle at, b\rangle + \langle at, ct\rangle = \langle at, b\rangle + \langle act, t\rangle$ to show that σ induces an isomorphism between coker ρ and $\Omega^1_{R/\mathbb{Z}}$; 5. Relations (D1) and (D2) show that the expressions $\langle at^i, t\rangle$ and $\langle bt^i, c\rangle$ are additive in a and b respectively; 6. The additivity of $\langle bt^i, c\rangle$ in c follows from $\langle bt^i, c\rangle + \langle bt^i, e\rangle = \langle bt^i, c + e - bcet^i\rangle =$
$= \langle bt^i, c + e\rangle + \langle bt^i, -bcet^i\rangle$ and $\langle bt^i, -bcet^i\rangle =$
$\langle bt^{2i}, -bce\rangle + i\langle -b^2cet^{2i-1}, t\rangle = 0$; 7. It follows from the last two remarks and relation (D3) that the formula given for τ defines a homomorphism.

8. A few more words about the inverse of τ. First note that there is a canonical isomorphism

$$\chi\colon R^{q-n} \oplus (\Omega^1_{R/\mathbb{Z}})^{\oplus(q-n)} \xrightarrow{\sim} \left[\frac{t^n\Omega^1_{R[t]/(t^{q+1})/\mathbb{Z}}}{t^q\Omega^1_{R/\mathbb{Z}} + Rt^q dt}\right]$$

Next we define a homomorphism

$$\nu: K_2(R[t]/(t^{q+1}), (t^n)) \to \left[\frac{t^n\Omega^1_{R[t]/(t^{q+1})/\mathbb{Z}}}{t^q\Omega^1_{R/\mathbb{Z}} + Rt^q dt}\right]$$

by

$$\nu\langle f, g\rangle = \begin{cases} \text{class of } fdg & \text{if } f \in t^nR[t]/(t^{q+1}) \\ \text{class of } -gdf & \text{if } g \in t^nR[t]/(t^{q+1}) \end{cases}$$

There is no ambiguity in this definition in case both f and g are in $t^nR[t]/(t^{q+1})$ and the relations (D1) - (D3) are also respected. So ν is a homomorphism indeed. Its kernel contains obviously the image of $K_2(R[t]/(t^{q+1}), (t^q))$. In view of (1.6) we get therefore a homomorphism

$$\bar{\nu}: K_2(R[t]/(t^q), (t^n)) \to \left[\frac{t^n\Omega^1_{R[t]/(t^{q+1})/\mathbb{Z}}}{t^q\Omega^1_{R/\mathbb{Z}} + Rt^q dt}\right]$$

One can check without difficulty that $\bar{\nu} \circ \tau = \chi$ and $\tau \circ \chi^{-1} \circ \bar{\nu} = 1$. □

Remark. The first part of the preceding theorem is actually Van der Kallen's result [12].

(1.12) Corollary. (1) For all $q > n$ the group $K_2(\mathbb{Z}[t]/(t^q), (t^n))$ is generated by the elements $\langle t^{m-1}, t\rangle$ with $n < m \leq q$.
(2) $K_2(\mathbb{Z}[t]/(t^{2n-1}), (t^n)) \cong \mathbb{Z}^{n-1}$, canonically.
Proof. (2) follows immediately from (1.11) and (1) follows from the theorem by (1.6) and induction. □

(1.13) Theorem. For every n there is a (canonical) surjective homomorphism

$$K_3(\mathbb{Z}[t]/(t^n), (t)) \twoheadrightarrow \mathbb{Z}^{n-1}$$

The kernel of this homomorphism consists of those elements which can be lifted to $K_3(\mathbb{Z}[t]/(t^q), (t))$ for every $q > n$.
Remark. For n = 2 Loday has a canonical way for splitting our surjection.

Proof. We use the exact sequences

$$K_3(\mathbb{Z}[t]/(t^q), (t)) \to K_3(\mathbb{Z}[t]/(t^n), (t)) \to K_2(\mathbb{Z}[t]/(t^q), (t^n)) \to K_2(\mathbb{Z}[t]/(t^q), (t))$$

From (1.12) one knows that $K_2(\mathbb{Z}[t]/(t^q), (t^n))$ is generated by the elements $<t^{m-1}, t>$ with $n < m \leq q$ and that for $q \geq 2n - 1$ the $<t^{m-1}, t>$ with $n < m < 2n$ generate a free abelian group of rank $n - 1$. On the other hand for $m \geq 2n$ one has $m < t^{m-1}, t> = 0$ by (1.9). Roberts and Geller showed in [19] that the order of the element $<t^{m-1}, t>$ in $K_2(\mathbb{Z}[t]/(t^q), (t))$ is exactly m for all $m \geq 2$. Thus we see that for all $q \geq 2n - 1$

$$\ker[K_2(\mathbb{Z}[t]/(t^q), (t^n)) \to K_2(\mathbb{Z}[t]/(t^q), (t))]$$

is a free abelian group of rank $n - 1$ with generators $m<t^{m-1}, t>$, $m = n + 1,\ldots,2n - 1$. □

(1.14) Remark. There is a remarkable alternative description of the isomorphism

$$\mathbb{Z}^{n-1} \cong \ker[K_2(\mathbb{Z}[t]/(t^q), (t^n)) \to K_2(\mathbb{Z}[t]/(t^q), (t))]$$

for $q \geq 2n - 1$, namely

$$(a_1,\ldots,a_{n-1}) \mapsto \sum_{i=1}^{n-1} (<t^{n+i}, a_i> + (n + i)<a_i t^{n+i-1}, t>)$$

To see that this formula gives the same isomorphism as

$$(a_1,\ldots,a_{n-1}) \mapsto \sum_{i=1}^{n-1} a_i(n + i) <t^{n+i-1}, t>$$

we proceed as follows. The formula gives a map from $\mathbb{Z}^{n-1}$ to $\ker[K_2(\mathbb{Z}[t]/(t^q), (t^n)) \to K_2(\mathbb{Z}[t]/(t^q), (t))]$, as (1.9) shows. We do not yet see that it is a homomorphism. Now compose this map with the canonical projection onto

$$\ker[K_2(\mathbb{Z}[t]/(t^{2n-1}), (t^n)) \to K_2(\mathbb{Z}[t]/(t^{2n-1}), (t))],$$

of which we know already that it is an isomorphism. Next note that the composite of the new map and the projection is the same as the composite of the old map and the projection, because in $K_2(\mathbb{Z}[t]/(t^{2n-1}), (t^n))$:

$$\langle at^{n+i-1}, t\rangle = a\langle t^{n+i-1}, t\rangle$$

and

$$\langle t^{n+i}, a\rangle = a\langle t^{n+i}, 1\rangle - \langle t^{n+i}, -\binom{a}{2}t^{n+i}\rangle$$

$$= -(n + i)\langle t, -\binom{a}{2}t^{2n+2i-1}\rangle = 0.$$

The new description fits exactly in the philosophy of the next sections; it shows the importance of the elements $\langle t^m, a\rangle + m\langle at^{m-1}, t\rangle$.

§2

In this section we take a close look at the system of groups

$$K_2(R[t]/(t^q), (t)), \ q \geqslant 2.$$

For every integer $m \geqslant 1$ the substitution $t \mapsto t^m$ induces a homomorphism

$$V_m: K_j(R[t]/(t^q), (t)) \to K_j(R[t]/(t^{qm}), (t))$$

called Verschiebung. There is also a transfer homomorphism associated to it, called Frobenius:

$$F_m: K_j(R[t]/(t^{qm}), (t)) \to K_j(R[t]/(t^q), (t)).$$

They both commute with the natural projections in the system. For some more details see [2] p. 219 - 221. The following relations hold

$$F_m F_r = F_{mr}, \ V_m V_r = V_{mr} \text{ for all } m, r$$
$$F_m V_m = \text{multiplication by } m$$

and for j = 1, 2 (at least) one can also show (for instance with (2.1) below)

$F_m V_r = V_r F_m$ if $(m, r) = 1$ (greatest common divisor)

$V_p F_p$ = multiplication by p, if R is an $\mathbb{F}_p$-algebra.

It is obvious how V_m acts on the generators given in (1.7). For F_m it is not immediately clear. The action of F_m is described explicitly in the following proposition.

(2.1) Proposition. The homomorphism

$$F_m: K_1(R[t]/(t^{qm}), (t)) \to K_1(R[t]/(t^q), (t))$$

acts on the generators $\langle at^i \rangle$ by:

$$F_m \langle at^i \rangle = d\langle a^{m/d} t^{i/d} \rangle$$

where $d = (m, i)$, the greatest common divisor. The homomorphism

$$F_m: K_2(R[t]/(t^{qm}), (t)) \to K_2(R[t]/(t^q), (t))$$

acts by

$$F_m \langle at^i, b \rangle = d\langle a^{m/d} b^{m/d-1} t^{i/d}, b \rangle$$

$$F_m \langle at^{i-1}, t \rangle = r\langle a^{m/d} t^{i/d-1}, t \rangle - s\langle a^{m/d-1} t^{i/d}, a \rangle +$$

$$+ (d - 1)\langle -a^{m/d} t^{i/d}, -1 \rangle$$

where $d = (m, i)$ and $rm + si = d$ (Note that although r and s are not uniquely determined, there is no ambiguity in the last formula in view of (1.5) (2)).

If i/d is invertible in R, then

$$F_m \langle at^{i-1}, t \rangle = -\frac{d}{i}\langle a^{m/d-1} t^{i/d}, a \rangle + (d - 1)\langle -a^{m/d} t^{i/d}, -1 \rangle$$

or, if m/d is invertible, then

$$F_m \langle at^{i-1}, t \rangle = \frac{d}{m}\langle a^{m/d} t^{i/d-1}, t \rangle + (d - 1)\langle -a^{m/d} t^{i/d}, -1 \rangle.$$

Moreover, if char R = 2 or $2 \in R^*$ or d is odd, the above formulas for $F_m\langle at^{i-1}, t\rangle$ simplify because $(d-1)\langle -a^{m/d}t^{i/d}, -1\rangle = 0$.

Proof. The statement about K_1 is very well known and can be found in any book on formal groups (e.g. [15]). In proving the statement for K_2 we will have the occasion to prove the statement for K_1, too. By functoriality it suffices to produce formulas in the universal case i.e. let

$$F_m:\ K_j(\mathbb{Z}[a, b]⟦t⟧, (t)) \to K_j(\mathbb{Z}[a, b]⟦t⟧, (t))$$

be the transfer homomorphism associated with the substitution $t \mapsto t^m$; we compute $F_m\langle at^i, b\rangle$ and $F_m\langle at^{i-1}, t\rangle$, where now $\langle at^i, b\rangle$ and $\langle at^{i-1}, t\rangle$ are elements of $K_2(\mathbb{Z}[a, b]⟦t⟧, (t))$. We need one more trick. We want Steinberg symbols instead of pointy brackets, because Steinberg symbols are products of elements in K_1 and there is a projection formula to reduce the problem then to a K_1-problem. In order to convert pointy brackets to Steinberg symbols we invert a, b, t.

(2.2) Lemma. The homomorphisms

$$K_1(\mathbb{Z}[a, b]⟦t⟧) \to K_1(\mathbb{Z}[a, b]⟦t⟧[a^{-1}, b^{-1}, t^{-1}])$$

$$\text{and}\quad K_2(\mathbb{Z}[a, b]⟦t⟧) \to K_2(\mathbb{Z}[a, b]⟦t⟧[a^{-1}, b^{-1}, t^{-1}])$$

are injective.

Let us accept this lemma for the moment and continue the proof of (2.1). The substitution $t \mapsto t^m$ induces a homomorphism

$$V_m:\ K_j(\mathbb{Z}[a, b]⟦t⟧[a^{-1}, b^{-1}, t^{-1}]) \circlearrowleft$$

and a transfer homomorphism

$$F_m:\ K_j(\mathbb{Z}[a, b]⟦t⟧[a^{-1}, b^{-1}, t^{-1}]) \circlearrowright$$

Both commute with the inclusions of the lemma. This reduces the problem to computing $F_m\langle at^i, b\rangle$ and $F_m\langle at^{i-1}, t\rangle$ in

$K_2(\mathbb{Z}[a, b][[t]][a^{-1}, b^{-1}, t^{-1}])$. The upshot of inverting b and t is, according to (1.5),

$$\langle at^i, b\rangle = \{1 - abt^i, b\}$$
$$\langle at^{i-1}, t\rangle = \{1 - at^i, t\}$$

Now it is time for the K_1 calculations. Recall that $\langle at^i\rangle$ is nothing but the unit $1 - at^i$. According to [17] p. 139 $F_m(1 - at^i)$ is equal to the determinant of the matrix which describes the multiplication by $1 - at^i$ with respect to the basis $1, t, t^2, \ldots, t^{m-1}$. If $m|i$ this matrix is the diagonal matrix

$$\begin{pmatrix} 1 - at^{i/m} & & \\ & \ddots & \\ & & 1 - at^{i/m} \end{pmatrix}$$

If $(m, i) = 1$ it is the matrix

$$\left.\begin{pmatrix} 1 & & & -at^{l+1} & & \\ & \ddots & & & \ddots & \\ & & & & & -at^{l+1} \\ -at^{l} & & & & & \\ & \ddots & & & & \\ & & -at^{l} & & & 1 \end{pmatrix}\right\} \begin{matrix} n \text{ rows} \\ \\ m-n \text{ rows} \end{matrix}$$

where $i = lm + n$ and $0 < n < m$. In both cases the determinant can easily be computed. The general case can be synthesized from these two special cases using $F_m = F_{m/d} \cdot F_d$. The result is

$$F_m(1 - at^i) = (1 - a^{m/d}t^{i/d})^d$$

Back to K_2. Using the projection formula [17] p. 137 one obtains

$$F_m\{1 - abt^i, b\} = \{F_m(1 - abt^i), b\}$$
$$= \{1 - a^{m/d}b^{m/d}t^{i/d}, b\}^d$$

Translated to $\langle , \rangle$'s this is:

$$F_m\langle at^i, b\rangle = d\langle a^{m/d}b^{m/d-1}t^{i/d}, b\rangle.$$

The computation for $F_m\langle at^{i-1}, t\rangle$ is slightly more complicated and can best be done by first considering the special cases $m|i$ and $(m, i) = 1$, respectively. Suppose $m|i$. Then

$$\begin{aligned} F_m\{1 - at^i, t\} &= F_m\{V_m(1 - at^{i/m}), t\} \\ &= \{1 - at^{i/m}, F_m t\} \text{ (projection formula)} \\ &= \{1 - at^{i/m}, t\}\{1 - at^{i/m}, (-1)^{m-1}\} \end{aligned}$$

because

$$F_m t = \det \begin{pmatrix} 0 & & & 0\,t \\ 1 & \ddots & & 0 \\ & \ddots & \ddots & \\ 0 & & 1 & 0 \end{pmatrix} = (-1)^{m-1}t.$$

Next suppose $(m, i) = 1$. Let r and s be such that $rm + si = 1$. Then

$$\begin{aligned} F_m\{1 - at^i, t\} &= F_m\{1 - at^i, t^m\}^r . F_m\{1 - at^i, t^i\}^s \\ &= F_m\{1 - at^i, t^m\}^r . F_m\{1 - at^i, a\}^{-s} \\ &= \{F_m(1 - at^i), t\}^r . \{F_m(1 - at^i), a\}^{-s} \\ &= \{1 - a^m t^i, t\}^r . \{1 - a^m t^i, a\}^{-s} \end{aligned}$$

The general case follows from the two special cases and $F_m = F_{m/d} \circ F_d$. When the result is rewritten in $\langle , \rangle$'s it reads as follows

$$\begin{aligned} F_m\langle at^{i-1}, t\rangle = {} & r\langle a^{m/d}t^{i/d-1}, t\rangle - s\langle a^{m/d-1}t^{i/d}, a\rangle \\ & + (d - 1)\langle -a^{m/d}t^{i/d}, -1\rangle \end{aligned}$$

This formula is without any change valid in $K_2(R[t]/(t^q), (t))$. If i/d is invertible in R we can make the following calculation.

$$\langle a^{m/d}t^{i/d-1}, t\rangle = \frac{d}{i}\langle a^{m/d}, t^{i/d}\rangle = \frac{m}{i}\langle a, a^{m/d-1}t^{i/d}\rangle;$$

whence

$$r\langle a^{m/d}t^{i/d-1}, t\rangle - s\langle a^{m/d-1}t^{i/d}, a\rangle = -\frac{d}{i}\langle a^{m/d-1}t^{i/d}, a\rangle$$

The formula for the case $m/d \in R^*$ is proved similarly. Finally, if 2 is invertible in R or d odd,

$$(d-1)\langle -a^{m/d}t^{i/d}, -1\rangle = 0$$

follows from (1.8) and

$$2\langle -a^{m/d}t^{i/d}, -1\rangle = \langle a^{m/d}t^{i/d}, (-1)^2\rangle = 0$$

In case char R = 2 it is trivial. This concludes the proof of proposition (2.1) modulo lemma (2.2).

Proof of lemma (2.2). We add a^{-1}, b^{-1}, t^{-1} in three steps and to each of these steps we apply the following lemma for j = 1, 2 (cf. [18])

(2.3) Lemma. Let A be a regular noetherian domain. Let $x \in A$ be such that A/xA is also regular. Suppose that the map $K_j(A) \to K_j(A/xA)$ is surjective. Then the map

$$K_j(A) \to K_j(A[x^{-1}])$$

is injective.

Proof. Because A and A/xA are regular noetherian rings, the K-theory of the category of finitely generated projective modules is equal to the K-theory of the category of all finitely generated modules. For the latter one has the localisation sequence [18]

$$\dots \to K_j(A/xA) \to K_j(A) \to K_j(A[x^{-1}]) \to \dots$$

We show that the map $K_j(A/xA) \to K_j(A)$ in this sequence is zero. That will prove the lemma. The composite map

$$K_j(A) \to K_j(A/xA) \to K_j(A) \quad \text{(the last map is the transfer map)}$$

is induced by the exact functor $- \otimes_A A/xA$ from the category of finitely generated projective A-modules to the category of all finitely generated A-modules. Since this functor fits in the short exact sequence

$$0 \to - \otimes_A A \xrightarrow{x} - \otimes_A A \to - \otimes_A A/xA \to 0$$

it induces the zero map on K-groups. So the composite $K_j(A) \to K_j(A/xA) \to K_j(A)$ is zero. Since by assumption $K_j(A) \to K_j(A/xA)$ is surjective, $K_j(A/xA) \to K_j(A)$ is zero. $\square$

For every ring R of characteristic $p > 0$ the expressions $\langle a^{p^{r-s}-1} t^{mp^r}, a\rangle + m\langle a^{p^r} t^{mp^{r+s}-1}, t\rangle$ with $a \in R$, $m, r, s \in \mathbb{Z}$, $0 \leq s \leq r$, $(m, p) = 1$, $mp^r \geq n$ and $mp^{r+s} \geq n + 1$ give elements of $K_2(R[t]/(t^q), (t^n))$ which vanish in $K_2(R[t]/(t^q), (t))$. The question is whether these elements (for varying a, m, r, s) generate

$$\ker[K_2(R[t]/(t^q), (t^n)) \to K_2(R[t]/(t^q), (t))].$$

A positive answer to this question has implications for $K_2(R[t]/(t^q), (t))$ as well as for $K_3(R[t]/(t^n), (t))$. In the next theorem we show that for certain rings the answer is positive. The first part of the proof is quite general. It shows, using the Frobenius operations, that the answer is yes if there exist a ring W and a W-algebra $\tilde{R}$ such that W/pW is perfect, $\tilde{R}/p\tilde{R} = R$ and for every $r \geq 0$ the only solutions for the equations

$$d\tilde{h} = p^r\tilde{\eta}, \quad \tilde{h} \in \tilde{R}, \ \tilde{\eta} \in \Omega^1_{\tilde{R}/W}$$

are $\tilde{h} = \tilde{e}_r^{p^r} + p\tilde{e}_{r-1}^{p^{r-1}} + p^2\tilde{e}_{r-2}^{p^{r-2}} + \ldots + p^{r-1}\tilde{e}_1^p + p^r\tilde{e}_0$ and

$\tilde{\eta} = \tilde{e}_r^{p^r-1}d\tilde{e}_r + \tilde{e}_{r-1}^{p^{r-1}-1}d\tilde{e}_{r-1} + \ldots + \tilde{e}_1^{p-1}d\tilde{e}_1 + d\tilde{e}_0$ with

$\tilde{e}_0, \ldots, \tilde{e}_r \in \tilde{R}$.

To get an example of this situation we let R be a smooth algebra over a perfect field k. Incidently, since k is perfect, R is smooth if and only if it is a k-algebra of finite type which is regular (in the sense of commutative algebra) (see [8]p. 99). It is well known that every smooth k-algebra can locally be lifted to a smooth algebra over W(k), the ring of Witt vectors of k (see [9] p. 69). We need however that R itself can be lifted and we must impose this as an additional condition. In lemma (2.7) we prove, using results of Cartier, that if R is liftable to a smooth W(k)-algebra $\tilde{R}$, the only solutions for the "differential equations" $d\tilde{h} = p^r\tilde{\eta}$ in $\Omega^1_{\tilde{R}/W(k)}$ are indeed those mentioned above.

(2.5) Theorem. Let R be a smooth algebra over a perfect field k of characteristic $p > 0$. Assume that there exists a smooth algebra $\tilde{R}$ over W(k) (the Witt vectors of k) such that $\tilde{R}/p\tilde{R} = R$. Then for all integers $q > n \geq 1$

$$\ker[K_2(R[t]/(t^q), (t^n)) \to K_2(R[t]/(t^q), (t))]$$

is the subgroup of $K_2(R[t]/(t^q), (t^n))$ which is generated by the elements

$$\langle a^{p^{r-s}-1} t^{mp^r}, a\rangle + m\langle a^{p^r} t^{mp^{r+s}-1}, t\rangle$$

with $a \in R$, $m, r, s \in \mathbb{Z}$ satisfying $0 \leq s \leq r$, $(m, p) = 1$, $mp^r \geq n$ and $mp^{r+s} \geq n + 1$.

Remarks. (1) Some of the above elements may be zero, as (1.10) shows. (2) If $mp^r \geq q$ or $mp^{r+s} \geq q + 1$, one of the two terms of the above element is zero. Neglect this, for it seems to suggest that there are two types of elements involved and it distorts the actual picture. There is just one crucial type of element! (3) As remarked in the introduction, we recover here a result of Bloch's ([2] p. 236, th. 4.1), namely: take $q = n + 1$ and recall that (1.11) gives an isomorphism

$$\tau: R \oplus \Omega^1_{R/\mathbb{Z}} \xrightarrow{\sim} K_2(R[t]/(t^{n+1}), (t^n))$$

Using this isomorphism, the result of (2.5) and (1.6) we see that

$$\ker[K_2(R[t]/(t^{n+1}), (t)) \to K_2(R[t]/(t^n), (t))]$$

is isomorphic to

$$\Omega^1_{R/\mathbb{Z}} \qquad \text{if } n \neq 0, -1 \bmod p$$

$$\Omega^1_{R/\mathbb{Z}} \oplus R/Rp^r \qquad \text{if } n = mp^r - 1, (m, p) = 1, r \geq 1$$

$$\Omega^1_{R/\mathbb{Z}}/D_{r,R} \qquad \text{if } n = mp^r, (m, p) = 1, r \geq 1$$

here $D_{r,R}$ is the subgroup of $\Omega^1_{R/\mathbb{Z}}$ generated by the forms $a^{p^l-1}da$ with $0 \leq l \leq r - 1$.

Proof of (2.5). First of all note that the element

$$\langle a^{p^{r-s}-1}t^{mp^r}, a\rangle + m\langle a^{p^r}t^{mp^{r+s}-1}, t\rangle$$

vanishes in $K_2(R[t]/(t^q), (t))$ by lemma (1.10) and (D1). So the group generated by these elements is contained in

$$\ker[K_2(R[t]/(t^q), (t^n)) \to K_2(R[t]/(t^q), (t))]$$

To prove the converse inclusion we use induction on $q - n$. We start with the induction step saving the more complicated $q = n + 1$ for later. Let us simplify notation for this proof and put

$$L_{q,n} = K_2(R[t]/(t^q), (t^n))$$

Fix q and n with $q - n \geq 2$ and assume that $\ker[L_{i,j} \to L_{i,1}]$ is generated by the elements described in the theorem if $i - j < q - n$. Consider the following commutative diagram with exact rows

$$\begin{array}{ccccccc} L_{q,q-1} & \to & L_{q,n} & \to & L_{q-1,n} & \to & 0 \\ \| & & \downarrow & & \downarrow & & \\ L_{q,q-1} & \to & L_{q,1} & \to & L_{q-1,1} & \to & 0 \end{array}$$

We apply the induction hypothesis to $L_{q-1,n} \to L_{q-1,1}$ and $L_{q,q-1} \to L_{q,1}$. Note that every element $<a^{p^{r-s}-1}t^{mp^r}, a> + m<a^{p^r}t^{mp^{r+s}-1}, t>$ of $L_{q-1,n}$ can be lifted to the element with the same name in $L_{q,n}$. It is now obvious how to finish the proof of the induction step.

We go on with our study of

$$\ker[K_2(R[t]/(t^{n+1}), (t^n)) \to K_2(R[t]/(t^{n+1}), (t))],$$

that is of relations

$$\sum_i <a_i t^n, b_i> + <ct^n, t> = 0$$

in $K_2(R[t]/(t^{n+1}), (t))$. Actually it turns out to be more efficient to study the more general relations

$$\sum_i <a_i t^n, b_i> + <ct^n, t> + <ft^{n-1}, t> = 0$$

in $K_2(R[t]/(t^{n+1}), (t))$. That is done in the following lemma.

(2.6) Lemma. Let $n = mp^r$ with $(m, p) = 1$ and $r \geq 0$. Suppose

$$\sum_i <a_i t^n, b_i> + <ct^n, t> + <ft^{n-1}, t> = 0$$

in $K_2(R[t]/(t^{n+1}), (t))$. Then there exist $e_0, e_1, \ldots, e_r \in R$ such that

$$\sum_i a_i db_i = \frac{1}{m} \sum_{l=0}^{r} e_l^{p^l-1} de_l \text{ in } \Omega^1_{R/\mathbb{Z}}$$

and $\quad f = e^{p^r}$ in R.

Assume for the moment this lemma is true. We finish the proof of (2.5). Take a general element of

$$\ker[K_2(R[t]/(t^{n+1}), (t^n)) \to K_2(R[t]/(t^{n+1}), (t))],$$

say $\sum_i \langle a_i t^n, b_i\rangle + \langle ct^n, t\rangle$.

Let $n = mp^r$ with $(m, p) = 1$, $r \geq 0$. By the lemma there are $e_0, e_1, \ldots, e_r \in R$ such that $e_r = 0$ and $\sum_i a_i db_i = \frac{1}{m} \sum_{l=0}^{r} e_l^{p^l - 1} de_l$ in $\Omega^1_{R/\mathbb{Z}}$. This implies by (1.11)

$$\sum_i \langle a_i t^n, b_i\rangle = \frac{1}{m} \sum_{l=0}^{r} \langle e_l^{p^l-1} t^n, e_l\rangle$$

in $K_2(R[t]/(t^{n+1}), (t^n))$. Next note that $e_r = 0$ and that for $l = 0, \ldots, r - 1$

$$\langle e_l^{p^l-1} t^n, e_l\rangle = \langle e_l^{p^l-1} t^{mp^r}, e_l\rangle + m\langle e_l^{p^r} t^{mp^{2r-l}-1}, t\rangle$$

Thus we see that $\sum_i \langle a_i t^n, b_i\rangle$ is a sum of elements of the desired form in $K_2(R[t]/(t^{n+1}), (t^n))$. Each of these elements vanishes in $K_2(R[t]/(t^{n+1}), (t))$. Therefore we find also

$$\langle ct^n, t\rangle = 0 \text{ in } K_2(R[t]/(t^{n+1}), (t)).$$

This implies that there is a relation

$$\langle ct^n, t\rangle + \sum_i \langle f_i t^{n+1}, g_i\rangle + \langle ht^{n+1}, t\rangle = 0$$

in $K_2(R[t]/(t^{n+2}), (t))$ for some $f_i, g_i, h \in R$. Let $n + 1 = m'p^{r'}$ with $(m', p) = 1$, $r' \geq 0$. Then by lemma (2.6) $c = e^{p^{r'}}$ for some $e \in R$. Thus we get

$$\langle ct^n, t\rangle = \frac{1}{m'}(\langle e^{p^{r'}-1} t^{m'p^{r'}}, e\rangle + m'\langle e^{p^{r'}} t^{m'p^{r'}-1}, t\rangle)$$

in $K_2(R[t]/(t^{n+1}), (t^n))$, again an element of the desired form. This concludes the proof of (2.5) modulo lemma (2.6).

<u>Proof of lemma</u> (2.6). At this point the W(k)-algebra $\tilde{R}$ with $\tilde{R}/p\tilde{R} = R$ appears on the scene. We claim that there exists a homomorphism

$$\overline{F}_n\colon K_2(\tilde{R}[t]/(t^{n+1}), (t)) \to \Omega^1_{\tilde{R}/W(k)}$$

such that the following square is commutative.

$$\begin{array}{ccc} K_2(\tilde{R}[t]/(t^{2n}),\ (t)) & \xrightarrow{F_n} & K_2(\tilde{R}[t]/(t^2),\ (t)) \\ \downarrow & & \downarrow \\ K_2(\tilde{R}[t]/(t^{n+1}),\ (t)) & \xrightarrow{\overline{F}_n} & \Omega^1_{\tilde{R}/W(k)} \end{array}$$

Here the vertical maps are the canonical projection and the map given in (1.11) composed with the obvious map, respectively. To prove our claim we show that every generator of

$$\ker[K_2(\tilde{R}[t]/(t^{2n}),\ (t)) \to K_2(\tilde{R}[t]/(t^{n+1}),\ (t))]$$

is killed by the composite of F_n and the map to $\Omega^1_{\tilde{R}/W(k)}$. From (1.6) and (1.7) one knows that this kernel is generated by the elements $\langle at^i, b\rangle$ and $\langle at^{j-1}, t\rangle$ with $a, b \in \tilde{R}$ and $n + 1 \leqslant i \leqslant 2n - 1$, $n + 2 \leqslant j \leqslant 2n$. By (2.1) we have

$$F_n\langle at^i, b\rangle = d\langle a^{n/d}b^{n/d-1}t^{i/d}, b\rangle$$

with $d = (n, i)$. Since $n + 1 \leqslant i \leqslant 2n - 1$ it must be that $i/d \geqslant 2$ and hence

$$F_n\langle at^i, b\rangle = 0 \text{ in } K_2(\tilde{R}[t]/(t^2),\ (t)).$$

Furthermore,

$$F_n\langle at^{j-1}, t\rangle = r\langle a^{n/e}t^{j/e-1}, t\rangle - s\langle a^{n/e-1}t^{j/e}, a\rangle$$

with $e = (n, j)$ and $rn + sj = e$. For $n + 2 \leqslant j \leqslant 2n - 1$ one has $j/e \geqslant 3$ and hence

$$F_n\langle at^{j-1}, t\rangle = 0 \text{ in } K_2(\tilde{R}[t]/(t^2),\ (t)).$$

For $j = 2n$ we have

$$F_n\langle at^{2n-1}, t\rangle = \langle at, t\rangle,$$

an element of $\ker[K_2(\tilde{R}[t]/(t^2),\ (t)) \to \Omega^1_{\tilde{R}/W(k)}]$. This proves the existence of $\overline{F}_n$.

Next write $n = mp^r$ with $(m, p) = 1$, $r \geq 0$. We claim that there is a homomorphism

$$\bar{\bar{F}}_n: K_2(R[t]/(t^{n+1}), (t)) \to \left[\frac{\Omega^1_{\tilde{R}/W(k)}}{p^{r+1}\Omega^1_{\tilde{R}/W(k)} + pd\tilde{R}}\right]$$

such that the following diagram commutes

$$\begin{array}{ccc} K_2(\tilde{R}[t]/(t^{n+1}), (t)) & \xrightarrow{\bar{F}_n} & \Omega^1_{\tilde{R}/W(k)} \\ \downarrow & & \downarrow \\ K_2(R[t]/(t^{n+1}), (t)) & \xrightarrow{\bar{\bar{F}}_n} & \left[\dfrac{\Omega^1_{\tilde{R}/W(k)}}{p^{r+1}\Omega^1_{\tilde{R}/W(k)} + pd\tilde{R}}\right] \end{array}$$

To prove this we must figure out what $\bar{F}_n$ does to the generators of

$$\ker[K_2(\tilde{R}[t]/(t^{n+1}), (t)) \to K_2(R[t]/(t^{n+1}), (t))]$$

By (1.6) and the obvious analogue of (1.7) we know that this kernel is generated by the elements $\langle at^i, b\rangle$ with $a, b \in \tilde{R}$, $ab \in p\tilde{R}$, $1 \leq i \leq n$ and the elements $\langle at^{i-1}, t\rangle$ with $a \in p\tilde{R}$, $2 \leq i \leq n+1$. From (2.1) and the definition of $\bar{F}_n$ one sees immediately

$$\bar{F}_n\langle at^i, b\rangle = \begin{cases} 0 & \text{if } i \nmid n \\ i\, a^{n/i} b^{n/i-1} db & \text{if } i \mid n \end{cases}$$

$$\bar{F}_n\langle at^{i-1}, t\rangle = \begin{cases} 0 & \text{if } i \nmid n \\ -\, a^{n/i-1} da & \text{if } i \mid n \end{cases}$$

Examine the 1-form $ia^{n/i}b^{n/i-1}db$ if $i \mid n$. Using $ab \in p\tilde{R}$ one can easily check that this form belongs to $p^{r+1}\Omega^1_{\tilde{R}/W(k)}$. Next examine the form $a^{n/i-1}da$ if $i \mid n$. It is given that $a = pc$ for some $c \in \tilde{R}$. Write $i = jp^s$ with $(j, p) = 1$ and compute

$$a^{n/i-1}da = p^{n/i}c^{n/i-1}dc = \frac{j}{m}\, p^{n/i-r+s}d(c^{n/i}).$$

Therefore $a^{n/i-1}da \in pd\tilde{R}$. All this proves our claim about the existence of $\overline{\overline{F}}_n$.

We are going to use this map

$$\overline{\overline{F}}_n : K_2(R[t]/(t^{n+1}), (t)) \to \left[\frac{\Omega^1_{\tilde{R}/W(k)}}{p^{r+1}\Omega^1_{\tilde{R}/W(k)} + pd\tilde{R}}\right]$$

to test when a relation

$$\sum_i \langle a_i t^n, b_i\rangle + \langle ct^n, t\rangle + \langle ft^{n-1}, t\rangle = 0$$

can hold in $K_2(R[t]/(t^{n+1}), (t))$.

Let $\tilde{a}_i$, $\tilde{b}_i$, $\tilde{c}$, $\tilde{f}$ be liftings of a_i, b_i, c, f respectively. The construction of $\overline{\overline{F}}_n$ shows immediately

$$\overline{\overline{F}}_n\langle a_i t^n, b_i\rangle = n\tilde{a}_i d\tilde{b}_i \mod p^{r+1}\Omega^1_{\tilde{R}/W(k)} + pd\tilde{R}$$

$$\overline{\overline{F}}_n\langle ct^n, t\rangle = 0 \qquad \text{mod idem}$$

$$\overline{\overline{F}}_n\langle ft^{n-1}, t\rangle = -d\tilde{f} \qquad \text{mod idem}$$

So from the relation given in $K_2(R[t]/(t^{n+1}), (t))$ one obtains

$$p^r m \sum_i \tilde{a}_i d\tilde{b}_i - d\tilde{f} = pd\tilde{g} - p^{r+1}\tilde{\omega}$$

in $\Omega^1_{\tilde{R}/W(k)}$ for some $\tilde{g} \in \tilde{R}$ and $\tilde{\omega} \in \Omega^1_{\tilde{R}/W(k)}$. Put $\tilde{\eta} = m \sum_i \tilde{a}_i d\tilde{b}_i + p\tilde{\omega}$ and $\tilde{h} = \tilde{f} + p\tilde{g}$. Then we have

$$\sum_i a_i db_i = \frac{1}{m}\tilde{\eta} \mod p\Omega^1_{\tilde{R}/W(k)}$$

$$f = \tilde{h} \mod p\tilde{R}$$

$$d\tilde{h} = p^r\tilde{\eta}$$

The proof of (2.6) is concluded with the following lemma in which we study the "differential equation" $d\tilde{h} = p^r\tilde{\eta}$.

(2.7) <u>Lemma</u>. Let $\tilde{R}$ and R be as in (2.5). Suppose $\tilde{h} \in \tilde{R}$ and $\tilde{\eta} \in \Omega^1_{\tilde{R}/W(k)}$ satisfy the equation

$$d\tilde{h} = p^r\tilde{\eta}$$

Then there exist $\tilde{e}_0, \tilde{e}_1, \ldots, \tilde{e}_r \in \tilde{R}$ such that

$$\tilde{h} = \tilde{e}_r^{p^r} + p\tilde{e}_{r-1}^{p^{r-1}} + \ldots + p^{r-1}\tilde{e}_1^p + p^r\tilde{e}_0$$

and hence

$$\tilde{\eta} = \tilde{e}_r^{p^r-1}d\tilde{e}_r + \tilde{e}_{r-1}^{p^{r-1}-1}d\tilde{e}_{r-1} + \ldots + \tilde{e}_1^{p-1}d\tilde{e}_1 + d\tilde{e}_0$$

<u>Proof</u>. (Compare with [2] p. 234), by induction on r. The case $r = 0$ is a triviality. Now assume $r \geqslant 1$ and the conclusion of the lemma holds for $r - 1$. Suppose $d\tilde{h} = p^r\tilde{\eta}$. The induction hypothesis implies that there exist $\tilde{g}_1, \tilde{g}_2, \ldots, \tilde{g}_r \in \tilde{R}$ such that

$$\tilde{h} = \sum_{j=1}^{r} p^{r-j}\tilde{g}_j^{p^{j-1}}$$

and

$$p\tilde{\eta} = \sum_{j=1}^{r} \tilde{g}_j^{p^{j-1}-1}d\tilde{g}_j$$

Let $g_1, \ldots, g_r \in R$ be the images of $\tilde{g}_1, \ldots, \tilde{g}_r$ respectively. The second equation above yields by reduction mod p

$$g_r^{p^{r-1}-1}dg_r + \ldots + g_2^{p-1}dg_2 + dg_1 = 0$$

Let K be the field of fractions of R. Cartier has defined an operator

$$C: \Omega^1_{K/k,\text{closed}} \to \Omega^1_{K/k}$$

which has the properties

$$C(f^p\omega) = fC\omega \qquad \text{for } f \in K,\ \Omega^1_{K/k,\text{closed}}$$

$$C(f^{p-1}df) = df \qquad \text{for } f \in K$$

$$C(df) = 0 \qquad \text{for } f \in K$$

(see [3], [20]). Applying $C^{r-1}, C^{r-2}, \ldots, C^2$, C to the relation above we find successively

$$dg_r = 0,\ dg_{r-1} = 0, \ldots, dg_2 = 0,\ dg_1 = 0.$$

This in turn implies $g_j \in K^p$ as is shown in [3] p. 196. Since R is integrally closed in K we get even

$$g_j \in R^p, \text{ say } g_j = e_j^p, \text{ for } j = 1,\dots,r-1$$

Let $\tilde{e}_j \in \tilde{R}$ be a lifting of e_j. Then $\tilde{g}_j = \tilde{e}_j^p + p\tilde{z}_j$ for certain $\tilde{z}_j \in \tilde{R}$. A well known lemma, based on the observation that the number $\binom{p^s}{l}p^l$ is divisible by p^{s+1} for all s,l ⩾ 1, states, that $\tilde{g}_j = \tilde{e}_j^p + p\tilde{z}_j$ implies

$$\tilde{g}_j^{p^{j-1}} = \tilde{e}_j^{p^j} + p^j\tilde{x}_j$$

for appropriate $\tilde{x}_j$. Finally we put $\tilde{e}_0 = \tilde{x}_1 + \tilde{x}_2 + \dots + \tilde{x}_r$. The result is

$$\tilde{h} = \sum_{j=0}^{r} p^{r-j}\tilde{e}_j^{p^j}$$

This concludes the proof of (2.7), and also of (2.6) and (2.5). □□□

Theorem (2.5) has the following counterpart in characteristic zero.

(2.8) <u>Theorem</u>. Let R be a $\mathbb{Q}$-algebra. Then for all $q > n \geq 2$

$$\ker[K_2(R[t]/(t^q), (t^n)) \to K_2(R[t]/(t^q), (t))]$$

is the subgroup of $K_2(R[t]/(t^q), (t^n))$ which is generated by the elements

$$\langle t^m, a\rangle + m\langle at^{m-1}, t\rangle$$

with $a \in R$, $m \in \mathbb{Z}$, $m \geq n+1$.

<u>Proof</u>. (Rough sketch). As in the proof of theorem (2.5) one uses induction on $q - n$. The induction step is made exactly as before. As for the case $q = n + 1$ one uses a lemma like (2.6) saying that the relation

$$\Sigma \langle a_i t^n, b_i\rangle + \langle ct^n, t\rangle + \langle ft^{n-1}, t\rangle = 0$$

in $K_2(R[t]/(t^{n+1}), (t))$ implies $\sum_i a_i db_i = \frac{1}{n}df$ in $\Omega^1_{R/\mathbb{Z}}$. To prove that lemma one plays with Frobenius as in the proof of (2.6) and one tests things in $\Omega^1_{R/\mathbb{Z}}$. Having $\sum_i a_i db_i = \frac{1}{n} df$ in $\Omega^1_{R/\mathbb{Z}}$ one gets with (1.11) $\sum_i \langle a_i t^n, b_i\rangle = \frac{1}{n}\langle t^n, f\rangle$ in $K_2(R[t]/(t^{n+1}), (t^n))$. Further details are left to the reader. □

(2.9) Corollary. For rings R as in (2.5) or (2.8) the projective system

$$\{K_3(R[t]/(t^n), (t))\}_{n \geq 2}$$

has the Mittag-Leffler property. More precisely, if n is given, put $q_0 = 2n - 1$ if char R = 0, $q_0 = 4n^2 - 1$ if char R = 2 or 3, $q_0 = pn^2 - 1$ if char R = p ⩾ 5. Then for all $q \geq q_0$

$$\text{image}[K_3(R[t]/(t^q), (t)) \to K_3(R[t]/(t^n), (t))]$$
$$=$$
$$\text{image}[K_3(R[t]/(t^{q_0}), (t)) \to K_3(R[t]/(t^n), (t))]$$

Proof. We will prove the equivalent statement: for $q \geq q_0$ the canonical map

$$\ker[K_2(R[t]/(t^q), (t^n)) \to K_2(R[t]/(t^q), (t))]$$
$$\downarrow$$
$$\ker[K_2(R[t]/(t^{q_0}), (t^n)) \to K_2(R[t]/(t^{q_0}), (t))]$$

is injective.
The kernel of this map is the image in $K_2(R[t]/(t^q), (t^n))$ of $\ker[K_2(R[t]/(t^q), (t^{q_0})) \to K_2(R[t]/(t^q), (t))]$. Therefore it is generated by the elements

$$\langle a^{p^{r-s}-1} t^{mp^r}, a\rangle + m\langle a^{p^r} t^{mp^{r+s}-1}, t\rangle$$

with $a \in R$, $0 \leq s \leq r$, $(m, p) = 1$, $mp^r \geq q_0$ and $mp^{r+s} \geq q_0 + 1$, if char R = p > 0, and by the elements

$$\langle t^m, a\rangle + m\langle at^{m-1}, t\rangle$$

with $a \in R$, $m \geq q_0 + 1$, if char $R = 0$. Lemmas (1.9) and (1.10) and our choice of q_0 guarantee that these elements are zero in $K_2(R[t]/(t^q), (t^n))$. □

(2.10) Corollary. If R is a $\mathbb{Q}$-algebra and $q \geq 2n - 1$, there is an isomorphism

$$R^{n-1} \xrightarrow{\sim} \ker[K_2(R[t]/(t^q), (t^n)) \to K_2(R[t]/(t^q), (t))]$$

given by

$$(a_1,\dots,a_{n-1}) \mapsto \sum_{i=1}^{n-1} (\langle t^{n+i}, a_i\rangle + (n+i)\langle a_i t^{n+i-1}, t\rangle)$$

Proof. (cf. (1.14)). The formula does define a map $R^{n-1} \to \ker[\dots]$ as (1.9) shows. Next looking at the proof of the previous corollary we see that it suffices to show that the formula defines an isomorphism in case $q = 2n - 1$. The map is surjective by (2.8). So we are left with showing it is an injective homomorphism. For that, recall (1.11):

$$R^{n-1} \oplus (\Omega^1_{R/\mathbb{Z}})^{\oplus(n-1)} \xrightarrow{\sim} K_2(R[t]/(t^{2n-1}), (t^n))$$

Via this isomorphism the map here corresponds to

$$R^{n-1} \to R^{n-1} \oplus (\Omega^1_{R/\mathbb{Z}})^{\oplus(n-1)}$$

$$(a_1,\dots,a_{n-1}) \mapsto ((n+1)a_1, (n+2)a_2,\dots,(2n-1)a_{n-1}; da_1, da_2,\dots,da_{n-1})$$

and this is clearly an injective homomorphism. □

§3

In this section we want to organise the elements of

$$\ker[K_2(R[t]/(t^q), (t^n)) \to K_2(R[t]/(t^q), (t))]$$

if char $R > 0$, similar to what we did in (1.14) for $R = \mathbb{Z}$ and in (2.10) for $\mathbb{Q}$-algebras. Actually we are interested in the subgroup of

$K_2(R[t]/(t^q), (t^n))$ which is generated by the elements

$$<a^{p^{r-s}-1} t^{mp^r}, a> + m<a^{p^r} t^{mp^{r+s}-1}, a>$$

with $a \in R$, $0 \leq s \leq r$, $(m, p) = 1$, $mp^r \geq n$ and $mp^{r+s} \geq n + 1$. For that we can work with any $\mathbb{F}_p$-algebra R; smoothness and liftability are not needed as in (2.5). We will construct a homomorphism from $\varprojlim_l K_1(R[t]/(t^l), (t))$ onto this group and show that its kernel contains the elements $<at^i>$ with $a \in R$ and $i \geq 2n$ or $i = n$. Thus we get a homomorphism Δ_n from $K_1(R[t]/(t^{2n}), (t))/\{<at^n>| a \in R\}$ onto the group we want to understand. The subgroup of $K_1(R[t]/(t^{2n}), (t))/$ $/\{<at^n>| a \in R\}$ generated by the elements $<at^i>$ with $a \in R$ and $n < i < 2n$ is clearly isomorphic to R^{n-1}. The restriction of Δ_n gives exactly the same map $R^{n-1} \to K_2(R[t]/(t^q), (t^n))$ as in characteristic zero. But its image is too small and to cover the remaining part of the group we are studying, we have to go through all the troubles hereafter.

(3.1). We start with recalling some known facts (cf. [2]). Following Bloch we write (for every ring R)

$$CK_j(R) = \varprojlim_l K_j(R[t]/(t^l), (t))$$

The Verschiebung and Frobenius operators V_m and F_m, defined in §2, pass to the limit and thus give endomorphisms V_m and F_m of $CK_j(R)$. For $j = 1, 2$ we copy the notation for elements of $CK_j(R)$ from the finite level i.e. $<f> \in CK_1(R)$ is the element which for every l projects onto $<f> \in K_1(R[t]/(t^l), (t))$; similarly for $<f, g> \in CK_2(R)$. Proposition (1.7) shows that $CK_1(R)$ is topologically generated by the elements $<at^i>$ with $a \in R$, $i \geq 1$ (note that this statement gives nothing but the well known fact that every power series with constant term 1 is equal to an infinite product $\prod_i(1 - a_i t^i)$). And $CK_2(R)$ is topologically generated by the $<at^i, b>$ and $<at^{i-1}, t>$ with $a, b \in R$,

$i \geq 1$. Proposition (2.1) describes the action of F_m in terms of these generators.

Now assume that R contains a field of characteristic $p \geq 0$. Then one can distinguish inside $CK_j(R)$ the group of p-typical elements

$$TCK_j(R) = \bigcap_{\substack{m \in I(p) \\ m \neq 1}} \ker F_m$$

where $I(p) = \mathbb{N} \setminus p\mathbb{N}$. Let μ be the Möbius function i.e. $\mu(1) = 1$, $\mu(m) = (-1)^r$ if m is a product of r distinct primes, $\mu(m) = 0$ otherwise. Recall (1.8) that for $m \in I(p)$ multiplication by m gives an automorphism of $CK_j(R)$ (at least for $j = 1, 2$). It is one of the well known facts that the expression

$$E = \sum_{m \in I(p)} \frac{\mu(m)}{m} V_m F_m$$

defines a projection operator on $CK_j(R)$ with image $TCK_j(R)$ (see [2] for instance). We need the following explicit formulas, which may be proved using (2.1) and the familiar identities for μ:

(3.1.1) If i is not a power of p then

$$E\langle at^i\rangle,\ E\langle at^i, b\rangle,\ E\langle at^{i-1}, t\rangle \text{ are all zero}$$

$$E\langle at^{p^r}\rangle = \sum_{m \in I(p)} \frac{\mu(m)}{m} \langle a^m t^{mp^r}\rangle$$

$$E\langle at^{p^r-1}, t\rangle = \sum_{m \in I(p)} \frac{\mu(m)}{m} \langle a^m t^{mp^r-1}, t\rangle.$$

Another well known fact is that there is an isomorphism

$$(3.1.2) \qquad CK_j(R) \xrightarrow{\sim} \prod_{m \in I(p)} TCK_j(R),$$

a product of infinitely many copies of $TCK_j(R)$ indexed by the set I(p); this isomorphism is given by

$$(3.1.2a) \qquad \alpha \mapsto \left(\tfrac{1}{m} E \circ F_m \alpha\right)_{m \in I(p)}$$

and its inverse is

(3.1.2b) $(\omega_m)_{m \in I(p)} \mapsto \sum_{m \in I(p)} V_m \omega_m .$

In [2] Bloch defined a homomorphism

$$d: CK_1(R) \to CK_2(R)$$

by (in our notation)

$$d\langle a_1 t + a_2 t^2 + \ldots + a_q t^q \rangle = \langle a_1 + a_2 t + \ldots + a_q t^{q-1}, t\rangle$$

on the finite levels (and then passing to the limit).

Remark. $TCK_1(R)$ is actually a very classical thing; namely for $p > 0$ it is the ring of Witt vectors of R, usually denoted W(R); for $p = 0$ the whole thing degenerates to $TCK_1(R) = R$. For $p = 0$ the element $E\langle at\rangle$ of $CK_1(R) = (1 + tR[\![t]\!])^\times$ is the exponential function exp(at) which by the identification $TCK_1(R) = R$ corresponds to a. For $p > 0$ the element $E\langle at\rangle$ of $CK_1(R)$ is the socalled Artin-Hasse exponential of at:

$$E\langle at\rangle = \prod_{m \in I(p)} (1 - a^m t^m)^{\mu(m)/m}$$

Although $TCK_2(R)$ started out as a piece of K-theory in Bloch's paper [2], it has been axiomatised and stripped of K-theory. Nowadays one writes, in case $p > 0$, $W\Omega^1_R$ instead of $TCK_2(R)$ and $W(R) \xrightarrow{d} W\Omega^1_R$ is the beginning of the so-called De Rham-Witt complex [10]. If $p = 0$ one has

$$TCK_2(R) \cong \Omega^1_{R/\mathbb{Z}}$$

with $E\langle at, b\rangle$ corresponding to the form adb, and d is (up to sign) the ordinary differentiation $R \to \Omega^1_{R/\mathbb{Z}}$.

Let $p > 0$. Fix n. We are ready to define a homomorphism

$$\Delta_n: CK_1(R) \to K_2(R[t]/(t^q), (t^n))$$

for every $q > n$; actually q does not really appear in the formulas and Δ_n may be considered as a map into the projective system. We decompose

$CK_1(R)$ as $\prod_{m\in I(p)} TCK_1(R)$ and define for each $m \in I(p)$ a homomorphism

$$\delta_m: TCK_1(R) \to K_2(R[t]/(t^q), (t^n)),$$

which itself will be the difference of two maps ϕ_m and ψ_m.

(3.2) <u>Construction</u> of ϕ_m, ψ_m, δ_m. Fix $m \in I(p)$ and define

$$r_m = \text{least integer} \geq 0 \text{ such that } mp^{r_m} > n.$$

To simplify notation we write $r = r_m$. The substitution $t \mapsto t^{mp^r}$ defines a homomorphism

$$CK_2(R) \to K_2(R[t]/(t^q), (t^n))$$

We compose it with the homomorphism

$$F_{p^r} \circ d: CK_1(R) \to CK_2(R)$$

and restrict the composite map to $TCK_1(R)$. Thus we get

$$\phi_m: TCK_1(R) \to K_2(R[t]/(t^q), (t^n)).$$

The map ψ_m is defined as follows. Consider the homomorphism

$$K_1(R[t]/(t^{q+1}), (t)) \to K_2(R[t]/(t^q), (t^n))$$

defined by

$$\langle a_1t+\dots+a_qt^q\rangle \mapsto \langle a_1t^{mp^r-1} + a_2t^{2mp^r-1}+\dots+a_qt^{qmp^r-1}, t\rangle$$

Compose it with the homomorphisms

$$\text{proj}: CK_1(R) \to K_1(R[t]/(t^{q+1}), (t)) \text{ and}$$

$$m.F_{p^r}: CK_1(R) \to CK_1(R)$$

and restrict the composite to $TCK_1(R)$. Thus we get

$$\psi_m: TCK_1(R) \to K_2(R[t]/(t^q), (t^n))$$

Finally we put $\delta_m = \psi_m - \phi_m$

(3.3) Proposition. Notation as above. For $s \leq r$ one has

$$\delta_m E\langle at^{p^s}\rangle =$$

$$= \sum_{l\in I(p)} \frac{\mu(l)}{l^2}(\langle(a^l)^{p^{r-s}-1} t^{lmp^r}, a^l\rangle + ml\langle a^{lp^r} t^{lmp^{r+s}-1}, t\rangle)$$

Furthermore if $s > r$ or $s = r$ and $mp^r \geq 2n$ or $s = r - 1$ and $mp^r = pn$ then $\delta_m E\langle at^{p^s}\rangle = 0$.

Proof. The formulas for δ_m are obtained from those for ψ_m and ϕ_m. Direct computation shows easily

$$\psi_m E\langle at^{p^s}\rangle = \sum_{l\in I(p)} \frac{\mu(l)}{l} m\langle(a^l)^{p^r} t^{lmp^{r+s}-1}, t\rangle$$

For ϕ_m it is slightly more complicated, but still straightforward: First one has

$$dE\langle at^{p^s}\rangle = \sum_{l\in I(p)} \frac{\mu(l)}{l} \langle a^l t^{lp^s-1}, t\rangle$$

Secondly, for $s > r$

$$F_{p^r}\langle a^l t^{lp^s-1}, t\rangle = \langle a^l t^{lp^{s-r}-1}, t\rangle$$

and for $s \leq r$

$$F_{p^r}\langle a^l t^{lp^s-1}, t\rangle = -\frac{1}{l}\langle(a^l)^{p^{r-s}-1} t^l, a^l\rangle$$

(see (2.1)). Finally we must replace t by t^{mp^r}. Thus we find for $s > r$

$$\begin{aligned}\phi_m E\langle at^{p^s}\rangle &= \sum_{l\in I(p)} \frac{\mu(l)}{l} \langle a^l (t^{mp^r})^{lp^{s-r}-1}, t^{mp^r}\rangle \\ &= \sum_{l\in I(p)} \frac{\mu(l)}{l} mp^r\langle a^l t^{mlp^s-1}, t\rangle \\ &= \sum_{l\in I(p)} \frac{\mu(l)}{l} m\langle a^{lp^r} t^{mlp^{s+r}-1}, t\rangle \\ &= \psi_m E\langle at^{p^s}\rangle\end{aligned}$$

and for $s \leq r$

$$\phi_m E\langle at^{p^s}\rangle = - \sum_{l\in I(p)} \frac{\mu(l)}{l^2} < (a^l)^{p^{r-s}-1} t^{lmp^r}, a^l>$$

The formulas for $\delta_m E\langle at^{p^s}\rangle$ now follow immediately. The special statements for the case $s = r$, $mp^r \geq 2n$ and for $s = r - 1$, $mp^r = pn$ are immediate consequences of the general formula and (1.10). □

(3.4) Construction of $\tilde{\Delta}_n$. Putting all δ_m's together we get a map

$$\prod_{m\in I(p)} TCK_1(R) \to K_2(R[t]/(t^q), (t^n))$$

$$(\alpha_m)_{m \in I(p)} \mapsto \sum_{m\in I(p)} \delta_m\alpha_m$$

Note that by (3.3) $\delta_m = 0$ for $m \geq 2n$ and that hence the sum $\sum_{m\in I(p)} \delta_m\alpha_m$ is actually a finite sum. Recall also from (3.1.2) that there is an isomorphism

$$CK_1(R) \xrightarrow{\sim} \prod_{m\in I(p)} TCK_1(R)$$

$$\alpha \mapsto (\tfrac{1}{m}E \circ F_m\alpha)_{m \in I(p)}$$

Composing we get the map

$$\tilde{\Delta}_n = \sum_{m\in I(p)} \frac{1}{m}\, \delta_m EF_m : CK_1(R) \to K_2(R[t]/(t^q), (t^n)).$$

The main properties of the map $\tilde{\Delta}_n$ are given in the following theorem

(3.5) Theorem. Let R be an $\mathbb{F}_p$-algebra. Let $q > n \geq 1$ be integers. Then the map

$$\tilde{\Delta}_n : CK_1(R) \to K_2(R[t]/(t^q), (t^n)),$$

constructed in (3.4), has the following properties

(1) The image of $\tilde{\Delta}_n$ is the subgroup of $K_2(R[t]/(t^q), (t^n))$ which is generated by the elements

$$\langle a^{p^{r-s}-1} t^{mp^r}, a\rangle + m\langle a^{p^r} t^{mp^{r+s}-1}, t\rangle$$

with $a \in R$, $m, r, s \in \mathbb{Z}$ such that $0 \leq s \leq r$, $(m, p) = 1$, $mp^r \geq n$ and

$mp^{r+s} \geq n+1$.

(2) The kernel of $\tilde{\Delta}_n$ contains all elements $\langle at^i \rangle$ with $a \in R$, and $i \geq 2n$ or $i = n$

(3) If $n < i < 2n$ then

$$\tilde{\Delta}_n \langle at^i \rangle = \langle t^i, a \rangle + i \langle at^{i-1}, t \rangle$$

Proof. (1) The group $CK_1(R)$ is generated (topologically) by the elements $V_m E \langle at^{p^s} \rangle$ with $a \in R$, $m \in I(p)$, $s \geq 0$. By construction $\tilde{\Delta}_n V_m E \langle at^{p^s} \rangle = \delta_m E \langle at^{p^s} \rangle$. The right-hand side is a sum of elements of the prescribed form by (3.3). This proves one inclusion for (1). The converse inclusion follows easily from the observation that $\tilde{\Delta}_n V_m E \langle at^{p^s} \rangle$ is equal to $\langle a^{p^{r-s}-1} t^{mp^r}, a \rangle + m \langle a^{p^r} t^{mp^{r+s}-1}, t \rangle$ plus higher order terms of the prescribed form.

(2) and (3) Let $i = jp^s$ with $(p, j) = 1$. By definition one has

$$\tilde{\Delta}_n \langle at^i \rangle = \sum_{m \in I(p)} \frac{1}{m} \delta_m E F_m \langle at^i \rangle$$

The formulas of (2.1) and (3.1.1) show $EF_m \langle at^i \rangle = 0$ if m is not a multiple of j and $EF_{lj} \langle at^i \rangle = jE \langle a^l t^{p^s} \rangle$ for all $l \in I(p)$. Thus we see

$$\tilde{\Delta}_n \langle at^i \rangle = \sum_{l \in I(p)} \frac{1}{l} \delta_{lj} E \langle a^l t^{p^s} \rangle.$$

Next note that $i \geq n$ implies $ljp^s \geq 2n$ for all $l \geq 2$ and hence by (3.3)

$$\tilde{\Delta}_n \langle at^i \rangle = \delta_j E \langle at^{p^s} \rangle$$

If $i = n$ or $i \geq 2n$ proposition (3.3) shows $\tilde{\Delta}_n \langle at^i \rangle = 0$. If $n < i < 2n$ then (3.3), (1.10) and (1.5) show

$$\tilde{\Delta}_n \langle at^i \rangle = \langle t^i, a \rangle + i \langle at^{i-1}, t \rangle \qquad \square$$

(3.6) Corollary. Let R be an $\mathbb{F}_p$-algebra. Let $q > n \geq 1$ be integers. Then $\tilde{\Delta}_n$ induces a homomorphism

$$\Delta_n : K_1(R[t]/(t^{2n}),(t))/\{\langle at^n \rangle | a \in R\} \to K_2(R[t]/(t^q),(t^n)) \qquad \square$$

(3.7) <u>Corollary</u>. Let R be a smooth algebra over a perfect field k of characteristic $p > 0$ which can be lifted to a smooth W(k)-algebra. Then for all integers $q > n \geq 1$ there is an exact sequence

$$[K_1(R[t]/(t^{2n}),(t))/\{\langle at^n\rangle | a \in R\}] \xrightarrow{\Delta_n} K_2(R[t]/(t^q),(t^n)) \to$$

$$\to K_2(R[t]/(t^q),(t)) \to K_2(R[t]/(t^n),(t)) \to 0.$$

<u>Proof</u>. (1.6), (2.5), (3.5). □

§4

(4.1) <u>Conjecture</u>. Let R be a domain of characteristic $p > 0$. Let n and q be integers with $q \geq \max(pn^2 - 1,\ 4n^2 - 1)$. Then the homomorphism

$$\Delta_n\colon K_1(R[t]/(t^{2n}),\ (t))/\{\langle at^n\rangle | a \in R\} \to K_2(R[t]/(t^q),\ (t^n))$$

is injective.

(4.2) <u>Comments</u>. 1. Every domain can be embedded in a perfect field, say $R \subset k$. If the conjecture can be proved for k, it will follow automatically for R.
If k is a perfect field, then

$$K_2(k[t]/(t^q),\ (t)) = 0$$

for all $q \geq 2$. This can be shown as follows. The group is generated by the elements $\langle at^i, b\rangle$ and $\langle at^{i-1}, t\rangle$ with $a, b \in k$, $i \geq 1$. Consider $\langle at^i, b\rangle$. Choose r such that $p^r i \geq q$. Since k is perfect, there is a $c \in k$ such that $b = c^{p^r}$. Now compute, using (1.5),

$$\langle at^i, c^{p^r}\rangle = p^r\langle ac^{p^r-1}t^i, c\rangle = \langle a^{p^r}c^{p^{2r}-1}t^{p^r i}, c\rangle = 0.$$

Generators of type $\langle at^{i-1}, t\rangle$ are zero, as one can easily show using (1.8), (1.10) and the preceding computation. So $K_2(k[t]/(t^q),\ (t))$ vanishes. Therefore Δ_n is surjective (see (2.5) and (3.5)). Hence for a perfect field conjecture (4.1) is equivalent to the statement that Δ_n is an isomorphism. One might try to prove this by constructing an inverse for Δ_n.

2. Let R be a domain of characteristic $p > 0$. For every positive integer s we denote, as is usual, by $W_s(R)$ the group of Witt vectors over R of length s i.e. in the notation used till now

$$W_s(R) = TCK_1(R)/V_p^s TCK_1(R).$$

In particular $W_1(R) \cong R$ with $a \in R$ corresponding to $E\langle at\rangle$. The isomorphism (3.1.2.b) induces an isomorphism

$$\prod_{m\in I(p)} W_{s_m}(R) \tilde{\to} K_1(R[t]/(t^{2n}), (t))/\{\langle at^n\rangle | a \in R\}$$

where s_m is the least integer ≥ 0 such that $mp^{s_m} = n$ or $mp^{s_m} \geq 2n$. The restriction of Δ_n to the component $W_{s_m}(R)$ which corresponds to m, is induced by the map δ_m of (3.2) (actually we are undoing the procedure of (3.4)). Let us denote this restriction by $\bar{\delta}_m$. Proposition (3.3) shows that for $s < s_m$

$$\bar{\delta}_m E\langle at^{p^s}\rangle = \sum_{l\in I(p)} \frac{\mu(l)}{l}(\langle a^{lp^{r-s}-1}t^{lmp^r}, a\rangle + mp^s\langle a^{lp^{r-s}}t^{lmp^r-1}, t\rangle)$$

where r is the least integer ≥ 0 such that $mp^r > n$. Instead of showing that Δ_n is injective one can also try to show that each $\bar{\delta}_m$ is injective.

3. Combining the ideas of the two preceding comments we would like to construct for a perfect field k and for every $m \in I(p)$ a homomorphism

$$K_2(k[t]/(t^q), (t^n)) \to W_{s_m}(k)$$

which is left inverse to $\bar{\delta}_m$. However, we can only find for every $m \in I(p)$ with $s_m \geq 1$ a homomorphism ε'_m: $K_2(k[t]/(t^q), (t^n)) \to k$ such that $\varepsilon'_m \circ \bar{\delta}_m$ is the canonical projection $W_{s_m}(k) \to k$. For those m for which $s_m = 1$, this is of course enough, but we are unable to treat cases with $s_m \geq 2$. It may be that the techniques which Coleman developed in [4] can do a better job.

4. In what follows, we will construct for every domain R of characteristic $p > 0$ and for every $m \in I(p)$ with $s_m \geq 1$ a homomorphism

$$\varepsilon_m: K_2(R[t]/(t^q), (t^n)) \to R$$

such that $\varepsilon_m \circ \overline{\delta}_m$ is the canonical projection $W_{s_m}(R) \to R$ followed by the r_m-th power of Frobenius on R; here r_m is the least integer ≥ 0 such that $mp^{r_m} > n$.

(4.3) <u>Lemma</u>. For every domain R and all n and q as in (4.1) there is a homomorphism

$$\nu: K_2(R[t]/(t^q), (t^n)) \to \Omega^1_{R[t]/(t^{2n})/R} \Big/ Rt^{2n-1}dt$$

given by $\nu\langle f, g\rangle = fdg$ if $f \in (t^n)$
$-gdf$ if $g \in (t^n)$.

<u>Proof</u>. This was shown in the proof of (1.11). □

(4.4) <u>Lemma</u>. Let R be a domain of characteristic $p > 0$. For all n and q as in (4.1) the assignment

$$\langle f, g\rangle \mapsto \sum_{i=1}^{p-1} \frac{1}{i} f^i g^{i-1} dg$$

defines a homomorphism

$$\pi: K_2(R[t]/(t^q), (t^n)) \to \Omega^1_{R[t]/(t^{np})/R} \Big/ B(R)$$

where B(R) is the R-module generated by the exact 1-forms and $(t^n)^{p-1}d(t^n)$.

<u>Proof</u>. One must check that the assignment respects relations (D1), (D2), (D3). For (D1) and (D3) this is easy, but (D2) requires more care. Let $f, h, g \in R[t]/(t^q)$ be such that $g \in (t^n)$ or f and $h \in (t^n)$. We must prove $\pi\langle f + h - fhg, g\rangle - \pi\langle f, g\rangle - \pi\langle h, g\rangle = 0$. Let k be a perfect field containing R. The R-module $\Omega^1_{R[t]/(t^{np})/R} \Big/ B(R)$ is free with basis

$\{t^{lp-1}dt \mid l = 1,\dots,n-1\}$ if $p|n$ and $\{t^{lp-1}dt \mid l = 1,\dots,n\}$ if $p \nmid n$. From this remark one sees immediately that the natural map

$$\Omega^1_{R[t]/(t^{np})/R}\big/B(R) \to \Omega^1_{k[t]/(t^{np})/k}\big/B(k)$$

is injective. Therefore, instead of checking $\pi\langle f + h - fhg, g\rangle - \pi\langle f, g\rangle - \pi\langle h, g\rangle = 0$ in $\Omega^1_{R[t]/(t^{np})/R}\big/B(R)$ we can do it in $\Omega^1_{k[t]/(t^{np})/k}\big/B(k)$. Let $W = W(k)$ be the ring of Witt vectors over k, K its field of fractions and K((t)) the field of Laurent series over K. Let $\tilde f, \tilde h, \tilde g \in W[[t]]$ be lifings of f, g and h respectively, such that order $\tilde f$ = order f, order $\tilde h$ = order h and order $\tilde g$ = = order g. Put $\tilde e = \tilde f + \tilde h - \widetilde{fhg}$. Define the forms $\omega_1, \omega_2, \omega_3 \in \Omega^1_{K((t))/K}$ by

$$-\log(1 - \widetilde{fg})\frac{d\tilde g}{\tilde g} = \sum_{i=1}^{p-1} \frac{1}{i}\tilde f^i \tilde g^{i-1} d\tilde g + \frac{1}{p}\tilde f^p \tilde g^{p-1} d\tilde g + \omega_1$$

$$-\log(1 - \widetilde{hg})\frac{d\tilde g}{\tilde g} = \sum_{i=1}^{p-1} \frac{1}{i}\tilde h^i \tilde g^{i-1} d\tilde g + \frac{1}{p}\tilde h^p \tilde g^{p-1} d\tilde g + \omega_2$$

$$-\log(1 - \widetilde{eg})\frac{d\tilde g}{\tilde g} = \sum_{i=1}^{p-1} \frac{1}{i}\tilde e^i \tilde g^{i-1} d\tilde g + \frac{1}{p}\tilde e^p \tilde g^{p-1} d\tilde g + \omega_3$$

Note that actually ω_1, ω_2 and ω_3 are elements of $t^{np}\Omega^1_{K[[t]]/K}$. Since $1 - \widetilde{eg} = (1 - \widetilde{fg})(1 - \widetilde{hg})$, we have in $\Omega^1_{K[[t]]/K}$

$$\sum_{i=1}^{p-1} \frac{1}{i}\tilde e^i \tilde g^{i-1} d\tilde g - \sum_{i=1}^{p-1} \frac{1}{i}\tilde f^i \tilde g^{i-1} d\tilde g - \sum_{i=1}^{p-1} \frac{1}{i}\tilde h^i \tilde g^{i-1} d\tilde g$$

$$= -\frac{1}{p}(\tilde e^p - \tilde f^p - \tilde h^p)\tilde g^{p-1} d\tilde g - \omega_3 + \omega_1 + \omega_2.$$

It is clear that the left-hand side is in $\Omega^1_{W[[t]]/W}$. We will show that the right-hand side belongs to $t^{np}\Omega^1_{K[[t]]/K} + Wt^{n(p-1)}dt^n$. The conclusion will be that both sides lie in $t^{np}\Omega^1_{W[[t]]/W} + Wt^{n(p-1)}dt^n$. It was noted already that $-\omega_3 + \omega_1 + \omega_2$ is in $t^{np}\Omega^1_{K[[t]]/K}$. So we must check $\frac{1}{p}(\tilde e^p - \tilde f^p - \tilde h^p)\tilde g^{p-1} d\tilde g$. This form is equal to

$$\frac{1}{p} \sum_{l,j} \frac{p!}{l!\,j!\,(p-l-j)!}(-1)^{p-l-j}\tilde f^{p-j}\tilde h^{p-l}\tilde g^{2p-l-j-1} d\tilde g$$

where the sum runs over all pairs (l, j) with $0 \leq l < p$, $0 \leq j < p$ and $l + j \leq p$. One sees immediately that the order of each term is at least np, except when order $\tilde{f}$ = order $\tilde{h}$ = 0 and order $\tilde{g}$ = n. Now assume order $\tilde{f}$ = order $\tilde{h}$ = 0 and order $\tilde{g}$ = n. In this case all terms $\tilde{f}^{p-j}\tilde{h}^{p-l}\tilde{g}^{2p-l-j-1}d\tilde{g}$ with $l + j < p$ have order $\geq$ np. So we are left with

$$\frac{1}{p}\sum_{j=1}^{p-1}\binom{p}{j}\tilde{f}^{p-j}\tilde{h}^{j}\tilde{g}^{p-1}d\tilde{g}$$

Define $\tilde{f}_0, \tilde{h}_0, \tilde{g}_0 \in W$ by $\tilde{f} = \tilde{f}_0 \bmod t$, $\tilde{h} = \tilde{h}_0 \bmod t$ and $\tilde{g} = \tilde{g}_0 t^n \bmod t^{n+1}$. Then modulo terms of order $\geq$ pn the above form is equal to

$$\left(\sum_{j=1}^{p-1}\frac{(p-1)!}{j!(p-j)!}\tilde{f}_0^{p-j}\tilde{h}_0^{j}\tilde{g}_0^{p}\right)t^{n(p-1)}dt^n$$

Thus we have shown that in all cases

$$\sum_{i=1}^{p-1}\frac{1}{i}\tilde{e}^{i}\tilde{g}^{i-1}d\tilde{g} - \sum_{i=1}^{p-1}\frac{1}{i}\tilde{f}^{i}\tilde{g}^{i-1}d\tilde{g} - \sum_{i=1}^{p-1}\frac{1}{i}\tilde{h}^{i}\tilde{g}^{i-1}d\tilde{g}$$

lies in $t^{np}\Omega^1_{W[[t]]/W} + Wt^{n(p-1)}dt^n$. So its image in $\Omega^1_{k[t]/(t^{np})/k}/B(k)$ is zero, which is exactly what we needed. □

Remark. It is clear that (4.4) and (4.3) are based on the expression $\log(1 - fg)\frac{dg}{g}$. When one tries to show that relation (D1) is respected, one tries to show more or less that $\log(1 - g)\frac{dg}{g}$ is an exact 1-form. At this point the classical dilogarithm function

$$\operatorname{dilog} x = \int \log(1 - x)\frac{dx}{x} = -\sum_{m\geq 1}\frac{x^m}{m^2}$$

appears.

(4.5) Let R be a domain of characteristic $p > 0$. Let n and q be integers with $q \geq \max(pn^2 - 1, 4n^2 - 1)$. Take $m \in I(p)$ and define as before s_m and r_m as the least non-negative integers which satisfy $mp^{s_m} = n$ or $mp^{s_m} \geq 2n$ and $mp^{r_m} > n$. Take m such that $s_m \geq 1$. We are going to define a homomorphism

$$\varepsilon_m\colon K_2(R[t]/(t^q),\ (t^n)) \to R$$

and check that the composite $\varepsilon_m \circ \bar{\delta}_m$ is the canonical projection $W_{s_m}(R) \to R$ followed by the r_m-th power of Frobenius on R. Three cases must be distinguished: either $1 \leqslant m < n$ or $n + 1 \leqslant m < 2n$ or $p|n$ and $mp^{s_m} = n$. First we consider the case $n + 1 \leqslant m < 2n$. To define ε_m we use (4.3). The free R-module $\Omega^1_{R[t]/(t^{2n})/R} / Rt^{2n-1}dt$ has a basis $dt, tdt, t^2dt, \ldots, t^{2n-2}dt$. Define

$$\zeta_m\colon \Omega^1_{R[t]/(t^{2n})/R} / Rt^{2n-1}dt \to R$$

by $\zeta_m(t^{m-1}dt) = \frac{1}{m}$ and $\zeta_m(t^{i-1}dt) = 0$ for $i \neq m$. Then we define

$$\varepsilon_m = \zeta_m \circ \nu$$

Note that in this case $s_m = 1$ and $r_m = 0$. Using (4.2.2) one checks immediately that $\varepsilon_m \circ \bar{\delta}_m$ is the identity map.

Next assume $1 \leqslant m < n$ or $p|n$ and $mp^{s_m} = n$. In these cases we use (4.4). The free R-module $\Omega^1_{R[t]/(t^{np})/R} / B(R)$ has a basis $\{t^{lp-1}dt \mid l = 1,\ldots,n-1\}$ if $p\nmid n$ and $\{t^{lp-1}dt \mid l = 1,\ldots,n\}$ if $p|n$.

Define

$$\xi_m\colon \Omega^1_{R[t]/(t^{np})/R} / B(R) \to R$$

by $\xi_m(t^{mp^{r_m}-1}dt) = \frac{1}{m}$ and $\xi_m(t^{lp-1}dt) = 0$ for $lp \neq mp^{r_m}$. And then we define

$$\varepsilon_m = \xi_m \circ \pi$$

Finally we compute $\varepsilon_m \circ \bar{\delta}_m$ using (4.2.2). Clearly $\varepsilon_m \circ \bar{\delta}_m E\langle at^{p^s}\rangle = 0$ for $s > 0$. And, say $r = r_m$,

$$\varepsilon_m \circ \bar{\delta}_m E\langle at\rangle = \varepsilon_m[m \sum_{l\in I(p)} \frac{\mu(l)}{l} \langle a^{lp^r} t^{lmp^r-1},\ t\rangle]$$

$$= \xi_m[m \sum_{l\in I(p)} \sum_{i=1}^{p-1} \frac{\mu(l)}{li}\, a^{ilp^r} t^{ilmp^r-1}dt]$$

$$= \xi_m[m \sum_{w=1}^{p-1} \frac{1}{w}(\sum_{l|w} \mu(l))a^{wp^r} t^{wmp^r-1}dt]$$

$$= \xi_m(ma^{p^r} t^{mp^r-1}dt)$$

$$= a^{p^r}$$

We conclude this paper with a theorem which summarizes our results in the case where for every $m \in I(p)$ s_m is 0 or 1.

(4.6) <u>Theorem</u>. Let R be a domain of characteristic $p > 0$. Let n and q be integers such that $1 < n < \frac{p}{2}$ or $n = p$ and $q \geq \max(pn^2 - 1, 4n^2 - 1)$. Then the homomorphism

$$\Delta_n: K_1(R[t]/(t^{2n}), (t))/\{<at^n> | a \in R\} \to K_2(R[t]/(t^q), (t^n))$$

is injective. Moreover, there is an isomorphism

$$R^{2n-2} \tilde{\to} K_1(R[t]/(t^{2n}), (t))/\{<at^n> | a \in R\}$$

Furthermore, if R satisfies the hypothesis of (2.5), there is a surjection

$$K_3(R[t]/(t^n), (t)) \to R^{2n-2},$$

and the kernel of this homomorphism consists of all elements of $K_3(R[t]/(t^n), (t))$ which can be lifted to $K_3(R[t]/(t^q), (t))$ for every $q > n$. □

<u>References</u>

[1] Bass, H. & Murthy, P., Grothendieck groups and Picard groups of abelian group rings. Ann. Math. 86 (1967), p. 16-73.

[2] Bloch, S., Algebraic K-theory and crystalline cohomology. Publ. Math. I.H.E.S. 47 (1978), p. 187-268.

[3] Cartier, P., Questions de rationalité des diviseurs en géométrie algébrique. Bull. de la Soc. Math. de France 86 (1958), p. 177-251.

[4] Coleman, R., The dilogarithm and the norm residue symbol. Preprint.

[5] Evens, L. & Friedlander, E., $K_r(\mathbb{Z}/p^2)$ and $K_r(\mathbb{Z}/p[\varepsilon])$ for $p \geq 5$ and $r \leq 4$. Bull. Amer. Math. Soc. 2 (1980), p. 440-443.

[6] Evens, L. & Friedlander, E., On $K_*(\mathbb{Z}/p^2\mathbb{Z})$ and related homology groups, preprint, Evanston 1980.

[7] Graham, J., Continuous symbols on fields of formal power series in: Algebraic K-theory II, Lecture Notes in Math. 342, Springer Verlag.

[8] Grothendieck, A., Eléments de géométrie algébrique IV. Publ. Math. I.H.E.S. 32 (1967).

[9] Grothendieck, A., S.G.A. I, Lecture Notes in Math. 224, Springer Verlag.

[10] Illusie, L., Complexe de De Rham-Witt, Asterisque 63 (1979), p. 83-112.

[11] Illusie, L., Complexe de De Rham-Witt et cohomologie cristalline, Ann. Scient. Ec. Norm. Sup. 12 (1979), p. 501-661.

[12] Van der Kallen, W., Sur le K_2 des nombres duaux. C.R. Acad. Sc. Paris t. 273, 1971, Serie A, p. 1204-1207.

[13] Keune, F., The relativization of K_2. J. of Alg. 54 (1978), p. 159-177.

[14] Keune, F., On the equivalence of two higher algebraic K-theories. Preprint, Nijmegen 1979.

[15] Lazard, M., Commutative formal groups. Lecture Notes in Math. 443, Springer Verlag.

[16] Maazen, H. & Stienstra, J., A presentation for K_2 of split radical pairs. J. of pure and applied alg. 10 (1977), p. 271-294.

[17] Milnor, J., Introduction to algebraic K-theory. Annals of Math. Study 72, Princeton Univ. Press.

[18] Quillen, D., Higher algebraic K-theory I, in: Algebraic K-theory I, Lecture Notes in Math. 341, Springer Verlag.

[19] Roberts, L. & Geller, S., K_2 of some truncated polynomial rings, in: Ring theory Waterloo 1978, Lecture Notes in Math. 734, Springer Verlag.

[20] Seshadri, C., L'opération de Cartier. Applications, Séminaire C. Chevalley 1958/1959, Ec. Norm. Sup.

[21] Snaith, V., in preparation.

[22] Soulé, C., Rational K-theory of the dual numbers of a ring of algebraic integers, preprint.

[23] Stienstra, J., Deformations of the second Chow group, a K-theoretic approach; thesis, Utrecht 1978.

Jan Stienstra
Mathematisch Instituut
Budapestlaan 6
3508 TA Utrecht
Netherlands.

ON THE NORMAL SUBGROUPS OF GL_n OVER A RING *

L.N.Vaserstein

University of Chicago and Cornell University

Let A be an associative ring with 1, GL_nA the group of invertible n by n matrices over A. For any (two sided) ideal B of A, let E_nB denote the subgroup of GL_nA generated by all elementary matrices $b^{i,j} = 1_n + e_{i,j}(b)$ with $b \in B$, where $1 \leq i \neq j \leq n$, and 1_n is the identity matrix.

We write $sr(A) \leq n$, if for any $b_1, \ldots, b_{n+1}$ in A satisfying $\sum_{i=1}^{n+1} Ab_i = A$ there exist c_i in A such that $\sum_{i=1}^{n} A(b_i + c_i b_{n+1}) = A$. This means (see Vaserstein [5]) that "n defines a stable range for GLA" in the sense of H.Bass [1].

THEOREM 1 (H.Bass [1, 2]). Let $sr(A) \leq n-1$ and $n \geq 3$. Then (2) a subgroup H of GL_nA is normalized by E_nA if and only if, for a unique ideal B of A, the image of H in GL_nA/B lies in the center and $H \supset [E_nA, E_nB]$.

THEOREM 3 (I.Golubchik [3], A.Suslin [4]). Let A be commutative and $n \geq 3$. Then (2) holds.

We hybridize Th.1, 3, i.e. prove (2) with $n \geq 3$ under a weaker condition than the condition $sr(A) < n$ or the commutativity of A:

THEOREM 4. Suppose that $n \geq 3$ and for every maximal ideal D of the center C of A there exists a multiplicative set $S \subset C - D$ such that $sr(S^{-1}A) \leq n-1$. Then (2) holds.

Th.4 implies Th.1 (take $S = \{1\}$ for any D), and Th.3 (take $S = C - D = A - D$). It also implies the following generalization of Th.3:

---

* The talk was given at Northwestern University in May, 1979.

COROLLARY 5. Let $n \geq 3$ and A be finitely generated as module over its center. Then (2) holds.

Indeed, then $sr((C - D)^{-1}A) \leq 1$ for any maximal ideal D of the center C, see [1].

Statement (2) will be contained in the following two statements, where, for any $g = (g_{i,j})$ in GL_nA, $J(g)$ denotes the ideal of A generated by $g_{i,j}$ and $ag_{i,i} - g_{j,j}a$ with $i \neq j$, $a \in A$:

(6) $E_n(A,B) = [E_nA, E_nB] = [GL_nA, E_n(A,B)] = [E_nA, G_n(A,B)]$ for any ideal B of A, where $E_n(A, B)$ is the subgroup generated by $a^{i,j}b^{j,i}(-a)^{i,j}$ with $a \in A$, $b \in B$, $i \neq j$, and where $G_n(A,B) = \{ g \in GL_nA : J(g) \subset B \}$;

(7) every subgroup H of GL_nA normalized by E_nA contains $E_n(A, J(H))$, where $J(H)$ is the ideal of A generated by all $J(g)$ with $g \in H$.

We will prove (6) and (7) under weaker conditions than that of Th.4. I know no counter example to (6) or (7) with $n \geq 3$.

Remark. Suslin's proof of the normality of E_nA in GL_nA for a commutative ring A, $n \geq 3$, is based on the fact, that for any $b_i \in A$ with $\sum_{i=1}^{n} Ab_i = A$, the A-module $P = \{(a_i)_{1 \leq i \leq n} : \sum_{i=1}^{n} a_ib_i = 0\}$ is generated by the rows of the form $b_ie_j - b_je_i$. This fact was proved by me in [7], Ch.1 for $n = 3$, and claimed in Ch.2 for any $n \geq 3$. A.Suslin has generalized this fact for A as in Cor.5 and got a similar proof of the normality of E_nA for such A. Our approach here is different.

The results of this paper can be generalized to the automorphism groups of projective A-modules (cf.[6]), and to the orthogonal groups in the sense of [6].

Proof of (6)

LEMMA 8 . Let B be an ideal of A , and $n \geq 3$. Then

$$E_n(A , B) = [E_nA , E_nB] .$$

PROOF. Let $a \in A$, $b \in B$, $1 \leq i \neq j \neq k \neq i \leq n$. The equality $(ab)^{i,j} = [a^{i,k} , b^{k,j}]$ shows that $[E_nA, E_nB] \supset E_nB$.

Now, we have to prove that $h\, a^{i,j}\, b^{j,i}\, (-a)^{i,j}\, h^{-1} \in E_n(A , B)$ for any $h \in E_nA$. It is enough to consider the case of elementary $h = c^{s,t}$.

The case $(s,t) = (i, j)$ is trivial.

If $(s,t) \neq (i, j), (j, i)$, then $[h, a^{i,j}\, b^{j,i}\, (-a)^{i,j}] \in E_nB$ and $g' := h\, a^{i,j}\, b^{j,i} (-a)^{i,j}\, h^{-1} \in E_n(A , B)$.

At last, if $(s, t) = (j, i)$, then

$$g' = h\, a^{i,j}\, [1^{j,k}, b^{k,i}]\, (h\, a^{i,j})^{-1} = h\, [a^{i,k}\, 1^{j,k} , (-ba)^{k,j} b^{k,i}]\, h^{-1}$$

$$= [(ca)^{j,k}\, a^{i,k}\, 1^{j,k} , (bac)^{k,i}\, (-ba)^{k,j}\, b^{k,i}]$$

$$\in (ca + 1)^{j,k}\, a^{i,k}\, (bac - b)^{k,i}\, (-ba)^{k,j}\, (-a)^{i,k}\, (-ca - 1)^{j,k}\, E_nB$$

$\subset E_n(A , B)$ by the second case.

LEMMA 9. Let B be an ideal of A , and $n \geq 3$. Then $E_n(A, B)$ is normalized by $GL_{n-1}A = \begin{pmatrix} GL_{n-1}A & 0 \\ 0 & 1 \end{pmatrix} \subset GL_nA$.

PROOF. Let $b \in B$, $1 \leq i \neq j \leq n$, $g \in GL_{n-1}A$. We have to prove that $gb^{i,j}g^{-1} \in E_n(A , B)$.

Case $B = A$. If i or $j = n$, then $g\, b^{i,j}\, g^{-1} \in E_nA$. Otherwise, $g\, b^{i,j}\, g^{-1} = [g\, b^{i,n}\, g^{-1} , g\, 1^{n,j}\, g^{-1}] \in E_nA$.

General case. Since $GL_{n-1}A$ normalizes E_nA , the set $E_nA\, GL_{n-1}A$ is a subgroup of GL_nA , normalized by E_nA and by all permutation matrices. Therefore, we can write $g = g_1 g_2$ with g_1 in E_nA and g_2 having the same j-th row as 1_n. Then $g_2 b^{i,j} g_2^{-1} \in E_nB$ and $g\, b^{i,j}\, g^{-1} \in E_n(A , B)$ by Lemma 8.

LEMMA 10. Let B be an ideal of A, and $n \geq 3$. Then $E_n(A,BB) \subset E_nB$.

PROOF. Let $b_1, b_2 \in B$, $1 \leq i \neq j \leq n$, and $a \in A$. Take $k \neq i, j$ $(1 \leq k \leq n)$. Then $a^{i,j}(b_1 b_2)^{j,i}(-a)^{i,j}$

$= [a^{i,j}\, b_1^{\,j,k}\, (-a)^{i,j}\ ,\ a^{i,j}\, b_2^{\,k,i}\, (-a)^{i,j}] \in E_nB$.

LEMMA 11. Let $n \geq 3$, S a central multiplicative set in A, $g \in GL_nA$, and assume that the image of g in $GL_nS^{-1}A$ lies in $E_nS^{-1}A\, GL_{n-1}S^{-1}A$. Then there exists $s \in S$ such that $g\,(E_nsA)\,g^{-1} \subset E_nA$.

PROOF. Consider the ring $A' = A[x, y]$, where x commutes with the center of A but does not commute with A, and y commutes with A and x. The localization $A' \longrightarrow S^{-1}A'$ induces the group morphism $F : GL_nA' \longrightarrow GL_nS^{-1}A'$. Let $1 \leq i \neq j \leq n$, and $h(y) = g\,(xy^2)^{i,j}\,g^{-1} \in GL_nA'$.

Applying Lemmas 9, 10 (with $S^{-1}A'$ instead of A) we get: $F(h(y)) \in E_n(S^{-1}A', S^{-1}A'y^2) \subset E_nS^{-1}A'y$. Therefore, there exists $s_1 \in S$ such that $F(h(s_1y)) \in F(E_nA'y)$. Let $F(h(s_1y)) = F(h'(y))$ with $h'(y) \in E_nA'y$. Since $F(h(s_1y)\,h'(y)^{-1}) = 1_n$, there exists $s_2 \in S$ such that $h(ys_1s_2)\,h'(s_2y)^{-1} = 1_n$.

Thus, $g\,(xs_{i,j})^{i,j}\,g^{-1} = h(s_1s_2) = h'(s_2) \in E_nA'$ for $s_{i,j} = (s_1s_2)^2 \in S$. It follows that $g\,(As_{i,j})^{i,j}\,g^{-1} \subset E_nA$.

Take s = the product of all $s_{i,j}$. Then $g\,(As)^{i,j}\,g^{-1} \subset E_nA$ for all $i \neq j$, hence $g\,(E_nAs)\,g^{-1} \subset E_nA$.

PROPOSITION 12. Let $n \leq 3$, and G_A the subgroup of $g \in GL_nA$ such that the image of g in $GL_nS^{-1}A$ lies in $E_nS^{-1}A\, GL_{n-1}S^{-1}A$ for every maximal ideal D of the center C of A, where $S = C - D$. Then E_nA is normal in G_A.

PROOF. Let $g \in G_A$. The set $X = \{ c \in C : g\,(E_ncA)\,g^{-1} \subset E_nA \}$ is an ideal of C. By Lemma 11, it contains some s outside any maximal ideal of C. Thus, $X = C$, i.e. E_nA is normalized by g.

THEOREM 13. Let $n \geq 3$. For any ideal B of A, let G_B be the subgroup of $g \in GL_nB := \{h \in GL_nA : h \equiv 1_n \bmod B\}$ such that the image of g in $GL_nS^{-1}A$ lies in $E_n(S^{-1}A, S^{-1}B)\, GL_{n-1}S^{-1}B$ for every maximal ideal D of the center C of A, where $S = C - D$. Then $[G_B, E_nA] = E_n(A, B)$. If $G_B = GL_nB$, then $[G_n(A, B), E_nA] = E_n(A, B)$. In particular, (6) holds provided $G_B = GL_nB$ for every ideal B of A.

PROOF. Consider the ring $A' = \{(a, b) \in A^2 : a + B = b + B\}$, and the ideal $B' = (B, 0)$ of A'.

Let $g \in G_B$, $h \in E_nA$. Put $g' = (g,0) \in GL_nB'$, $h' = (h,h) \in E_nA'$.

Let C' be the center of A', D' a maximal ideal of C', D a maximal ideal of C containing the ideal $\{c \in C : (c, c) \in D'\}$, $S = C - D$, $T = \{(s, s) : s \in S\} \subset S' = C' - D'$.

Then the image of g' in $GL_nT^{-1}A'$ lies in $E_nT^{-1}A'\, GL_{n-1}T^{-1}A'$. So, the image of g' in $GL_nS'^{-1}A'$ lies in $E_nS'^{-1}A'\, GL_{n-1}S'^{-1}A'$.

By Prop.12, $[g', h'] \in E_nA'$. On the other hand, $[g', h'] \in GL_nB'$.

As A' is the semidirect product of the ideal B' and the subring $\{(a,a) : a \in A\}$, isomorphic to A, we have $E_nA' \cap GL_nB' = E_n(A', B')$.

Thus, $[g', h'] \in E_n(A', B')$, hence $[g, h] \in E_n(A, B)$.

In the view of Lemma 8, only the inclusion $E_n(A,B) \supset [G_n(A,B), E_nA]$ is left to prove (provided $G_A = GL_nA$ and $G_B = GL_nB$). Following [2], we take any $g \in G_n(A, B)$ and for every h in E_nA define $f(h) := [h, g]\, E_n(A, B) \in (E_nA \cap GL_nB)/E_n(A, B)$.

Since E_nA and GL_nB commute mod $E_n(A, B)$, the last quotient group is commutative, and f is a group morphism from $E_nA = [E_nA, E_nA]$ to an abelian group. Therefore, f is trivial, i.e. $[h, g] \in E_n(A, B)$ for every $h \in E_nA$.

COROLLARY 14. Under the condition of Th. 4, statement (6) holds. In particular, every subgroup H of GL_nA, containing $E_n(A, J(H))$, is normalized by E_nA.

Indeed, the condition $sr(S^{-1}A) < n$ implies [1] that $GL_nS^{-1}B = E_n(S^{-1}A, S^{-1}B)\, GL_{n-1}S^{-1}B$ for any ideal B of A. Hence, the image of

GL_nB in $GL_nT^{-1}B$ lies in $E_n(T^{-1}A, T^{-1}B)\, GL_{n-1}T^{-1}B$ for $T = C - D \supset S$. Thus, the condition of Th.13 holds.

Proof of (7)

Let $n \geq 2$, $a \in A$, $1 \leq i \neq j \leq n$. A matrix $g = (g_{i,j}) \in GL_nA$ commutes with $a^{i,j}$ if and only if $ag_{j,k} = 0$ for all $k \neq j$, $g_{k,i}a = 0$ for all $k \neq i$, and $ag_{j,j} = g_{i,i}a$. Therefore, g commutes with all $a^{i,j}$, $1 \leq i \neq j \leq n$, if and only if $ag_{i,j} = g_{i,j}a = ag_{i,i} - g_{j,j}a = 0$ for all $i \neq j$. In particular, $G_n(A, 0)$ is the centralizator of E_nA in GL_nA, and, for any ideal B of A, the inverse image of the center under the homomorphism $GL_nA \longrightarrow GL_nA/B$ is exactly $G_n(A, B)$.

LEMMA 15. Let $n \geq 3$, $0 \neq a \in A$, a not a zero divisor in A, and let H be a subgroup of GL_nA normalized by all $a^{i,j}$ $(i \neq j)$ and containing $g = (g_{i,j})$ such that $g_{n,1} = 0$ and g does not commute with $a^{i,j}$ for some $i \neq j$. Then H contains an elementary matrix $\neq 1_n$.

PROOF. Case 1 : $g = \prod_{i=1}^{n-1} (b_i)^{i,n}$ with $b_i \in A$, $b_j \neq 0$ for some j. Take k, such that $1 \leq k \leq n-1$, $k \neq j$. Then

$$H \ni [a^{k,j}, g] = (ab_j)^{k,n} \neq 1_n .$$

Case 2 : $g_{n,i} = 1 - g_{n,n} = 0$ for all $i \leq n-1$. If g is not as in Case 1, then g does not commute with $a^{i,n}$ for some $i < n$. So $[g, a^{i,n}] \in H$ is as g in Case 1.

General case. If g does not commute with $a^{1,i}$ for some i $(2 \leq i \leq n)$, then $[g, a^{1,i}] \in H$ satisfies the same condition as g in Case 2.

Otherwise, if g does not commute with $a^{i,1}$ for some i $(2 \leq i \leq n-1)$, then $[a^{i,1}, g] \in H$ satisfies the same condition as g in Case 2, and we are again reduced to Case 2.

At last, if g commutes with all $a^{1,i}$ and with $a^{2,1}$, then g is a diagonal matrix and $[g, a^{2,3}] \in H$ is an elementary matrix $\neq 1_n$.

LEMMA 16 . *Let* S *be a central multiplicative set in* A , $n \geq 3$, $F : GL_nA \to GL_nS^{-1}A$ *the group morphism induced by the localization* $A \to S^{-1}A$. *Let* $h \in E_nS^{-1}A\, GL_{n-1}S^{-1}A$. *Then for any* $s \in S$ *there exists* $s' \in S$ *such that* $h\, F(E_ns'A)\, h^{-1} \subset F(E_nsA)$.

PROOF. Let $A' = A[x, y]$ be as in the proof of Lemma 11. Let $1 \leq i \neq j \leq n$, and $g(y) = h\, (xy^2)^{i,j}\, h^{-1} \in GL_nS^{-1}A'$. By Lemmas 9 and 10 (with $S^{-1}A'$ instead of A), we have : $g(y) \in E_nS^{-1}A'y$. Therefore, there exists $s_o \in S$ such that $g(s_oy) \in F(E_nA'y)$, hence $g(s_os) \in F(E_nA\, x\, s)$. It follows that $h\, F((xs_{i,j})^{i,j})\, h^{-1} \in F(E_nsA)$ for $s_{i,j} = (s_os)^2 \in S$ and all $x \in A$. Now we can take $s' = \prod_{i \neq j} s_{i,j} \in S$.

LEMMA 17 . *Let* A, S, n, F *be as in Lemma 16. Let* H *be a subgroup of* $GL_nS^{-1}A$ *normalized by* $F(E_nAs)$ *for some* $s \in S$. *Suppose that* $\sum_{i=1}^{n-1} S^{-1}A(g_{i,1} + c_ig_{n,1}) = S^{-1}A$ *for some* $g = (g_{i,j}) \in H - G_n(S^{-1}A,0)$ *and* $c_i \in S^{-1}A$. *Then* H *contains an elementary matrix* $\neq 1_n$.

PROOF . *Case 1* : $g_{n,2} = 0$. Then we can apply Lemma 15 (with $S^{-1}A$ instead of A) .

Case 2 : $S^{-1}Ag_{1,2} \ni g_{n,2}$. Let $ag_{1,2} = g_{n,2}$ where $a \in S^{-1}A$. Then $H' = (-a)^{n,1}\, H\, a^{n,1} \ni (-a)^{n,1}\, g\, a^{n,1} = (h_{i,j}) = h$ with $h_{n,2} = 0$, and $h \notin G_n(S^{-1}A, 0)$. By Lemma 16, H' is normalized by $F(E_ns'A)$ for some $s' \in S$. Therefore, by Case 1, $H' \ni b^{i,j}$ for some non-zero b in $S^{-1}A$, $1 \leq i \neq j \leq n$. It follows that $H' \ni c^{n,1}$ for some $0 \neq c \in S^{-1}A$ (we can take $c = bs'^3$). Thus,

$$H = a^{n,1}\, H'\, (-a)^{n,1} \ni a^{n,1}\, c^{n,1}\, (-a)^{n,1} = c^{n,1} \neq 1_n .$$

Case 3: $S^{-1}Ag_{1,1} \ni g_{n,1}$. If $g_{n,1} = 0$, we can apply Lemma 15 . Otherwise, $H \ni [g\, ,\, s^{1,n}] = h = (h_{i,j})$ with $S^{-1}Ah_{1,2} \ni h_{n,2}$, and $h \notin G_n(S^{-1}A\, ,\, 0)$. So, we are reduced to Case 2 .

Case 4 : $\sum_{i=1}^{n-1} S^{-1}Ag_{i,1} = S^{-1}A$. Find $c_i \in S^{-1}A$ such that $\sum_{i=1}^{n-1} c_i g_{i,1} = -g_{n,1}$. Then, putting $f = \prod_{i=2}^{n-1} (c_i)^{n,i}$, we have: $H' = f H f^{-1} \ni f g f^{-1} = h = (h_{i,j})$ with $c_1 h_{1,1} = -h_{n,1}$, $h \notin G_n(S^{-1}A, 0)$. By Lemma 16, H' is normalized by $F(E_n As')$ for some $s' \in S$. By Case 3 , H' contains an elementary matrix $\neq 1_n$. It follows that $H' \ni c^{n,1}$ for some $c \neq 0$. Thus ,

$$H = f^{-1} H' f \ni f^{-1} c^{n,1} f = c^{n,1} .$$

General case. Let c_i be as in the condition of the lemma . Then , putting $f = \prod_{i=1}^{n-1} (c_i)^{i,n} \in E_n S^{-1}A$, we have : $H' = f H f^{-1} \ni f g f^{-1} = h = (h_{i,j})$ with $\sum_{i=1}^{n-1} S^{-1}Ah_{i,1} = S^{-1}A$, and $h \notin G_n(S^{-1}A, 0)$. By Lemma 15 and Case 4 , $H' \ni c^{1,n}$ for some $0 \neq c \in S^{-1}A$. Thus , $H \ni f c^{1,n} f^{-1} = c^{1,n}$.

PROPOSITION 18 . Let A be a ring with 1 , $n \geq 3$, H a subgroup of GL_nA . Suppose that for every $g = (g_{i,j}) \in GL_nJ(H)$ and any maximal ideal D of the center C of A there exist $b_i \in A$, $a \in C - D$ such that $\sum_{i=1}^{n-1} A(g_{i,1}s + b_i g_{n,1}) \ni s^2$. Then H is normalized by E_nA if and only if $H \supset E_nJ(H)$.

PROOF. Assume that $H \supset [H, E_nA]$, and let $B = \{ b \in A : b^{1,2} \in H \}$. Then B is an ideal of A and $B \subset J(H)$. We have to prove that $B = J(H)$.

Let D be a maximal ideal of C , and $S = C - D$. The image of H in $GL_nS^{-1}(A/B)$ is normalized by the image of E_nA/B there. By Lemma 17 (with A/B instead of A), either the image of H contains $x^{i,j}$ with $1 \leq i \neq j \leq n$, $0 \neq x \in S^{-1}(A/B)$, or the image of H lies in the center.

In the first case we have $g \in H$ of the form $a^{i,j} h$, where the image of $a \in A$ in $S^{-1}(A/B)$ equals $x \neq 0$ and the image of h in $GL_nS^{-1}(A/B)$ equals 1_n. Take $k \neq i, j$ and $s \in S$ such that

$1 \leq k \leq n$ and $[h, s^{j,k}] \in GL_n B$. Then
$H \ni g' = [g, s^{j,k}] = (as)^{i,k} h'$ with $h' \in GL_n B$.

The condition of Prop.18 implies that $GL_n B = G_B$ (for any ideal $B \subset J(H)$).(Indeed, let $g \in GL_n B$. Find $a_i, b_i \in A$, $s \in S = C - D$ such that $\sum_{i=1}^{n-1} a_i(g_{i,n}s + b_i g_{n,n}) = s^2$. Let $f = \prod_{i=1}^{n-1} (b_i/s)^{i,n} \in E_n S^{-1}A$, and $h = \prod_{i=1}^{n-1} ((a_i - g_{n,n}a_i)/s)^{n,i} \in E_n S^{-1}B$. Then, for the image $F(g)$ of g in $GL_n S^{-1}A$, we have: $(f^{-1} h f F(g))_{n,n} = (hf F(g))_{n,n} = 1$, hence $f^{-1} h f F(g) \in E_n S^{-1}B \, GL_{n-1} S^{-1}B$, and
$F(g) \in E_n(S^{-1}A, S^{-1}B) \, GL_{n-1} S^{-1}B$.)

By Theorem 13, $[h'^{-1}, 1^{k,j}] \in E_n(A, B) \subset H$. Therefore, $H \ni g' [h'^{-1}, 1^{k,j}] (1^{k,j} g'^{-1} (-1)^{k,j}) = [g'h'^{-1}, 1^{k,j}] = (as)^{i,j}$. This contradicts the definition of B. So the first case is impossible.

In the second case, for any $g \in H$ there exists $s \in S$ such that $sJ(g) \subset B$. Since we can take such s outside any maximal ideal D of C, we have $J(g) \subset B$. Since this holds for every $g \in H$, we have $J(H) \subset B$. Thus, $J(H) = B$.

Now, assume that $H \supset E_n(A, J(H))$. Then, by Th.13 (with $B = J(H)$), we have $[H, E_n A] \subset [G_n(A, J(H)), E_n A] = E_n(A, J(H))$, so H is normalized by $E_n A$.

Prop.18 implies Th.4 with a weaker condition :

THEOREM 19. Let $n \geq 3$. Suppose that for every maximal ideal D of the center C of A and any $g = (g_{i,j}) \in GL_n A$ there exist a_i, $b_i \in A$, $s \in C - D$ such that $\sum_{i=1}^{n-1} b_i(g_{i,1}s + a_i g_{n,1}) = s^2$. Then a subgroup H of $GL_n A$ is normalized by $E_n A$ if and only if $H \supset E_n(A, J(H))$.

REFERENCES

1 H.Bass, K-theory and stable algebra, Publ.Math. IHES . No. 22 (1964), 5-60.

2 H.Bass, Algebraic K-theory, New York, 1968.

3 I.Golubchik, On the general linear group over an associative ring, Uspekhi Mat.Nauk, 28:3 (1973, 179-180 (in Russian).

4 A.Suslin, On the structure of the special linear group over polynomial rings, Izv.Akad.Nauk,ser.mat. 41:2 (1977),235-252 = Math.USSR Izvestjia 11:2, 221-238.

5 L.N.Vaserstein, The stable range of rings and the dimension of topological spaces, Funct.Anal.Appl. 5:2 (1971),102-110.

6 L.N.Vaserstein, The stabilization for classical groups over rings, Math.USSR Sb.22 (1974), 271-303.

7 L.N.Vaserstein, A.A.Suslin, Serre's problem on projective modules over polynomial rings and algebraic K-theory, Math.USSR Izv. 10:5 (1976), 937-1001.

MEYER VIETORIS SEQUENCES AND

MODULE STRUCTURES ON NK_*

C. A. Weibel

The purpose of this essay is to point out a number of consequences of some module structures on the nilgroups NK_*, particularly in characteristic p. For example, if we ignore p-torsion and only consider rings in which p is nilpotent, then Quillen K-theory satisfies excision and has Mayer-Vietoris exact sequences. We can deduce similar, but weaker results in characteristic 0 .

In §1 we introduce the notation we need about Witt vectors. In §2 we remind the reader of the $End_0(R)$-module structure on $NK_*(\Lambda)$. It is Stienstra's observation that this extends to a W(R)-module structure, and we explore some elementary consequences of this in §3. In §4 we compute this module structure on $NK_2(k[\varepsilon],\varepsilon)$ and show that it is <u>not</u> the "usual" k-module structure. In §5 we prove the above mentioned results about $K_*(\Lambda) \otimes \mathbb{Z}[\frac{1}{p}]$ using known facts about Karoubi-Villamayor theory. Finally, in §6 we discuss the problem of localization of the ring, and show for example that for a finite group π the groups $NK_*(\mathbb{Z}\pi)$ are torsion groups, the only torsion being at primes dividing the order of π.

I would like to thank J. Stienstra for showing me that I was talking about Witt vectors, and for explaining his point of view to me. I would also like to thank R. G. Swan for motivating this essay with his work on localization in [Sw].

Finally, I would like to thank L. G. Roberts and Queen's University for their hospitality during the final stage of work.

§1. Witt Vectors

In this section, R denotes any commutative ring with 1. By $W(R)$ we will mean the ring of all (big) Witt vectors. That is, the underlying additive group of $W(R)$ is the multiplicative group $1 + tR[[t]]$ of power series. The multiplication on $W(R)$ is the unique continuous functorial operation $*$ for which $(1-at) * (1-bt) = (1-abt)$; for example, if $d = \text{g.c.d.}(m,n)$ then

$$(1.1) \quad (1-at^m) * (1-bt^n) = (1-a^{n/d}b^{m/d}t^{mn/d})^d \ .$$

A very quick and readable introduction to this point of view is [Bl 1,§I.1]; other points of view are discussed in [Cl], [Gr], and [SGA6]. Note that the "zero" and "one" of the ring $W(R)$ are 1 and $1-t$ in our convention, which differs from that of [Bl 1] by a minus sign but agrees with [Stl].

A quick way to check multiplicative formulas in $W(R)$ is to use the ghost map $gh: W(R) \to \prod_{1}^{\infty} R$. It is obtained from the abelian group homomorphism

$$-t \frac{d}{dt}(\log) : (1 + tR[[t]])^{\times} \to (tR[[t]])^{+}$$

$$\alpha(t) \mapsto \frac{-t}{\alpha(t)} \frac{d\alpha}{dt}$$

by identifying the left side with $W(R)$ and the right side with $\prod R$ (via $\Sigma a_n t^n \leftrightarrow (a_1, a_2, \ldots)$). For example, $gh(1-t)^n = (n,n,\ldots)$.

The map gh is a ring homomorphism (for the product structure on ΠR), and is an injection if R has no $\mathbb{Z}$-torsion (q.v. [Bl 1,p.195]). If $\mathbb{Q} \subseteq R$, gh is a ring isomorphism, so that $W(R) \approx \Pi R$.

Now suppose R is one of the following rings: $S^{-1}\mathbb{Z}$ for some set S of primes, $\hat{\mathbb{Z}}_p$, or a (commutative) $\mathbb{Q}$-algebra. Then for $r \in R$ the coefficients of the power series

$$\lambda_t(r) = (1-t)^r = 1-rt + \binom{r}{2} t^2 - \binom{r}{3} t^3 + \dots$$

all belong to R. This is obvious for $\mathbb{Q}$-algebras, and may be proven for $\hat{\mathbb{Z}}_p$ by an easy convergence argument. Here is a proof for $S^{-1}\mathbb{Z}$ that I learned from R. G. Swan: let $r=k/s$, and suppose by induction that $\binom{r}{n} \in R$ for $n < N$. The coefficients of t^N in the equation

$$[(1-t)^r]^s = [\sum_{i=0}^{\infty} \binom{r}{i} (-t)^i]^s = \sum_{i=0}^{\infty} \binom{k}{i} (-t)^i = (1-t)^k$$

are all in R, with the possible exception of $s\binom{r}{N}$. Hence $\binom{r}{N} \in R$ as well.

<u>Proposition 1.2</u> If R is $S^{-1}\mathbb{Z}$, $\hat{\mathbb{Z}}_p$, or a $\mathbb{Q}$-algebra, the map $\lambda_t : r \mapsto (1-t)^r$ defines a ring injection from R to $W(R)$.

Proof. Composition with the ghost map, which is a ring injection, gives the set map $gh(\lambda_t(r)) = (r,r,\ldots)$. Since $gh(\lambda_t)$ is a ring injection, λ_t must also be a ring injection.

Remark 1.3 We can formulate Proposition 1.2 in the language of λ-rings (see [SGA6]). A binomial ring is a commutative ring R with no $\mathbb{Z}$-torsion, such that $\binom{r}{n} \in R$ for every $r \in R$, $n \in \mathbb{N}$. The above discussion shows that $S^{-1}\mathbb{Z}$, $\hat{\mathbb{Z}}_p$, and $\mathbb{Q}$-algebras are binomial rings. A λ-ring is a commutative ring R with a given ring homomorphism $\lambda_t : R \to W(R)$. Proposition 1.2 shows that every binomial ring is a λ-ring in such a way that λ_t is an injection, and in this guise is proven on p.322 of [SGA6].

(1.4) In order to perform brute force computations, it is useful to introduce the following three abelian group endomorphisms of $W(R)$: the homothety $[r] : \alpha(t) \mapsto \alpha(rt)$, the Verschiebung $V_m : \alpha(t) \mapsto \alpha(t^m)$, and the Frobenius transfer

$$F_m : \alpha(t) \mapsto \sum_{\zeta^m=1} \alpha(\zeta t^{1/m}) .$$

F_m is a ring endomorphism, and $F_m(\lambda_t(r)) = \lambda_t(r)$ for the ring map λ_t of Proposition 1.2. Since every Witt vector can be written uniquely as a product $\alpha(t) = \Pi(1-r_m t^m)$, we can think of multiplication by $\alpha(t)$ as the endomorphism $\Sigma V_m[r_m] F_m$. The theoretical foundations of this viewpoint are developed in [C1] and [C2]. This viewpoint is applied to K-theory computations in [Bl 1], [LR], and [St1].

(1.5) The subgroups $I_N = (1+t^N R[[t]])$ of $W(R)$ are actually ideals, as is clear from the formula (1.1). We will call the resulting topology on $W(R)$ the <u>t-adic topology</u>. $W(R)$ is separated and complete in this topology, and the quotient rings $W_N(R) = W(R)/I_{N+1}$ are the rings of <u>truncated Witt vectors.</u> In particular, $W_1(R) \cong R$, so the ring maps λ_t of Proposition 1.2 are split injections.

When $p^m = 0$ in R and N is fixed, $(1-t)^{p^n} \in I_N$ for large n . Thus "p" is nilpotent in each $W_N(R)$, but "$p^n \neq 0$" in $W(R)$. From this, we see that the composite $\hat{\mathbb{Z}}_p \to W(\hat{\mathbb{Z}}_p) \to W(R)$ is a ring injection, which is continuous with respect to the p-adic and t-adic topologies.

§2. $\mathrm{End}_0(R)$

In this section, we recall some facts about the ring $\mathrm{End}_0(R)$. A very readable survey of the relation of $\mathrm{End}_0(R)$ to $W(R)$ may be found in [Gr].

We can define $\mathrm{End}_0(\Lambda)$ for any ring Λ with 1. Let <u>End</u> (Λ) denote the exact category of endomorphisms of finitely generated projective right Λ-modules. This is the category denoted $\underline{\underline{P}}(\Lambda)^{\mathbb{N}}$ on p.5 of [Ba]: objects are pairs (M,f) with $f \in \mathrm{End}(M)$, and morphisms $(M_1,f_1) \to (M_2,f_2)$ are maps $\alpha: M_1 \to M_2$ with $f_2\alpha = \alpha f_1$.

There are two interesting subcategories of $\underline{\underline{\mathrm{End}}}(\Lambda)$. One is the full exact subcategory $\underline{\underline{\mathrm{Nil}}}(\Lambda)$ of nilpotent endomorphisms, and the other is the reflective subcategory of zero endomorphims, which is naturally equivalent to $\underline{\underline{P}}(\Lambda)$, the category of f.g. Λ-projectives. We define $\mathrm{End}_n(\Lambda)$ and $\mathrm{Nil}_n(\Lambda)$ by the splittings

$$K_n \underline{\underline{\mathrm{End}}}(\Lambda) = K_n(\Lambda) \oplus \mathrm{End}_n(\Lambda)$$

$$K_n \underline{\underline{\mathrm{Nil}}}(\Lambda) = K_n(\Lambda) \oplus \mathrm{Nil}_n(\Lambda) .$$

Now suppose that Λ is an R-algebra for some commutative ring R. Then there are exact pairings

$$\otimes : \underline{\underline{\mathrm{End}}}(R) \times \underline{\underline{\mathrm{End}}}(\Lambda) \to \underline{\underline{\mathrm{End}}}(\Lambda)$$

$$\otimes : \underline{\underline{\mathrm{End}}}(R) \times \underline{\underline{\mathrm{Nil}}}(\Lambda) \to \underline{\underline{\mathrm{Nil}}}(\Lambda)$$

$$(M,f) \otimes (N,g) = (M \otimes_R N, f \otimes g) .$$

These induce maps $K_0 \underline{\underline{\mathrm{End}}}(R) \otimes K_* \underline{\underline{\mathrm{End}}}(\Lambda) \to K_* \underline{\underline{\mathrm{End}}}(\Lambda)$, $K_0 \underline{\underline{\mathrm{End}}}(R) \otimes K_* \underline{\underline{\mathrm{Nil}}}(\Lambda) \to K_* \underline{\underline{\mathrm{Nil}}}(\Lambda)$ by the usual generators-and-relations tricks on K_0. It is easy to see that $(0,0)$ and $(R,1) \in K_0(\underline{\underline{\mathrm{End}}}(R)$ act as the zero and identity maps. If we take $R = \Lambda$, we see that K_0 End (R) is a commutative ring with 1. $K_0(R)$ is an ideal, generated by the idempotent (R,0), and the quotient ring is $End_0(R)$. Since (R,0)⊗ reflects End (Λ) into P (Λ), $K_0(R)$ acts as zero on End_* (Λ) and Nil_* (Λ). The following is immediate (and well-known):

Proposition 2.1 If Λ is an R-algebra with 1, End_* (Λ) and Nil_* (Λ) are graded modules over the ring End_0 (R) .

Remark 2.2 Of course, we can use the construction in [Wa, §9] to see that End_* (Λ) and Nil_* (Λ) are graded modules over the graded ring End_* (R) . We will not use this, except to make the following observation: there is another embedding of P(R) into End(R), namely as the subcategory of identity endomorphisms. It is not hard to see that this induces a ring homomorphism $K_*(R) \to End_*(R)$ preserving "one". The resulting $K_*(R)$-module structure on Nil_* (Λ) agrees with the "usual" one, obtained by identifying Nil_* (Λ) with a $K_*(R)$-submodule of K_{*+1} (Λ[y]). We will return to this point in §3.

There is a well-defined map $\chi: End_0(R) \to W(R)$ given by taking characteristic polynomials : $\chi(M,f) = \det(1-tf)$. Note that $\chi(R,0)=1$ and $\chi(R,1)=1-t$. It is easy to see that χ is a ring homomorphism, and that the image of χ is the set of all rational functions in W(R), i.e., quotients of polynomials in 1+tR[t]. The induced t-adic topology on $End_0(R)$ is defined by the ideals $I_N = \{f \in End_0(R) \mid \chi(f) \equiv 1 \pmod{t^N}\}$, and End_0 (R) is separated in this topology. The key fact is:

<u>Theorem 2.3</u> (Almkvist [A]). The map $\chi : End_0(R) \to W(R)$ is a ring <u>injection</u>, and W(R) is the t-adic completion of End_0 (R).

The operations on End_0 (R) inducing the homothety, Verschiebung, and Frobenius of (1.4) are discussed in [Gr]. Grayson also points out in [Gr] that End_0 (R) is a λ-ring via $\lambda_t(M,f) = \Sigma(\Lambda^n M, \Lambda^n f)t^n$, and that χ is a λ-ring homomorphism. The λ-ring structure on W(R) is given in [SGA6, p.319].

<u>Exercise 2.4</u> Show that $\chi(f) = 1 + a_1 t + \ldots + a_m t^m$ for the endomorphism

$$f = \begin{bmatrix} 0 & & & & -a_m \\ 1 & 0 & & & \vdots \\ & 1 & \ddots & & \vdots \\ & & \ddots & 0 & -a_2 \\ & & & 1 & -a_1 \end{bmatrix}$$

of R^m. Then show that if $\nu^N = 0$ then $f \otimes \nu$ represents 0 in Nil_0 (Λ) whenever $\chi(f) \equiv 1$ modulo t^N.

§3. <u>The W(R)-modules $NK_*(\Lambda)$.</u>

We keep the notation that R is a commutative ring with 1 and that Λ is an R-algebra with 1. We take $NK_n(\Lambda)$ to be the kernel of "y=0" : $K_n(\Lambda[y]) \to K_n(\Lambda)$. $NK_n(\Lambda)$ is isomorphic to $Nil_{n-1}(\Lambda)$ via the composite

$$NK_n(\Lambda) \subset K_n(\Lambda[y]) \subset K_n(\Lambda[x,y]/(xy=1)) \xrightarrow{\partial} K_{n-1} \underline{\underline{Nil}}(\Lambda)$$

(this is proven on p.237 of [GQ]). Thus the groups $NK_n(\Lambda)$ are $End_0(R)$ modules. For $n \geq 1$, this is just Proposition 2.1; for $n = 0$ (and $n < 0$) this follows from the functoriality of the module structure and the fact that $NK_0(\Lambda)$ is the "contracted functor" of $NK_1(\Lambda)$, q.v. [Ba, XII §7].

<u>Theorem 3.1</u> (Stienstra [St2]). For every $\gamma \in Nil_*(\Lambda)$ there is an N so that γ is annihilated by the ideal $I_N = \{f | \chi(f) \equiv 1 \bmod t^N\}$ of $End_0(R)$. Consequently, $NK_*(\Lambda)$ is a module over the t-adic completion W(R) of $End_0(R)$.

The second sentence follows from the first by Theorem 2.3. The first sentence may be proven in the spirit of Exercise 2.4, but we will refer the reader to [St2] for a careful proof.

<u>Exercise 3.2</u> Use the sign convention that $[1-\nu y] \in NK_1(\Lambda)$ corresponds to $(N,\nu) - (N,0) \in Nil_0(\Lambda)$ and Exercise 2.4 to show that the W(R)-module structure on $NK_1(\Lambda)$ is completely determined by the formula $\alpha(t) * [1-\nu y] = [\alpha(\nu y).]$

Show that the ring map $K_*(R) \to End_*(R)$ induces the "usual" $K_*(R)$-module structure on $NK_*(\Lambda)$, i.e., that coming from the $K_*(R)$-module structure on $K_*(\Lambda[y])$. Use this to show that for $\gamma \in K_{n-1}(R)$, $(N,\nu) \in$ <u>Nil</u> (Λ), and $\alpha(t) \in W(R)$ we have the formula

$$\alpha(t) * \{\gamma, 1-\nu y\} = \{\gamma, \alpha(\nu y)\} \in NK_n(\Lambda).$$

This formula was first proven by Bloch on p.238 of [Bl 1], and is especially useful in determining the W(R)-module structure on $NK_2(\Lambda)$. (See Example (4.4).)

<u>Corollary 3.3</u> Fix an integer p and a ring Λ with 1.

(a) If Λ is an $S^{-1}\mathbb{Z}$- algebra, $NK_*(\Lambda)$ is an $S^{-1}\mathbb{Z}$-module.

(b) If Λ is a Q-algebra, $NK_*(\Lambda)$ is a center(Λ) -module.

(c) If Λ is a $\hat{\mathbb{Z}}_p$-algebra, $NK_*(\Lambda)$ is a $\hat{\mathbb{Z}}_p$-module.

(d) If $p^m=0$ in Λ, $NK_*(\Lambda)$ is a p-group.

<u>Proof.</u> The first three parts follow from (1.2) and (3.1). In case (d), note by (1.5) and (3.1) that every element of $NK_*(\Lambda)$ is annihilated by some $p^n \in \hat{\mathbb{Z}}_p$, i.e., that $NK_*(\Lambda)$ is a p-group.

<u>Historical Remark (3.4)</u> The observation that NK_0, NK_1 are p-groups for $\mathbb{Z}/p^m\mathbb{Z}$-algebras is due to Chase, and may be found

on p.646 of [Ba]. Chase asked in [Ge, Problem 18] if the same were true for all NK_*. The affirmative answer of (3.3) is implicit in [Bl 1] (as well as [Bl 2], [vdK, p.310], and [St1]). For $\mathbb{Q}$-algebras, it was remarked on pp.13,51 of [Ba2] that NK_0, NK_1 are divisible groups. For $S^{-1}\mathbb{Z}$-algebras, Swan proved in [Sw] that NU, NPic are $S^{-1}\mathbb{Z}$-modules, and observed that the same was true for NK_0, NK_1 . The $End_0(R)$ approach is due to Stienstra, mentioned on p.68 of [St1], and will appear in [St2].

<u>Corollary 3.5</u> If I is an ideal in Λ, the relative groups $NK_*(\Lambda,I)$ are $W(R)$-modules, and there is an exact sequence of $W(R)$-modules

$$(*)\quad NK_{*+1}(\Lambda) \to NK_{*+1}(\Lambda/I) \to NK_*(\Lambda,I) \to NK_*(\Lambda) \to NK_*(\Lambda/I).$$

In particular, if Λ is an R-algebra for R one of $S^{-1}\mathbb{Z}$, $\hat{\mathbb{Z}}_p$, $\mathbb{Q}$-algebra, then $NK_*(\Lambda,I)$ is an R-module. If $1 \in \Lambda$ and $p^m = 0$, then $NK_*(\Lambda,I)$ is a p-group.

<u>Proof</u>. The commutative diagram

$$\begin{array}{ccc} \underline{\underline{Nil}}(\Lambda) \times \underline{\underline{End}}(R) & \longrightarrow & \underline{\underline{Nil}}(\Lambda) \\ \downarrow & & \downarrow \\ \underline{\underline{Nil}}(\Lambda/I) \times \underline{\underline{End}}(R) & \longrightarrow & \underline{\underline{Nil}}(\Lambda/I) \end{array}$$

is a special case of diagram (5.5) of [We3], with $A = C = \underline{\underline{Nil}}(\Lambda)$, $B = \underline{\underline{End}}(R)$, etc. The discussion following (5.5) - especially the

the penultimate paragraph of Section 5 - applies here to prove that (after discarding the summand $K_*(R)$ of $K_*\underline{\text{End}}(R)$) the sequence (*) is an exact sequence of $\text{End}_0(R)$-modules. By Stienstra's theorem (3.1), (*) is actually a sequence of $W(R)$-modules. The rest of the Corollary follows as in (3.3) above.

Corollary 3.6 (Murthy-Pedrini [MP]) Let Λ be an algebra over a field $k \supseteq \mathbb{Q}$. Then $NK_n(\Lambda)$ is either zero or a torsionfree divisible group of rank at least $[k:\mathbb{Q}]$. In particular, if k is uncountable, then $NK_n(\Lambda)$ is either zero or of uncountable rank. The same is true of $NK_n(\Lambda,I)$.

Proof Each group $NK_n(\Lambda)$, $NK_n(\Lambda,I)$ is a k-vector space. We remark that the rank of these groups is **always** infinite - see Proposition 4.1 below.

Proposition 3.7 Let $\Lambda = \Lambda_0 \oplus \Lambda_1 \oplus \ldots$ be a graded ring, and write $K_*(\Lambda) = K_*(\Lambda_0) \oplus \tilde{K}_*(\Lambda)$. Then:

(a) if $\mathbb{Q} \subset \Lambda$ (resp. Λ is an $S^{-1}\mathbb{Z}$-module), $\tilde{K}_*(\Lambda)$ is a $\mathbb{Q}$-vector space (resp. an $S^{-1}\mathbb{Z}$-module)

(b) if Λ is a $\mathbb{Z}/p^m\,\mathbb{Z}$-algebra, $\tilde{K}_*(\Lambda)$ is a p-group.

Proof There is a ring homomorphism $\varphi:\Lambda \to \Lambda[y]$ defined on Λ_n by $\varphi(\lambda) = \lambda y^n$. Apply K_* to the following commutative diagram:

$$\begin{array}{ccc} & & \Lambda \\ & \diagup\!\!\diagup & \uparrow (y=1) \\ \Lambda & \xrightarrow{\varphi} & \Lambda[y] \\ \downarrow & & \downarrow (y=0) \\ \Lambda_0 & \longrightarrow & \Lambda \ . \end{array}$$

We see that $K_*(\varphi)$ maps $\tilde{K}_*(\Lambda)$ into $NK_*(\Lambda)$ as a summand. Now apply (3.3).

§4. Other methods

There are two other ways to obtain a $W(R)$-module structure on $NK_*(\Lambda)$. For completeness, we mention them here. Proof that they agree with the $End_0(R)$ pairing will appear in [St2].

The first is to use the Dieudonné ring $D = End(W)$. Cartier proved in [Cl] that every endomorphism of the functor W :(R-algebras)→ (abelian groups) can be written uniquely as $\Sigma\, V_m\, [r_{mn}]\, F_n$, where the operations $[r]$, V_m, F_n are those of (1.5). There are three pertinent types of endomorphism of $NK_*(\Lambda)$: The homothety $[r] = \rho_r^*$, the Verschiebung $V_m = (\iota_m)^*$, and the Frobenius transfer $(\iota_m)_*$. These are induced by the endomorphisms $\rho_r(y) = ry$ and $\iota_m(y) = y^m$ of $\Lambda[y]$. It is proven on p.317 of [Bl 2] that F_m corresponds to the endofunctor

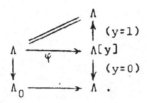

$\theta(N,\nu) = (N,\nu^m)$ of $\underline{\underline{\text{Nil}}}(\Lambda)$, so for every γ in $NK_*(\Lambda)$ we have $F_m(\gamma) = 0$ for large m. Thus the expression $\Sigma V_m[r_{mn}]F_n(\gamma)$ makes sense. Cartier has given in [C2] the necessary relations for this to define a D-module structure on NK_*, hence a W(R)-module structure using the map $W(R) \to D$ of (1.5). This brute force approach is the one taken in [Bl 1], [LR], and [Stl]. A nice application of this method is the following, which van der Kallen attributes to Farrell:

<u>Proposition 4.1</u>([vdK],p.310) For every i the group $NK_i(\Lambda)$ is either zero or is not finitely generated.

<u>Proof</u> If not, then there is an M such that $F_m \equiv 0$ on $NK_i(\Lambda)$ for all $m > M$. Pick $\alpha \neq 0$ in $NK_i(\Lambda)$ and choose an integer $m > M$ with $m\alpha \neq 0$, and note that $F_m(V_m\alpha) = m\alpha$, a contradiction.

The second method is given on p.315 of [Bl 2]. Bloch considers the biexact functor

$$\underline{\underline{P}}(R[[t]]) \times \underline{\underline{\text{Nil}}}(\Lambda) \to \underline{\underline{P}}(\Lambda),$$
$$M \otimes (N,\nu) = M \underset{R[[t]]}{\otimes} N$$

where N is considered to be an R-module with t acting via ν. Waldhausen's machinery (in §9.2 of [Wa]) produces a map

$$W(R) \otimes NK_*(\Lambda) \subseteq K_1(R[[t]]) \otimes K_{*-1}\ \underline{\underline{\mathrm{Nil}}}\ (\Lambda) \to K_*(\Lambda) .$$

Bloch then injects $NK_*(\Lambda)$ as a summand in $NK_*(\Lambda[x])$ via $y \mapsto xy$ and obtains a pairing $W(R) \otimes NK_*(\Lambda) \to NK_*(\Lambda)$. Work is then needed to show that this defines a module structure. Note that the original map $W(R) \otimes NK_*(\Lambda) \to K_*(\Lambda)$ is the composite of the module map and the projection "y=1" : $NK_*(\Lambda) \to K_*(\Lambda)$. Bloch also shows in [Bl 1,p.224] that

(4.2) The relative groups $K_*(R[\varepsilon]/(\varepsilon^{m+1}), \varepsilon) = C_m K_*(R)$ have the structure of W(R)-modules.

We warn the reader that there are two <u>different</u> W(R)-module structures on $NK_*(R[\varepsilon]/(\varepsilon^{m+1}),\varepsilon)$: one from the $End_0(R)$ pairing, and one induced from the pairing (4.2) on $K_*(R[\varepsilon,y],\varepsilon)$. We will not consider Bloch's pairing (4.2), except to give the following two examples of the difference in module structure:

<u>Example 4.3</u> Let k be a Q-algebra, and set $R = k[\varepsilon]/(\varepsilon^{n+1}=0)$. If we give $K_1(R,\varepsilon)$ the R-module structure induced from the W(R)-module structure of (4.2), then there is an R-module isomorphism $\varepsilon R \cong K_1(R,\varepsilon)$, $f \mapsto \exp(f)$. As a sample of the W(R)-module structure, we note that $(1-rt) * \exp(f(\varepsilon)) = \exp(f(r\varepsilon))$. The R-module structure on $NK_1(R,\varepsilon)$ induced from the $End_0(R)$ pairing agrees with this : $\lambda_t(r) * \exp(f) = \exp(rf)$ for $f = f(\varepsilon,y)$ in $\varepsilon yR[y]$. The W(R)-structure on $NK_1(R,\varepsilon)$ gives $(1-rt) * \exp(f(\varepsilon,y)) = \exp(f(\varepsilon,ry))$ however, which is different. The map

"y = 1": $NK_1(R,\varepsilon) \to K_1(R,\varepsilon)$ is an R-module map but not a W(R)-module map.

Example (4.4) We describe the R-module structure on $NK_2(R,\varepsilon)$ for $R = k[\varepsilon]$, $\varepsilon^2 = 0$, k a Q-algebra. It was a surprise to me that we cannot replace "Q-algebra" by "field of characteristic $\neq$ 2,3", since the formulas require Ω_k to be a Q-module. Recall that there is a well-known group isomorphism $\Omega_k \cong K_2(R,\varepsilon)$ given by $a\,db \mapsto \langle a\varepsilon, b\rangle$. (We are using Stienstra's <,> notation in order to make use of the computations in [St1].) Using Bloch's pairing in (4.2), $K_2(R,\varepsilon)$ is a $k = W_1(k)$-module and $\Omega_k \cong K_2(R,\varepsilon)$ is a k-module isomorphism: this observation is stated on p.62 of [St1].

If we use the k[y]-module structure on $NK_2(R,\varepsilon)$ coming from the (4.2) pairing on $K_2(R[y],\varepsilon)$, we obtain the k[y]-module isomorphism $(y\Omega_k[y]) \otimes k[y] \cong NK_2(R,\varepsilon)$ which (for b in k and f_i in k[y]) associates $yf_1 db \otimes f_2$ and $\langle \varepsilon y f_1, b\rangle\langle \varepsilon f_2, y\rangle$. On the other hand, we will show that under the k-module structure induced from the $End_0(R)$-pairing we have the formulas:

$$(4.4.1) \qquad \lambda_t(r) * \langle \varepsilon y f_1, b\rangle = \langle r\varepsilon y f_1, b\rangle \qquad (b \text{ in } k,$$

$$(4.4.2) \qquad \lambda_t(r) * \langle \varepsilon, y f_2\rangle = \langle \varepsilon, r y f_2\rangle. \qquad f_i(y) \text{ in } k[y])$$

Thus (for example) the two actions of r on $\langle \varepsilon, y^i\rangle$ are different if

$dr \neq 0$, so that "$y=1$" : $NK_2(R,\varepsilon) \to K_2(R,\varepsilon)$ is not a k-module homomorphism. Using these formulas, it is easy to see that there is a k-module isomorphism

$$(y\Omega_k[y]) \oplus k[y] \simeq NK_2(R,\varepsilon)$$

$$yf_1 db \oplus f_2 \mapsto \langle \varepsilon y f_1, b\rangle \langle \varepsilon, y f_2\rangle .$$

We now derive the formulas (4.4.1)and (4.4.2). By functoriality of the $W(R)$-module structure, we can assume that b is a unit of k. Formula (4.4.1) then follows from Exercise 3.2 and Example 4.3, given the identification $\langle f\varepsilon, b\rangle = \{1-bf\varepsilon, b\}$. To establish (4.4.2), we first consider the case $f_2(y) = b$. Using the formulas on p.62 of [St1], which are also valid for the module structure on $NK_2(R)$, we find that

$$(1-rt^m) * \langle \varepsilon, by\rangle = \begin{cases} \langle \varepsilon, rby\rangle, & m = 1 \\ 0, & m \neq 1 . \end{cases}$$

The formula (4.4.2) for $\langle \varepsilon, by\rangle$ is immediate. For $\langle \varepsilon, by^i\rangle$ we use the following formula of Bloch, found on p.316 of [Bl2] (note the missing V_m in [Bl2]):

(4.4.3) $\quad V_m[F_m(\omega) * \gamma] = \omega * V_m(\gamma) \quad (\omega \text{ in } W(R), \gamma \text{ in } NK_*(\Lambda)).$

We can now compute:

$$\lambda_t(r)*\langle\varepsilon,by^m\rangle = \lambda_t(r)*V_m\langle\varepsilon,by\rangle = V_m[F_m\lambda_t(r)*\langle\varepsilon,by\rangle]$$
$$= V_m[\lambda_t(r)*\langle\varepsilon,by\rangle] = V_m\langle\varepsilon,rby\rangle = \langle\varepsilon,rby^m\rangle .$$

The formula (4.4.2) now follows from the fact that $\langle\varepsilon,yf_2\rangle$ is additive in f_2 .

Exercise 4.5 Show that for $R = k[\varepsilon]/(\varepsilon^{n+1})$, k a Q-algebra, there is an isomorphism of R-modules

$$(\Omega_k \otimes \varepsilon yR[y]) \oplus \varepsilon yR[y] \cong NK_2(R,\varepsilon)$$

$$\varepsilon yf_1 db \oplus \varepsilon yf_2 \mapsto \langle\varepsilon yf_1,b\rangle \langle\varepsilon, \frac{\exp(\varepsilon yf_2)-1}{\varepsilon}\rangle .$$

§5. Main Results

In order to pass from the NK_*-groups to K_*-groups, it is necessary to consider the Karoubi-Villamayor groups, whose key property is that $KV_*(\Lambda) = KV_*(\Lambda[y])$. Recall that the groups $N^pF(\Lambda)$ are defined by iteration from $N^0F = F$ using the formula $NF(\Lambda)$ = kernel of $F(\Lambda[y]) \to F(\Lambda)$. If $F = K_*$ and Λ is an R-algebra, we see that the $N^pK_*(\Lambda)$ are W(R)-modules for $p \neq 0$ by functoriality of the W(R)-module structure on NK_* . The result we need is this:

<u>Theorem 5.1</u> If Λ is a ring with unit, there is a first quadrant spectral sequence (defined for $p \geq 0$, $q \geq 1$)

$$E^1_{pq} = N^p K_q(\Lambda) \Rightarrow KV_{p+q}(\Lambda) .$$

If I is an ideal in Λ, there is a spectral sequence

$$E^1_{pq} = N^p K_q(\Lambda, I) \Rightarrow KV_{p+q}(I) .$$

Moreover, there is a long exact sequence

$$\ldots \to KV_{i+1}(\Lambda/I) \to H_i(C_*) \oplus KV_i(I) \to KV_i(\Lambda) \to \ldots$$

$$\ldots \to KV_1(\Lambda/I) \to K_0(I)/\mathrm{im} NK_1(\Lambda/I) \to K_0(\Lambda) \to K_0(\Lambda/I) .$$

Here C_* is a chain complex with C_p the cokernel of $N^p K_1(\Lambda) \to N^p K_1(\Lambda/I)$.

<u>Proof</u> See [We2], theorems 2.5, 2.6, and 3.2. The spectral sequence was originally discovered by Gersten and Anderson.

Since p-groups form a Serre subcategory of all abelian groups (and since $K_0(\Lambda) = KV_0(\Lambda)$ by definition), we immediately deduce the following result:

<u>Theorem 5.2</u> Let Λ be a $\mathbb{Z}/p^m\mathbb{Z}$-algebra with unit, and let I be an ideal in Λ. Then $K_*(\Lambda) \otimes \mathbb{Z}[\frac{1}{p}] \xrightarrow{\cong} KV_*(\Lambda) \otimes \mathbb{Z}[\frac{1}{p}]$ and $K_*(\Lambda, I) \otimes \mathbb{Z}[\frac{1}{p}] \xrightarrow{\cong} KV_*(I) \otimes \mathbb{Z}[\frac{1}{p}]$.

<u>Corollary 5.3</u> ("Excision") If $f = \Lambda_1 \to \Lambda_2$ is a map of $\mathbb{Z}/p^m\mathbb{Z}$-algebras with unit, and I is an ideal of Λ_1 with $I \cong f(I)$, then

$$K_*(\Lambda_1, I) \otimes Z[\tfrac{1}{p}] \xrightarrow{\cong} K_*(\Lambda_2, I) \otimes Z[\tfrac{1}{p}] .$$

<u>Corollary 5.4</u> If I is a nilpotent ideal in a $\mathbb{Z}/p^m\mathbb{Z}$-algebra Λ with unit, then $K_*(\Lambda, I)$ is a p-group.

<u>Proof:</u> We have $KV_*(I) = 0$ by [We 1, Theorem 2.3].

<u>Theorem 5.5</u> ("Mayer-Vietoris"). Let $\begin{array}{ccc} \Lambda_1 & \to & \Lambda_2 \\ \downarrow & & \downarrow \\ \Lambda_3 & \to & \Lambda_4 \end{array}$ be a pullback square of $\mathbb{Z}/p^m\mathbb{Z}$-algebras with $\Lambda_2 \to \Lambda_4$ onto. Then there is a long exact sequence

$$\cdots K_{*+1}(\Lambda_4)\otimes Z[\tfrac{1}{p}] \to K_*(\Lambda_1)\otimes Z[\tfrac{1}{p}] \to [K_*(\Lambda_2) \oplus K_*(\Lambda_3)]\otimes Z[\tfrac{1}{p}] \to K_*(\Lambda_4)\otimes Z[\tfrac{1}{p}] \cdots$$

valid for all integers * . The same is true if we replace K_* by KV_* .

<u>Proof:</u> We splice together the long exact ideal sequences for $\Lambda_1 \to \Lambda_3$ and $\Lambda_2 \to \Lambda_4$ in the familar way.

Example 5.6 Let k be a finite field of characteristic p. Then $K_*(k[x,y]/(x^3=y^2)) = K_*(k) \oplus (p\text{-group})$ follows either from the fact that the ring is graded and (3.7), or from the conductor pullback square. For the node $\Lambda_1 = k[x,y]/(y^2=x^2+x^3)$ it is well-known that $KV_*(\Lambda_1) = K_*(k) \oplus K_{*+1}(k)$ from the conductor pullback square. Since $K_*(k)$ is a group of order prime to p (for $* \neq 0$), we must have $K_*(\Lambda_1) = K_*(k) \oplus K_{*+1}(k) \oplus (p\text{-group})$.

If we try to use the same technique for our other λ-rings, the best we can do is this:

Theorem 5.7 Let Λ be an R-algebra with unit, and let I be an ideal in Λ, where $R = S^{-1}\mathbb{Z}$ or $\mathbb{Q}$. Let η be either $K_*(\Lambda) \to KV_*(\Lambda)$ or $K_*(\Lambda,I) \to KV_*(I)$. Then:

(a) If $R = S^{-1}\mathbb{Z}$, ker (η) is divisible by the primes in S and coker (η) has no S-torsion.

(b) If $R = \mathbb{Q}$, ker (η) is divisible and coker (η) is torsionfree.

Remark From Example (4.4) we know that $d_1 : E^1_{12} \to E^1_{02}$ is not a k-module homomorphism for a $\mathbb{Q}$-algebra k, so it is not reasonable to expect a k-module version. In [We4], I will give an example to show that ker (η) and coker (η) need not be $S^{-1}\mathbb{Z}$-modules.

Proof We prove (a); (b) is a special case. In the spectral sequence of (5.1), the E^1_{pq} are R-modules for $p \geq 1$. By induction, we observe that E^n_{pq} is an R-module for $p \geq n$, S-torsionfree for $0 < p < n$, and $K_q(\Lambda)$/(S-divisible group) for $p = 0$. The key observation here is that, if h is any map from an $S^{-1}\mathbb{Z}$-module to an abelian group, im (h) is S-divisible and ker (h) is S-torsionfree. For $n = \infty$ we see that $K_q(\Lambda)$/(S-divisible group) is a subgroup of $KV_q(\Lambda)$, and that the quotient is filtered by S-torsionfree groups, whence the result.

Corollary 5.8 Let I be a nilpotent ideal in a ring Λ. Then:

(a) If Λ is an $S^{-1}\mathbb{Z}$-algebra, $K_*(\Lambda,I)$ is S-divisible.

(b) If $\mathbb{Q} \subset \Lambda$, $K_*(\Lambda,I)$ is divisible.

Proof Again, $KV_*(I) = 0$ by [We 1, Theorem 2.3].

In a subsequent paper [We4], I expect to show that $K_*(\ ,I)$ is an R-module in Corollary 5.8 by the method of K-theory with mod p coefficients.

§6. Localization of R

In this section, we consider the effect of localizing Λ on the resulting map $NK_*(\Lambda) \to NK_*(S^{-1}\Lambda)$.

<u>Lemma 6.1</u> Let R be one of $S^{-1}\mathbb{Z}$, $\hat{\mathbb{Z}}_p$, or a $\mathbb{Q}$-algebra. Then (for each nonnilpotent s in R and integer i) $\binom{1/s}{i}$ is in $R[1/s]$ and $\binom{s^n}{i}$ is in sR for large n.

<u>Proof</u> The first part follows from (1.1). For the second part, choose n large enough so that g.c.d. $(i!,s^{n-1})$ = g.c.d.$(i!s^n)$.

<u>Proposition 6.2</u> Let R be one of $S^{-1}\mathbb{Z}$, $\hat{\mathbb{Z}}_p$, or a $\mathbb{Q}$-algebra. Then for each nonnilpotent s in R and integer N there is an isomorphism

$$W_N(R)\left[\frac{1}{\lambda_t(s)}\right] \cong W_N(R)\left[\frac{1}{1-st}\right] \cong W_N(R\left[\frac{1}{s}\right]) .$$

<u>Proof</u> For a,b in any commutative ring A we have $A[\frac{1}{a}] = A[\frac{1}{b}]$ just in case there are α,β in A and integers m,n so that $a\alpha = b^m$, $b\beta = a^n$. We take $a = \lambda_t(s) = (1-t)^s$, $b = (1-st)$ and choose m,n so that

$$\alpha(t) = (1-s^m t)^{1/s} \quad , \quad \beta(t) = (1-t/s)^{s^n}$$

belong to $W_N(R)$; this is possible by lemma 6.1. We then have $a*\alpha = 1-s^m t = b^{*m}$ and $b*\beta = (1-t)^{s^n} = a^{*n}$, proving the first isomorphism. The second isomorphism is easy to establish from the equation $(1-rs^{-i}t^m)*(1-st) = (1-rs^{m-i}t^m)$.

Remark For the (big) Witt vectors, $W(R)[\lambda_t(s)^{-1}]$ and $W(R)[(1-st)^{-1}]$ are distinct subrings of $W(R[s^{-1}])$, at least for $R = S^{-1}\mathbb{Z}$ or $\hat{\mathbb{Z}}_p$. This is because $\alpha(t), \beta(t) \notin W(R)$ when s is a prime of R.

Theorem 6.3 (Vorst) Let R be a commutative ring with unit, and let Λ be an R-algebra with unit. Then for every multiplicative set $S \subset R$ of nonzerodivisors on Λ there is an isomorphism

$$\{(1-st)\,|\,s\in S\}^{-1}W(R) \underset{W(R)}{\otimes} NK_*(\Lambda) \to NK_*(S^{-1}\Lambda) .$$

Proof This is (1.4) of [V], as interpreted in Remark (1.8) of [V].

Corollary 6.4 Let R be one of $T^{-1}\mathbb{Z}, \hat{\mathbb{Z}}_p$, or a $\mathbb{Q}$-algebra, and let Λ be an R-algebra. Then for every multiplicative set $S \subset R$ of nonzerodivisors on Λ there is an isomorphism of $S^{-1}R$-modules:

$$S^{-1}R \underset{R}{\otimes} NK_*(\Lambda) \to NK_*(S^{-1}\Lambda) .$$

Proof $NK_*(\Lambda)$ is the direct colimit over the family of finitely generated W(R)-submodules M. By Stienstra's Theorem (3.1) we have

$$S^{-1}R\underset{R}{\otimes}M = \{\lambda(s)\}^{-1}W_N(R)\underset{W(R)}{\otimes} M = \{(1-st)\}^{-1}W(R)\underset{W(R)}{\otimes} M .$$

The corollary now follows from (6.3) by taking colimits of both sides.

Consequence (6.5) Let π be a finite group of order n, and let p be a prime dividing n. Then $NK_*(\Lambda)$ is a p-group for $\Lambda = \mathbb{Z}_{(p)}[\pi]$ and $\hat{\mathbb{Z}}_p[\pi]$. $NK_*(\mathbb{Z}[\pi])$ is an n-torsion group, and $\mathbb{Z}_{(p)} \otimes NK_*(\mathbb{Z}[\pi]) = NK_*(\mathbb{Z}_{(p)}\pi)$. Similar statements hold for $NK_*(R[\pi])$, where R is a finite extension of $S^{-1}\mathbb{Z}$ or $\hat{\mathbb{Z}}_p$.

Proof $\mathbb{Z}[\frac{1}{n}] \otimes NK_*(\Lambda) = NK_*(\Lambda[\frac{1}{n}]) = 0$, since $\Lambda[\frac{1}{n}]$ is a regular ring. (See [Ba], pp.648, 695 for previous results.)

Consequence (6.6) Let Λ be a ring with no $\mathbb{Z}$-torsion. Then the rank of the abelian group $NK_*(\Lambda)$ is the dimension of the $\mathbb{Q}$-vector space $NK_*(\mathbb{Q}\Lambda)$, and the p-torsion subgroup of $NK_*(\Lambda)$ is the torsion subgroup of $NK_*(\mathbb{Z}_{(p)}\Lambda)$.

In order to apply these results when S contains zerodivisors, we can use the following result of Vorst:

Theorem 6.7 (Vorst [V],(1.7)) Let A be a reduced commutative ring. Then for every multiplicative set $S \subset A$ there is an isomorphism

$$\{(1-st) | s \in S\}^{-1} W(A) \otimes NK_*(A) \to NK_*(S^{-1}A) .$$

<u>Corollary 6.8</u> Let R be one of $T^{-1}\mathbb{Z}$, $\hat{\mathbb{Z}}_p$, or a $\mathbb{Q}$-algebra, and let A be a reduced commutative R-algebra. Then for every multiplicative set $S \subset R$ there is an $S^{-1}R$-module isomorphism $S^{-1}R \otimes NK_*(A) \to NK_*(S^{-1}A)$.

<u>Consequence (6.9)</u> Let R be the coordinate ring of an affine variety over a field of characteristic zero. Then $NK_*(R)$ is a torsion R-module supported only at the singular locus of the variety.

<u>Consequence (6.10)</u> ([Ba],p.648) Let R be a reduced commutative ring, and n an integer with $R[\frac{1}{n}]$ regular. Then $NK_*(R)$ is an n-torsion abelian group.

<u>Consequence (6.11)</u> Vorst's Theorem 1.9 of [V] in characteristic zero merely states that the R-module $NK_i(R)$ is zero iff it is locally zero.

<u>Remark</u> In [Sw], Swan proved that for every commutative ring A and for every multiplicative set $S \subset \mathbb{Z}$ there are isomorphisms $S^{-1}NU(A) \to NU(S^{-1}A)$, $S^{-1}NPic(A) \to NPic(S^{-1}A)$. The possibility of generalizing this result to the higher nilgroups was the original motivation for this paper.

References

[A] G. Almkvist, The Grothendieck group of the category of endomorphisms, J. Alg. 28 (1974), 375-388.

[Ba] H. Bass, Algebraic K-theory, Benjamin, New York, 1968.

[Ba2] H. Bass, Introduction to some methods of algebraic K-theory, CBMS Regional Conf. series, No. 20, AMS, Providence, 1974.

[Bl1] S. Bloch, Algebraic K-theory and Crystalline Cohomology, Publ. Math. I.H.E.S. 47 (1978), 188-268.

[Bl2] S. Bloch, Some formulas pertaining to the K-theory of commutative groupschemes, J. Alg. 53 (1978), 304-326.

[C1] P. Cartier, Groupes formels associés aux anneaux de Witt généralisés, C.R. Acad. Sci. Paris 265 (1967), 49-52.

[C2] P. Cartier, Modules associés à un groupe formel commutatif, courbes typiques, C.R. Acad. Sci. Paris 265 (1967), 129-132.

[Ge] S. Gersten, Problems about higher K-functors, Lecture Notes in Math. 341, Springer-Verlag, New York, 1973.

[GQ] D. Grayson, Higher algebraic K-theory: II (after Daniel Quillen), Lecture Notes in Math. 551, Springer-Verlag, New York, 1976.

[Gr] D. Grayson, Grothendieck rings and Witt vectors, Comm. in Alg. 6 (1978), 249-255.

[LR] J. Labute and P. Russell, On K_2 of truncated polynomial rings, J.P.A.A. 6 (1975), 239-251.

[MP] M. P. Murthy and C. Pedrini, K_0 and K_1 of polynomial rings, Lecture Notes in Math. 342, Springer-Verlag, New York, 1973.

[SGA6] P. Berthelot, S.G.A. VI, Exposé V, Lecture Notes in Math. 225, Springer-Verlag, New York, 1971.

[St1] J. Stienstra, Deformations of the second Chow group, a K-theoretic approach, thesis, Utrecht, 1978.

[St2] J. Stienstra, to appear.

[Sw] R. Swan, On seminormality, preprint (1979).

[V] T. Vorst, Localization of the K-theory of polynomial equations, Math. Ann. 244 (1979), 33-53.

[vdK] W. van der Kallen, Generators and relations in algebraic K-theory, Proc. I.C.M. (Helsinki,1978), 305-310.

[Wa] F. Waldhausen, Algebraic K-theory of generalized free products, Ann. Math. 108 (1978), 135-256.

[We1] C. Weibel, Nilpotence in K-theory, J. Alg. 61 (1979), 298-307.

[We2] C. Weibel, KV-theory of categories, Trans. Amer. Math. Soc., to appear.

[We3] C. Weibel, A survey of products in algebraic K-theory, these Proceedings.

[We4] C. Weibel, Meyer-Vietoris Sequences and mod p K-theory, Oberwolfach (1980), Lect. Notes in Math., to appear.

A SURVEY OF PRODUCTS IN ALGEBRAIC K-THEORY

C. A. Weibel

In the last ten years, several authors have defined various kinds of product structures in algebraic K-theory, the idea being that $K_*(R)$ should be a graded commutative ring whenever R is a commutative ring. Such a structure should coincide with the products given by Milnor [Mi] for $* \leq 1$. The purpose of this paper is to survey the two main constructions: Waldhausen's product via the "double Q" construction [Wa] and the Loday product via the $BG\ell^+$ construction [L], as generalized by May in [May 1,2,3]. We also include a discussion of the product in KV-theory, since it is more tractable and so appears useful in understanding products for K-theory (see [S] for example).

It should be pointed out that the material in this survey is well-known to "the experts." It is my hope that condensing the technical matter into the Bibliography will result in a readable introduction for "non-experts." In particular, I have tried to show just how various results are implicit in Waldhausen's fundamental paper [Wa].

It goes without saying that I am enormously grateful to "the experts" for many useful discussions. I would like to thank Fiedorowicz, Loday and May in particular. I would also like to thank Queen's University for their hospitality during the writing stage.

1. Bicategories.

The key idea in Waldhausen's approach to multiplicative structures is that one should consider bicategories (or double categories) as well as categories in K-theory. The best introduction to bicategories is [K-S]; [E] and [Mac, p. 44] also give formal definitions. We will recount and use the viewpoint of §5 of [Wa] in this section.

Associated to every small category S is a simplicial set $S_{\bullet}$, called its nerve; S_0 is the set of objects of S, S_1 is the set of morphisms, S_2 is the set of pairs of composable morphisms (i.e., $S_2 = S_1 \times_{S_0} S_1$), etc. The category S can be completely recovered from its nerve, and it is possible to write down axioms that describe which simplicial sets are nerves of categories. Identifying small categories and their nerves, we will think of a small category as a special kind of simplicial set.

Thinking of a simplicial set as a functor $S_{\bullet}: \Delta^{op} \to (\text{Sets})$, where Δ is the category of finite ordinal numbers ([Mac, p. 171]), a bisimplicial set is a functor $C_{\bullet\bullet}: \Delta^{op} \times \Delta^{op} \to (\text{Sets})$. We may visualize $S_{\bullet\bullet}$ as a lattice of sets:

$$\begin{array}{ccccc}
\vdots & & \vdots & & \\
S_{01} & \overset{\longleftarrow}{\underset{\longrightarrow}{\longleftarrow}} & S_{11} & \cdots & \\
\uparrow\downarrow\downarrow & & \uparrow\downarrow\downarrow & & \\
S_{00} & \overset{\longleftarrow}{\underset{\longrightarrow}{\longleftarrow}} & S_{10} & \cdots &
\end{array}$$

A (small) bicategory $S_{\bullet\bullet}$ is then a special kind of bisimplicial set, namely one for which each of the simplicial sets $S_{m\bullet}$ and

$S_{\cdot n}$ are categories. The 'interchange law' is automatic, as it merely states that $d_1^h d_1^v = d_1^v d_1^h : S_{22} \to S_{11}$. We call the sets S_{11}, S_{01} and S_{10}, S_{00} the sets of <u>bimorphisms</u>, <u>horizontal</u> and <u>vertical</u> <u>morphisms</u>, and <u>objects</u>, respectively.

In the notation of [K-S], bimorphisms are called <u>squares</u>. The horizontal source and target maps $d_0^h, d_1^h : S_{11} \to S_{01}$ strip off the left and right edges of the squares, while the vertical source and target maps $d_0^v, d_1^v : S_{11} \to S_{10}$ strip off the top and bottom edges.

Two pertinent constructions of bicategories from categories A, B, C are bi(C), whose bimorphisms are commutative squares in C

$$\begin{array}{ccc} \bullet & \longrightarrow & \bullet \\ \downarrow & & \downarrow \\ \bullet & \longrightarrow & \bullet \end{array}$$

and $A \otimes B$, whose objects are pairs (A,B) in $\mathrm{Obj}(A) \times \mathrm{Obj}(B)$ and bimorphisms are pairs $(a:A_0 \to A_1, b:B_0 \to B_1)$ in $\mathrm{Mor}(A) \times \mathrm{Mor}(B)$. We can represent (a,b) as the square

$$\begin{array}{ccc} (A_0,B_0) & \xrightarrow{(a,1)} & (A_1,B_0) \\ {\scriptstyle (1,b)}\downarrow & & \downarrow{\scriptstyle (1,b)} \\ (A_0,B_1) & \xrightarrow{(a,1)} & (A_1,B_1). \end{array}$$

The content of Lemma 3 on p. 170 of [Wa] is: if we think in terms of bisimplicial sets, then $A \otimes B$ is the 'product' of A and B in the sense that $(A \otimes B)_{mn} = A_m \times B_n$.

There is a geometric realization functor B from bicategories (through bisimplicial sets) into topological spaces, described on [Wa, p. 164]. There is also a diagonalization functor 'diag' from bisimplicial sets into simplicial sets, and it is well known that $B \circ \mathrm{diag} \simeq B$. Since $\mathrm{diag}(A \otimes B)$ is the usual product of categories $A \times B$, we have $(BA) \times (BB) \simeq B(A \otimes B)$. Since there is a map $\mathrm{diag} \circ \mathrm{bi}(C) \to C$, BC is a retract of $B(\mathrm{bi}(C))$. From these remarks it is clear that for every functor $A \times B \to C$ there is a commutative diagram

(1.1)
$$\begin{array}{ccc} BA \times BB & \longrightarrow & BC \\ \downarrow \simeq & & \uparrow \\ B(A \otimes B) & \longrightarrow & B(\mathrm{bi}C). \end{array}$$

2. Waldhausen's Product.

When A is a small exact category, Waldhausen defines (on p. 194 of [Wa]) a bicategory QQA as follows. The bimorphisms are equivalence classes of commutative diagrams

(2.1)
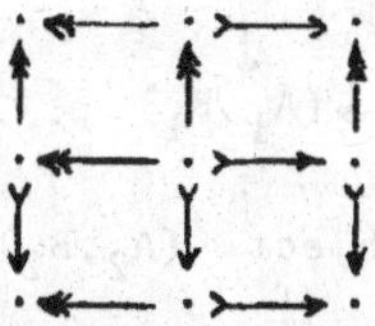

in which the four little squares can be embedded in a 3×3 diagram with short exact rows and columns. Two diagrams are equivalent if they are isomorphic by an isomorphism which restricts to the identity on each corner object. Waldhausen proves (on p. 196) that the loop space $\Omega BQQA$ is homotopy equivalent to BQA (the category QA is defined on [Q, p.100]). Thus by definition ([Q, p.103]) we have $K_pA = \pi_{p+1}BQA = \pi_{p+2}BQQA$.

If A, B, C are small exact categories, a functor $\boxtimes : A \times B \to C$ is called <u>biexact</u> if (i) each partial functor $a \boxtimes (-) : B \to C$, $(-) \boxtimes b : A \to C$ is exact and if (ii) $A \boxtimes 0 = 0 \boxtimes B = 0$ for distinguished zero objects 0 of A, B, C. Note that we can assume C skeletal if necessary to obtain the technical condition (ii). Given a biexact functor $\boxtimes$, there is an induced bicategory factorization

$$QA \otimes QB \to QQC \to bi(QC)$$

of the map of §1. The right-hand map is given on bimorphisms by "forgetting" the middle object in the diagram (2.1). The left-hand map is given on the bimorphism

$$\begin{array}{ccccc}
(A_0,B_0) & \twoheadleftarrow & (A_2,B_0) & \rightarrowtail & (A_1,B_0) \\
\uparrow & & \uparrow & & \uparrow \\
(A_0,B_2) & \leftarrow & (A_2,B_2) & \rightarrow & (A_1,B_2) \\
\downarrow & & \downarrow & & \downarrow \\
(A_0,B_1) & \twoheadleftarrow & (A_2,B_1) & \rightarrowtail & (A_1,B_1)
\end{array}$$

of $QA \otimes QB$ by adding the middle object (A_2,B_2) as shown, and then applying $\boxtimes$. This factorization is pointed out in Proposition 9.2 of [Wa], where Waldhausen notes that the resulting map $BQA \times BQB \to BQQC$ of realizations vanishes on the subspace $BQA \vee BQB$ (because of the technical condition (ii)), and hence induces a map of topological spaces

(2.2) $$BQA \wedge BQB \to BQQC.$$

A lucid account of the product map (2.2) is also given in [Gr].

If we take homotopy groups, we obtain (using, e.g., [Br,(1.6)]) a map

$$K_p(A) \otimes K_q(B) \to K_{p+q}(C).$$

(2.3) In the special case that $A = C$ and there is an object b_0 of B so that $(-) \boxtimes b_0$ is the identity on A, there is a commutative diagram (Lemma 9.2.4 of [Wa]):

$$\begin{array}{ccc} BQA = BQA \wedge S^0 & \xrightarrow{\simeq} & \Omega BQQB \\ \downarrow & & \uparrow \\ BQA \wedge \Omega BQB & \to & \Omega(BQA \wedge BQB). \end{array}$$

The left vertical map comes from the inclusion of S^0 into ΩBQB by selection of the loop $[b_0]: 0 \rightarrowtail b_0 \twoheadrightarrow 0$. The fact that the top composite is the natural map is stated on p. 199, line 18 of [Wa].

When there is an associative pairing $B \times B \to B$, $K_*(B)$ becomes a graded ring; $K_*(B)$ has unit $[b_0]$ if $(-)\boxtimes b_0 = b_0 \boxtimes(-) = id(B)$. The map $\boxtimes : A \times B \to A$ induces a right $K_*(B)$-module structure on K_*A when the two evident functors $A \times B \times B \to A$ agree up to natural isomorphism. These remarks apply notably to the case $B = \underline{P}(k)$, the category of fin. gen. projective k-modules for a commutative ring k: tensor product makes $K_*(k)$ a graded commutative ring with unit, and for every k-algebra A the group $K_*(A)$ is a 2-sided $K_*(k)$-module.

3. Loday's Product

Another approach to products in K-theory is to deal with symmetric monoidal categories, and invoke the "+=Q" theorem. This approach was first used by Loday in [L], using the category $\underset{\sim}{F}(A) = \coprod_{n=0}^{\infty} G\ell_n(A)$, and later generalized by May in [May 1,2]. For simplicity, we first describe Loday's method, and then present the more complicated approach used by May.

The choice of an isomorphism $\otimes_{p,q}: A^p \otimes B^q \to (A\otimes B)^{pq}$ for every p and q gives a pairing $\otimes: \underset{\sim}{F}(A) \times \underset{\sim}{F}(B) \to \underset{\sim}{F}(A\otimes B)$. Since $0 \otimes \underset{\sim}{F}(B) = \underset{\sim}{F}(A) \otimes 0 = 0$, it induces a map of topological spaces, which is the top row of the following diagram:

$$
\begin{array}{ccc}
B\underset{\sim}{F}(A)\wedge B\underset{\sim}{F}(B) & \xrightarrow{\quad B\otimes \quad} & B\underset{\sim}{F}(A\otimes B) \\
\| & & \| \\
* \amalg \coprod_{p,q\geq 1} BGl_p(A)\times BGl_q(B) & \xrightarrow{\coprod B\otimes_{pq}} & \coprod_{r\geq 0} BGl_r(A\otimes B) \\
\downarrow & & \downarrow \\
* \amalg \coprod_{p,q\geq 1} BGl_p^+(A)\times BGl_q^+(B) & \xrightarrow{\coprod f_{pq}=f} & \coprod_{r\geq 0} BGl_r^+(A\otimes B) \\
\| & & \Big\downarrow i_{A\otimes B} \\
\coprod_{p\geq 0} BGl_p^+(A)\wedge \coprod_{q\geq 0} BGl_q^+(B) & & \\
\Big\downarrow i_A\wedge i_B & & \\
[\mathbb{Z}\times BGl^+(A)]\wedge[\mathbb{Z}\times BGl^+(B)] & \overset{\hat{\gamma}}{\dashrightarrow} & \mathbb{Z}\times BGl^+(A\otimes B).
\end{array}
\tag{3.1}
$$

The convention is that $BGl_p^+(A)$ denotes $BGl_p(A)$ for $p \leq 2$; for $p \geq 3$ it denotes the result of the plus construction relative to $E_p(A)$. The maps f_{pq} are the universal maps determined uniquely by the $B\otimes_{pq}$. Loday's idea is to define a map $\hat{\gamma}$ making the diagram (3.1) commute up to homotopy. As before, taking homotopy groups yields a map

$$K_p(A) \otimes K_q(B) \longrightarrow K_{p+q}(A\otimes B).$$

If k is a commutative ring, the map $k \otimes k \to k$ makes $K_*(k)$ into a graded ring; the maps $k \otimes A \to A$ make $K_*(A)$ into a $K_*(k)$-algebra for every k-algebra A.

Loday first observes that $B\underset{\sim}{F}(A) = \coprod BGl_n(A)$ is an H-space (see [Wh]) under direct sum, and that $x\otimes(-)$, $(-)\otimes y$ are H-space maps for each $x \in B\underset{\sim}{F}(A)$, $y \in B\underset{\sim}{F}(B)$. He next observes that $\coprod BGl_p^+$ is an

H-space (etc.) and that $f(x,)$, $f(,y)$ are H-space maps; this is (2.1.2)(ii) of [L]. Now $\mathbb{Z}\times BG\ell^+(A)$ is the "group completion" of $\coprod BG\ell_p^+(A)$ in the very strong sense that for every x in $\mathbb{Z}\times BG\ell^+(A)$ there is a positive integer n such that $x+n$ is in the image of $\coprod BG\ell_p^+(A)$. Since $\mathbb{Z}\times BG\ell^+(A\otimes B)$ is an H-group, there is a unique extension of f to an H-space map

$$\gamma:[\mathbb{Z}\times BG\ell^+(A)]\times[\mathbb{Z}\times BG\ell^+(B)] \to \mathbb{Z}\times BG\ell^+(A\otimes B).$$

Specifically, if $x \in \mathbb{Z}\times BG\ell^+(A)$, $y \in \mathbb{Z}\times BG\ell^+(B)$, we choose $m,n \in \mathbb{N}$, $x_0 \in \coprod BG\ell_p^+(A)$, $y_0 \in \coprod BG\ell_q^+(B)$ such that $x+m = i_A(x_0)$, $y+n = i_B(y_0)$, and define

$$\begin{aligned}\gamma(x,y) &= \gamma(i_A(x_0)-m, i_B(y_0)-n)\\ &= i\circ f(x_0,y_0) - i\circ f(*_m,y_0) - i\circ f(x_0,*_n) + i\circ f(*_m,*_n).\end{aligned}$$

Here $*_m, *_n$ are the basepoints of $BG\ell_m^+(A)$ and $BG\ell_n^+(B)$, and we have used i for $i_{A\otimes B}$. If we take x and y to be in the respective basepoint components, we recover the map

$$\operatorname{colim} \gamma_{pq}: BG\ell^+(A)\times BG\ell^+(B) \to BG\ell^+(A\otimes B)$$

on the top of p. 332 of [L]. Since γ is homotopically trivial on $[\mathbb{Z}\times BG\ell^+(A)]\vee[\mathbb{Z}\times BG\ell^+(B)]$, γ factors through the smash product to give the map $\hat{\gamma}$ of diagram (3.1). The choices used to define γ mean that γ is only well-defined up to weak homotopy type.

In May's generalization, one considers "pairings" $\otimes: A\times B \to C$ of symmetric monoidal categories. This means that $A\otimes 0 = 0\otimes B = 0$ and that there is a coherent natural bidistributivity axiom

$$(a+a')\otimes(b+b') \cong (a\otimes b) + (a\otimes b') + (a'\otimes b) + (a'\otimes b').$$

Instead of making the technical notion of coherence precise, we refer the reader to §2 of [May 2] and content ourselves with the remark that $\otimes : \underset{\sim}{F}(A) \times \underset{\sim}{F}(B) \to \underset{\sim}{F}(A \otimes B)$ is such a pairing.

At this stage, we need to introduce the "group completion" map $B\mathcal{A} \to E_0 B\mathcal{A}$, defined for every symmetric monoidal category $\mathcal{A}$. For example, $E_0 B\underset{\sim}{F}(A)$ is the space $\mathbb{Z} \times B\mathcal{G}\ell^{+}(A)$. One way to construct the group completion is to use the $S^{-1}S$ construction of [GQ]. Another way is to use an infinite loop space machine; for example, one can first obtain a Γ-space $\widetilde{B\mathcal{A}}$, use Segal's machine to obtain a spectrum $EB\mathcal{A}$, and take the zeroth space $E_0 B\mathcal{A}$. This latter approach has both the advantages and disadvantages inherent in infinite loop space machinery.

The point is that a pairing of symmetric monoidal categories functorially determines a pairing $E_0 B\mathcal{A} \wedge E_0 B\mathcal{B} \to E_0 B\mathcal{C}$ of infinite loop spaces. This follows, for example, from Theorems 1.6 and 2.1 of [May 2]. More is true: a pairing $EB\mathcal{A} \wedge EB\mathcal{B} \to EB\mathcal{C}$ is determined in the stable category of infinite loop spectra, allowing spectrum level work to be performed. There is also a commutative diagram:

(3.2)
$$\begin{array}{ccccc}
B\mathcal{A} \times B\mathcal{B} & \longrightarrow & B\mathcal{A} \wedge B\mathcal{B} & \xrightarrow{B\otimes} & B\mathcal{C} \\
\downarrow & & \downarrow \quad \uparrow \gamma & & \downarrow \\
E_0 B\mathcal{A} \times E_0 B\mathcal{B} & \longrightarrow & E_0 B\mathcal{A} \wedge E_0 B\mathcal{B} & \longrightarrow & E_0 B\mathcal{C}
\end{array}$$

in which the bottom composite is an infinite loop space map. Commutativity of the diagram is Corollary (6.5) of [May 2]. As remarked in the introduction of [May 2], it is immediate from (3.1) and (3.2) that the product defined by May specializes to Loday's product.

Here is an example of the usefulness of Loday's product. Let Ens denote the skeletal category $\coprod \Sigma_n$ of finite sets and their isomorphisms. For this symmetric monoidal category we have E_0B Ens $= \mathbb{Z}\times B\Sigma^+ = \Omega^\infty\Sigma^\infty$, and a pairing Ens×Ens → Ens induced by multiplication, which is discussed on [May 1, p. 161]. Consequently, K_*Ens $= \pi_*\Omega^\infty\Sigma^\infty = \pi_*^s$ is a graded commutative ring. The map Ens → $\underset{\sim}{F}(\mathbb{Z})$ embedding the symmetric group Σ_n into $Gl_n(\mathbb{Z})$ induces a map of pairings in the senses of [May 1, p. 155] and [May 2, §2], hence a ring map $\pi_*^s \to K_*(\mathbb{Z})$.

4. Agreement of Product Structures

In order to directly compare Waldhausen's product and Loday's product, consider a pairing $A\times B \to C$ of exact categories. The subcategories $Is(A)$, $Is(B)$, $Is(C)$ of isomorphisms are all symmetric monoidal categories, and the induced functor $Is(A)\times Is(B) \to Is(C)$ is a pairing of symmetric monoidal categories. Waldhausen's Lemma 9.2.6 in [Wa] states that the following diagram commutes up to basepoint preserving homotopy:

(4.1)
$$\begin{array}{ccc}
BIs(A)\wedge BIs(B) & \longrightarrow & BIs(C) \\
\downarrow & & \downarrow \\
(\Omega BQA)\wedge(\Omega BQB) & \overset{\hat{\gamma}}{\dashrightarrow} & \Omega BQC \\
\downarrow & & \downarrow\simeq \\
\Omega\Omega(BQA\wedge BQB) & \longrightarrow & \Omega\Omega(BQQC).
\end{array}$$

The maps $BIs(A) \to \Omega BQA$, etc., are described on p. 198 of [Wa]. The top arrow in (4.1) is induced from the composite

$$BIs(A)\times BIs(B) \cong B(Is(B)\otimes Is(B)) \to B(bi(Is(C))) \to BIs(C),$$

which by (1.1) is the natural map $B\boxtimes$. The bottom map is the double looping of Waldhausen's map (2.2), and we have already remarked on the lower right-hand homotopy equivalence (of H-spaces!). There is a unique way, up to homotopy, to fill in the broken arrow so that the diagram remains homotopy commutative.

We point out that the broken arrow is induced from an H-space map $(\Omega BQA)\times(\Omega BQB) \to \Omega BQC$. To see this, note that the functor $QA\otimes QB \to QQC$ is a map of symmetric monoidal bicategories (the operation being slotwise direct sum), so that $BQA\times BQB \simeq B(QA\otimes QB) \to BQQC$ is an H-space map (in fact it is an infinite loop space map).

Now suppose that all exact sequences in A split. Then there is a basepoint preserving homotopy equivalence $E_0BIs(A) \to \Omega BQA$ so that

$$\begin{array}{ccc} & BIs(A) & \\ \swarrow & & \searrow \\ E_0BIs(A) & \xrightarrow{\simeq} & \Omega BQA \end{array}$$

commutes (up to basepoint preserving homotopy). This is proven as (9.3.2) of [Wa], modulo the observation that (by (6.3) of [Wa]) the space $\Omega BN_\Gamma(Is(A))$ in (9.3.2) is just $E_0BIs(A)$. Having said this, it follows that the top part of (4.1) induces the following homotopy commutative diagram of <u>H-spaces</u>:

(4.2)
$$\begin{array}{ccc} BIs(A)\times BIs(B) & \longrightarrow & BIs(C) \\ \downarrow & & \downarrow \\ E_0BIs(A)\times E_0BIs(B) & \xrightarrow{\gamma} & E_0BIs(C). \end{array}$$

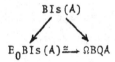

The H-space map γ in (4.2) is uniquely determined, so it must be the same as the map γ in Loday's construction, as generalized by May. Comparing (4.1) and (4.2), we see that the broken arrow $\hat{\gamma}$ in (4.1) must be the same as the $\hat{\gamma}$ in (3.1) and (3.2). We summarize this as follows:

Theorem 4.3 (Waldhausen). If $A \times B \to C$ is a biexact pairing of exact categories for which all exact sequences split, then the groups $K_*(A) = \pi_*(\Omega BQA)$ agree with the groups $K_*Is(A) = \pi_* E_0 BIs(A)$. There are commutative diagrams

$$\begin{array}{ccc} E_0BIs(A) \wedge E_0BIs(B) & \xrightarrow{\hat{\gamma}} & E_0BIs(C) \\ \downarrow & & \downarrow \simeq \\ \Omega BQA \wedge \Omega BQB & \longrightarrow & \Omega\Omega BQQC \end{array}$$

$$\begin{array}{ccc} K_*Is(A) \otimes K_*Is(B) & \longrightarrow & K_*Is(C) \\ \cong \downarrow & & \cong \downarrow \\ K_*(A) \otimes K_*(B) & \longrightarrow & K_*(C) \end{array}$$

in which the top maps are the Loday-May pairings, and the bottom maps are the Waldhausen pairings.

We can apply this result to the exact category $\underline{\underline{P}}(A) = A$ of fin. gen. projective A-modules, etc. Since $E_0BIs\underline{\underline{P}}(A)$ is $K_0(A) \times BG\ell^+(A)$, we obtain:

Corollary 4.4 (Waldhausen). Let A, B be rings with unit. There is a homotopy commutative diagram

$$\begin{array}{ccc} [K_0(A) \times BG\ell^+(A)] \wedge [K_0(B) \times BG\ell^+(B)] & \xrightarrow{\hat{\gamma}} & [K_0(A \otimes B) \times BG\ell^+(A \otimes B)] \\ \downarrow & & \downarrow \\ [\Omega BQ\underline{P}(A)] \wedge [\Omega BQ\underline{P}(B)] & \longrightarrow & \Omega\Omega[BQQ\underline{P}(A \otimes B)], \end{array}$$

where the top arrow is Loday's pairing and the bottom arrow is Waldhausen's product. Thus the two pairings agree on homotopy to give the same graded map

$$K_*(A) \otimes K_*(B) \longrightarrow K_*(A\otimes B).$$

Remark. Waldhausen gave the argument for Theorem (4.3) on p.235 of [Wa] for the special case $A = \mathbb{Z}$, $B = \mathbb{Z}G$, in order to show that the map $\pi_*K(BG;\mathbb{Z}) \longrightarrow K_*(\mathbb{Z}G)$ constructed on [Wa, p. 227] agrees with Loday's map on [L, p. 226].

If we had used the skeletal subcategories $\underline{\underline{F}}(A) \subseteq \underline{\underline{P}}(A)$ of free modules, we would get $\underset{\sim}{F}(A) = Is(\underline{\underline{F}}(A)$ when A is commutative, or more generally when $A^m \not\cong A^n$ for $m \neq n$. In this case we could write (4.4) as the commutative diagram

$$\begin{array}{ccc} [\mathbb{Z}\times BG\ell^+(A)]\wedge[\mathbb{Z}\times BG\ell^+(B)] & \xrightarrow{\hat{\gamma}} & [\mathbb{Z}\times BG\ell^+(A\otimes B)] \\ \downarrow & & \downarrow \\ [\Omega BQ\underline{F}(A)]\wedge[\Omega BQ\underline{F}(B)] & \longrightarrow & \Omega\Omega[BQQ\underline{F}(A\otimes B)] \end{array}$$

5. Relative K-theory.

When I is an ideal in a ring A, we will construct a pairing $K_*(A,I) \otimes K_*(B) \to K_*(A\otimes B, I\otimes B)$ so that:

(5.1) If B is commutative and A is a B-algebra, the map $K_*(A,I) \otimes K_*(B) \to K_*(A\otimes B, I\otimes B) \to K_*(A,I)$ makes $K_*(A,I)$ a graded $K_*(B)$-module.

There are two approaches to this problem, corresponding to the two types of product. We will first describe May's approach, which is more conceptually straightforward, and then describe Waldhausen's more subtle method.

In May's approach, the basic object is the category $\underset{\sim}{F}(B)$ of free B-modules. One has a "morphism of pairings of symmetric monoidal categories" (q.v. §2 of [May 2]):

$$\begin{array}{ccc} \underset{\sim}{F}(A)\times\underset{\sim}{F}(B) & \longrightarrow & \underset{\sim}{F}(A\otimes B) \\ \downarrow & & \downarrow \\ \underset{\sim}{F}(A/I)\times\underset{\sim}{F}(B) & \longrightarrow & \underset{\sim}{F}(A/I\otimes B). \end{array}$$

The machinery described in Theorem 1.6 of [May 2] produces a "morphism of pairings in the stable category" which is adequately represented by commutativity of the right-hand portion of the following diagram:

$$(5.2)\qquad \begin{array}{ccccc} \mathrm{Fiber}(A,I)\wedge E_0B\underset{\sim}{F}B & \longrightarrow & E_0B\underset{\sim}{F}A\wedge E_0B\underset{\sim}{F}B & \longrightarrow & E_0B\underset{\sim}{F}(A/I)\wedge E_0B\underset{\sim}{F}B \\ \vdots & & \downarrow & & \downarrow \\ \mathrm{Fiber}(A\otimes B,I\otimes B) & \longrightarrow & E_0BF(A\otimes B) & \longrightarrow & E_0B\underset{\sim}{F}(A/I\otimes B). \end{array}$$

In (5.2) we have written Fiber(A,I) for the homotopy fiber of the map $E_0B\underset{\sim}{F}(A) \to E_0B\underset{\sim}{F}(A/I)$, and similarly for Fiber(A⊗B,I⊗B). This being said, it is standard that there is a broken arrow making (5.2) a map of (infinite loop space) fibrations. Since $K_p(A,I) = \pi_p\mathrm{Fiber}(A,I)$, the homotopy groups of (5.2) yield a map of long exact sequences:

$$\begin{array}{ccccc} \cdots K_{p+1}(A/I)\times K_q(B) & \longrightarrow & K_p(A,I)\times K_q(B) & \longrightarrow & K_p(A)\times K_q(B)\cdots \\ \downarrow & & \vdots & & \downarrow \\ \cdots K_{p+q+1}(A/I\otimes B) & \longrightarrow & K_{p+q}(A\otimes B,I\otimes B) & \longrightarrow & K_{p+q}(A\otimes B)\cdots \end{array} .$$

The problem with this approach is that the broken arrow in (5.2) is not unique and, unless care is taken, will <u>not</u> make (5.1) hold.

Happily, there is enough structure in the categories involved to save the day. The functoriality of May's approach makes the right-hand square in (5.2) commute on the nose, and the details of associativity can be checked directly. This is done in detail in [May 3].

The second approach to relative pairings is due to Waldhausen and is implicit in §7 of [Wa]. The idea is to use simplicial exact categories (SEC's) to produce a model for $\Omega^{-2}\mathrm{Fiber}(A,I)$, and do all work at the category-theoretic level.

To an exact category A, Waldhausen associates an SEC denoted $S_\bullet A$, and proves in [Wa, (7.1)] that $\Omega BQS_\bullet A \simeq BQA$. An object of the exact category $S_n A$ is a sequence

$$A: A_{01} \rightarrowtail A_{02} \rightarrowtail \cdots \rightarrowtail A_{0n}$$

of admissible monics in A, together with choices of objects A_{ij} $(i>j)$ and isomorphisms $A_{ij} \cong A_{0i}/A_{0j}$. The ith face map $d_i: S_n A \to S_{n-1} A$ is induced by "dropping the index i". For example, the sequence for $d_0(A)$ is

$$A_{12} \rightarrowtail A_{13} \rightarrowtail \cdots \rightarrowtail A_{1n}.$$

Next, we suppose given an exact functor $f: A \to A'$. Waldhausen constructs on [Wa, p. 182] an SEC denoted $F_\bullet(f)$ fitting into a sequence of SEC's:

(5.3) $$A \to A' \to F_\bullet(f) \to S_\bullet A \to S_\bullet A'.$$

Of course, in (5.3) we consider $\mathcal{A}$ and $\mathcal{A}'$ to be constant simplicial exact categories.

Briefly, an object of the exact category $F_p(f)$ is a triple $(A',A,\cong)$: an object A' of $S_{p+1}(\mathcal{A}')$, an object A of $S_p(\mathcal{A})$, and an isomorphism $f(A) \cong d_0(A')$ in $S_p(\mathcal{A}')$. The map $\mathcal{A}' \to F_p(f)$ sends A' in $\mathcal{A}'$ to the object $(A'=A'=\cdots=A',0,0)$ in $F_p(f)$, and the map $F_p(f) \to S_p(\mathcal{A})$ sends $(A',A,\cong)$ to A.

The content of Propositions (7.1) and (7.2) of [Wa] is that

$$BQ\mathcal{A} \longrightarrow BQ\mathcal{A}' \longrightarrow BQF_{\bullet}(f) \longrightarrow BQS_{\bullet}\mathcal{A} \longrightarrow BQS_{\bullet}\mathcal{A}'$$

is a fibration sequence up to homotopy. Thus $\Omega BQF_{\bullet}(f)$ is a model for the fiber of $BQ\mathcal{A} \to BQ\mathcal{A}'$; if we set $K_p(f) = \pi_{p+2}BQF_{\bullet}(f)$, there is a long exact sequence

$$\text{(5.4)} \qquad \cdots K_{*+1}\mathcal{A} \longrightarrow K_{*+1}\mathcal{A}' \longrightarrow K_*(f) \longrightarrow K_*\mathcal{A} \longrightarrow K_*\mathcal{A}' \cdots$$

In particular: if $\mathcal{A} = \underline{P}(A)$, $\mathcal{A}' = \underline{P}(A/I)$ then $K_p(A,I) = \pi_{p+2}BQF_{\bullet}(f)$

Given this category-theoretic encoding of the relative term in K-theory, we can construct relative pairings with ease. One starts with a commutative diagram

$$\text{(5.5)} \qquad \begin{array}{ccc} \mathcal{A} \times \mathcal{B} & \xrightarrow{\boxtimes} & \mathcal{C} \\ {\scriptstyle f\times \mathcal{B}}\downarrow & & \downarrow{\scriptstyle f'} \\ \mathcal{A}' \times \mathcal{B} & \xrightarrow{\boxtimes} & \mathcal{C}' \end{array}$$

in which the horizontal arrows are biexact. The functors $\boxtimes$ induce simplicial biexact functors $S_{\bullet}\mathcal{A}\times\mathcal{B} \to S_{\bullet}\mathcal{C}$ and $F_{\bullet}(f)\times\mathcal{B} \to F_{\bullet}(f')$ in an obvious manner. The result is the commutative diagram of SEC's, analogous to (5.3):

$$\begin{array}{ccccccccc} A\times B & \longrightarrow & A'\times B & \longrightarrow & F_\bullet(f)\times B & \longrightarrow & S_\bullet A\times B & \longrightarrow & S_\bullet A'\times B \\ \downarrow & & \downarrow & & \downarrow & & \downarrow & & \downarrow \\ C & \longrightarrow & C' & \longrightarrow & F_\bullet(f) & \longrightarrow & S_\bullet C & \longrightarrow & S_\bullet C'. \end{array}$$

Following §2 above, there is a commutative diagram of simplicial bicategories, the middle of which is

$$(5.6)\qquad \begin{array}{ccccc} QA' \otimes QB & \longrightarrow & QF_\bullet(f) \otimes QB & \longrightarrow & QS_\bullet A \otimes QB \\ \downarrow & & \downarrow & & \downarrow \\ QQC' & \longrightarrow & QQF_\bullet(f') & \longrightarrow & QQS_\bullet(C). \end{array}$$

Geometric realization yields a map of fibration sequences, the middle of which is

$$\begin{array}{ccccc} BQA'\wedge BQB & \longrightarrow & BQF_\bullet(f)\wedge BQB & \longrightarrow & BQS_\bullet A\wedge BQB \\ \downarrow & & \downarrow & & \downarrow \\ BQQC' & \longrightarrow & BQQF_\bullet(f') & \longrightarrow & BQQS_\bullet(C). \end{array}$$

Taking homotopy groups yields the map of long exact sequences

$$\begin{array}{ccccc} \cdots K_{p+1}A'\times K_q(B) & \longrightarrow & K_p(f)\times K_q(B) & \longrightarrow & K_p(A)\times K_q(B)\cdots \\ \downarrow & & \downarrow & & \downarrow \\ \cdots K_{p+q+1}(C') & \longrightarrow & K_{p+q}(f') & \longrightarrow & K_{p+q}(C)\cdots, \end{array}$$

the middle vertical arrow being the desired pairing.

Now suppose that we are in the situation of (2.3) above, i.e., that $A=C$ and $A'=C'$. We assume that there is an associative pairing $B\times B \to B$, $(-)\otimes b_0$ is the identity functor on A, A' and B, and that (5.5) fits into an "associativity axiom" cube (going from $A\times B\times B$ to A'). Then the two evident functors $F_\bullet(f)\times B\times B \to F_\bullet(f)$ agree (up to natural isomorphism), so that $K_*(f)$ is a graded

$K_*(B)$-module in such a way that (5.4) is a sequence of $K_*(B)$-modules.

The above paragraph applies to the situation of (5.1). We take (5.5) to be the diagram

$$\begin{array}{ccc} \underline{P}(A)\times\underline{P}(B) & \xrightarrow{\otimes} & \underline{P}(A\otimes B) \\ \downarrow & & \downarrow \\ \underline{P}(A/I)\times\underline{P}(B) & \xrightarrow{\otimes} & \underline{P}(A/I\otimes B). \end{array}$$

All hypotheses are met, and $K_*(A,I)$ is a graded $K_*(B)$-module.

6. Products for KV-theory

There is another type of K-theory with product: the Karoubi-Villamayor groups $KV_*(A)$. This theory makes sense for any ring (with or without "one"), and is uniquely determined by the axioms given in [K-V]. One way to define them is to set $KV_0(A) = K_0(A)$, and for $p > 0$ to define $KV_p(A)$ by exactness of

$$0 \longrightarrow KV_p(A) \longrightarrow K_0(\Omega^p A) \longrightarrow K_0(E\Omega^{p-1}A).$$

Here we have used the notation $\Omega A = (t^2-t)A[t] \subset tA[t] = EA$ for any ring A, and defined $\Omega^p A$ by iteration: $\Omega^p A = \Pi(t_i^2-t_i)A[t_1,\ldots,t_p]$.

In [K], Karoubi constructs a pairing for KV-theory from the following "usual" pairing for K_0: When A is a ring without "one" we can define $K_0(A) = K_0(R\oplus A, A)$ for any ring R with "one" with an R-algebra structure on A; this definition is independent of the choice of R. Since $(\mathbb{Z}\oplus A)\otimes(\mathbb{Z}\oplus B) = R\oplus(A\otimes B)$ for $R = \mathbb{Z}\oplus A\oplus B$, we can define the bilinear map $\theta_{0,0}$ as the composite:

$$K_0(A)\otimes K_0(B) \longrightarrow K_0(\mathbb{Z}\oplus A)\otimes K_0(\mathbb{Z}\oplus B) \longrightarrow K_0(R\oplus(A\otimes B)) \longrightarrow K_0(A\otimes B).$$

One interpretation of Karoubi's construction of pairings is this: It is easy to show that $K_0(EA\otimes\Omega B) \to K_0(EA\otimes EB)$ and $K_0(\Omega A\otimes EB) \to K_0(EA\otimes EB)$ are injections for every A,B. We then have the following commutative square with exact rows (for $p,q\geq 1$):

(6.1)
$$\begin{array}{ccccccc} 0 \to & KV_p(A)\times KV_q(B) & \to & K_0(\Omega^p A)\times K_0(\Omega^q B) & \to & K_0(E\Omega^{p-1}A)\times K_0(E\Omega^{q-1}B) \\ & \downarrow{\scriptstyle \theta_{p,q}} & & \downarrow & & \downarrow \\ 0 \longrightarrow & KV_{p+q}(A\otimes B) & \longrightarrow & K_0(\Omega^p A\otimes \Omega^q B) & \longrightarrow & K_0(E\Omega^{p-1}A\otimes E\Omega^{q-1}B). \end{array}$$

It follows that the broken arrow $\theta_{p,q}$ is defined when $p, q \geq 1$ When $p = 0, q \geq 1$ and $p \geq 1, q = 0$ it is also easy to induce maps $\theta_{p,q}$. Karoubi then proves the following theorem in [K, p. 78]:

Theorem (6.2) The maps $\theta_{p,q}:KV_p(A) \otimes KV_q(B) \to KV_{p+q}(A\otimes B)$ are the unique natural bilinear maps satisfying the following axioms:

(i) $\theta_{0,0}$ is the "usual" product $K_0(A) \otimes K_0(B) \to K_0(A\otimes B)$

(ii) Every map of GL-fibrations (see [K-V])

(6.2a)
$$\begin{array}{ccccccccc} 0 & \longrightarrow & A\times B & \longrightarrow & A\times B & \longrightarrow & A''\times B & \longrightarrow & 0 \\ & & \downarrow & & \downarrow & & \downarrow & & \\ 0 & \longrightarrow & C' & \longrightarrow & C & \longrightarrow & C'' & \longrightarrow & 0 \end{array}$$

gives rise to a commutative diagram

$$\begin{array}{ccccc} KV_{p+1}(A'') \otimes KV_q(B) & \longrightarrow & KV_{p+q+1}(C'') & \longleftarrow & KV_p(B) \otimes KV_{q+1}(A'') \\ \downarrow{\scriptstyle \partial\times 1} & & \downarrow{\scriptstyle \partial} & & \downarrow{\scriptstyle (-1)^p\times\partial} \\ KV_p(A') \otimes KV_q(B) & \longrightarrow & KV_{p+q}(C') & \longleftarrow & KV_p(B) \otimes KV_q(A'). \end{array}$$

In the remainder of this section, we show that the natural map $K_*A \to KV_*(A)$ sends the K_*-pairing to the KV_*-pairing. In order to do this, it is necessary to recall the construction of the map $K_* \to KV_*$.

For a ring A with "one" we can define a simplicial ring $A_\bullet$, which in degree n is the coordinate ring $A[x_0,\ldots,x_n]/(\Sigma x_i=1)$ of the "standard n-simplex," the face and degeneracy maps being dictated by the geometry. Applying $BG\ell^+$ gives the simplicial topological space $p \mapsto BG\ell^+(A_p)$, and we have for $* \geq 1$ that $KV_*(A) = \pi_*|BG\ell^+(A_\bullet)|$. This is proven in [A, p. 65]. The map of simplicial spaces $BG\ell^+(A) \to BG\ell^+(A_\bullet)$ induces the map $K_*(A) \to KV_*(A)$ of homotopy groups.

Loday's pairing now induces a pairing in KV-theory in a completely canonical way: the choice of isomorphisms $A^p \otimes B^q \cong (A\otimes B)^{pq}$ completely determines a simplicial pairing $\underline{F}(A_\bullet)\times\underline{F}(B_\bullet) \to \underline{F}(A\otimes B)_\bullet$, and this in turn yields a map of simplicial topological spaces

$$\hat{\gamma}_\bullet : [\mathbb{Z}\times BG\ell^+(A_\bullet)]\wedge[\mathbb{Z}\times BG\ell^+(B_\bullet)] \to \mathbb{Z}\times BG\ell^+(A\otimes B)_\bullet .$$

Applying geometric realization yields a map

$$|\hat{\gamma}_\bullet| = [\mathbb{Z}\times|BG\ell^+(A_\bullet)|]\wedge[\mathbb{Z}\times|BG\ell^+(B_\bullet)|] \to \mathbb{Z}\times|BG\ell^+(A\otimes B)_\bullet| .$$

Another way to proceed is to use Waldhausen's pairing and a Q-version of the map $K_* \to KV_*$. Define the simplicial subcategory $\underline{P}_\bullet(A)$ of $\underline{P}(A_\bullet)$ by letting $\underline{P}_p(A)$ denote the full subcategory of $\underline{P}(A_p)$ of projective A_p-modules extended from A (i.e., isomorphic to some $P\otimes_A A_p$). We then have

(6.3)
$$K_0(A)\times|BG\ell^+(A_\bullet)| \simeq \Omega|BQ\underline{P}_\bullet(A)| .$$
$$KV_*(A) = \pi_{*+1}|BQ\underline{P}_\bullet(A)| .$$

This is Theorem 2.1 of [We]; the technical reason for using $\underline{\underline{P}}_\bullet(A)$ instead of $\underline{\underline{P}}(A_\bullet)$ is that $\Omega|BQ\underline{\underline{P}}(A_\bullet)| \simeq |K_0(A_\bullet)|\times|BG\ell^+(A_\bullet)|$, and the space $|K_0(A_\bullet)|$ need not be $K_0(A)$. We now induce a pairing from Waldhausen's product:

External $\otimes$ gives a biexact functor $\underline{\underline{P}}_\bullet(A)\times\underline{\underline{P}}_\bullet(B) \to \underline{\underline{P}}_\bullet(A\otimes B)$, and so a morphism of simplicial bicategories

$$Q\underline{\underline{P}}_\bullet(A) \otimes Q\underline{\underline{P}}_\bullet(B) \longrightarrow QQ\underline{\underline{P}}_\bullet(A\otimes B),$$

which realizes to a map of topological spaces:

$$|BQ\underline{\underline{P}}_\bullet(A)|\wedge|BQ\underline{\underline{P}}_\bullet(B)| \longrightarrow |BQQ\underline{\underline{P}}_\bullet(A\otimes B)|. \tag{6.4}$$

Since each $BQQ\underline{\underline{P}}_p$ is connected, we have that

$$\Omega|BQQ\underline{\underline{P}}_\bullet| \simeq |\Omega BQQ\underline{\underline{P}}_\bullet| \simeq |BQ\underline{\underline{P}}_\bullet|,$$

and therefore we have an induced map of homotopy groups

$$\begin{array}{ccc} \pi_{p+1}|BQ\underline{\underline{P}}_\bullet(A)| \otimes \pi_{q+1}|BQ\underline{\underline{P}}_\bullet(B)| & \longrightarrow & \pi_{p+q+2}|BQQ\underline{\underline{P}}_\bullet(A\otimes B)| \\ \| & & \| \\ KV_p(A) \otimes KV_q(B) & \longrightarrow & KV_{p+q}(A\otimes B), \end{array} \tag{6.5}$$

defined for $p, q \geq 0$. In view of (6.3) and (4.4), it is clear that the map (6.4) agrees with Loday's $|\hat{\gamma}_\bullet|$.

In view of Theorem (6.2), we can show that the pairing (6.5) agrees with Karoubi's pairing (6.1) by checking the two axioms. Since Waldhausen's map $K_0(A) \otimes K_0(B) \to K_0(A\otimes B)$ agrees with the classical external product used by Karoubi, we only have to check axiom (ii). We can assume that A, A'', C, C'', and B have a "one", so that the commutative diagram (6.2a) gives rise to a commutative diagram of bisimplicial bicategories analogous to (5.6):

$$\begin{array}{ccccc}
Q\underline{\underline{P}}_{\bullet}(A'') \otimes Q\underline{\underline{P}}_{\bullet}(B) & \longrightarrow & QF_{\bullet}(f_{\bullet}) \otimes Q\underline{\underline{P}}_{\bullet}(B) & \longrightarrow & QS_{\bullet}\underline{\underline{P}}_{\bullet}(A) \otimes Q\underline{\underline{P}}_{\bullet}(B) \\
\downarrow & & \downarrow & & \downarrow \\
QQ\underline{\underline{P}}_{\bullet}(A''\otimes B) & \longrightarrow & QQF_{\bullet}(f_{\bullet}\otimes B) & \longrightarrow & QQS_{\bullet}\underline{\underline{P}}_{\bullet}(A\otimes B).
\end{array}$$

Applying geometric realization gives a map of fibrations <u>at each level</u>, and the assumption that the exact sequences of rings were Gl-fibrations implies that we have <u>global</u> fibrations, i.e., that the rows in the following diagram are fibrations of topological spaces:

$$\begin{array}{ccccc}
|BQ\underline{\underline{P}}_{\bullet}(A'')| \wedge |BQ\underline{\underline{P}}_{\bullet}(B)| & \longrightarrow & |BQF_{\bullet}(f_{\bullet})| \wedge |BQ\underline{\underline{P}}_{\bullet}(B)| & \longrightarrow & |BQS_{\bullet}\underline{\underline{P}}_{\bullet}(A)| \wedge |BQ\underline{\underline{P}}_{\bullet}(B)| \\
\downarrow & & \downarrow & & \downarrow \\
|BQQ\underline{\underline{P}}_{\bullet}(A''\otimes B)| & \longrightarrow & |BQQF_{\bullet}(f_{\bullet}\otimes B)| & \longrightarrow & |BQQS_{\bullet}\underline{\underline{P}}_{\bullet}(A\otimes B)|.
\end{array}$$

The fact that $\pi_{*+2}(BQF_{\bullet}(f_{\bullet})) = KV_{*}(A')$, $\pi_{*+3}(BQQF_{\bullet}(f_{\bullet}\otimes B)) = KV_{*}(A'\otimes B)$ follows from consideration of the long exact homotopy sequences of the rows (and §5), and the commutative diagram

$$\begin{array}{ccc}
\pi_{p+1}\Omega|BQ\underline{\underline{P}}_{\bullet}A''| \otimes \pi_{q}\Omega|BQ\underline{\underline{P}}_{\bullet}B| & \longrightarrow & \pi_{p+1}\Omega|BQF_{\bullet}(f_{\bullet})| \otimes \pi_{q}\Omega|BQ\underline{\underline{P}}_{\bullet}B| \\
\downarrow & & \downarrow \\
\pi_{p+q+1}\Omega\Omega|BQQ\underline{\underline{P}}_{\bullet}(A''\otimes B)| & \longrightarrow & \pi_{p+q+1}\Omega\Omega|BQQF_{\bullet}(f_{\bullet}\otimes B)| \\
\uparrow & & \uparrow \\
\pi_{p}\Omega|BQ\underline{\underline{P}}_{\bullet}B| \otimes \pi_{q+1}\Omega|BQ\underline{\underline{P}}_{\bullet}A''| & \xrightarrow{(-1)^{p}} & \pi_{p}\Omega|BQ\underline{\underline{P}}_{\bullet}B| \otimes \pi_{q+1}\Omega|BQF_{\bullet}(f_{\bullet})|
\end{array}$$

translates into the diagram of axiom (ii). We summarize this:

<u>Proposition (6.6)</u> The pairing (6.5) on KV_*-theory induced from the pairing on K_*-theory satisfies the axioms of Theorem (6.2), and so agrees with Karoubi's product.

References

[A] D.W. Anderson, Relationship among K-theories, Lecture Notes in Math. 341, Springer-Verlag, New York, 1973.

[Br] W. Browder, Algebraic K-theory with coefficients $\mathbb{Z}/p$, Lecture Notes in Math. 657, Springer-Verlag, New York, 1978.

[E] C. Ehresmann, Catégories et Structures, Dunod, Paris, 1965.

[GQ] D. Grayson, Higher algebraic K-theory: II (after D. Quillen), Lecture Notes in Math. 551, Springer-Verlag, New York, 1976.

[Gr] D. Grayson, Products in K-theory and intersecting algebraic cycles, Inv. Math. 47 (1978), 71-84.

[K] M. Karoubi, La périodicité de Bott en K-theorie générale, Ann. Scient. Éc. Norm. Sup. (Paris), t. 4 (1971), 63-95.

[K-V] M. Karoubi and O. Villamayor, Foncteurs K^n en algèbre et en topologie, C.R. Acad. Sci. (Paris) 269 (1969), 416-419.

[K-S] G.M. Kelly and R. Street, Review of the elements of 2-categories, Lecture Notes in Math. 420, Springer-Verlag, New York, 1974.

[L] J.L. Loday, K-theorie algébrique et représentations des groupes, Ann. Sc. Éc Norm. Sup. (Paris), t. 9 (1976), 309-377.

[Mac] S. MacLane, Categories for the Working Mathematician, Springer-Verlag, New York, 1971.

[May 1] J.P. May, E_∞ Ring Spaces and E_∞ Ring Spectra, Lecture Notes in Math. 577, Springer-Verlag, New York, 1977.

[May 2] J.P. May, Pairings of Categories and Spectra, J.P.A.A. (to appear).

[May 3] J.P. May, Multiplicative Infinite Loop Space Theory (to appear).

[Mi] J. Milnor, Introduction to Algebraic K-theory, Annals of Math. Studies, No. 72, Princeton University Press, Princeton, 1971.

[Q] D. Quillen, Higher Algebraic K-theory: I, Lecture Notes in Math. 341, Springer-Verlag, New York, 1973.

[S] A. Suslin, Milnor K-theory injects into Quillen K-theory modulo torsion, to appear.

[Wa] F. Waldhausen, Algebraic K-theory of generalized free products, Ann. Math. 108 (1978), 135-256.

[We] C. Weibel, KV-theory of categories, preprint (1979).

[Wh] G. Whitehead, Elements of Homotopy Theory, Springer-Verlag, New York, 1978.

Vol. 700: Module Theory, Proceedings, 1977. Edited by C. Faith and S. Wiegand. X, 239 pages. 1979.

Vol. 701: Functional Analysis Methods in Numerical Analysis, Proceedings, 1977. Edited by M. Zuhair Nashed. VII, 333 pages. 1979.

Vol. 702: Yuri N. Bibikov, Local Theory of Nonlinear Analytic Ordinary Differential Equations. IX, 147 pages. 1979.

Vol. 703: Equadiff IV, Proceedings, 1977. Edited by J. Fábera. XIX, 441 pages. 1979.

Vol. 704: Computing Methods in Applied Sciences and Engineering, 1977, I. Proceedings, 1977. Edited by R. Glowinski and J. L. Lions. VI, 391 pages. 1979.

Vol. 705: O. Forster und K. Knorr, Konstruktion verseller Familien kompakter komplexer Räume. VII, 141 Seiten. 1979.

Vol. 706: Probability Measures on Groups, Proceedings, 1978. Edited by H. Heyer. XIII, 348 pages. 1979.

Vol. 707: R. Zielke, Discontinuous Čebyšev Systems. VI, 111 pages. 1979.

Vol. 708: J. P. Jouanolou, Equations de Pfaff algébriques. V, 255 pages. 1979.

Vol. 709: Probability in Banach Spaces II. Proceedings, 1978. Edited by A. Beck. V, 205 pages. 1979.

Vol. 710: Séminaire Bourbaki vol. 1977/78, Exposés 507-524. IV, 328 pages. 1979.

Vol. 711: Asymptotic Analysis. Edited by F. Verhulst. V, 240 pages. 1979.

Vol. 712: Equations Différentielles et Systèmes de Pfaff dans le Champ Complexe. Edité par R. Gérard et J.-P. Ramis. V, 364 pages. 1979.

Vol. 713: Séminaire de Théorie du Potentiel, Paris No. 4. Edité par F. Hirsch et G. Mokobodzki. VII, 281 pages. 1979.

Vol. 714: J. Jacod, Calcul Stochastique et Problèmes de Martingales. X, 539 pages. 1979.

Vol. 715: Inder Bir S. Passi, Group Rings and Their Augmentation Ideals. VI, 137 pages. 1979.

Vol. 716: M. A. Scheunert, The Theory of Lie Superalgebras. X, 271 pages. 1979.

Vol. 717: Grosser, Bidualräume und Vervollständigungen von Banachmoduln. III, 209 pages. 1979.

Vol. 718: J. Ferrante and C. W. Rackoff, The Computational Complexity of Logical Theories. X, 243 pages. 1979.

Vol. 719: Categorial Topology, Proceedings, 1978. Edited by H. Herrlich and G. Preuß. XII, 420 pages. 1979.

Vol. 720: E. Dubinsky, The Structure of Nuclear Fréchet Spaces. V, 187 pages. 1979.

Vol. 721: Séminaire de Probabilités XIII. Proceedings, Strasbourg, 1977/78. Edité par C. Dellacherie, P. A. Meyer et M. Weil. VII, 647 pages. 1979.

Vol. 722: Topology of Low-Dimensional Manifolds. Proceedings, 1977. Edited by R. Fenn. VI, 154 pages. 1979.

Vol. 723: W. Brandal, Commutative Rings whose Finitely Generated Modules Decompose. II, 116 pages. 1979.

Vol. 724: D. Griffeath, Additive and Cancellative Interacting Particle Systems. V, 108 pages. 1979.

Vol. 725: Algèbres d'Opérateurs. Proceedings, 1978. Edité par P. de la Harpe. VII, 309 pages. 1979.

Vol. 726: Y.-C. Wong, Schwartz Spaces, Nuclear Spaces and Tensor Products. VI, 418 pages. 1979.

Vol. 727: Y. Saito, Spectral Representations for Schrödinger Operators With Long-Range Potentials. V, 149 pages. 1979.

Vol. 728: Non-Commutative Harmonic Analysis. Proceedings, 1978. Edited by J. Carmona and M. Vergne. V, 244 pages. 1979.

Vol. 729: Ergodic Theory. Proceedings, 1978. Edited by M. Denker and K. Jacobs. XII, 209 pages. 1979.

Vol. 730: Functional Differential Equations and Approximation of Fixed Points. Proceedings, 1978. Edited by H.-O. Peitgen and H.-O. Walther. XV, 503 pages. 1979.

Vol. 731: Y. Nakagami and M. Takesaki, Duality for Crossed Products of von Neumann Algebras. IX, 139 pages. 1979.

Vol. 732: Algebraic Geometry. Proceedings, 1978. Edited by K. Lønsted. IV, 658 pages. 1979.

Vol. 733: F. Bloom, Modern Differential Geometric Techniques in the Theory of Continuous Distributions of Dislocations. XII, 206 pages. 1979.

Vol. 734: Ring Theory, Waterloo, 1978. Proceedings, 1978. Edited by D. Handelman and J. Lawrence. XI, 352 pages. 1979.

Vol. 735: B. Aupetit, Propriétés Spectrales des Algèbres de Banach. XII, 192 pages. 1979.

Vol. 736: E. Behrends, M-Structure and the Banach-Stone Theorem. X, 217 pages. 1979.

Vol. 737: Volterra Equations. Proceedings 1978. Edited by S.-O. Londen and O. J. Staffans. VIII, 314 pages. 1979.

Vol. 738: P. E. Conner, Differentiable Periodic Maps. 2nd edition, IV, 181 pages. 1979.

Vol. 739: Analyse Harmonique sur les Groupes de Lie II. Proceedings, 1976-78. Edited by P. Eymard et al. VI, 646 pages. 1979.

Vol. 740: Séminaire d'Algèbre Paul Dubreil. Proceedings, 1977-78. Edited by M.-P. Malliavin. V, 456 pages. 1979.

Vol. 741: Algebraic Topology, Waterloo 1978. Proceedings. Edited by P. Hoffman and V. Snaith. XI, 655 pages. 1979.

Vol. 742: K. Clancey, Seminormal Operators. VII, 125 pages. 1979.

Vol. 743: Romanian-Finnish Seminar on Complex Analysis. Proceedings, 1976. Edited by C. Andreian Cazacu et al. XVI, 713 pages. 1979.

Vol. 744: I. Reiner and K. W. Roggenkamp, Integral Representations. VIII, 275 pages. 1979.

Vol. 745: D. K. Haley, Equational Compactness in Rings. III, 167 pages. 1979.

Vol. 746: P. Hoffman, τ-Rings and Wreath Product Representations. V, 148 pages. 1979.

Vol. 747: Complex Analysis, Joensuu 1978. Proceedings, 1978. Edited by I. Laine, O. Lehto and T. Sorvali. XV, 450 pages. 1979.

Vol. 748: Combinatorial Mathematics VI. Proceedings, 1978. Edited by A. F. Horadam and W. D. Wallis. IX, 206 pages. 1979.

Vol. 749: V. Girault and P.-A. Raviart, Finite Element Approximation of the Navier-Stokes Equations. VII, 200 pages. 1979.

Vol. 750: J. C. Jantzen, Moduln mit einem höchsten Gewicht. III, 195 Seiten. 1979.

Vol. 751: Number Theory, Carbondale 1979. Proceedings. Edited by M. B. Nathanson. V, 342 pages. 1979.

Vol. 752: M. Barr, *-Autonomous Categories. VI, 140 pages. 1979.

Vol. 753: Applications of Sheaves. Proceedings, 1977. Edited by M. Fourman, C. Mulvey and D. Scott. XIV, 779 pages. 1979.

Vol. 754: O. A. Laudal, Formal Moduli of Algebraic Structures. III, 161 pages. 1979.

Vol. 755: Global Analysis. Proceedings, 1978. Edited by M. Grmela and J. E. Marsden. VII, 377 pages. 1979.

Vol. 756: H. O. Cordes, Elliptic Pseudo-Differential Operators – An Abstract Theory. IX, 331 pages. 1979.

Vol. 757: Smoothing Techniques for Curve Estimation. Proceedings, 1979. Edited by Th. Gasser and M. Rosenblatt. V, 245 pages. 1979.

Vol. 758: C. Năstăsescu and F. Van Oystaeyen; Graded and Filtered Rings and Modules. X, 148 pages. 1979.

Vol. 759: R. L. Epstein, Degrees of Unsolvability: Structure and Theory. XIV, 216 pages. 1979.

Vol. 760: H.-O. Georgii, Canonical Gibbs Measures. VIII, 190 pages. 1979.

Vol. 761: K. Johannson, Homotopy Equivalences of 3-Manifolds with Boundaries. 2, 303 pages. 1979.

Vol. 762: D. H. Sattinger, Group Theoretic Methods in Bifurcation Theory. V, 241 pages. 1979.

Vol. 763: Algebraic Topology, Aarhus 1978. Proceedings, 1978. Edited by J. L. Dupont and H. Madsen. VI, 695 pages. 1979.

Vol. 764: B. Srinivasan, Representations of Finite Chevalley Groups. XI, 177 pages. 1979.

Vol. 765: Padé Approximation and its Applications. Proceedings, 1979. Edited by L. Wuytack. VI, 392 pages. 1979.

Vol. 766: T. tom Dieck, Transformation Groups and Representation Theory. VIII, 309 pages. 1979.

Vol. 767: M. Namba, Families of Meromorphic Functions on Compact Riemann Surfaces. XII, 284 pages. 1979.

Vol. 768: R. S. Doran and J. Wichmann, Approximate Identities and Factorization in Banach Modules. X, 305 pages. 1979.

Vol. 769: J. Flum, M. Ziegler, Topological Model Theory. X, 151 pages. 1980.

Vol. 770: Séminaire Bourbaki vol. 1978/79 Exposés 525–542. IV, 341 pages. 1980.

Vol. 771: Approximation Methods for Navier-Stokes Problems. Proceedings, 1979. Edited by R. Rautmann. XVI, 581 pages. 1980.

Vol. 772: J. P. Levine, Algebraic Structure of Knot Modules. XI, 104 pages. 1980.

Vol. 773: Numerical Analysis. Proceedings, 1979. Edited by G. A. Watson. X, 184 pages. 1980.

Vol. 774: R. Azencott, Y. Guivarc'h, R. F. Gundy, Ecole d'Eté de Probabilités de Saint-Flour VIII-1978. Edited by P. L. Hennequin. XIII, 334 pages. 1980.

Vol. 775: Geometric Methods in Mathematical Physics. Proceedings, 1979. Edited by G. Kaiser and J. E. Marsden. VII, 257 pages. 1980.

Vol. 776: B. Gross, Arithmetic on Elliptic Curves with Complex Multiplication. V, 95 pages. 1980.

Vol. 777: Séminaire sur les Singularités des Surfaces. Proceedings, 1976-1977. Edited by M. Demazure, H. Pinkham and B. Teissier. IX, 339 pages. 1980.

Vol. 778: SK1 von Schiefkörpern. Proceedings, 1976. Edited by P. Draxl and M. Kneser. II, 124 pages. 1980.

Vol. 779: Euclidean Harmonic Analysis. Proceedings, 1979. Edited by J. J. Benedetto. III, 177 pages. 1980.

Vol. 780: L. Schwartz, Semi-Martingales sur des Variétés, et Martingales Conformes sur des Variétés Analytiques Complexes. XV, 132 pages. 1980.

Vol. 781: Harmonic Analysis Iraklion 1978. Proceedings 1978. Edited by N. Petridis, S. K. Pichorides and N. Varopoulos. V, 213 pages. 1980.

Vol. 782: Bifurcation and Nonlinear Eigenvalue Problems. Proceedings, 1978. Edited by C. Bardos, J. M. Lasry and M. Schatzman. VIII, 296 pages. 1980.

Vol. 783: A. Dinghas, Wertverteilung meromorpher Funktionen in ein- und mehrfach zusammenhängenden Gebieten. Edited by R. Nevanlinna and C. Andreian Cazacu. XIII, 145 pages. 1980.

Vol. 784: Séminaire de Probabilités XIV. Proceedings, 1978/79. Edited by J. Azéma and M. Yor. VIII, 546 pages. 1980.

Vol. 785: W. M. Schmidt, Diophantine Approximation. X, 299 pages. 1980.

Vol. 786: I. J. Maddox, Infinite Matrices of Operators. V, 122 pages. 1980.

Vol. 787: Potential Theory, Copenhagen 1979. Proceedings, 1979. Edited by C. Berg, G. Forst and B. Fuglede. VIII, 319 pages. 1980.

Vol. 788: Topology Symposium, Siegen 1979. Proceedings, 1979. Edited by U. Koschorke and W. D. Neumann. VIII, 495 pages. 1980.

Vol. 789: J. E. Humphreys, Arithmetic Groups. VII, 158 pages. 1980.

Vol. 790: W. Dicks, Groups, Trees and Projective Modules. IX, 127 pages. 1980.

Vol. 791: K. W. Bauer and S. Ruscheweyh, Differential Operators for Partial Differential Equations and Function Theoretic Applications. V, 258 pages. 1980.

Vol. 792: Geometry and Differential Geometry. Proceedings, 1979. Edited by R. Artzy and I. Vaisman. VI, 443 pages. 1980.

Vol. 793: J. Renault, A Groupoid Approach to C*-Algebras. III, 160 pages. 1980.

Vol. 794: Measure Theory, Oberwolfach 1979. Proceedings 1979. Edited by D. Kölzow. XV, 573 pages. 1980.

Vol. 795: Séminaire d'Algèbre Paul Dubreil et Marie-Paule Malliavin. Proceedings 1979. Edited by M. P. Malliavin. V, 433 pages. 1980.

Vol. 796: C. Constantinescu, Duality in Measure Theory. IV, 197 pages. 1980.

Vol. 797: S. Mäki, The Determination of Units in Real Cyclic Sextic Fields. III, 198 pages. 1980.

Vol. 798: Analytic Functions, Kozubnik 1979. Proceedings. Edited by J. Ławrynowicz. X, 476 pages. 1980.

Vol. 799: Functional Differential Equations and Bifurcation. Proceedings 1979. Edited by A. F. Izé. XXII, 409 pages. 1980.

Vol. 800: M.-F. Vignéras, Arithmétique des Algèbres de Quaternions. VII, 169 pages. 1980.

Vol. 801: K. Floret, Weakly Compact Sets. VII, 123 pages. 1980.

Vol. 802: J. Bair, R. Fourneau, Etude Géometrique des Espaces Vectoriels II. VII, 283 pages. 1980.

Vol. 803: F.-Y. Maeda, Dirichlet Integrals on Harmonic Spaces. X, 180 pages. 1980.

Vol. 804: M. Matsuda, First Order Algebraic Differential Equations. VII, 111 pages. 1980.

Vol. 805: O. Kowalski, Generalized Symmetric Spaces. XII, 187 pages. 1980.

Vol. 806: Burnside Groups. Proceedings, 1977. Edited by J. L. Mennicke. V, 274 pages. 1980.

Vol. 807: Fonctions de Plusieurs Variables Complexes IV. Proceedings, 1979. Edited by F. Norguet. IX, 198 pages. 1980.

Vol. 808: G. Maury et J. Raynaud, Ordres Maximaux au Sens de K. Asano. VIII, 192 pages. 1980.

Vol. 809: I. Gumowski and Ch. Mira, Recurences and Discrete Dynamic Systems. VI, 272 pages. 1980.

Vol. 810: Geometrical Approaches to Differential Equations. Proceedings 1979. Edited by R. Martini. VII, 339 pages. 1980.

Vol. 811: D. Normann, Recursion on the Countable Functionals. VIII, 191 pages. 1980.

Vol. 812: Y. Namikawa, Toroidal Compactification of Siegel Spaces. VIII, 162 pages. 1980.

Vol. 813: A. Campillo, Algebroid Curves in Positive Characteristic. V, 168 pages. 1980.

Vol. 814: Séminaire de Théorie du Potentiel, Paris, No. 5. Proceedings. Edited by F. Hirsch et G. Mokobodzki. IV, 239 pages. 1980.

Vol. 815: P. J. Slodowy, Simple Singularities and Simple Algebraic Groups. XI, 175 pages. 1980.

Vol. 816: L. Stoica, Local Operators and Markov Processes. VIII, 104 pages. 1980.